ELECTRONIC CIRCUITS AND APPLICATIONS

ELECTRONIC CIRCUITS AND APPLICATIONS

STEPHEN D. SENTURIA
BRUCE D. WEDLOCK
Massachusetts Institute of Technology

John Wiley & Sons, Inc.
New York London Sydney Toronto

Library of Congress Cataloging in Publication Data:

Senturia, Stephen D 1940–
Electronic circuits and applications.

1. Electronics. 2. Integrated circuits. I. Wedlock, Bruce D., 1934–
joint author. II. Title.
TK816.S43 621.381 74-7404

Printed in the United States of America

10 9 8 7 6 5 4 3 2 1

To our students

PREFACE

This book has been written for the person who wishes to learn the principles of modern electrical engineering and use those principles to make intelligent and practical use of electronic components, circuits, and instruments. Designed for students who have no previous background in circuit theory or electronics, the text provides a sufficiently broad and thorough exposure to practical electronics to permit the immediate application of electronic circuits and instruments to laboratory and research work. Because these applications involve increasingly sophisticated concepts in signal processing, we have included practical introductions to network theory, linear system theory, modulation and detection, noise, guarding and shielding, and analog and digital instrumentation. Thus, this book can be used either as a textbook for an introductory first course in electrical engineering, as a textbook for a one-semester or two-semester "Electronics for Scientists and Engineers" survey course or as a self-study primer for the professional scientist or engineer who needs additional background in the theory and practice of electronics.

The development of integrated circuits during the last decade has led to many important changes in electronic-circuit technology. A scientist or engineer can now work with inexpensive, flexible, integrated-circuit function modules to design and build surprisingly sophisticated circuits without first having to master all of the intricacies and subtleties of the transistor. Furthermore, the manufacturers of commercial equipment have used these same modules to build sophisticated signal-processing circuits into successively lower-cost packages, making available to the user of electronics a wealth and diversity of commercial electronic instruments that were unheard of a decade ago.

This textbook evolved from a one-semester introductory electronics course taught by the authors at the Massachusetts Institute of Technology. The students in this course (and there have been more than a thousand since the course became organized in its present format) come from all fields—physics, biology, chemistry, mechanical, civil, chemical, and aeronautical engineering—and range from freshmen to graduate students, including several medical and law students. The course is used by many freshmen as a precursor to the MIT Electrical Engineering Core program.

In order to prepare this diverse audience for dealing with modern integrated circuit electronics, we have developed a pedagogical approach that differs in several ways from the approaches found in other introductory texts. We stress the close intermingling of:

(*1*) theoretical results based on ideal elements;
(*2*) facts about real devices, and practical circuit models that represent the devices; and
(*3*) applications that exploit both the devices and the key theoretical concepts.

Network theory is treated as a learnable skill instead of a mathematical discipline. The student is encouraged to learn how to put network theory to work with the aid of text examples and carefully graded exercises and problems based on actual applications. Design problems, which represent challenging applications of the text material, are identified by boldface problem numbers. Equivalent circuit concepts are heavily emphasized in both purely resistive and single-time-constant networks. The operational amplifier (Chapter 5) is the first active element introduced because it is simpler to understand and apply than the transistor and because it has largely replaced the transistor as the basic building block of analog circuits. Diodes and transistors (including both bipolar and field-effect transistors) are then introduced in a context that stresses those circuit applications that cannot be performed by op-amps alone. In the process, of course, the student encounters many of the circuit configurations that are employed in op-amps. Thus the student is "bootstrapped" to an understanding of op-amp characteristics and limitations in two steps; first, with the op-amp itself treated as an ideal element, and, second, with the op-amp reappearing as a complex circuit containing transistors and diodes.

With the exception of step responses in single-time-constant circuits (which is introduced in Chapter 6 to permit the early discussion of many important applications examples), the subject of ac circuits is postponed until after op-amps, transistors, and diodes have been fully discussed on a dc or quasi-static basis. During the presentation of small-signal models (Chapter 11), it becomes apparent that techniques are needed for dealing with circuits containing more than one energy storage element, and for dealing with time-varying wave-forms of more complexity than a step function. The discussion of amplifier frequency response thus becomes the vehicle for a three-chapter introduction to linear system theory, including complex impedance and s-plane techniques at a level that requires no advanced mathematics. Superposition in both the time and frequency domain is then discussed, the latter leading into the concept of a frequency spectrum. The spectrum concept is immediately applied in the discussion of modulation (Chapter 15), and is also exploited in the later chapters on noise and instrumentation applications. The material on digital signals and circuits (Chapter 16) includes a review of logic families, gates, flip-flops, and their applications to counters, shift registers, and combinational logic problems. Chapters 17 and 18 contain information on noise and instrumentation of a kind not usually encountered in elementary texts, including shielding and guarding, device noise, analog signal processing, analog-to-digital conversion, sampled data, and digital signal processing. This material has become essential for the modern user of electronics, particularly in the age of the minicomputer, and is now made accessible to beginning students.

The prerequisite background for this textbook has been kept to a minimum: elementary calculus and the equivalent of high-school physics. The course at MIT includes a modest laboratory program to supplement lecture and recitation classes, and makes extensive use of lecture demonstrations. The laboratory program, suggested course outlines, and sample lecture demonstrations are described in the *Teacher's Manual*.

The preparation of a book of this size has drawn on the contributions of many people. The concept of teaching network theory and electronics as a single unified subject derives from Professor Campbell Searle, who taught the introductory electronics course when one of us (S.D.S.) was a first-year physics graduate student trying to learn electronics. In addition, Professor Searle has provided invaluable constructive criticism throughout the writing of this text. Several members of the MIT faculty and nearly 40 graduate teaching assistants have participated in the teaching of this material over the past five years, many of whom have made important contributions through their suggestions and examples. Among these, we especially wish to thank O. R. Mitchell, Irvin Englander, George Lewis, Ernest Vincent, David James, Kenway Wong, Gim Hom, Tom Davis, James Kirtley, and Robert Donaghey. The chairman of the MIT Department of Electrical Engineering, Professor Louis D. Smullin, has provided support and encouragement during this project, as have many colleagues throughout the department. To Mrs. Thalia P. Stone, who prepared the original manuscript and nurtured it through several revisions, our thanks for her skill and good humor throughout a long job. Mr. Max Byer was particularly helpful in the production of preliminary editions for student use at MIT. To the legion of students who prodded us to do it over until we did it right, our thanks for their encouragement and witheringly honest criticism. Finally, to our wives, Alice Senturia and Mary Ann Wedlock, goes that special kind of thanks reserved for those who understand when no one else does.

We have tried to make this an accurate and useful book. To the extent that we have succeeded, we owe a great deal to the people listed above and to others. Any errors and flaws that remain are our own, and for them we take full responsibility.

Cambridge, Massachusetts
August 1974

Stephen D. Senturia
Bruce D. Wedlock

CONTENTS

CHAPTER ONE
PROLOGUE

1.0 SCOPE AND GOALS

To say that we live in an age of electronics is an understatement. From the omnipresent transistor radio to the equally omnipresent digital computer, we encounter electronic devices and systems on a daily basis. In every aspect of our increasingly technological society—whether it be science, engineering, medicine, music, maintenance, or even espionage—the role of electronics is large, and it is growing.

The primary goal of this textbook is to provide a broad introduction to the concepts and uses of modern electronics. We shall discuss the language, the ideas, and the techniques of this field with reference to a variety of applications. These applications have been chosen primarily to suit the needs of persons in scientific and engineering fields who must learn to *use* electronics in intelligent and creative ways. Thus, although we will begin with the most fundamental aspects of electrical networks, our ultimate purpose is to teach the student to understand and to synthesize interconnections of electrical and electronic components into systems that perform useful and interesting tasks.

In general, all of the tasks with which we shall be concerned can be classified as "signal-processing" tasks. Let us explore the meaning of this term.

1.1 SIGNAL PROCESSING

1.1.1 What Is a Signal?

A *signal* is any physical variable whose magnitude or variation with time contains *information*. This information might involve speech and music, as in radio broadcasting, a physical quantity such as the temperature of the air in a room, or numerical data, such as the record of stock market transactions. The physical variables that can carry information in an electrical system are voltage and current.[1] When we speak of "signals" in this book, therefore, we refer implicitly to voltages or currents. However, most of the concepts we discuss can be applied directly to systems with different information-carrying variables. Thus, the behavior of a mechanical system (in which force and velocity are the variables) or a hydraulic system (in which pressure and flow rate are the variables) can often be *modeled* or represented by an equivalent electrical system. An understanding of the behavior of electrical systems, therefore, provides a basis for understanding a much broader range of phenomena.

1.1.2 Analog and Digital Signals

A signal can carry information in two different forms. In an *analog signal* the continuous variation of the voltage or current with time carries the information. An example, in Fig. 1.1, is the voltage produced by a thermocouple pair when the two junctions are at different temperatures.[2] As the temperature difference between the two junctions varies, the magnitude of the voltage across the thermocouple pair also varies. The voltage thus provides an analog representation of the temperature difference.

The other kind of signal is a *digital signal*. A digital signal is one that can take on values within two *discrete* ranges. Such signals are used to represent ON–OFF or YES–NO information. An ordinary household thermostat delivers a digital signal to control the furnace. When the room temperature drops below a preset value, the thermostat switch closes turning ON the furnace. Once the room temperature rises high enough, the switch opens turning

[1] Readers totally unfamiliar with these terms may wish to look ahead briefly to Section 2.1 of the following chapter.

[2] A thermocouple is a junction between dissimilar metals, such as copper and constantan. The voltage generated by a pair of thermocouples is used to measure the temperature difference between the two junctions.

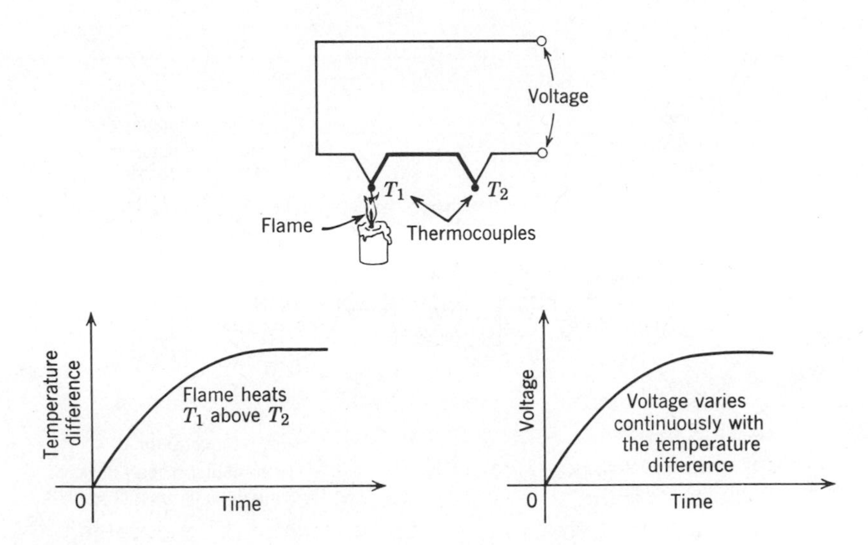

Figure 1.1
An example of an analog signal. One thermocouple is heated above the temperature of the other, giving rise to an analog output voltage.

OFF the furnace. The current through the switch provides a digital representation of the temperature variation: ON equals "too cold" while OFF equals "not too cold."

1.1.3 Signal-Processing Systems

A *signal-processing system* is an interconnection of components and devices that can accept an input signal or a group of input signals, operate on the signals in some fashion either to extract or improve the quality of the information, and present the information as an output in the proper form at the proper time.

Figure 1.2 illustrates the components in such a system. The central circles represent the two types of signal processing (digital and analog), while the block between the two signal-processing blocks represents the conversion of an analog signal to equivalent digital form (A/D = *A*nalog-to-*D*igital) and the reverse conversion of a digital signal to the corresponding analog form (D/A = *D*igital-to-*A*nalog). The remaining blocks involve inputs and outputs—getting signals into and out of the processing system.

Many electrical signals derived from physical systems are obtained from devices called *transducers.* We have already encountered an example of

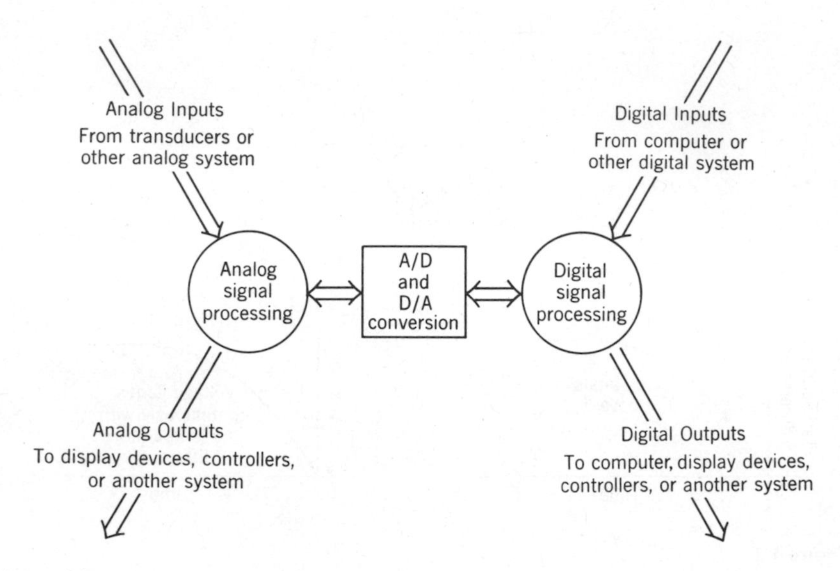

Figure 1.2
Components of a signal-processing system.

an analog transducer, the thermocouple pair. It converts temperature difference (the physical variable) to a voltage (the electrical variable). Generally, a transducer is a device that converts a physical or mechanical variable to an equivalent voltage or current signal. Unlike the thermocouple example, however, most transducers require some form of electrical excitation to operate.

The output from a system can be in many forms, depending on the use to be made of the information contained in the input signals. One can seek to display the information, either in analog form (using a meter, for example, in which the needle position indicates the size of the variable of interest) or in digital form (using a set of digital display elements that are lit up with a number corresponding to the variable of interest). Other possibilities are to convert the output to sound energy (with a loudspeaker), or to use the output as an input signal to another system, or to use the output as a control signal to initiate some action. The examples presented in Section 1.2 below illustrate some of these cases.

1.2 ILLUSTRATIVE SYSTEMS

1.2.1 A Communications System

The first system we shall illustrate is the *communications system.* The input can be either speech, music, or data that is produced at one location and transmitted efficiently over long distances, permitting faithful recovery of the original input. The example chosen, the familiar AM broadcast system is shown schematically in Fig. 1.3.

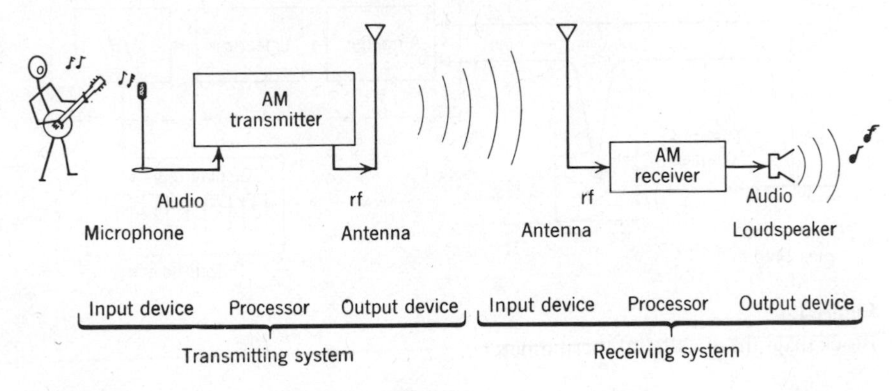

Figure 1.3
An AM broadcast communication system.

The letters *AM* stand for *A*mplitude *M*odulation. They mean that the amplitude or magnitude of a radio frequency (rf) signal is caused to vary according to the magnitude of a lower-frequency signal (audio, corresponding to audible frequencies). The function of the transmitter in an AM broadcast system is to accept the input signal from an input device (microphone), use this signal to control the amplitude of a radio frequency signal (each broadcast station is assigned its own radio frequency), and drive the output device (the antenna) with a radio frequency current to produce electromagnetic waves radiating into space. The receiving system consists of an input device (the antenna), a processor (the receiver), and an output device (the loudspeaker). The functions of the receiver are to amplify or increase the strength of the relatively weak signal obtained from the antenna, to filter or select the desired radio frequency signal from the signals of all other broadcast stations, to recover the audio signal from the amplitude variations of the radio frequency signal, and to drive a loudspeaker with this audio signal.

1.2.2 A Measurement System

A second system is a *measurement system.* The purpose of this system is to acquire information from suitable transducers about the behavior of some physical system and to display this information to the observer. An example of such a system, a digital thermocouple thermometer, is shown in Fig. 1.4.

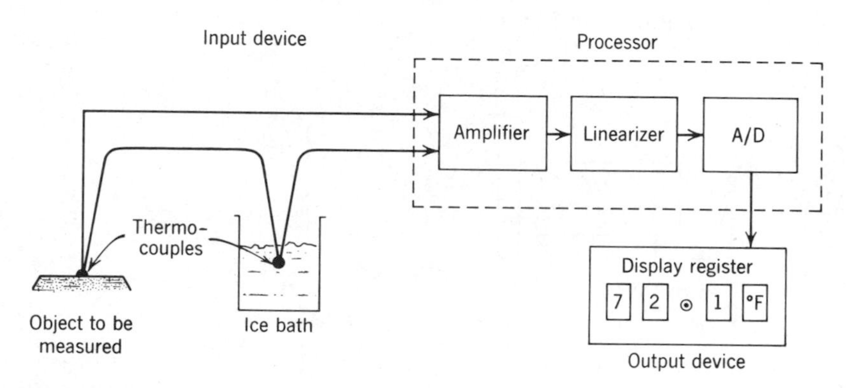

Figure 1.4
Block diagram of a digital thermometer.

The pair of thermocouple junctions, one attached to the object whose temperature is to be measured, the other submerged in an ice bath (to establish a stable reference point), presents to the processor a voltage that depends on the temperature difference between the object to be measured and the ice bath. Because the thermocouple voltage is never exactly proportional to the temperature difference, a small correction must be applied to the thermocouple voltage to produce an analog voltage that is exactly proportional to the temperature. This correction is the role of the linearizer. The analog voltage from the thermocouple is first amplified (i.e., made larger) then linearized, and then converted to digital form. Finally, it is displayed in a digital display register as the output of the thermometer.

Although a major goal of the communications system is to transmit a faithful reproduction of the source signal, a major goal of the measurement system is to produce numerically accurate data. In a measurement system, therefore, one expects to be concerned with locating and removing any small errors that might be added to the signal at each step of the processing sequence.

1.2.3 A Feedback Control System

The third system is the *feedback control system*, in which information about the behavior of the output modifies the signals driving the system. In Fig. 1.5 a thermostat is used to control the temperature of a room. In this case, the thermostat contains the input device for determining the room temperature (normally a bimetallic strip that bends as its temperature is varied), a mechanism for setting the desired temperature (the set point dial), and mechanical switches, activated by the bimetallic strip, which control the furnace.

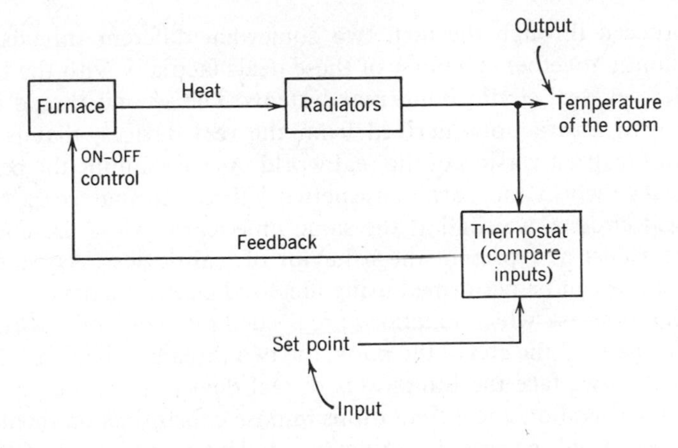

Figure 1.5
A feedback system.

This familiar example, which in fact includes *no* electrical components other than a switch, was chosen to emphasize the feedback concept. Suppose the feedback line were disconnected in Fig. 1.5. That is, suppose there were no mechanism for turning the furnace ON or OFF. The temperature of the room would either rise to some maximum (if the furnace were on all the time) or drop to some minimum (if the furnace were off all the time). Presuming that the maximum temperature is too hot for comfort while the minimum temperature is too cool, some "controller" is needed to turn the furnace ON and OFF. The "controller" might be a person who turns the heat ON when he feels cold and turns it OFF when he feels warm. Even at this level, the system (*including* the person) comprises a *feedback* control system, because information about the output (i.e., the temperature of the room) is used to modify one of the driving signals (the on-off switch on the furnace). The thermostat is simply a piece of hardware that performs automatically what the temperature-sensitive person would do; namely, it turns the furnace ON when the temperature drops below the point where the person set the thermostat, and turns it OFF otherwise.

There are many other feedback systems, including systems in which the processing of signals is done electronically. We shall encounter several electronic feedback systems later.

1.3 MODELING

As we proceed through the text, two somewhat different threads will be found running together. The first of these deals factually with the behavior of real devices, real circuits, and real systems. The second thread involves working with abstractions derived from the real devices, that is, with a somewhat idealized version of the real world. As we examine the basic laws of electrical networks and learn to use network theory to analyze the behavior of idealized circuits, we shall at the same time learn to use these idealized circuits to *model* or represent the behavior of real devices. As we examine what functions can be performed using idealized electronic devices, we shall at the same time ask which functions are needed for signal processing in the real world. Toward the end of the book, the two threads will get more tightly interwound as we face the limitations of real devices (e.g., the presence of noise) and discuss how these limitations impose constraints on the design of real systems. It will nevertheless help the student to be aware of this dual approach from the outset.

At each step of the development we shall encounter a conflict between accuracy and simplicity. If the model of a device is too simple, it will fail to portray essential device characteristics. If the model is too detailed, unnecessarily cumbersome calculations may obscure our understanding of the essential issues. As a result, the models we shall ultimately employ are compromises, accurate enough to represent the essential features of device and circuit performance, yet simple enough to permit rapid analysis and to enhance our understanding of system performance.

QUESTIONS

Q1.1 Describe a doorbell circuit as a communications signal-processing system. What is the information communicated? Identify the input device, the output device, and the processor. Are the signals analog or digital?

Q1.2 Describe an electric toaster that "pops up" the toast when finished as a feedback system.

Q1.3 How many different kinds of transducers can you think of?

Q1.4 How many different output devices can you think of?

PART A

LEARNING THE LANGUAGE

CHAPTER TWO
FUNDAMENTAL CONCEPTS

2.0 INTRODUCTION

Learning electronics is in many ways like learning a new language. The richly varied combinations of devices and circuits form a body of literature that becomes accessible to those conversant with the fundamentals of device behavior (vocabulary) and with the laws governing network behavior (grammar). Although the analogy can be overstressed, we urge the student to approach this subject as if it were a language, seeking to acquire not merely a modicum of understanding, but a conversational skill in working with the new terms. We begin with the definitions of basic electrical quantities, statements of fundamental network laws, and a first example of how the behavior of a real device is represented or modeled by an idealized network element.

2.1 ELECTRICAL QUANTITIES AND UNITS

2.1.1 Charge

In electrical circuits, the physical phenomenon of central importance is the motion of electric *charge*. Charge can exist in nature with either positive or negative polarity. Strong forces of attraction exist between opposite charges, while strong forces of repulsion exist between like charges. As a result, there is a tendency for positive and negative charges to associate together in equal amounts. An atom, for example, consists of a positively charged nucleus surrounded by a cloud of negatively charged *electrons.* The total amount of negative charge is equal in magnitude to the positive charge of the nucleus. The atom, therefore, has zero total charge and is said to be electrically neutral.

In the materials that can conduct electricity, some of the electrons break free of the atoms and are able to move through the conducting material. These electrons are called the *mobile charge*, or the *charge carriers.* Since each of the atoms was initially neutral, an atom becomes positively charged when a negatively charged electron breaks free. These positively charged atomic cores are not free to move, and form a background of immobile, *fixed charge.* In the important class of materials called *semiconductors*, the mobile electrons can have two types of motion, one in which the electrons behave as if they were ordinary negatively charged carriers, the other a complex collective motion of many electrons that behaves just as if there were positively charged mobile carriers in the material. Fixed charges with both signs are also encountered. (See Chapter 7 for further discussion of the details of the conduction process.) For the present, it is sufficient to recognize that conducting materials can be represented as containing mobile charge carriers (which might be of either sign) together with fixed charges of the opposite polarity.

There are many materials, called *insulators*, that do not conduct electricity. All of the charges in insulators are fixed. Examples of insulators include air, mica, glass, the thin oxides that form on many metallic surfaces and, of course, a vacuum (which has no charges at all).

To discuss charge and its motion in precise terms, we shall need an internally consistent system of units in which to express all quantities. Throughout this book we shall use the MKS units, in which distance is expressed in meters, mass in kilograms, and time in seconds. In the MKS unit system, *charge* is measured in *coulombs* (Table 2.1), and is usually denoted by Q.

TABLE 2.1

Symbols and MKS Units for Basic Electrical Quantities

Quantity	Symbol	Unit	Abbreviation	Relations
Mass	m	Kilogram	kg	
Length		Meter	m	
Time	t	Second	s	
Charge	Q, q	Coulomb	C	
Energy	W	Joule	J	

TABLE 2.1—continued

Quantity	Symbol	Unit	Abbreviation	Relations
Voltage (potential difference)	V, v	Volt	V	J/C
Electric field	E	Volt/Meter	V/m	
Current	I, i	Ampere	A	C/s
Power	P	Watt	W	J/s
Frequency	f	Hertz	Hz	Cycle/s
Angular frequency	ω	Radians/Second	rad/s	$2\pi f$
Resistance	R	Ohm	Ω	
Conductance	G	Siemens (or Mho)	S (or Ω^{-1})	
Capacitance	C	Farad	F	
Inductance	L	Henry	H	
Impedance	Z	Ohm	Ω	
Admittance	Y	Siemens (or Mho)	S (or Ω^{-1})	

2.1.2 Voltage

Because of the strong force of attraction between opposite charges, most materials are electrically neutral. Energy is required to separate the positive and negative charges. Consider Fig. 2.1, in which two initially neutral conducting plates are separated by a distance d. The space between the plates is presumed to contain an insulator like air or a vacuum. In Fig. 2.1*a*, both plates are neutral, the zero total charge on the upper plate being schematically represented as a charge $+Q$ together with a charge of $-Q$.

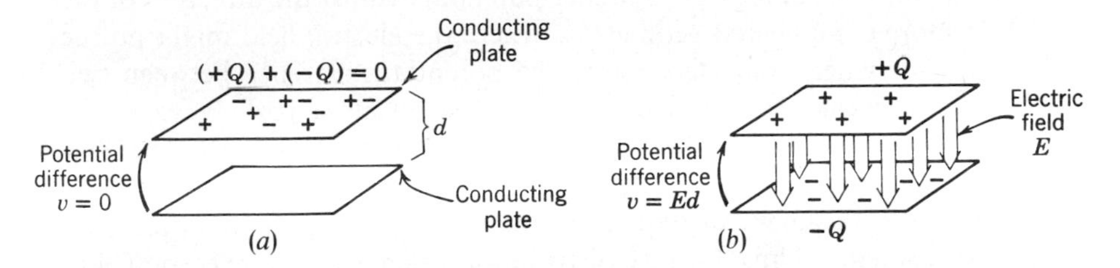

Figure 2.1
In (*a*) both plates are electrically neutral. In (*b*) a charge of $-Q$ has been transferred to the lower plate. An electric field and a potential difference exist between these plates.

In Fig. 2.1*b*, the charge of $-Q$ has been moved from the upper plate to the lower plate. (We shall discuss *how* that charge got moved in Section 2.3.) If we were to provide a conducting path between the plates, the attractive forces between unlike charges would produce a spontaneous charge transfer back to the uncharged situation of Fig. 2.1*a*. Positive charges would move toward the negatively charged plate, while negative charges would move toward the positively charged plate.

In Fig. 2.1*b* we say that a *potential difference* exists between the charged plates, with the positive upper plate being at a higher potential than the negative lower plate. More generally, a potential difference exists between two points whenever the introduction of a conducting path between the two points would result in spontaneous charge transfer. The direction of motion for positive charges is from higher potential to lower potential, while the direction of motion for negative charges is the opposite, from lower to higher potential.

The unit for measuring potential difference is the *volt*; for this reason, potential difference is also called *voltage*, and is usually denoted by v. To specify a numerical value for the voltage between two points, the concept of an *electric field* is useful. In the situation of Fig. 2.1*b*, a uniform electric field E exists between the plates, directed from the point of higher potential (positive plate) to the point of lower potential (negative plate). The strength of the field, expressed in volts/meter, is proportional to the charge on the plates, and can always be calculated from the laws of physics if the distribution of charges is known.[1] The relation between the magnitude of the electric field and the voltage v between the plates is

$$\begin{array}{ccccc} v & = & E & \times & d \\ \text{volts} & = & (\text{volts/meter}) & \times & \text{meters} \end{array} \tag{2.1}$$

where one moves from lower to higher potential against the direction of the field. In more complicated geometries, where the electric field might not be uniform everywhere, one determines the potential difference between two points by a simple extension of Eq. 2.1. One selects a path between the two points of interest, and breaks the path into many segements, each segment short enough so that the field is uniform near that segment. One then applies Eq. 2.1 to each segment in sequence, summing up the potential difference for each segment.[2] Thus, for any distribution of charges and electric fields, it is possible to find the potential difference between any two points.

[1] For the example of Fig. 2.1, if there is a vacuum in the space between the plates, the strength of the field is $E = Q/A\varepsilon_0$ volts/meter, where A is the area of the plates in square meters, and ε_0 is a fundamental constant, called the *permittivity*, having the value 8.9×10^{-12} coulombs/volt-meter.

[2] More precisely, the contribution to the potential difference is the *projection* of the electric field along the path segment.

In specifying potential difference, it is necessary to indicate not only the magnitude of the voltage between two points but also which point has the higher potential. However, in electric networks containing several different elements, it is not always possible to determine *in advance* which point has the higher potential. To avoid confusion, therefore, we must adopt consistant sign conventions (see Fig. 2.2). A two-terminal *network element* is represented as a box with two terminals attached (Fig. 2.2*a*). The solid lines leading from the box to the terminals are assumed to be perfect conductors of electricity. One terminal is labeled with a + sign, while the other terminal has a − sign. These signs fix the *reference polarity*. The voltage v in Fig. 2.2*a* is defined by convention to be

$$v = (\text{potential at + terminal}) - (\text{potential at − terminal}) \qquad (2.2)$$

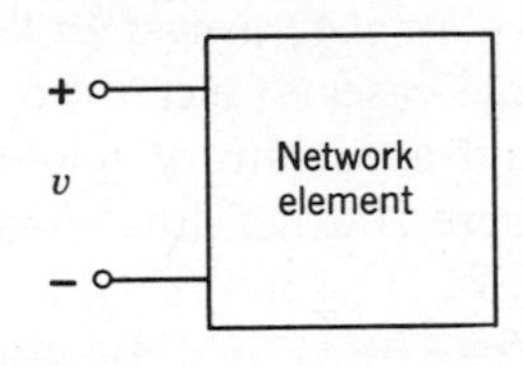

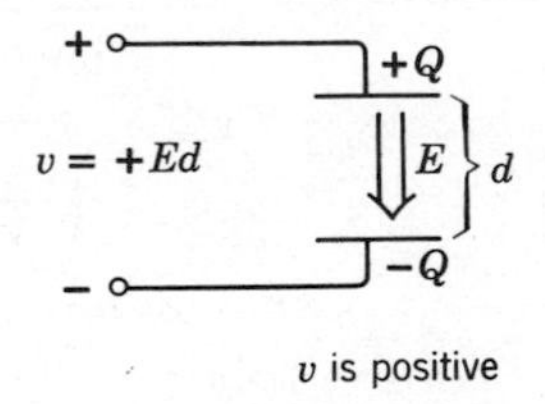

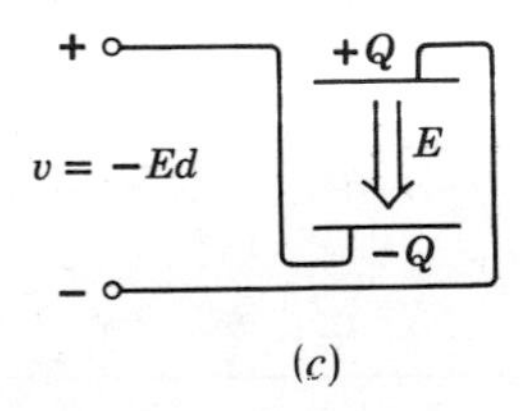

Figure 2.2
Voltage is an algebraic quantity. It can be positive or negative.

Fig. 2.2*b* shows the charged plates of Fig. 2.1 connected so that the + terminal is connected to the plate with the higher potential. Here the voltage v is a positive number. In Fig. 2.2c the + terminal is connected to the plate with the lower potential. If we apply the definition, Eq. 2.2, the voltage comes out negative. It is important to remember the algebraic nature of voltage. Once the reference polarity is specified, a positive voltage means that the + terminal has the higher potential, while a negative voltage means that the − terminal has the higher potential.

2.1.3 Current

We have already remarked that positive mobile charge moves from a region of high potential to a region of low potential, while negative mobile charge moves from a region of low potential to a region of high potential. Whenever this charge transport occurs, an *electric current* flows.

In Fig. 2.3, several simple cases of electric current flow are illustrated. A surface S is selected, and an arbitrary reference direction is assigned. If in a time interval Δt, there is a net flow of charge ΔQ across S in the

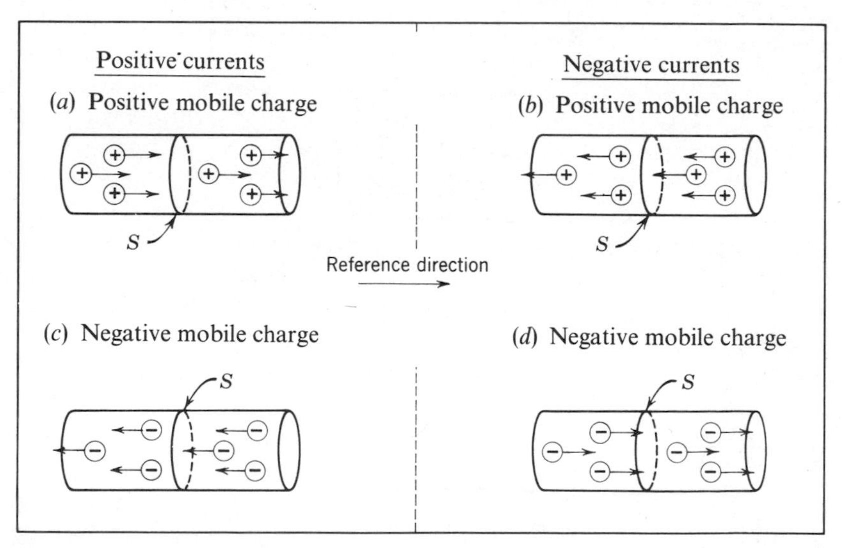

Figure 2.3
The relation between the algebraic sign of the current and the flow of mobile charge. The current is positive (a and c) if there is a net flow of positive charge across some surface S in the direction of the reference arrow. The current is negative (b and d) if there is a net flow of positive charge across the surface in the direction opposite to the reference arrow.

direction of the reference arrow, the current i through S is defined as

$$i = \frac{\Delta Q}{\Delta t} \tag{2.3}$$

The unit of current is the *ampere* (one ampere = one coulomb/second).

The algebraic sign of i is often a problem for beginning students. When the mobile charge is positive (Figs. 2.3*a* and 2.3*b*), a positive current describes the physical flow of mobile charge *along* the reference direction, while a negative current describes the flow of mobile charge *opposite* to the reference direction. When the mobile charge is negative, (Figs. 2.3*c* and 2.3*d*), the definition of current (Eq. 2.3) must be applied with care. Consider Fig. 2.3*d*, in which negative mobile charge crosses S along the reference direction. Suppose each carrier has a charge $-q$ and suppose the rate of flow across S is n carriers/second. In a time Δt, the total charge crossing S along the reference direction is $\Delta Q = -nq\Delta t$, which corresponds to a current

$$i = \frac{\Delta Q}{\Delta t} = -nq \tag{2.4}$$

Thus the current of Fig. 2.3*d* is negative. Furthermore, this current is identical to what one would obtain if positive carriers with charge $+q$ were crossing S at the rate of n carriers/second in a direction *opposite* to the reference direction (Fig. 2.3*b*). Thus the dual nature of charge is reflected in the dual nature of current.

In most of the examples one encounters in electronic circuits it is the algebraic sign of the *current* that is important; one is not normally concerned with the sign of the charge carriers responsible for the current. As a result, one often talks about electric currents as if the mobile charge were always positive. However, in semiconductor devices, the distinction between positive and negative charge carriers is critical to the operation of the device. When discussing in detail how these devices operate, the signs of the mobile charges must be clearly specified.

The concept of current crossing a specific area is easily extended to the concept of current *through* a network element. In Fig. 2.4, a two-terminal network element is shown. The reference direction for positive current is indicated by the short arrow next to terminal A. If a positive current i is flowing through the network element, positive charge is entering terminal A at the rate of i coulombs/second. But as we have remarked, materials (and network elements) of electrical importance remain electrically neutral overall. (Even the "charged" element of Fig. 2.1 has zero *total* charge.) Therefore, if charge flows into the element at terminal A, an equal amount of charge

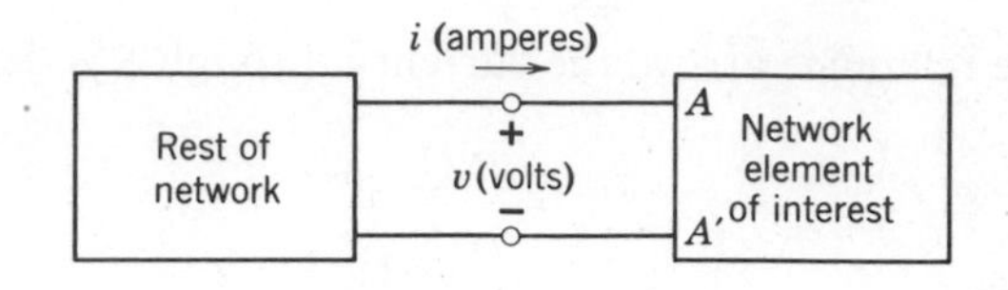

Figure 2.4
Current through a network element. Charge enters the element through terminal A at the rate of i coulombs/second, and leaves the element through terminal A' at the same rate.

must simultaneously flow out of the element from terminal A'. This continuity of electric current through network elements follows from the neutrality of the element as a whole.

2.1.4 Power

Every two-terminal network element can have a voltage between its terminals as well as a current flowing through it. Both voltage and current could be defined according to separate, independent sign conventions; however, an important physical relation exists between the polarities of voltage and current, and an additional convention is usually adopted to help to clarify this relationship. Figure 2.4 shows how we define the relative polarities for voltage and current. The reference direction for current *enters* the + voltage terminal. When this convention is adopted consistently, it is possible to define precisely a fourth electrical quantity of importance, *electric power*.

Consider the network element of Fig. 2.4. If the voltage is positive *and* the current is positive, there is continuous flow of positive charge from a point of high potential to a point of low potential. To maintain this flow, positive charge must be continuously separated from negative charge and fed into the + terminal. This continuous separation requires a continuous supply of energy. As the charge flows through the element, it gives up this energy. Since energy must be conserved, the energy appears in the network element either as heat (as in a toaster) or as stored energy (as in the charging of a car battery). The *rate* at which this energy transfer takes place is called the *power*, and is defined as

$$P = v \times i \tag{2.5}$$

$$\text{watts} = \text{volts} \times \text{amperes}$$

As indicated, the unit of power is the watt, which corresponds to an energy transfer of one joule/second.

Power, which is equal to the product of the voltage and the current when defined with the polarities of Fig. 2.4, is an algebraic quantity. When $P > 0$, as in the case discussed above, the element either dissipates or absorbs power. Suppose, however, that the product of v and i is negative. In that case, the element is actually supplying power to the rest of the network in which it is contained (see Section 2.3).

2.1.5 Prefixes and Powers of Ten

Engineering and scientific usage involves numerical quantities with vastly differing magnitudes. Instead of writing the explicit powers of 10 associated with each number (e.g., 1456 watts = 1.456×10^3 watts), a set of prefixes has been adopted (Table 2.2). In the example above, we would write 1.456 kW, meaning 1.456 kilowatts.

TABLE 2.2

Prefixes

Prefix	Power of Ten	Abbreviation
Giga-	10^9	G
Mega-	10^6	M
Kilo-	10^3	k
Milli-	10^{-3}	m
Micro-	10^{-6}	μ
Nano-	10^{-9}	n
Pico-	10^{-12}	p
Femto-	10^{-15}	f

2.2 NETWORKS AND KIRCHHOFF'S LAWS

2.2.1 Networks

When two or more network elements are connected together, a network is formed. Figure 2.5 shows a network containing four elements. Points of connection labeled 1, 2, and 3 in Fig. 2.5 are called *nodes*. The very fact of these interconnections among the elements imposes constraints on the

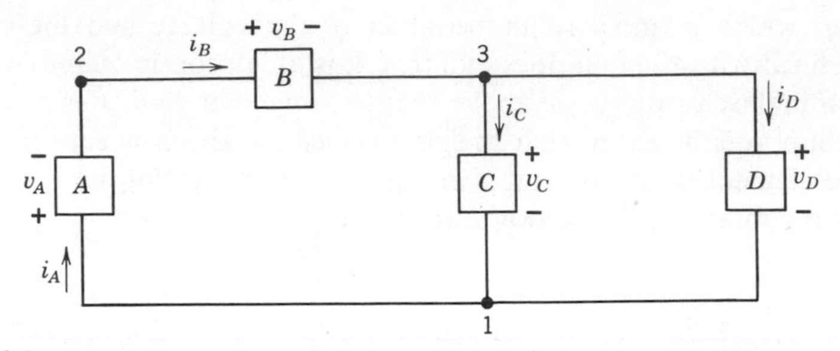

Figure 2.5
A four-element network. The points of connection between elements (labeled 1, 2, 3) are called nodes.

possible values of voltage and current that can exist in the network. These constraints are summarized by Kirchhoff's laws.

2.2.2 Kirchhoff's Voltage Law (KVL)

The first of Kirchhoff's laws arises from a fundamental physical property of voltage: the voltage between any two points in a network is independent of the path through the network. In Fig. 2.5, for example, elements C and D are connected in *parallel* (side by side). The voltage between nodes 3 and 1 cannot depend on whether the path goes through element C or element D. Therefore, the voltage v_C must equal the voltage v_D. As a rule, *elements connected in parallel always have the same voltage across them.*

In more general network situations, this property of voltage is stated as *Kirchhoff's voltage law* (abbreviated KVL):

> The algebraic sum of voltage drops taken around any loop in a network is equal to zero.

Refering again to Fig. 2.5, consider the closed loop 1 → 2 → 3 → 1. Between 1 and 2 the voltage drops by v_A. Between 2 and 3, the voltage drops by v_B. Between 3 and 1 the voltage drops by v_C (which is, of course, equal to v_D). Summing these voltage drops, we obtain

$$v_A + v_B + v_C = 0 \tag{2.6}$$

Notice that for Eq. 2.6 to be satisfied, at least one of the three voltages must be algebraically negative.

It is important to keep track of the signs of the voltage drops. A convenient method is to write each variable with that sign encountered first as one proceeds around a given loop. In the loop $1 \rightarrow 2 \rightarrow 3 \rightarrow 1$, the plus sign is encountered first at each network element, leading to Eq. 2.6 written with plus signs. If instead, for example, v_B had been defined with opposite polarity, it would have entered Eq. 2.6 with a minus sign.

2.2.3 Kirchhoff's Current Law (KCL)

The strong attractive force between opposite charges assures that an unneutralized charge will not build up spontaneously anywhere in a network. In particular, therefore, the *net mobile charge* entering a network node during any given time interval must be zero. This fact is expressed as *Kirchhoff's current law* (KCL).

> The algebraic sum of currents entering any node must be zero.

Referring again to Fig. 2.5 and looking at node 3, we see that i_B is defined with its reference direction entering node 3 (remember that i_B flows right *through* element B), while i_C and i_D are defined with their reference directions leaving node 3. When writing KCL, currents with reference directions leaving a node enter the equation with a minus sign. Therefore, the total current *entering* node 3 is

$$i_B - i_C - i_D = 0 \tag{2.7}$$

In Fig. 2.5, elements A and B are connected in *series* (end to end). A simple application of KCL at node 2 tells us that the currents i_A and i_B must be equal. As a rule, *elements connected in series always carry the same current.*

These two network laws and their simple applications apply to *every* electrical network regardless of the detailed characteristics of the elements. As we proceed, several concepts will be introduced that are much more limited in their range of validity. But KCL and KVL, which arise out of the basic physics of the forces between charges, will always be valid.

2.3 SOURCES AND WAVEFORMS

2.3.1 Batteries

When discussing voltage in Section 2.1, we stated that a source of energy is required to accomplish the separation of charge associated with a potential difference. One example of such a source is a *battery*, in which an electrochemical reaction between metallic electrodes and a suitable acid solution or salt paste establishes a potential difference between the electrodes.[3] Figure 2.6 shows the circuit symbol for a battery. Every battery carries + and − labeling, with the + or "positive" terminal being connected to the electrode having the higher potential.

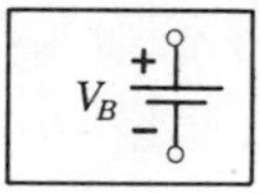

Figure 2.6
Circuit symbol for a battery.

A battery is identified primarily by the numerical value of the voltage between the terminals, denoted by the symbol V_B in Fig. 2.6. For example, an automobile storage battery is a 12 volt battery, while the common battery used in flashlights is a 1.5 volt battery.

A second important battery characteristic is its ability to deliver power to a load. When a conducting path is introduced between the battery terminals, a positive current will flow through that conducting path from the positive terminal (higher potential) to the negative terminal (lower potential), as shown in Fig. 2.7. The conducting path is called the *load*. It might be a motor, a

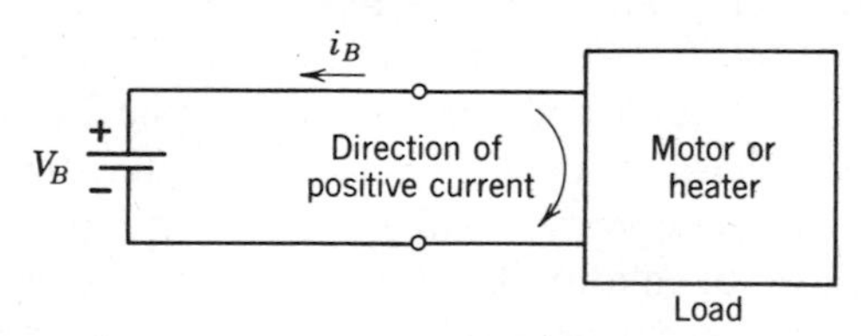

Figure 2.7
A battery connected to a load.

[3] Batteries convert the chemical energy of the constituents to electrical energy. Other sources of electrical energy are generators, which convert mechanical energy to electrical energy, and solar cells, which convert electromagnetic radiant energy to electrical energy.

heater, a radio, or any other component that requires a voltage to operate. The direction of positive current flow is labeled in the illustration. Notice that by the sign conventions of the previous section, the "battery current," labeled with the reference direction i_B, is actually *negative* when the battery is connected to a load. Since V_B is positive and i_B is negative, the power "absorbed by the battery" is negative. Alternatively, we say that the battery is delivering power to the load.

In all real batteries, this process of delivering power to a load can only last for a finite time. As the current is drawn from the battery, the chemical components in the battery are gradually used up. As the components are consumed, the battery voltage decreases toward zero, and we say that the battery is *discharged*. Figure 2.8 shows two discharging curves for a battery. The rate of discharge depends on how much current is drawn from the battery. At higher currents, the battery lifetime is shorter.

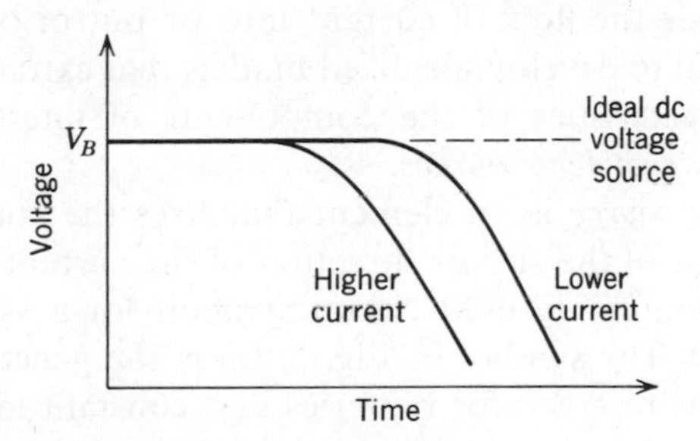

Figure 2.8
Battery discharge curves for two different currents.

The total energy that a battery can store in chemical form depends primarily on how big it is. Since chemical constituents characteristically store a fixed amount of electrochemical energy per unit volume, the larger the battery volume the longer it can supply energy at a given rate before its stored energy is used up. For the larger batteries, such as the automobile batteries, this lifetime is often specified in ampere-hours. The lifetime in hours is estimated by dividing the ampere-hour rating by the size of the current being drawn. Thus a 100 ampere-hour battery can be expected to deliver 5 amperes of current for about 20 hours.

Several kinds of batteries, particularly the lead-acid automobile batteries and nickel-cadmium cells, can be *recharged*, as illustrated in Fig. 2.9. The battery is connected to a battery charger, a source circuit that forces a positive current into the + terminal of the battery. Some of this power produces heating of the battery, but some of it is converted to chemical energy,

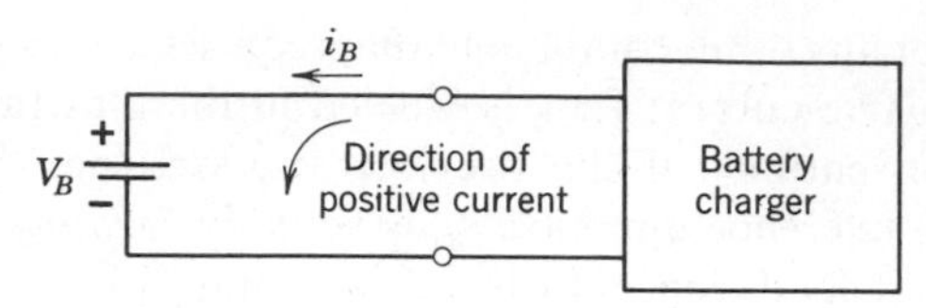

Figure 2.9
Charging of a battery.

available for delivery to a load. Only batteries with reversible chemical reactions can be recharged.[4]

2.3.2 Ideal Independent Sources

In analyzing further the flow of current into or out of batteries or other elements, it is useful to develop idealized models that extract and emphasize the essential characteristics of the components of interest. This section introduces ideal *independent sources.*

An ideal *voltage source* is an element that fixes the voltage between its terminals regardless of the size or direction of the current flowing through the source. Two commonly used circuit symbols for a voltage source are shown in Fig. 2.10. The symbol in Fig. 2.10*a* is the general symbol, while that of Fig. 2.10*b* is reserved for batteries and constant ideal sources. The voltage source has a labeled polarity (+ and −), and has a label (v_0 in Fig. 2.10*a*) that indicates the value of the voltage source. Considering the voltage source as a network element with voltage v and current i, the voltage at the

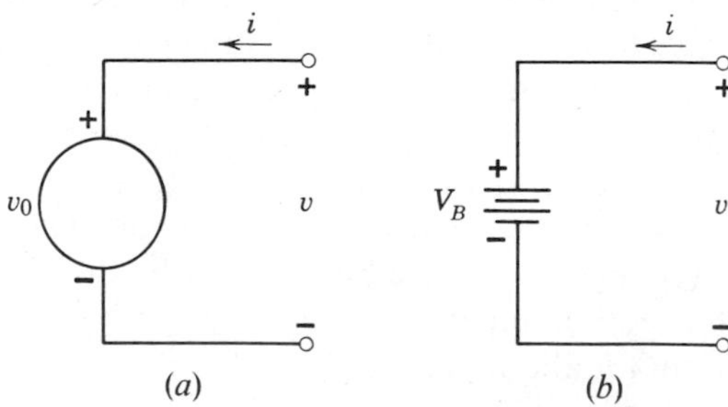

Figure 2.10
Symbols for independent voltage sources.

[4] Do not attempt to recharge flashlight or other dry-cell batteries. They may explode.

terminals v is equal (by our definition of the voltage source) to the source voltage v_0, regardless of the size or sign of the current i.

A battery, in which the voltage between its terminals is fixed by means of a chemical reaction, can be approximated, or *modeled* by an ideal voltage source (see Fig. 2.8). Notice that in the definition of the ideal element we are not concerned with the physical mechanism for maintaining a constant voltage. We simply postulate a "constant voltage property" for an ideal voltage source. Since the battery has a nearly constant voltage within its operating limits, we can model it with the ideal voltage source.

An ideal *current source* is an element that maintains a fixed current regardless of the voltage between its terminals. Figure 2.11 shows two circuit symbols for a current source. Each has a reference direction indicated by the arrow and a label indicating the value of the current. If we consider the current source as a network element with voltage v and current i, the current i is equal to $-i_0$ regardless of the value of the voltage v. (Notice that the reference directions defining positive i and positive i_0 happen to be opposite in this example.)

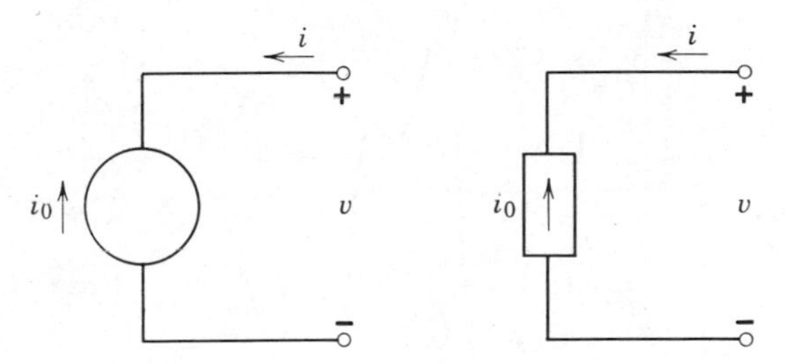

Figure 2.11
Current source symbols.

Examples of elements that can be approximated by ideal current sources are harder to find than for voltage sources. There is no single physical box analogous to the battery. As we shall see, however, current sources are most useful in modeling the behavior of transistors. In addition, in dealing with many problems of network analysis it is convenient to have both sources at our disposal.

2.3.3 Waveforms

Many of the voltages and current we shall encounter change their values with time. A graph which shows this variation with time is called a *waveform.*

Figure 2.12 shows a time-varying voltage source along with a graph of its waveform. This particular waveform might be a speech waveform output from a tape recorder. The graph can be plotted only *after* the reference polarity of the voltage is specified. When the waveform of Fig. 2.12 is positive, the + terminal has the higher potential; when the waveform is negative, the − terminal has the higher potential. Similarly one can talk about waveforms for currents once a reference direction is selected. Current flow along the reference direction corresponds to the positive parts of the waveform; current flow opposite to the reference direction corresponds to the negative parts of the waveform.

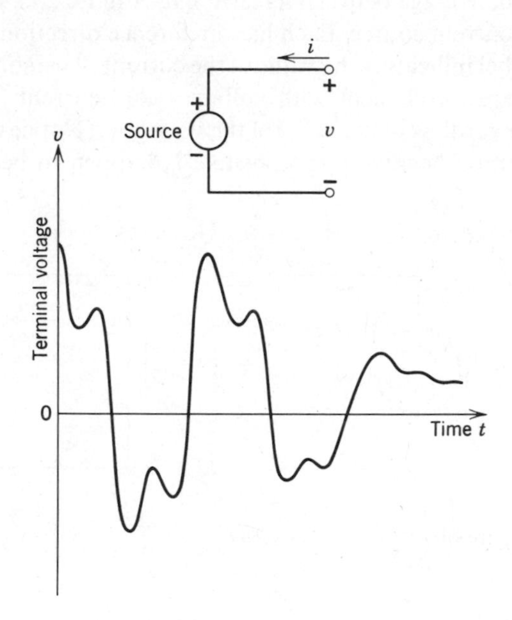

Figure 2.12
A time-varying voltage source with a graph of its waveform.

Several specific waveforms are so useful that they are given special names. The simplest of these is the constant waveform or *dc* waveform (for "*d*irect *c*urrent"). Sources that produce constant waveforms are called constant sources, or dc sources. The battery is a good example of a dc voltage source.

Three important time-varying waveforms are shown in Fig. 2.13. The first of these is a *step function.* It is used to represent a constant waveform that is "turned on" at some instant in time. The step function is given its own symbolic notation, $u_{-1}(x)$, which is defined such that for $x < 0$, the

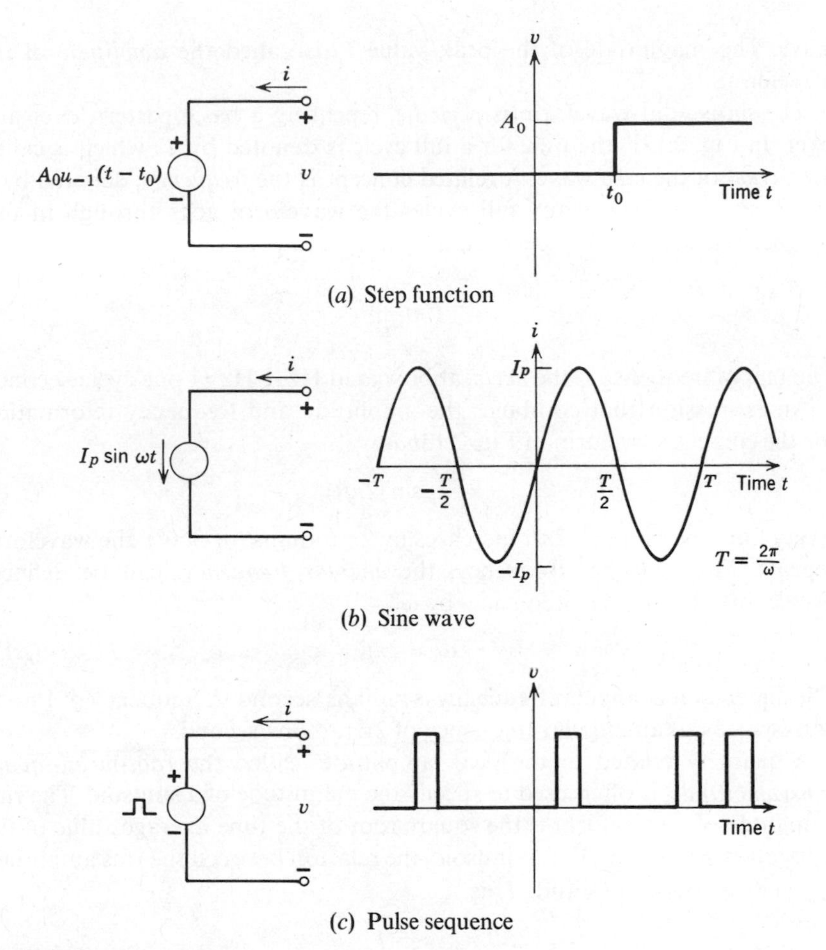

Figure 2.13
Time-varying waveforms.

function is zero, while for $x \geq 0$ the function has a value of $+1$. In Fig. 2.13*a*, the waveform is written

$$v = A_0 u_{-1}(t - t_0) \tag{2.8}$$

The use of $(t - t_0)$ as the argument for the function assures that the turn-on will occur at $t = t_0$, while the amplitude A_0 gives the waveform the correct value after turn-on has occurred.

The waveform shown in Fig. 2.13*b* is a *sinusoidal* or *ac* waveform (for "*a*lternating *c*urrent"). The waveform continually alternates between a peak positive value and a peak negative value, following a sine wave or cosine wave. The magnitude of the peak value I_p is called the *amplitude* of the waveform.

The sinusoidal waveform is *periodic*, repeating a basic pattern over and over. In Fig. 2.13*b*, the time for a full cycle is denoted by T, which is called the *period* of the sine wave. A related concept is the *frequency*, denoted by f, which specifies how many full cycles the waveform goes through in one second. The relationship between frequency and period is

$$f = \frac{1}{T} \tag{2.9}$$

The unit of frequency is the *hertz*, abbreviated Hz (1 Hz = one cycle/second).

An expression that combines the amplitude and frequency information for the current waveform of Fig. 2.13*b* is

$$i = I_p \sin(2\pi f t) \tag{2.10}$$

Every time the "angle" $2\pi f t$ increases by 2π radians (or 360°), the waveform repeats. Thus another frequency, the *angular frequency*, can be defined. We denote the angular frequency by ω.

$$\omega = 2\pi f \tag{2.11}$$

The dimension of angular frequency is radians/second. A frequency of 1 hertz corresponds to an angular frequency of 2π radians/second.

A quantity related to the peak amplitude, called the *root-mean-square (rms) amplitude*, is often used to specify the magnitude of a sinusoid. The rms amplitude of a waveform is the square root of the time average value of the square of a waveform. For a sinusoid, the relation between the rms amplitude I_{rms} and the peak amplitude I_p is

$$I_{rms} = \frac{I_p}{\sqrt{2}} \tag{2.12}$$

Thus the sinsusoid of Eq. 2.10 could also be written

$$i = \sqrt{2} I_{rms} \sin(2\pi f t) \tag{2.13}$$

It is conventional when dealing with ac voltages and currents to use the rms instead of the peak amplitude. For example, the voltage on the electric power lines supplying homes and industry is a 60 Hz sinusoidally varying voltage. The amplitude of this voltage is usually written as 115 Vac, which is an equivalent name for 115 volts rms. Thus the *peak amplitude* of the voltage on the 115 Vac power line is $\sqrt{2} \times 115 = 163$ volts.

The waveform of Fig. 2.13*c* consists of a sequence of positive-going pulses. This kind of waveform is found in digital systems and computers. The principal feature of these digital waveforms is the ON or OFF nature of the pulse sequence at a precise instant of time. The precise value of the amplitude is not critical as long as it falls within a rather wide predetermined range. Pulse sequences might be periodic, in which case they could be described with a frequency, just as with the sinusoidal waveforms.

2.3.4 Notation Conventions

To distinguish the different types of time-varying waveforms, a convention has been widely adopted regarding the use of capital or lower-case letters for network variables (voltage and current). Table 2.3 summarizes these conventions. Lower-case variables (i or v) with upper-case subscripts are used to represent general time-varying waveforms. Upper-case variables with upper-case subscripts are used to represent dc or constant waveforms. Upper-case variables with lower-case subscripts indicate the peak amplitude of sinusoidal waveforms. Finally, when a waveform consists of a dc part plus a small time-varying part, a lower-case variable with a lower-case subscript is used to represent the time-varying part that is then called the *incremental* component of the waveform.

TABLE 2.3

Notation Conventions for Waveforms

Waveform	Variable	Subscript	Examples
General network variable	Lower case	Upper case	v_A, i_C
DC component of a waveform	Upper case	Upper case	V_{CC}, I_B
Peak amplitude of a sinusoid	Upper case	Lower case	V_o, I_a
Incremental component of a waveform	Lower case	Lower case	v_a, i_c

QUESTIONS

Q2.1 Suppose atoms consisted of negatively charged nuclei surrounded by a cloud of positive charges, some of which could break free and become mobile charge carriers. How, if at all, would the definitions of voltage, current, and power be affected?

Q2.2 Why is a sign convention needed to define power?

Q2.3 Formulate Kirchhoff's laws for a fluid-flow system, in which pressure and flow rate are the "network" variables.

Q2.4 What form does the electrical energy absorbed by a light bulb take? By a battery being charged (Fig. 2.9)? By a loudspeaker? By a motor?

Q2.5 Does the voltage across an independent voltage source depend on the network to which the source is connected?

Q2.6 Does the voltage across an independent current source depend on the network to which the source is connected?

EXERCISES

E2.1 What is the value of the current in the network element in Fig. E2.1?

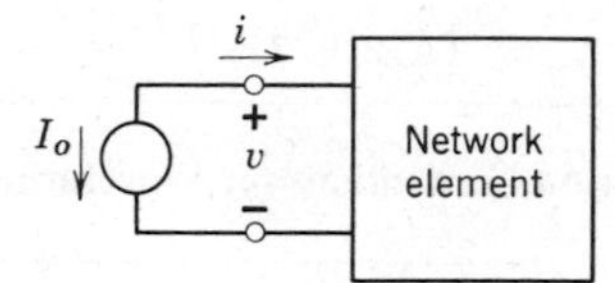

Figure E2.1

E2.2 A network element N is connected in a network such that

$$v = A \cos \omega t$$

and

$$i = B \cos \omega t$$

Find the instantaneous power dissipated in N (Fig. E2.2) and sketch a graph of power versus time

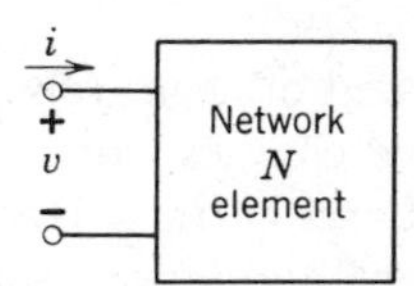

Figure E2.2

E2.3 Use the definitions of independent voltage and current sources to prove that $i = i_0$ in Fig. E2.3.

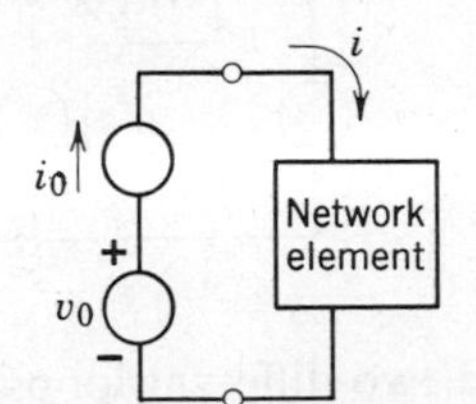

Figure E2.3

E2.4 Use the definition of independent voltage and current sources to prove that $v = -v_0$ in Fig. E2.4.

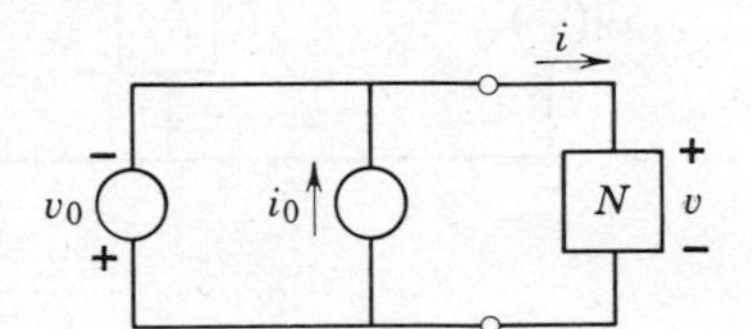

Figure E2.4

E2.5 What is the voltage across the element N in Fig. E2.5?

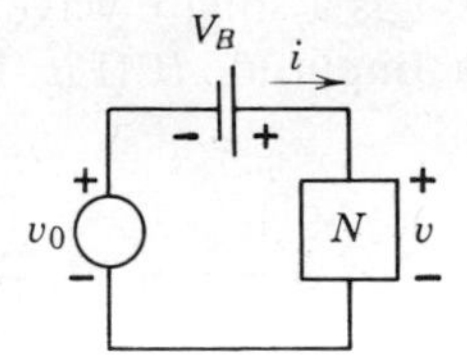

Figure E2.5

E2.6 What is the current through element N in Fig. E2.6?

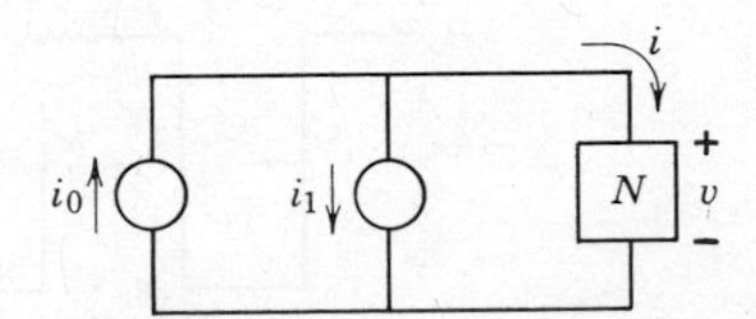

Figure E2.6

E2.7 Write KCL at node A of Fig. E2.7.

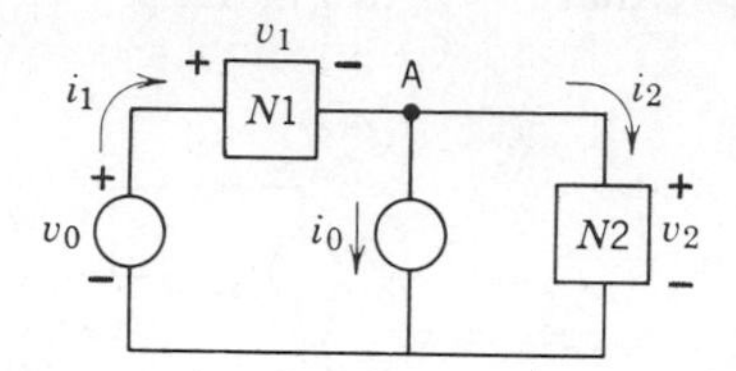

Figure E2.7

E2.8 Write KVL around two different loops and write KCL at node N of Fig. E2.8.

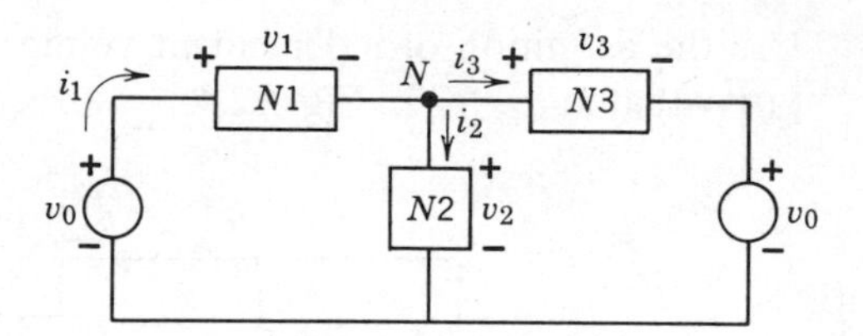

Figure E2.8

PROBLEMS

P2.1 For the example of Exercise E2.2, find the *average power* P_{av} by averaging the instantaneous power over one period of the power waveform.

P2.2 Repeat P2.1, where v is a square wave with amplitude A', and i is a square wave with amplitude B' (Fig. P2.2).

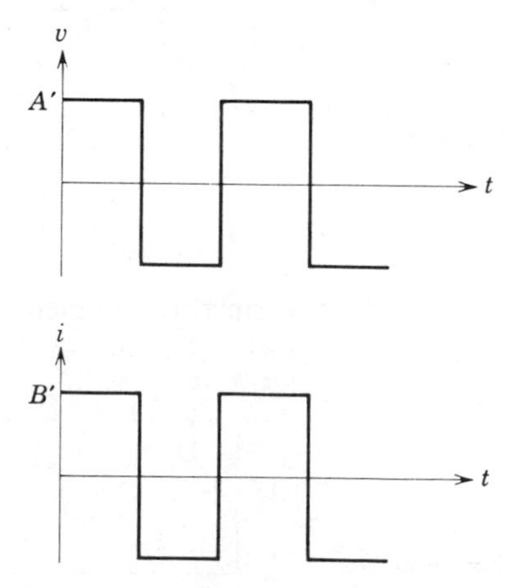

Figure P2.2

P2.3 Prove that Eq. 2.12 is correct.

P2.4 Use KVL and KCL to show that

$$v_1 = v_0$$

$$i_2 = -i_3$$

$$v_2 = v_3$$

Then show that $i_4 = -(i_1 + i_2)$, proving that for current flow a three-terminal element behaves like a single node (Fig. P2.4).

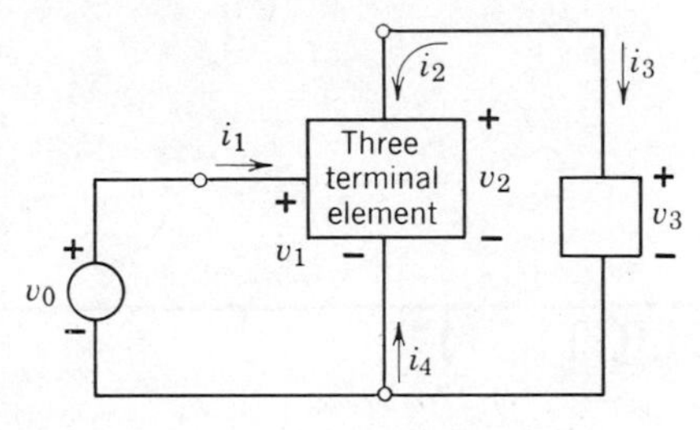

Figure P2.4

CHAPTER THREE
LINEAR RESISTIVE NETWORKS

3.0 INTRODUCTION

3.0.1 Resistive Elements

All network elements are characterized by some kind of relation between the voltage at the terminals and the current through the element. We have already seen that ideal sources are characterized either by a constant voltage, independent of current (the voltage source), or by a constant current, independent of voltage (the current source). All other elements of interest exhibit some interdependence between voltage and current.

A *resistive element* is an element for which the relation between voltage and current can be plotted as a graph. Such a graph is called, appropriately enough, a *$v - i$ characteristic*, an example of which is shown in Fig. 3.1. If one knows the voltage at the terminals of element D in Fig. 3.1, one can use the graph to determine the current through element D. Similarly if the current is known, the voltage can be determined.

Not all network elements are resistive elements. In Chapter 6 we shall encounter network elements for which the current at any instant of time depends on the rate of change of the voltage rather than on its value at that same instant. The class of resistive elements, however, is surprisingly large. Most of the familiar electronic devices (resistors, diodes, transistors, and vacuum tubes) behave as resistive elements over substantial parts of their operating ranges. We shall explore in this and the following chapters a rich variety of examples ranging from simple networks of resistors to complete

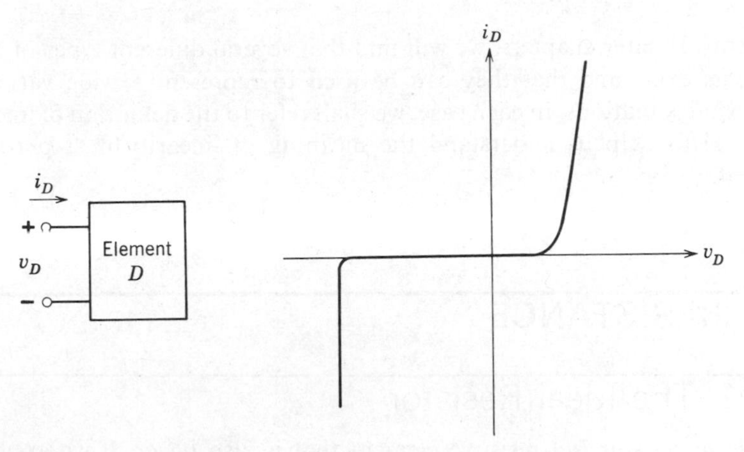

Figure 3.1
An example of a v-i characteristic for a resistive element.

amplifiers, all falling under the general heading of resistive networks. In the present chapter, however, our attention is limited to the subset of resistive elements that are linear.

3.0.2 Linearity

Linear elements have the following particular kind of interdependence between voltage and current. Suppose one has a function $f(x)$ that depends on a single variable x. In the context of electric circuits, x is an input or excitation, and $f(x)$ is an output or a response. In the case of the terminal characteristics of a network element, x might represent the voltage and $f(x)$ might represent the current. Alternatively, we might wish to identify x as the current and $f(x)$ as the voltage. The use of functional notation simply reminds us of the interdependence of the two quantities.

Suppose that when $x = x_1$, the function takes on the specific value $f(x_1)$, and when $x = x_2$, the function takes on a different value $f(x_2)$. The function $f(x)$ is called *linear* if for *any* two inputs, x_1 and x_2, the following relation is true:

$$f(x_1 + x_2) = f(x_1) + f(x_2) \tag{3.1}$$

Equation 3.1 says in mathematical language that the response to the sum of two excitations $(x_1 + x_2)$ must be equal to the sum of the responses to each excitation taken separately (first x_1, then x_2). We shall see in the following section how this definition of linearity is applied to resistive elements. In later chapters, we will find that several different types of linear elements exist, and that they can be used to represent a wide variety of interesting situations. In each case, we shall refer to the definition of linearity (Eq. 3.1) to help us understand the meaning of linearity in a particular context.

3.1 RESISTANCE

3.1.1 The Ideal Resistor

The *ideal resistor* is a resistive element that is also *linear*. If we apply the definition of linearity, the relation between voltage and current is such that if the current is doubled, the voltage is also doubled. More generally, the voltage must be proportional to the current.

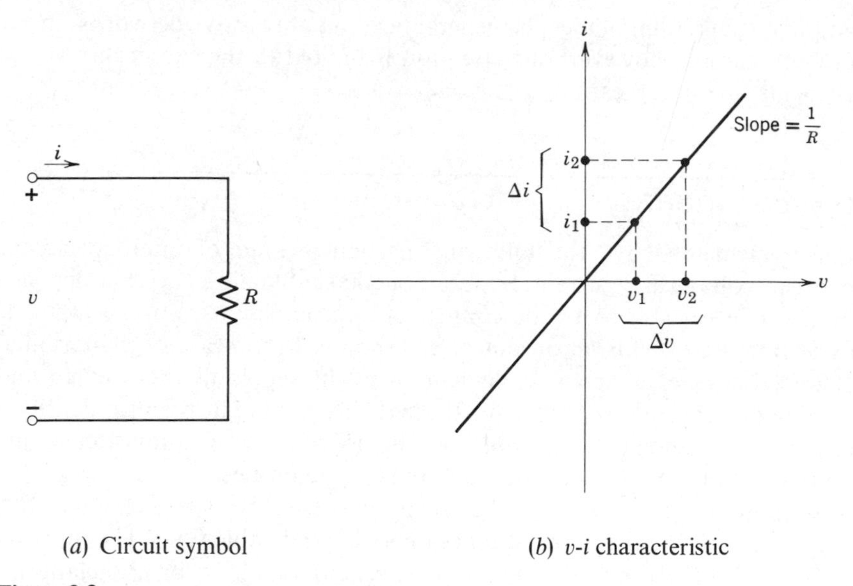

Figure 3.2
Circuit symbol and v-i characteristic of the ideal resistor.

The circuit symbol for an ideal resistor is shown in Fig. 3.2*a*, labeled R. The proportional relation between voltage and current, called *Ohm's law*, can be written in two ways. First,

$$v = iR \tag{3.2}$$

where R is the *resistance* of the element. The unit of resistance is the ohm (symbol Ω). Alternatively, one can express Ohm's law as

$$i = Gv \tag{3.3}$$

where $G = (1/R)$ is the *conductance* of the element. The unit of conductance is the siemens (or S). An older, widely used name for the siemens is the mho (or Ω^{-1}). The $v - i$ characteristic for the ideal resistor is shown in Fig. 3.2*b*. The plot is a straight line through the origin with slope $1/R$.[1]

Figure 3.3 can be used to demonstrate that the ideal resistor is a linear element. In Fig. 3.3*a* two voltage sources are connected in series with an ideal resistor, R. Writing KVL for the network yields

$$v_1 + v_2 - i_R R = 0 \tag{3.4}$$

or

$$i_R = \frac{v_1 + v_2}{R} \tag{3.5}$$

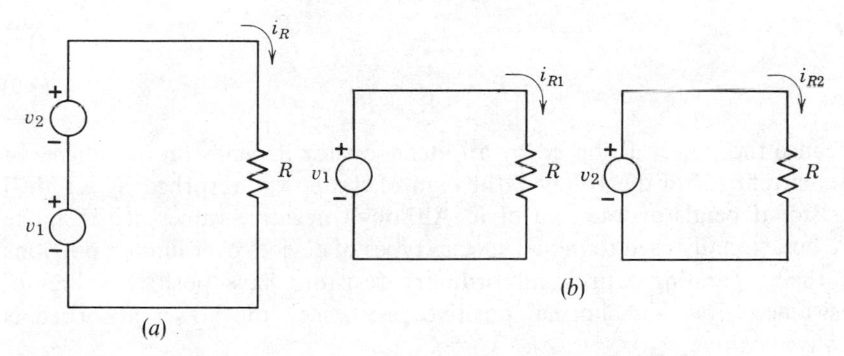

Figure 3.3
Linearity of the ideal resistor.

Figure 3.3*b* shows the two voltage sources connected separately to an ideal resistor R. The currents i_{R1} and i_{R2} are given by

$$i_{R1} = \frac{v_1}{R} \qquad i_{R2} = \frac{v_2}{R} \tag{3.6}$$

---

[1] The slope of a function plotted on x–y axes is defined as the ratio of a small change in y (denoted by Δy) to a corresponding small change in x (denoted by Δx) in the limit that Δx gets very small. In the notation of differential calculus, the slope is the same as the derivative, written

$$\text{Slope} = \frac{dy}{dx} = \underset{\Delta x \to 0}{\text{limit}} \frac{\Delta y}{\Delta x}$$

If the ideal resistor is linear, Eq. 3.1 requires that

$$i_R = i_{R1} + i_{R2} \tag{3.7}$$

Substitution from Eqs. 3.5 and 3.6 into 3.7 reduces to the algebraic identity

$$\frac{v_1 + v_2}{R} = \frac{v_1}{R} + \frac{v_2}{R} \tag{3.8}$$

Therefore, we conclude that the ideal resistor is a linear element.

3.1.2 Power in an Ideal Resistor

Noting that the sign conventions for the voltage and current polarities in Fig. 3.2 agree with those defined in Fig. 2.4, the power absorbed by an ideal resistor is given by

$$P = vi = (iR)i = i^2R$$

or

$$P = v\frac{v}{R} = \frac{v^2}{R} \tag{3.9}$$

Since the power absorbed by an ideal resistor depends on the *square* of the current (or of the voltage), the sign of the power absorbed by an ideal resistor depends on the sign of R. Although negative values of resistance are occasionally used to model special types of devices over limited portions of their operating ranges, all ordinary resistors have positive values of resistance. For these normal positive resistances, the power absorbed is always positive.

The electrical energy absorbed by a resistor must, by the law of conservation of energy, get converted to some other form. The most common form is heat energy, through the process known as *Joule heating*. The rate of Joule heat production in a resistor is identical to the rate of electrical energy absorption (Eq. 3.9) except for those resistive elements (e.g., light bulbs or loudspeakers) that convert a fraction of the absorbed energy to yet another form (e.g., radiant energy and sound energy).

3.1.3 Actual Resistors

The behavior of a wide range of real conducting materials can be modeled very accurately with ideal resistors. Metals and homogeneous semiconductors exhibit linearity between voltage and current over an enormous range of voltages and currents. Electrical devices, called *resistors*, are manufactured to have specific values of resistance. They are commonly made from pressed carbon powders, metal films, or fine wires. These real devices, the resistors, behave very nearly like ideal resistors. For this reason, we shall drop the distinction between "ideal" and "real" resistors and shall use the term "resistor" to refer both to the actual device and to the idealized model.

In actual resistors, however, Ohm's law is *not* valid for all ranges of current and voltage. The key to the breakdown of Ohm's law lies in the expression for the power absorbed by the resistors (Eq. 3.9). Joule heating raises the internal temperature of the resistors. The highest permissible internal temperature for a given resistor is determined by its composition and geometry. The *power rating* of a resistor is specified so that the internal temperature remains within permissible limits. If the power absorbed exceeds the power rating, Joule heating might heat the resistor beyond its melting or burning temperature. Ohm's law fails dramatically in this case. The simple *fuse*, which provides overload protection for electric circuits, operates on the basis of the Joule heat generated in the fuse material. The fuse link is designed so that the rated fuse current raises the link to its melting point, which causes the circuit to open and interrupt further current flow.

3.2 BASIC NETWORK CONFIGURATIONS

3.2.1 Simple Applications of Kirchhoff's Laws

We can now begin to build networks out of ideal sources and resistors. The simplest connections involve one source and one resistor, and are shown in Fig. 3.4.

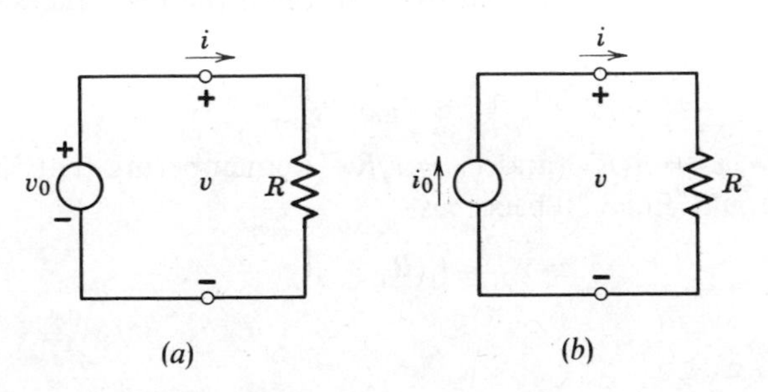

Figure 3.4
Simple source-resistor connections.

Figure 3.4*a* shows a voltage source connected to a resistor. By the definition of the voltage source, the voltage across the resistor must equal v_0. By Ohm's law, therefore, $i = v/R = v_0/R$. (Question: What is the current flowing through the v_0 source?)

Figure 3.4*b* shows the corresponding network involving a current source. By definition of the current source, the current is fixed at $i = i_0$. Ohm's law then yields $v = iR = i_0R$. (Question: What is the voltage across the i_0 source?)

3.2.2 Resistors in Series

Figure 3.5 shows a voltage source in series with two resistors, with the currents and voltages labeled for each resistor. KCL requires that i_1 and i_2 be equal.

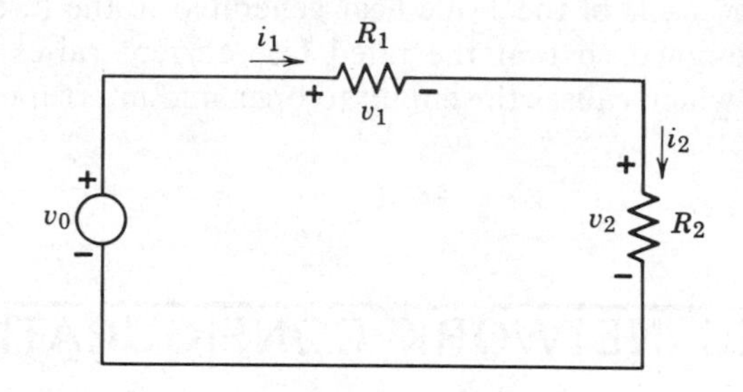

Figure 3.5
Resistors in series.

Applying KVL around the loop, the voltage drops by $-v_0$ through the source, and drops by v_1 and then by v_2 through the resistances. Therefore, by KVL

$$v_0 = v_1 + v_2 \tag{3.10}$$

By Ohm's law, $v_1 = i_1R_1$, and $v_2 = i_2R_2$. Remembering that the currents i_1 and i_2 are equal, Eq. 3.10 becomes

$$v_0 = i_1(R_1 + R_2) \tag{3.11}$$

Therefore,

$$i_1 = i_2 = \frac{v_0}{R_1 + R_2} \tag{3.12}$$

Once the current flowing in each resistance is known, the voltage across each resistance can be found from Ohm's law:

$$v_1 = \frac{R_1}{R_1 + R_2} v_0$$
$$v_2 = \frac{R_2}{R_1 + R_2} v_0 \qquad (3.13)$$

Examine Eq. 3.12 and compare it to Ohm's law. The current through the series combination of R_1 and R_2 is exactly the same as the current that would flow if a single resistance of value $(R_1 + R_2)$ were connected to the voltage source. Therefore, for purposes of calculating the current, *resistances in series add* (Fig. 3.6).

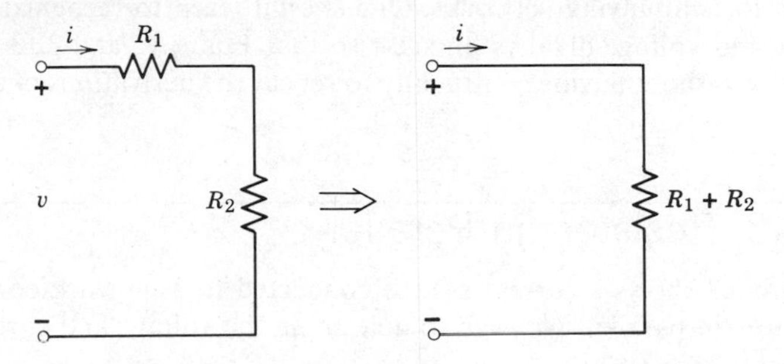

Figure 3.6
Resistances in series add.

The second result of importance is contained in Eq. 3.13. It states that the total voltage across resistors in series gets divided proportional to the resistance of each element. This relation, called the *voltage divider* relation, is illustrated in Fig. 3.7.

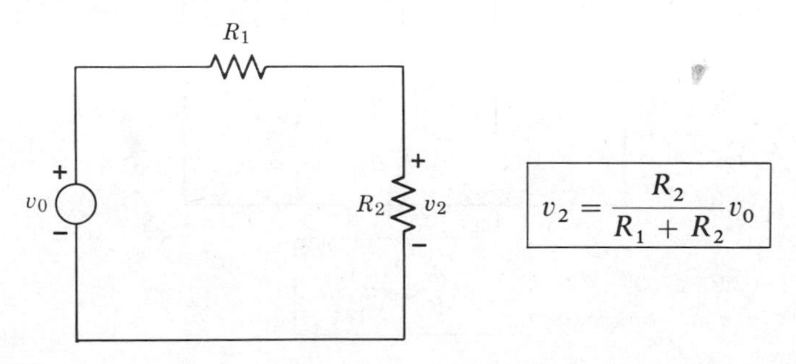

Figure 3.7
The voltage divider.

An Important Comment:

We have just used KCL, KVL, and the $v - i$ relations for each element in the network of Fig. 3.5 to determine the current and voltage for every element in the network. In general, a systematic application of Kirchhoff's laws together with the $v - i$ characteristics will always yield a complete solution for the voltages and currents. In most problems, however, one is not interested in such detailed information about every element in the network. Instead, one might be interested in the value of a single voltage or current. There is an advantage, therefore, in being able to concentrate on the particular variable of interest, and to simplify the rest of the network as much as possible. The example just completed illustrates two important results about linear resistive networks that will be employed again and again for simplifying networks. One should learn to recognize resistors in series and voltage dividers on sight so that Eqs. 3.12 and 3.13 can be used directly without having continually to repeat the derivations of these results.

3.2.3 Resistors in Parallel

Figure 3.8 shows a current source connected to a network containing two resistors in parallel. We wish to determine the voltage and current for each element.

Since all three elements are in parallel, they must all have the same voltage across them. Calling this voltage v and using Ohm's law, one finds

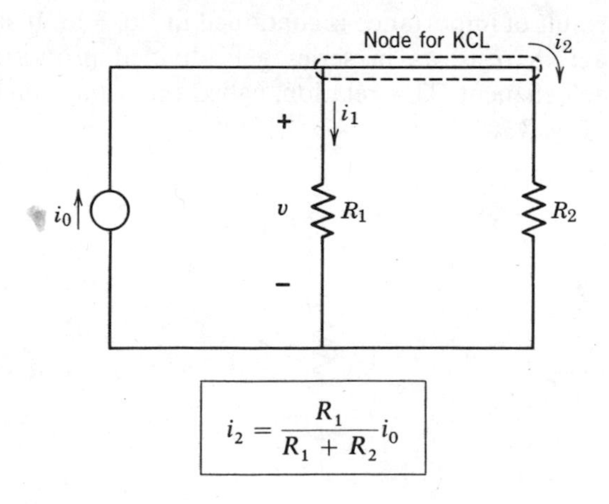

$$i_2 = \frac{R_1}{R_1 + R_2} i_0$$

Figure 3.8
Resistors in parallel; the current divider.

that $i_1 = v/R_1$ and $i_2 = v/R_2$. Kirchhoff's current law can be used at the node joining the three elements. The currents entering the node are, i_0, $-i_1$, and $-i_2$. By KCL, therefore,

$$i_0 - i_1 - i_2 = 0 \tag{3.14}$$

or

$$i_0 = i_1 + i_2 = v\left(\frac{1}{R_1} + \frac{1}{R_2}\right). \tag{3.15}$$

Equivalently,

$$v = \frac{R_1 R_2}{R_1 + R_2} i_0 \tag{3.16}$$

and

$$\begin{aligned} i_1 &= \frac{R_2}{R_1 + R_2} i_0 \\ i_2 &= \frac{R_1}{R_1 + R_2} i_0 \end{aligned} \tag{3.17}$$

The first result (Eq. 3.16) states that the total voltage across the parallel combination of R_1 and R_2 is the same as that which occurs across a single resistance of value $R_1 R_2/(R_1 + R_2)$. Because this expression for parallel resistance occurs so often, it is given a special notation $(R_1 \| R_2)$. That is, when R_1 and R_2 are in parallel (Fig. 3.9), the equivalent resistance is

$$(R_1 \| R_2) = \frac{R_1 R_2}{R_1 + R_2}. \tag{3.18}$$

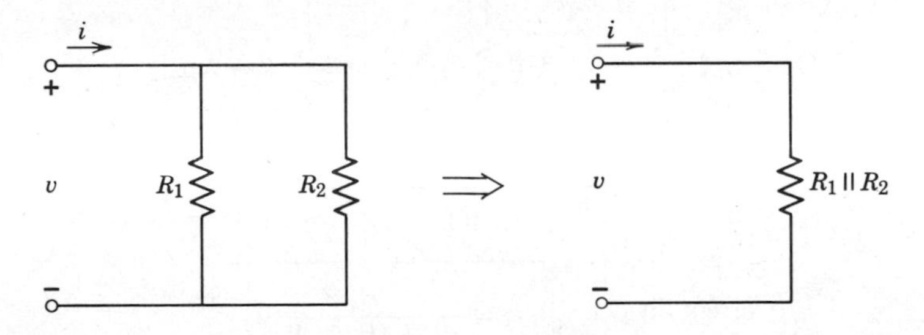

Figure 3.9
Combining resistors in parallel.

The second set of results (Eq. 3.17) indicates how the current from the source divides between the two conducting paths. Much like water in a pipe, one expects more flow down the pipe with the *smallest* resistance. Equation 3.17 has this characteristic. If R_1 and R_2 are exactly equal, $i_1 = i_2 = i_0/2$. But, if R_1 is larger than R_2, more of the current flows in R_2. This configuration of resistances is called a *current divider*, and is illustrated in Fig. 3.8. As with the voltage divider, one should learn to identify a current divider on sight and apply Eq. 3.17 directly.

3.2.4 A Network Example

Figure 3.10*a* shows a 6 mA source, i_0, driving a network of resistors.

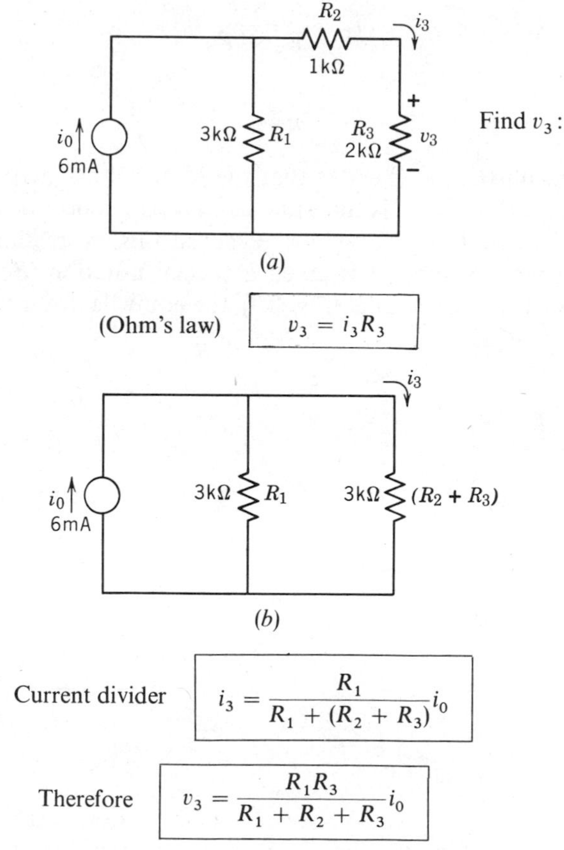

Figure 3.10
A network example.

Suppose that one needs to determine v_3, the voltage across R_3. One *could* write down KVL and KCL and Ohm's law and solve for all the voltages and currents in the network. *But this would be an enormous waste of effort.* Instead, let us break the network down into simple pieces.

First, $v_3 = i_3 R_3$ by Ohm's law. Second, we can find i_3 from the current divider relation. In order to use the current divider relation, however, we must recognize that R_2 and R_3 can be combined into a single 3 kΩ resistance for purposes of determining the total current through the series combination. The current divider relation (see Fig. 3.10*b*) yields

$$i_3 = \frac{R_1}{R_1 + (R_2 + R_3)} i_O \tag{3.19}$$

Therefore, we obtain the final result:

$$v_3 = i_3 R_3 = \frac{R_1 R_3}{R_1 + R_2 + R_3} i_O \tag{3.20}$$

With the numerical values given, we find that $i_3 = i_O/2 = 3$ mA, and that $v_3 = 3\text{ mA} \times 2\text{ k}\Omega = 6$ volts.

The student should mark well the method used to find the solution. Attention was focused on what was needed to obtain the required answer. Once it was determined how v_3 was related to i_3, *any* simplification that helped in finding i_3 was permitted. In this case, the current divider was used, after recognizing the presence of a series combination of resistances. This kind of network analysis does not proceed by formal rules. It depends instead on the student developing a working knowledge of the simple relations, the skill to recognize what is needed, and the ability to simplify wherever possible to reach the problem solution.

3.3 SUPERPOSITION

3.3.1 An Example

Consider the network of Fig. 3.11. It is identical to the network of Fig. 2.5, except that the four network elements have been specified as sources or resistors. There are many good ways to analyze this network, depending on which voltages or currents one might wish to determine. For now, let us find both i_1 and i_2 using KVL, KCL, and Ohm's law. The following relations

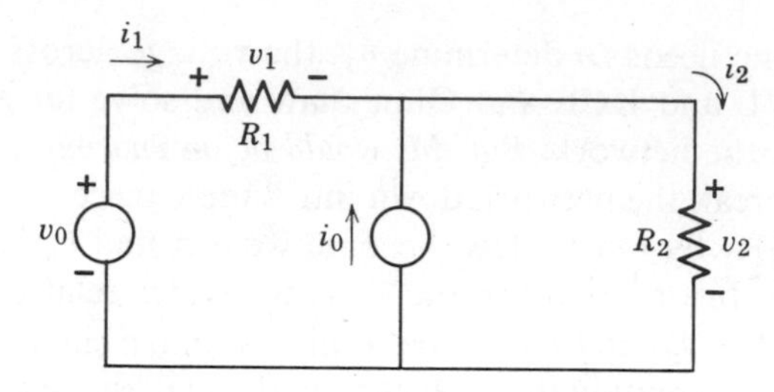

Figure 3.11
A superposition example.

are obtained for the network:

$$\begin{aligned} &\text{KVL} && v_0 = v_1 + v_2 \\ &\text{KCL} && i_0 + i_1 = i_2 \\ &\text{Ohm's law} && v_1 = i_1 R_1 \\ &\text{Ohm's law} && v_2 = i_2 R_2 \end{aligned} \tag{3.21}$$

Substituting in KVL for v_1 and v_2, one obtains

$$v_0 = i_1 R_1 + i_2 R_2 \tag{3.22}$$

which can be solved for i_2 to yield

$$i_2 = \frac{(v_0 - i_1 R_1)}{R_2} \tag{3.23}$$

Substituting for i_2 in KCL yields, after algebraic manipulation

$$i_1 = \frac{v_0}{R_1 + R_2} - \frac{R_2}{R_1 + R_2} i_0 \tag{3.24}$$

Finally, substitution of the result for i_1 back into Eq. 3.23 yields

$$i_2 = \frac{v_0}{R_1 + R_2} + \frac{R_1}{R_1 + R_2} i_0 \tag{3.25}$$

Let us examine these solutions for i_1 and i_2 carefully. First, each current is the sum of two terms, and each term is exactly proportional to one and only one independent source. This result illustrates an extension of the concept of linearity to cases with more than one independent source. If i_0 happened to be zero, i_1 and i_2 would be exactly proportional to v_0 (linearity), and if v_0 were zero, i_1 and i_2 would be exactly proportional to i_0 (linearity). We say then that the solutions for i_1 and i_2 are *superpositions* of linear responses, one response from each independent source. If the network had three independent sources, each response would consist of three terms, one term proportional to each independent source. This suggests a way of breaking up problems containing more than one independent source into several smaller problems. This method is called *superposition*:

> Determine the response to each independent source, one at a time, assuming that all other independent sources are zero. Sum the results to get the total response.

3.3.2 Suppression of Independent Sources

To use the superposition method, one must know what happens to a network when one sets an independent source to zero. The definitions of the independent sources provide the answer.

By definition, a voltage source maintains the value of the source voltage regardless of the current. If this maintained voltage is assumed to be zero, it means that any amount of current can flow and no voltage will result. We call this a *short circuit*. Schematically, a short circuit looks just like the solid lines we use to connect network elements (Fig. 3.12*a*).

To set a current source to zero, we again use the definition. If the current is to be zero regardless of the voltage, we have an *open circuit*. No current can flow in an open circuit. Schematically, an open circuit is represented by a broken connection, or more simply by the absence of any connection (Fig. 3.12*b*).

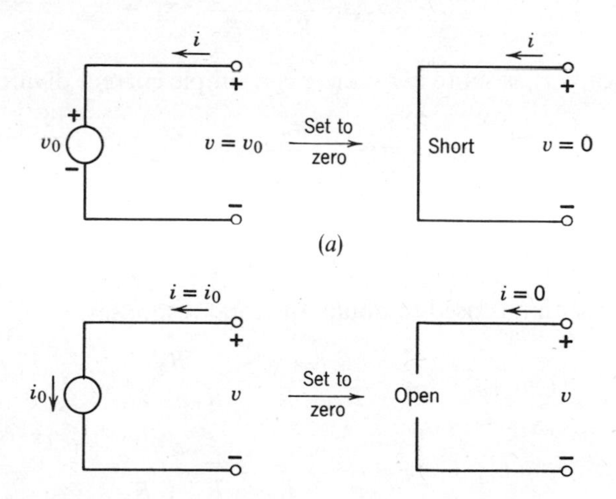

Figure 3.12
Suppression of independent sources.

3.3.3 An Example Revisited

Let us return to the network of Fig. 3.11 and determine the currents i_1 and i_2 using superposition. Figure 3.13 shows the two simpler networks that result, first, when i_0 is set to zero (Fig. 3.13*a*) and, second, when v_0 is set to zero (Fig. 3.13*b*).

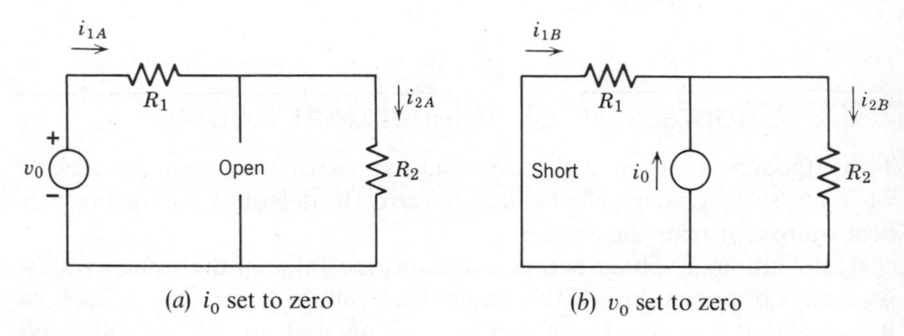

Figure 3.13
Superposition solution of the example of Fig. 3.11.

For the case when i_0 is set to zero, R_1 and R_2 appear in series, and both must carry the same current. Therefore,

$$i_{1A} = i_{2A} = \frac{v_0}{R_1 + R_2} \tag{3.26}$$

For the case when v_0 is set to zero, there is a simple current divider:

$$\begin{aligned} i_{1B} &= -\frac{R_2}{R_1 + R_2} i_0 \\ i_{2B} &= +\frac{R_1}{R_1 + R_2} i_0 \end{aligned} \tag{3.27}$$

Finally, superposition is used to obtain the total responses:

$$\begin{aligned} i_1 &= i_{1A} + i_{1B} = \frac{v_0}{R_1 + R_2} - \frac{R_2}{R_1 + R_2} i_0 \\ i_2 &= i_{2A} + i_{2B} = \frac{v_0}{R_1 + R_2} + \frac{R_1}{R_1 + R_2} i_0 \end{aligned} \tag{3.28}$$

These results are identical to Eqs. 3.24 and 3.25, which were obtained using KVL and KCL on the entire network. The superposition method reached the same solution, but with much less algebraic manipulation. The ability to set independent sources to zero and the ability to recognize basic configurations (e.g., series resistors and current divider) have enabled us to use superposition and skip most of the algebra.

3.3.4 A Cautionary Note

It is worth emphasizing that *superposition can be used only in linear networks.*

3.4 EQUIVALENT CIRCUITS

3.4.1 The Equivalent Network Concept

Figure 3.14*a* illustrates a problem we shall encounter frequently. To the left of the terminal pair (A-A') is a linear resistive network made up of sources and resistances. To the right of the A-A' pair is an unknown element, used to represent the fact that we might connect many different things there. (We might, for example, connect A-A' to a resistance, or perhaps to a diode, or perhaps to the input connection of a transistor amplifier.) It would be a terrible waste of effort to have to analyze the entire network to the left of A-A' each time we connected it to a different element. In many cases, the only feature of interest in the network to the left of A-A' is the relation between the voltage across the terminal pair v_A and the current i_A. If we could find a simple equivalent network that had the *same relation between* v_A *and* i_A, much repetitive analysis could be avoided. In the next section we shall examine the relationship between the terminal voltage and the terminal

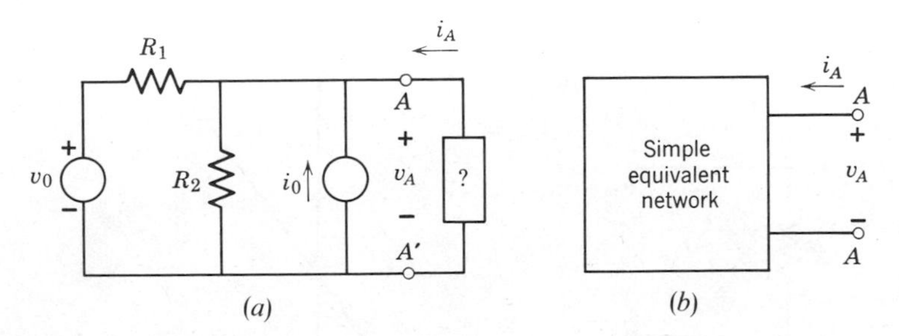

Figure 3.14
The equivalent network concept.

current for a few simple networks. Then, the principle of superposition will be used to develop two important general forms of equivalent networks. We shall then return to the network of Fig. 3.14*a* to illustrate the equivalent network concept.

3.4.2 Terminal Characteristics

The relation between terminal voltage and terminal current is called a *terminal characteristic*. We have already encountered three examples: the *v-i* characteristics of the ideal voltage source, the ideal current source, and

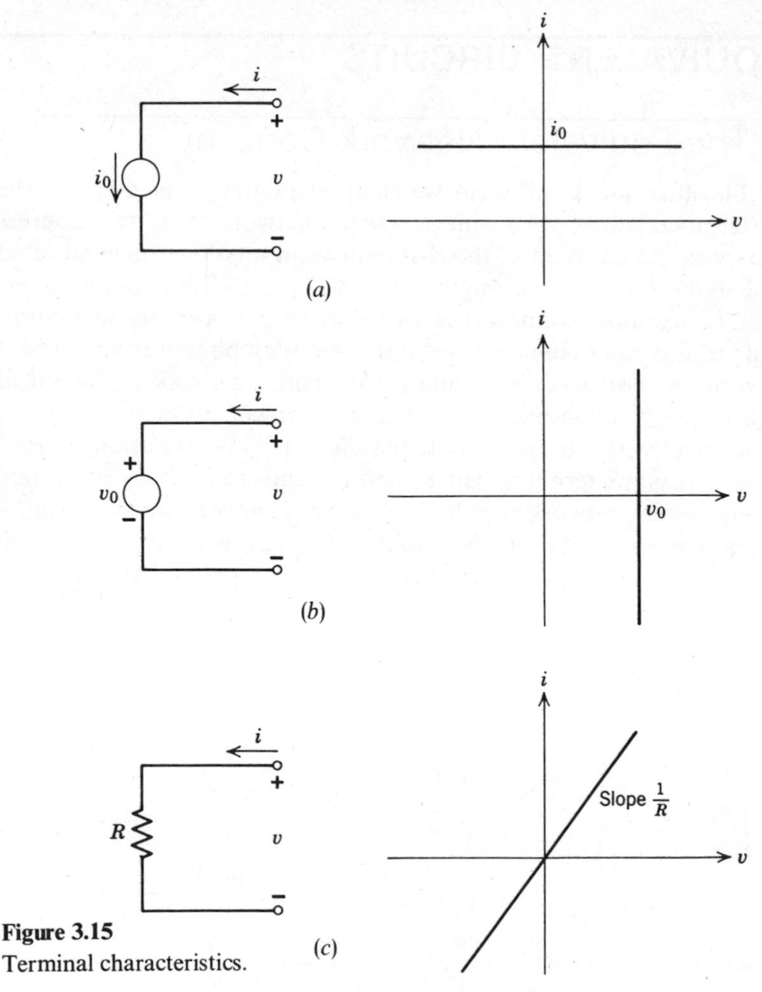

Figure 3.15
Terminal characteristics.

the ideal resistance (see Fig. 3.15). Let us now determine the terminal characteristics of a more general network.

Imagine that the network of Fig. 3.16*a* is connected to some other external network that has the ability to vary the terminal voltage v over a wide range. As the voltage v is varied, the current i must also vary. The terminal characteristic can be determined from KVL:

$$v = v_0 + iR \tag{3.29}$$

which, when solved for i, gives

$$i = \frac{v}{R} - \frac{v_0}{R} \tag{3.30}$$

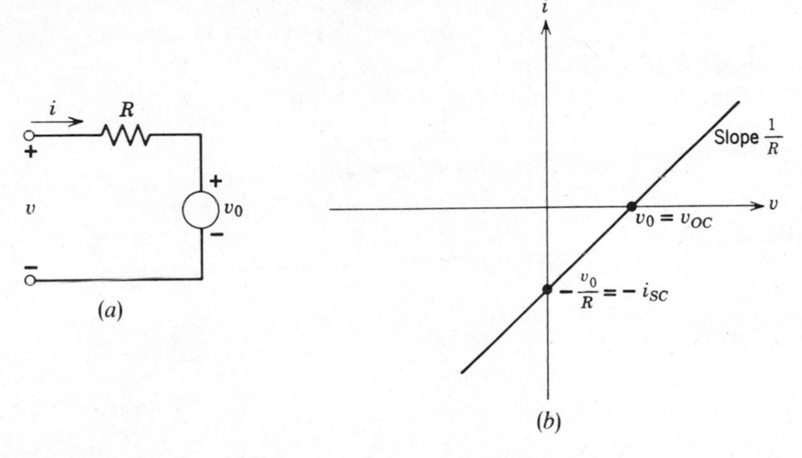

Figure 3.16
Terminal characteristic.

The graph of Eq. 3.30, shown in Fig. 3.16b, is a straight line that passes through $v = v_0$ when $i = 0$, and that passes through $i = -v_0/R$ when $v = 0$. The slope of the line is $1/R$, which is the same slope one would obtain for the resistance R taken alone.

The intercepts of this characteristic (at $v = 0$ or at $i = 0$) can be related to the behavior of the network under open- and short-circuit conditions (Fig. 3.17). If the network is not connected to anything else, no current can flow ($i = 0$). Whatever voltage appears between the terminals in this case is called the *open-circuit voltage*, denoted by v_{OC}. In the network of Fig. 3.16*a*, when $i = 0$, there is no voltage drop across the resistor. Therefore $v_{OC} = v_0$.

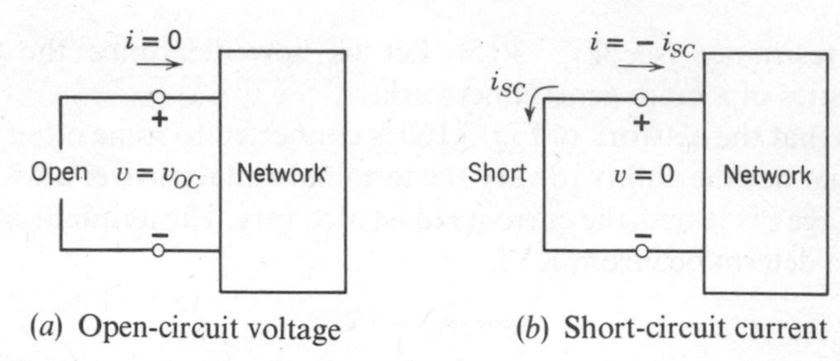

Figure 3.17
Open-circuit voltage and short-circuit current.

When a short circuit is connected between the terminals, the voltage v becomes zero regardless of the current. In this case, we define the current flowing out of the + terminal to be the *short-circuit current*, denoted by i_{SC} (Fig. 3.17*b*). With the polarity for i defined *into* the + terminal, $i = -i_{SC}$. For the network of Fig. 3.16*a*, $i_{SC} = v_0/R$.

Consider next the circuit of Fig. 3.18*a*. Using KCL, one obtains

$$i + i_0 = \frac{v}{R} \tag{3.31}$$

which transposes to

$$i = \frac{v}{R} - i_0 \tag{3.32}$$

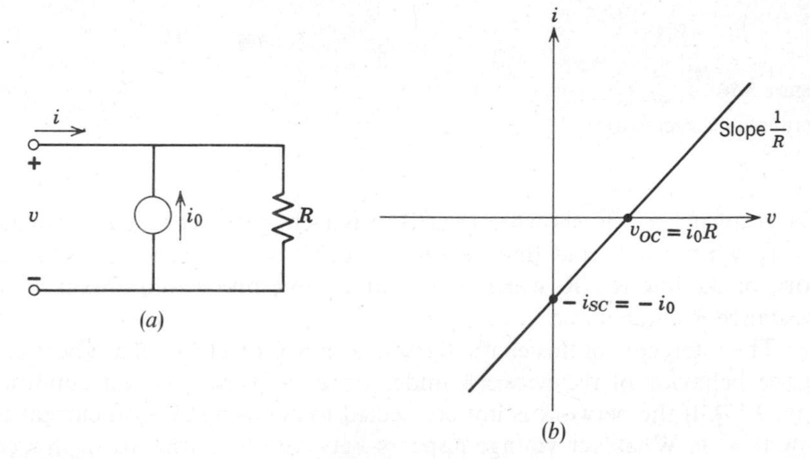

Figure 3.18
Terminal characteristic.

The open-circuit voltage is found by setting $i = 0$ in Eq. 3.32. The result is $v_{OC} = i_0 R$. Similarly, the short-circuit current ($v = 0$) is $i = -i_{SC} = -i_0$. Therefore, $i_{SC} = i_0$. The terminal characteristic for this network is plotted in Fig. 3.18*b*. Notice the similarity between this characteristic and that of Fig. 3.16*b*. Both have slopes of $1/R$; both have intercepts at v_{OC} and at $-i_{SC}$. In fact, the terminal characteristics of the two networks would be identical if v_0 happened to equal $i_0 R$. This idea is explored further in the next section.

3.4.3 Thévenin and Norton Equivalents

Consider the linear resistive network of Fig. 3.19. It may contain resistances and independent current and voltage sources. Imagine that it is connected to an external voltage source that fixes the terminal voltage v. If we use the superposition method of Section 3.3, the terminal current i must be a sum of terms, each term proportional to one independent source. Thus, by superposition

$$i = G_T v - [a_1 v_1 + a_2 v_2 + \ldots + a_N v_N + b_1 i_1 + \ldots + b_N i_N] \qquad (3.33)$$

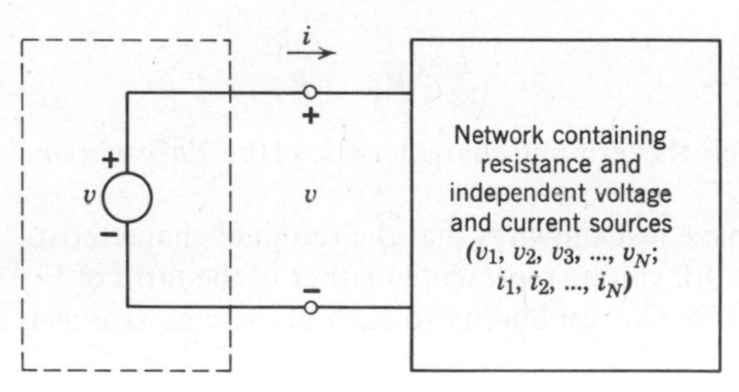

Figure 3.19
A general linear resistive network.

where $G_T, a_1, a_2, \ldots, b_N$ are constants determined from the simple superposition networks with all but one independent source suppressed. When $v = 0$ (short-circuit condition), we know the terminal current i must equal $-i_{SC}$. Therefore, the term in brackets, which depends only on the elements inside the network, is identified as i_{SC} and Eq. 3.33 becomes

$$i = G_T v - i_{SC} \qquad (3.34)$$

Since G_T must have the dimension of a conductance, G_T is equivalent to $1/R_T$, where R_T is the effective, or *equivalent* resistance for the network. That is,

$$i = \frac{v}{R_T} - i_{SC} \tag{3.35}$$

which is exactly the terminal characteristic of the network in Fig. 3.20*a*, called the *Norton equivalent network.*

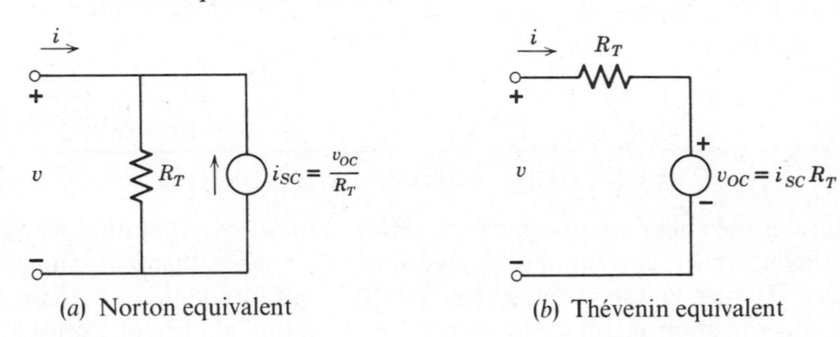

Figure 3.20
Thévenin and Norton equivalents.

Furthermore, the open-circuit voltage for the general network must from Eq. 3.35 be equal to i_{SC}/G_T, which is equal to $i_{SC}R_T$. Therefore, the terminal characteristic for the general network can also be written in the form below:

$$i = \frac{v}{R_T} - \frac{v_{OC}}{R_T} \tag{3.36}$$

This is precisely the terminal characteristic of the *Thévenin equivalent network* of Fig. 3.20*b*.

What we have just shown is that the terminal characteristic of *any* linear, resistive network can be represented either in the form of Eq. 3.35 or in the form of Eq. 3.36. Corresponding to each of these equations is a very simple network. The Norton equivalent network has a current source of magnitude i_{SC} in parallel with the equivalent resistance R_T, while the Thévenin equivalent network has a voltage source of magnitude v_{OC} in series with the same equivalent resistance R_T. The complete equivalence of these simple networks to each other is assured by the choice of R_T.

$$R_T = \frac{v_{OC}}{i_{SC}} \tag{3.37}$$

To determine the Thévenin or Norton equivalent network for any given network, either determine both v_{OC} and i_{SC} using standard network analysis methods and then determine R_T from Eq. 3.37, *or* find either v_{OC} or i_{SC} and then use a superposition argument to obtain R_T directly. Both methods are illustrated below.

Figure 3.21*a* shows a network with a single independent voltage source and two resistors. When $i = 0$, the relation between v and v_1 is a simple voltage divider. Therefore, the open-circuit voltage is

$$v_{OC} = \frac{R_2}{R_1 + R_2} v_1 \tag{3.38}$$

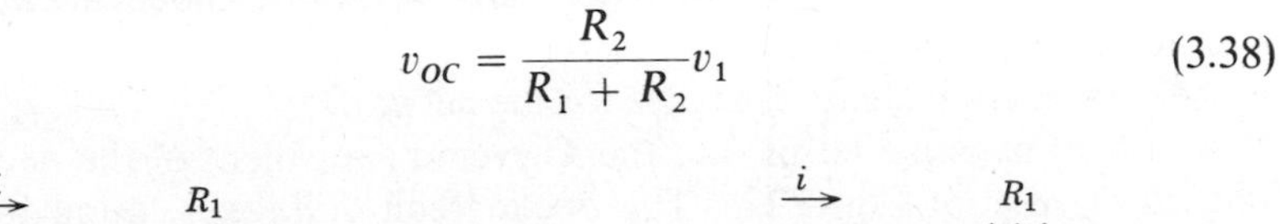

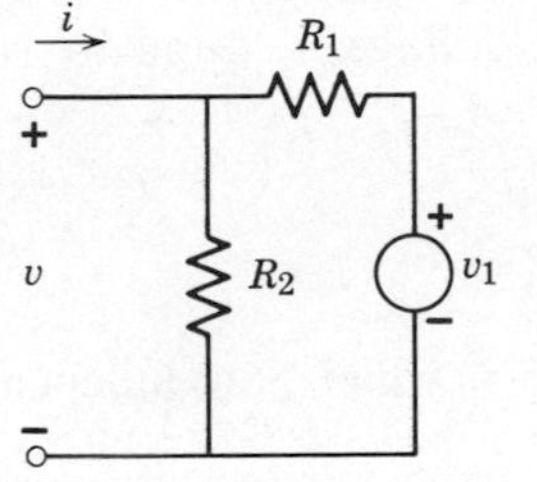

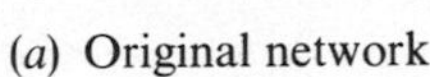

(*a*) Original network

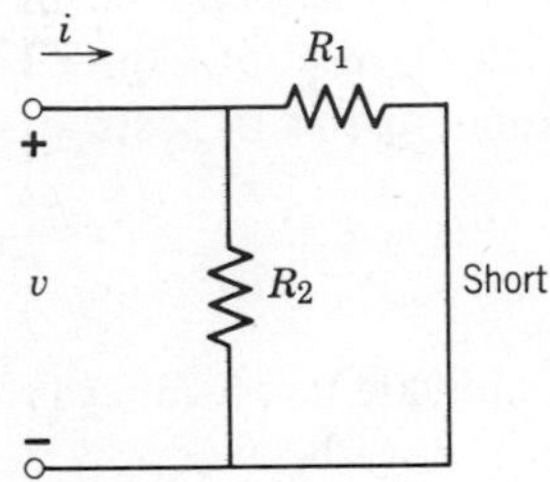

(*b*) Independent source suppressed

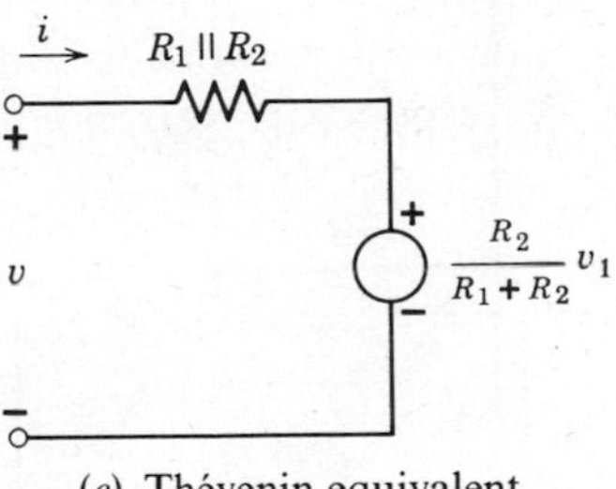

(*c*) Thévenin equivalent

Figure 3.21
Example of a Thévenin equivalent.

When $v = 0$, the current through R_1 is v_1/R_1, and all of this current must flow out the + terminal; otherwise, v could not be zero. Therefore, the short-circuit current is simply

$$i_{SC} = \frac{v_1}{R_1} \tag{3.39}$$

When we combine these results,

$$R_T = \frac{v_{OC}}{i_{SC}} = \frac{R_1 R_2}{R_1 + R_2} = (R_1 \| R_2) \tag{3.40}$$

The corresponding Thévenin equivalent network is shown in Fig. 3.21*c*.

An alternative method of finding the equivalent resistance is based on the fact that when *all* independent sources in the network are suppressed, the resistance seen between the terminals must be equal to R_T. This can be verified from Eq. 3.33. Suppression of all the independent sources makes the term in brackets zero, in which case $i = G_T v = v/R_T$. Alternatively, one can look at the general Thévenin network or the Norton network (Fig. 3.20). If the independent source in either is suppressed, the network reduces to a simple resistance R_T. Therefore, one can find R_T by suppression of the independent sources, as shown in Fig. 3.21*b*. With the independent source suppressed, the resistance between the terminals is simply $(R_1 \| R_2)$, which is the correct equivalent resistance for the network.

As a final example let us find the Thévenin equivalent of the network to the left of *A*-*A*′ in Fig. 3.22*a*. The open-circuit voltage is found by superposition, just as in Section 3.3.

$$v_{OC} = \frac{R_2}{R_1 + R_2} v_0 + \frac{R_1 R_2}{R_1 + R_2} i_0 \tag{3.41}$$

The equivalent resistance is found from Fig. 3.22*b*, in which both independent sources have been suppressed.

$$R_T = (R_1 \| R_2) \tag{3.42}$$

Finally, the Thévenin equivalent network is shown in Fig. 3.22*c*. As an exercise, construct the corresponding Norton equivalent network for this example.

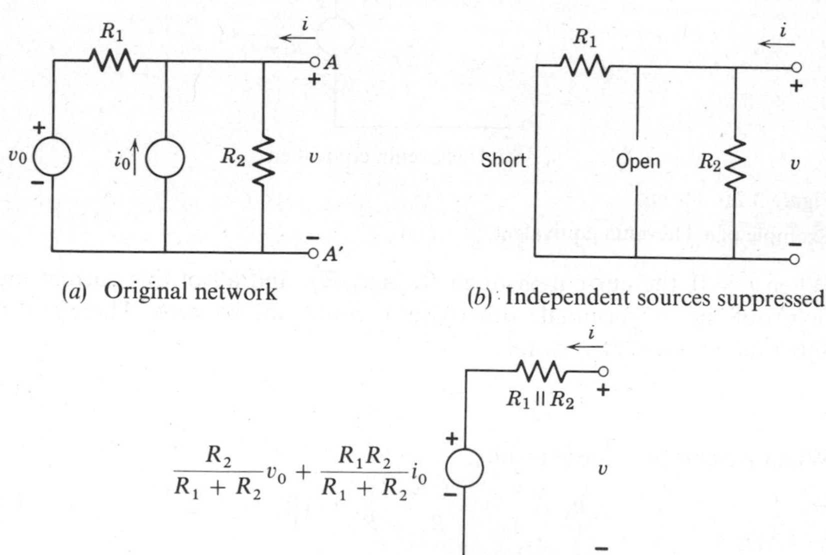

Figure 3.22
Thévenin equivalent of the network of Fig. 3.14.

3.5 SOURCES AND LOADS

3.5.1 Voltage and Current Transfer

We have seen that the terminal characteristics of any linear resistive network can be represented by a Thévenin or Norton equivalent circuit. This fact greatly simplifies the problem of determining what happens when networks are connected together. In this section, we consider a particular connection of networks: a source network is connected to a resistance called the load. Figure 3.23 illustrates the problem.

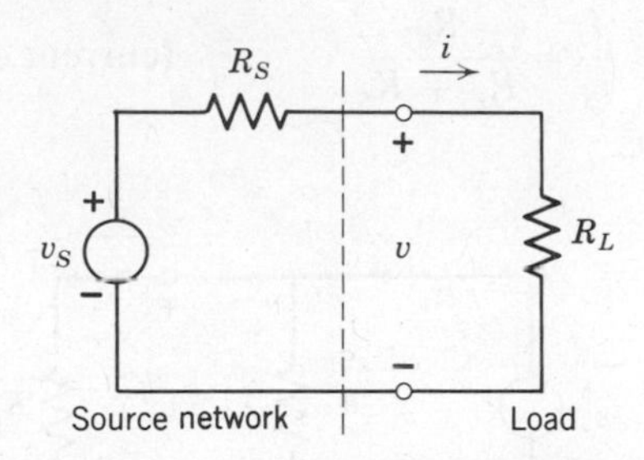

Figure 3.23
A Thévenin equivalent source network connected to a load.

The source network is shown in Thévenin equivalent form. Every real source, whether a battery, electronically regulated power supply, oscillator, or amplifier, can be modeled over at least a part of its operating range as a linear Thévenin network. The source resistance, R_S, indicates how much the terminal voltage drops when current is drawn from the source network. For example, the voltage across a battery decreases linearly as the current increases. Thus the battery could be modeled with the source network of Fig. 3.23.

The load might represent any network or element that we wanted the source to drive. The load for a high-fidelity amplifier source is the loudspeaker. Within the amplifier itself, the source might be the output of one stage of amplification, while the load might be the input of the next stage of amplification. There are many types of loads that can be accurately modeled as a simple resistance.

The relationship between the voltage across the load v and the open circuit source voltage v_S is called the voltage transfer. It is usually given as the ratio v/v_S. For the network of Fig. 3.23,

$$\frac{v}{v_S} = \frac{R_L}{R_L + R_S} \qquad \text{(voltage divider)} \qquad (3.43)$$

If R_L is very large compared to the source resistance R_S, then the voltage transfer ratio is approximately unity ($v = v_S$). Here the source network is acting like a voltage source. The voltage delivered to the load is not strongly dependent on the current in the load.

If R_L happens to be much smaller that R_S, the voltage transfer ratio becomes R_L/R_S. Here the current through the load is approximately equal to v_S/R_S, which is the short-circuit current of the source network. A convenient way to picture this example is to express the same source network in its Norton equivalent form (Fig. 3.24). Thus one can speak of a current transfer ratio:

$$\frac{i}{i_S} = \frac{R_S}{R_L + R_S} \qquad \text{(current divider)} \qquad (3.44)$$

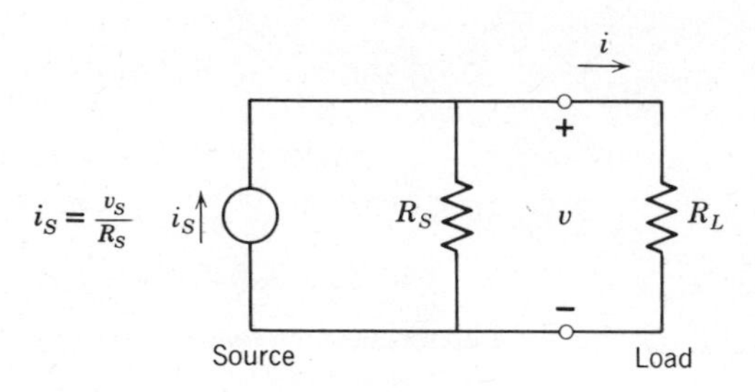

Figure 3.24
A Norton equivalent source network connected to a load.

When R_S is very large compared to R_L, the current divider relation tells us that most of the current goes through R_L. In this case, we say that the source network is acting like a current source. The current through the load does not depend strongly on the voltage across the load.

In summary, source networks with low resistances compared with the load behave like voltage sources; source networks with high resistances compared with the load behave like current sources.

3.5.2 Power Transfer

Power transfer is more complex than voltage or current transfer. Consider again the network of Fig. 3.23. The voltage across the load is

$$v = \frac{R_L}{R_L + R_S} v_S \qquad (3.45)$$

while the current through the load is

$$i = \frac{v}{R_L} = \frac{v_S}{R_L + R_S} \tag{3.46}$$

Therefore, the power dissipated in the load is

$$P_L = vi = \frac{R_L}{(R_L + R_S)^2} v_S^2 \tag{3.47}$$

There are some applications where one is concerned with the amount of power dissipated in a load. An example is an amplifier source and loudspeaker load, in which the power dissipated in the loudspeaker is converted to sound energy. Equation 3.47 indicates how the relative sizes of the source and load resistances should be chosen for effective power transfer.

If the source network is fixed, with both v_S and R_S given, the only way to optimize power transfer to the load is to vary R_L. Examination of Eq. 3.47 shows that if R_L is very small compared to R_S, very little power reaches the load. If R_L is very much larger than R_S, the power becomes v_S^2/R_L, which decreases as the load increases. Therefore, there is some choice of R_L comparable to R_S for which the power transfer to the load is a maximum. A simple calculus proof yields

$$R_L = R_S \tag{3.48}$$

for maximum power in the load. In this case, with source and load resistances equal, the power in the load is $v_S^2/4R_S$, while the power dissipated in the source resistance ($i^2 R_S$) also equals $v_S^2/4R_S$. Thus the condition for maximum power in the load when the source network is fixed corresponds to equal source and load resistances and equal amounts of power being dissipated in the source and load resistances. The quantity $v_S^2/4R_S$ is called the *available power*; it is the maximum power that can be delivered to a properly matched load.

A much more common situation is for the load resistance to be fixed and the source resistance variable. In this case examination of Eq. 3.47 shows that the power dissipated in the load for fixed R_L is maximized by making R_S as small as possible. In succeeding chapters we will see how feedback can be used to construct power source networks that have very small equivalent resistances. The power transfer in these cases is limited only by the maximum size of v_S that the source network is capable of providing.

3.6 CONSISTENT UNITS

Throughout the text we will numerically evaluate algebraic expressions in order to develop a feel for the actual circuit values encountered in typical applications. Since these algebraic expressions can be complex combinations of voltage, current and resistance, it is essential that a consistent set of units for the variables be employed.

The most obvious set is simply the basic MKS units of volts, amperes and ohms. However, in most solid-state electronics applications, a more useful set is volts, milliamps and kilohms, as these represent the multipliers most frequently encountered. Thus, when substituting numbers into algebraic expressions, *every* voltage is expressed in volts, *every* current in milliamps and *every* resistance in kilohms. If this procedure is followed, the multipliers of 10^{-3} and 10^{3} will exactly cancel. When later on we study time functions, the inclusion of the proper values of capacitance and inductance will also speed numerical evaluations. A summary of several consistent sets of units is given in Table 3.1.

TABLE 3.1

Quantity	Consistent Units				
Voltage	V	V	V	V	V
Current	A	mA	mA	mA	μA
Resistance	Ω	kΩ	kΩ	kΩ	MΩ
Conductance	S	mS	mS	mS	μS
Capacitance	F	μF	nF	pF	μF
Inductance	H	H	mH	μH	—
Time	s	ms	μs	ns	s
Frequency	Hz	kHz	MHz	GHz	Hz

3.7 ELECTRICAL MEASUREMENTS

We close this chapter on resistive circuits with a brief introduction to the subject of electrical measurements. Quantitative examples are presented in the problems at the end of the chapter (P3.1 to P3.8).

The simplest electrical measurement instrument is the *ammeter* (Fig. 3.25). An ammeter consists of a coil of wire suspended in a magnetic field. When current is passed through the coil, the magnetic field exerts a force on the coil that causes it to rotate. The amount of rotation depends on the amount of current in the coil. A meter needle attached to the coil enables the position of the coil to be observed. Ideally, an ammeter should have zero resistance. In practice, the coil resistance is never zero. Thus the simplest circuit model for an ammeter consists of an ideal ammeter in series with suitably chosen coil resistance (Fig. 3.25*b*).

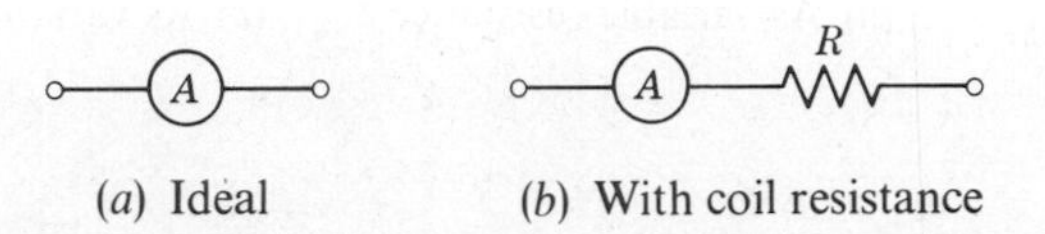

Figure 3.25
Ammeter.

An ideal *voltmeter* is an instrument in which a meter deflection is made proportional to voltage (Fig. 3.26). In order that the ideal voltmeter not change any of the voltages in the network to which it is connected, the ideal voltmeter should have an infinite resistance. In practice, all voltmeters draw a small amount of current. The simplest circuit model for an actual voltmeter, therefore, consists of an ideal voltmeter in parallel with a suitably chosen shunt resistance (Fig. 3.26*b*). It is possible to construct a voltmeter using an ammeter in series with a properly chosen large resistor (see Problem P3.6).

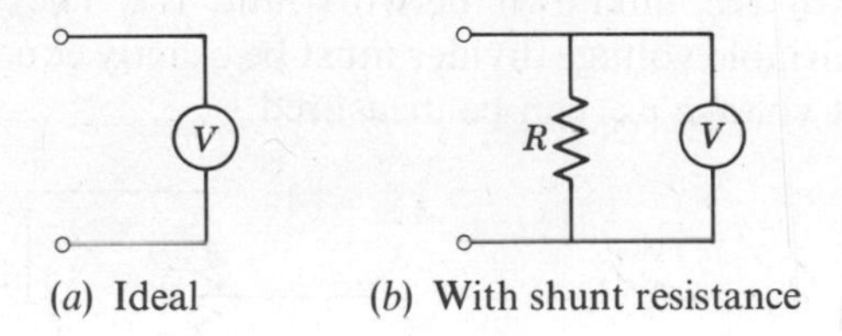

Figure 3.26
Voltmeter.

The voltmeter and ammeter discussed above are called direct-reading instruments because the electrical quantity of interest is converted directly into an observable meter motion. Many electrical measurements, however, use indirect or comparison methods. A network element of great value in such measurements is the *potentiometer* (Fig. 3.27). A potentiometer is a resistor constructed with a third contact whose position of contact can be varied. This third contact is called the *wiper* (W). As the wiper position is moved from terminal 1 to terminal 2, the resistance between the wiper and

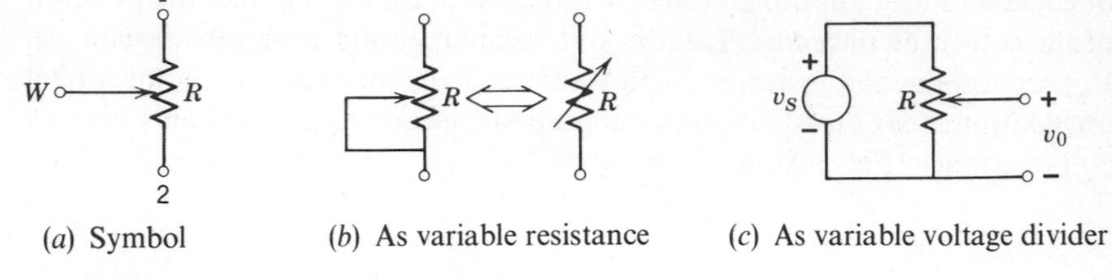

(*a*) Symbol (*b*) As variable resistance (*c*) As variable voltage divider

Figure 3.27
Potentiometer.

terminal 1 increases from zero to R, while the resistance between the wiper and terminal 2 decreases from R to zero. Figure 3.27*b* shows how a potentiometer can be used to make a variable resistor. Figure 3.27*c* shows how a potentiometer can be used in combination with a voltage source to make a variable voltage divider. As the wiper position in Fig. 3.27*c* is moved, both the Thévenin voltage *and* the Thévenin resistance of the variable voltage divider can change.

One example of the use of a potentiometer in a comparison measurement is shown in Fig. 3.28. Here an unknown network, represented by a Thévenin equivalent (v_{OC}, R_T) is connected through an ammeter to a variable voltage divider. (It is assumed that the relation between wiper position and open-circuit voltage has already been determined.) The wiper position is adjusted until the ammeter carries zero current. At this point the Thévenin open-circuit voltage of the unknown network and the Thévenin open-circuit voltage of the variable voltage divider must be exactly equal. In this manner, the open-circuit voltage v_{OC} can be measured.

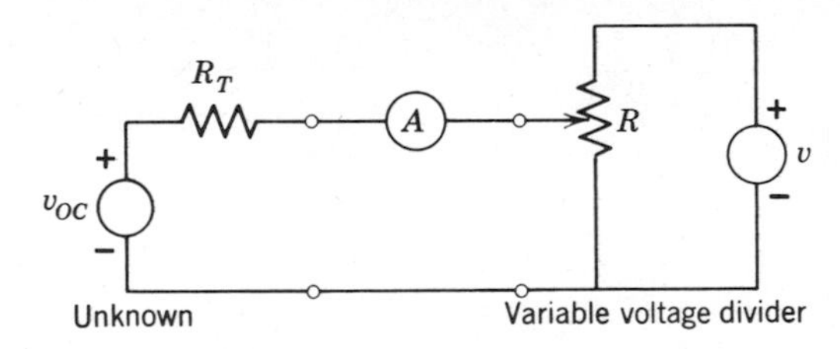

Figure 3.28
Potentiometric voltage measurement.

Resistance measurements can be made either directly or indirectly. A direct measurement requires both a voltmeter and an ammeter. One measures simultaneously the current through a resistance and the voltage across it, and then applies Ohm's law. Indirect methods use a comparison between the unknown resistance and known resistances. Problems P3.1 to P3.4 explore several indirect methods for measuring Thévenin resistances of networks that contain nonzero Thévenin sources. In networks that contain no Thévenin sources, that is, for pure resistances, a *Wheatstone bridge* (Fig. 3.29) can be used. The unknown resistance can be any one of the four resistances in the bridge. The other three resistance values are adjusted until the ammeter carries zero current. At this point, the ratio of R_1 to R_2 must be the same as the ratio of R_3 to R_4. If three of the resistance values are known, the fourth unknown resistance can be calculated. Problem P3.8 explores the Wheatstone bridge further.

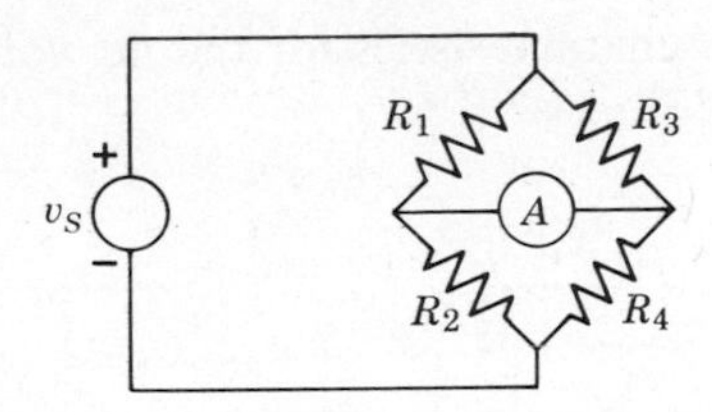

Figure 3.29
Wheatstone bridge.

QUESTIONS

Q3.1 Can you prove that a two-terminal element that is both linear and resistive must obey Ohm's law?

Q3.2 What is the equivalent circuit for an ideal current source connected in series with an ideal voltage source?

Q3.3 What is the equivalent circuit for an ideal current source connected in parallel with an ideal voltage source?

Q3.4 What happens when two unequal ideal voltage sources are connected in parallel? Does the introduction of a small Thévenin resistance in series with each source resolve the problem?

Q3.5 What happens when two unequal ideal current sources are connected in series? Does the introduction of a large Norton resistance in parallel with each source resolve the problem?

Q3.6 Is the power dissipated in a resistor a linear function of the voltage across the resistor?

Q3.7 Are ideal voltage sources and ideal current sources linear elements?

EXERCISES

E3.1 Starting with the definition of slope and with Ohm's law, prove that the slope of the v-i characteristic of an ideal resistor is $1/R$.

E3.2 Sketch the v-i characteristics for the networks in Fig. E3.2. Label intercepts and slopes.

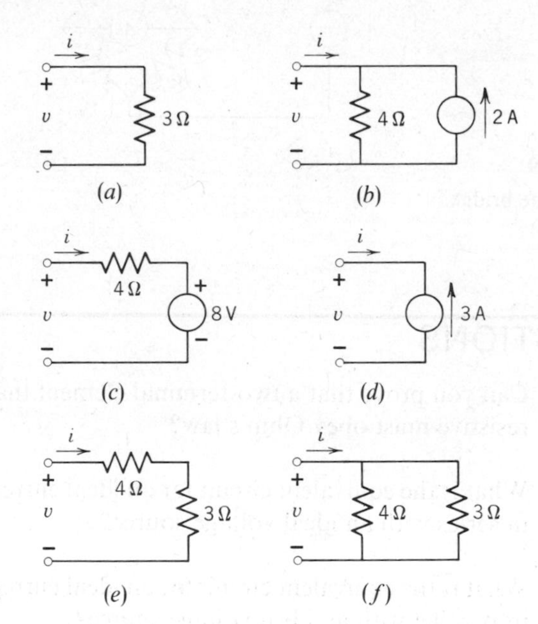

Figure E3.2

E3.3 For the circuit in Fig. E3.3, find values of R_1 to satisfy each of the following conditions:

(a) $v = 3$ V
(b) $v = 9$ V
(c) $i = 3$ A
(d) The power dissipated in R_1 is 12 watts.

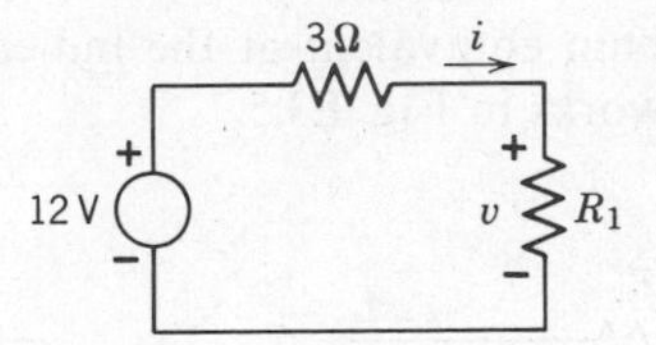

Figure E3.3

E3.4 Find the equivalent resistance R_T at the indicated terminals for each of the networks in Fig. E3.4.

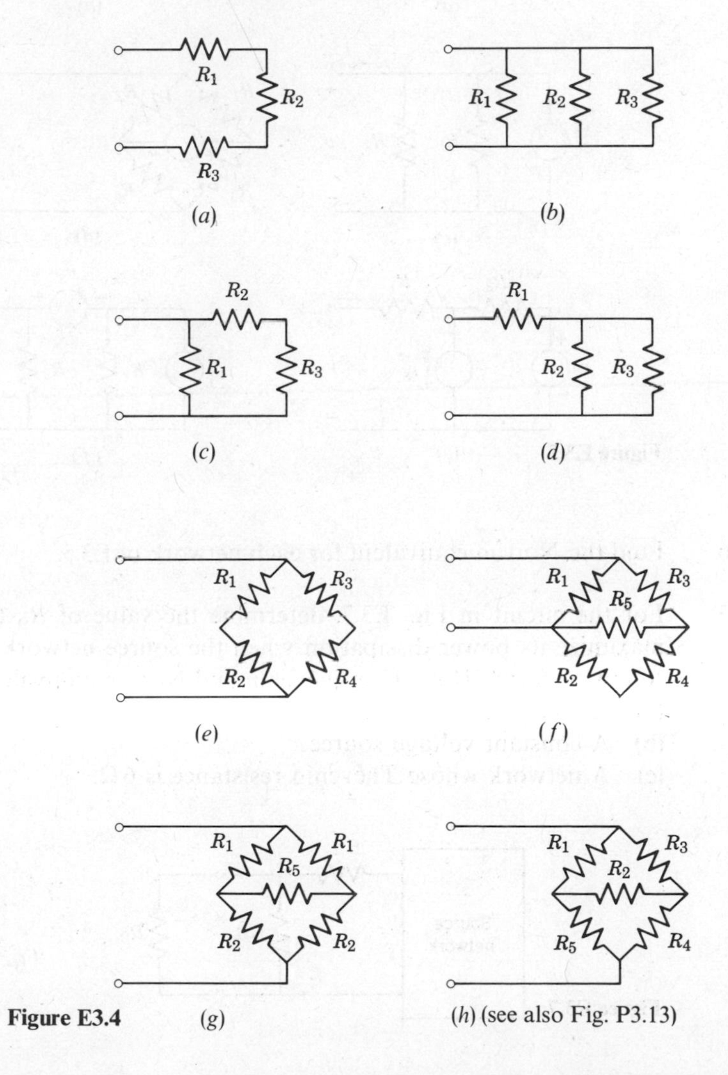

Figure E3.4

E3.5 Find the Thévenin equivalent at the indicated terminal pairs for each of the networks in Fig. E3.5.

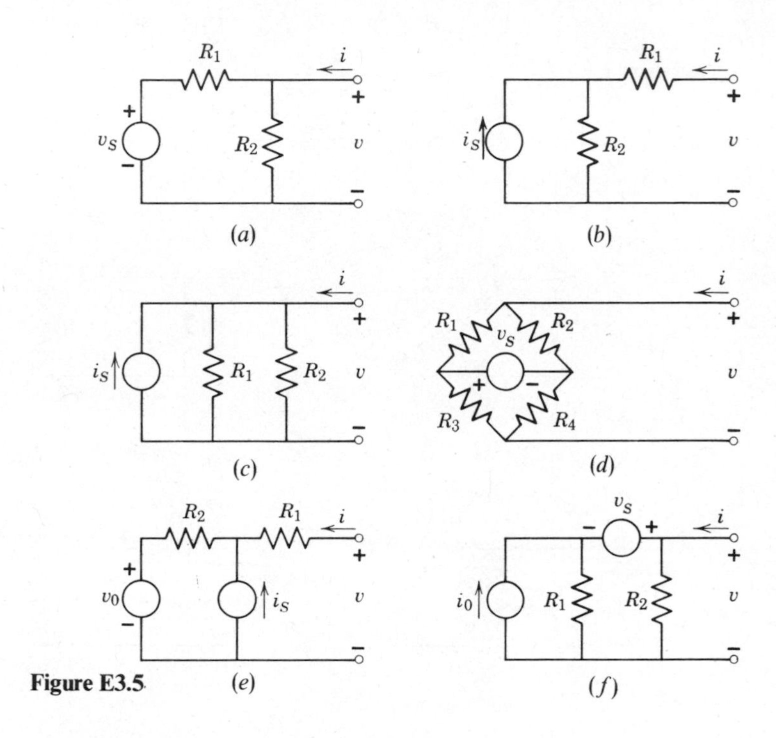

Figure E3.5

E3.6 Find the Norton equivalent for each network of E3.5.

E3.7 For the circuit in Fig. E3.7, determine the value of R_0 that will maximize its power dissipation when the source network has the following form. (*Hint*: Use Thévenin and Norton equivalents.)

(a) A constant current source.
(b) A constant voltage source.
(c) A network whose Thévenin resistance is 6 Ω.

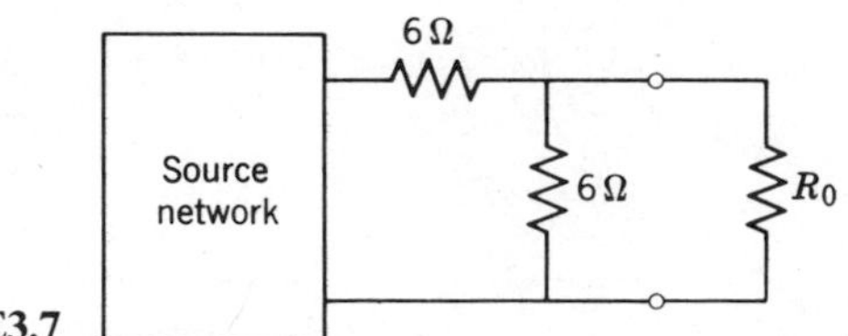

Figure E3.7

E3.8 Prove that energy is conserved in the network of Fig. 3.11 by finding the power dissipated or supplied by each element.

E3.9 Using Fig. E3.9 show that the power P_3 dissipated in R_3 is

$$P_3 = \frac{R_2^2 R_3}{(R_1 R_3 + R_2 R_3 + R_1 R_2)^2} v_0^2$$

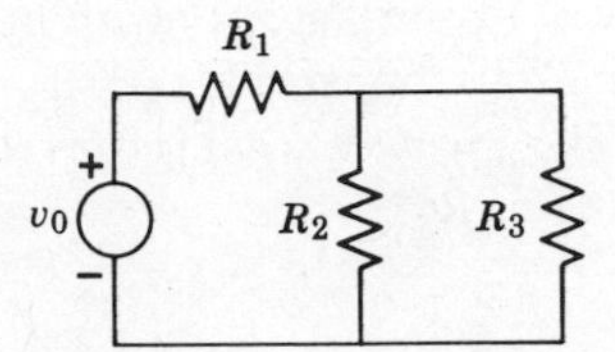

Figure E3.9

E3.10 (a) Find v_0 in the network of Fig. E3.10*a*.

(b) Solve for v_0 in Fig. E3.10*b* by superposition, using the results of part *a*.

(c) Find v_0 in the network of Fig. E3.10*c*. You may use superposition if you wish, but be careful.

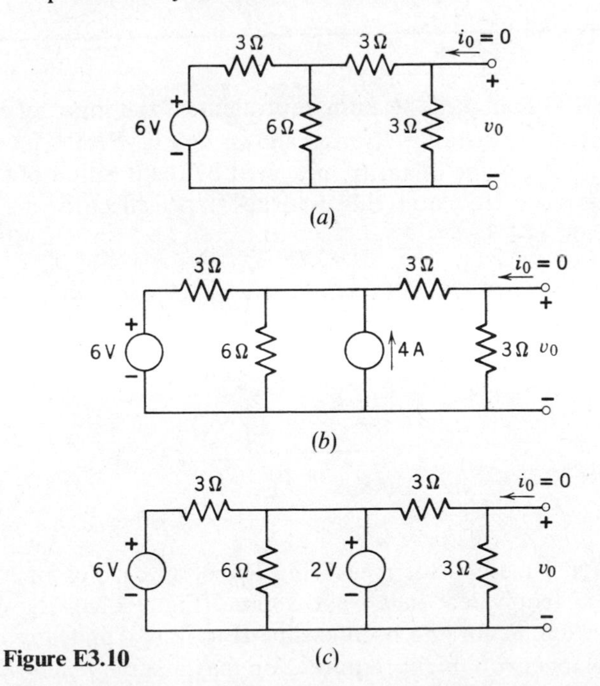

Figure E3.10

PROBLEMS

P3.1 A battery has an open-circuit voltage of 12 volts. When connected to a load and delivering 10 amps to that load, the battery voltage drops to 11.7 volts. What is the internal resistance of the battery?

P3.2 The Thévenin equivalent of the output circuit of an audio-frequency sine-wave generator is shown in Fig. P3.2*a*. The circuit of Fig. P3.2*b* illustrates one way to measure R_S with an oscilloscope. With the variable resistor R disconnected, $v_R = v_S$. When R is connected, and adjusted until $v_R = v_S/2$, what is the relation between the value of R and the value of R_S?

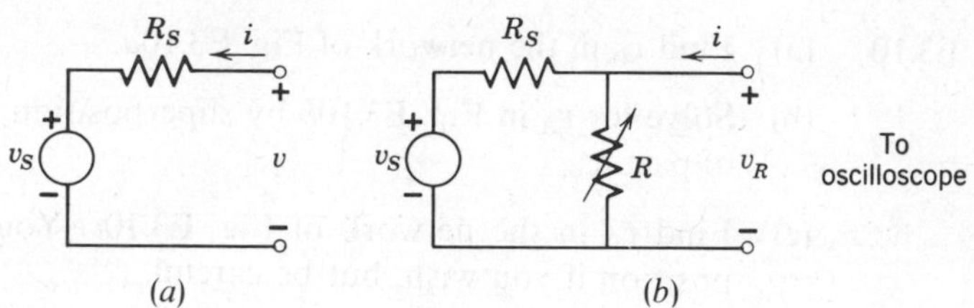

Figure P3.2

P3.3 Suppose that the Thévenin equivalent of the input of an oscilloscope is a resistance R_{IN}, as shown in Fig. P3.3. The voltage v_0 across R_{IN} is the quantity measured by the position of the oscilloscope trace. How does this input resistance affect the measurement method of P3.2?

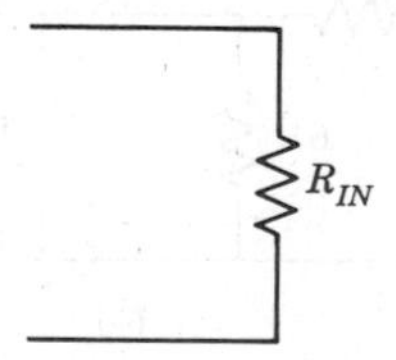

Figure P3.3

P3.4 Design a method for measuring R_{IN} of an oscilloscope. You may use an equivalent sine-wave source (Fig. P3.2*a*), the equivalent input circuit for the oscilloscope (Fig. P3.3), and any additional resistances you might require. You may assume $R_{IN} \gg R_S$.

P3.5 A voltmeter is an instrument that measures voltage. A nonideal voltmeter will have finite input resistance R_{IN}. One possible circuit model for the nonideal voltmeter is shown in Fig. P3.5*a*.

If a nonideal voltmeter with an input resistance of 100 kΩ is used to measure v_1 in the circuit in Fig. P3.5*b*, find the voltage that will be indicated on the meter scale. How does this compare with the voltage an ideal voltmeter would indicate?

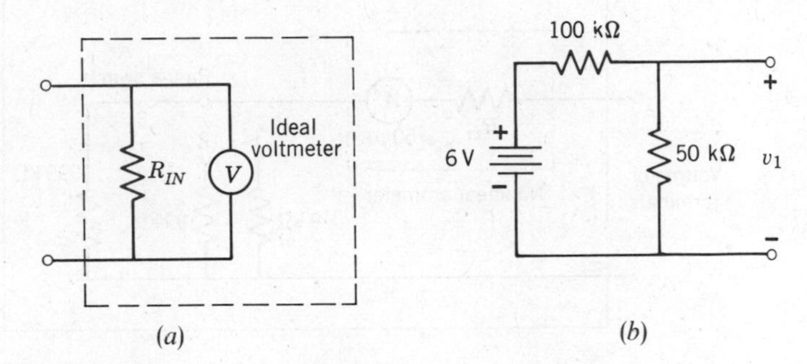

Circuit model for nonideal voltmeter.

Figure P3.5

P3.6 An ammeter is an instrument that measures current. A nonideal ammeter will have an internal resistance R_m because of the resistance of the wire used to wind the movement coil. This may be modeled by connecting R_m in series with an ideal ammeter, as shown in Fig. P3.6*a*.

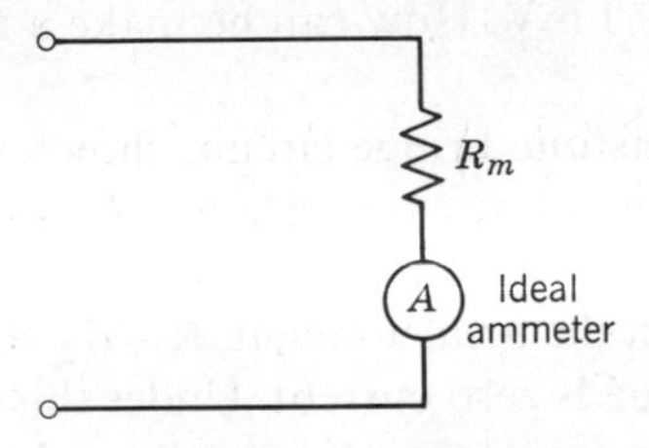

Figure P3.6*a*
Circuit model for nonideal ammeter.

Using an ammeter that has a full-scale deflection of 50 μA and a value of $R_m = 1$ kΩ, we construct a three-range voltmeter as shown in Fig. P3.6*b*.

(a) What will be the full-scale voltage reading on each of the three ranges in Fig. P3.6*b*?
(b) What will be the value of R_{IN} (see P3.5) for each of the three voltage ranges?

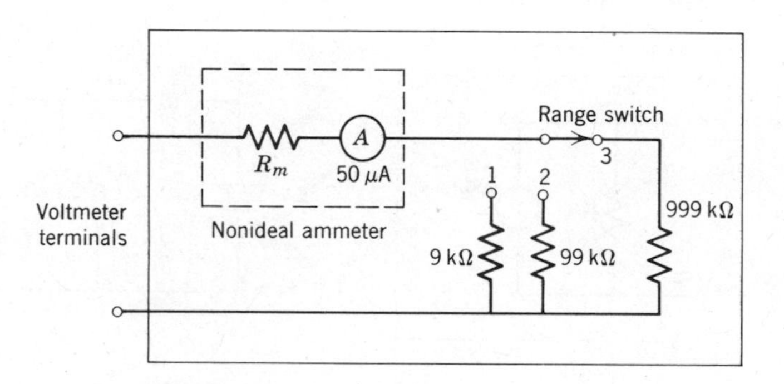

Figure P3.6*b*
Multirange voltmeter.

P3.7 A student is required to determine R_m of a dc microammeter that has a 10 μA full-scale deflection (see P3.6). He suspects R_m is on the order of 100 Ω, but does not have a voltmeter sufficiently sensitive to indicate 1 mV. How can he make a measurement of R_m?

P3.8 The Wheatstone bridge circuit, shown in Fig. P3.8, can be used to measure an unknown resistance R_x. Resistors R_1, R_2, and R_3 are adjustable resistors whose values can be read from calibrated dials. To perform the measurement, R_1, R_2, and R_3 are adjusted until the ammeter reads zero current. Under this condition the bridge is said to be balanced.

(a) Find R_x in terms of R_1, R_2, and R_3 when the bridge is balanced.
(b) Find the Thévenin equivalent of the bridge as seen at the ammeter terminals for an unbalance caused by a small change ΔR_x in the unknown resistance.
(c) If an ammeter with a resistance $R_m = 2$ kΩ and a minimum detectable current of 1 μA is used in the bridge, what is the minimum change ΔR_x that can be detected under the following

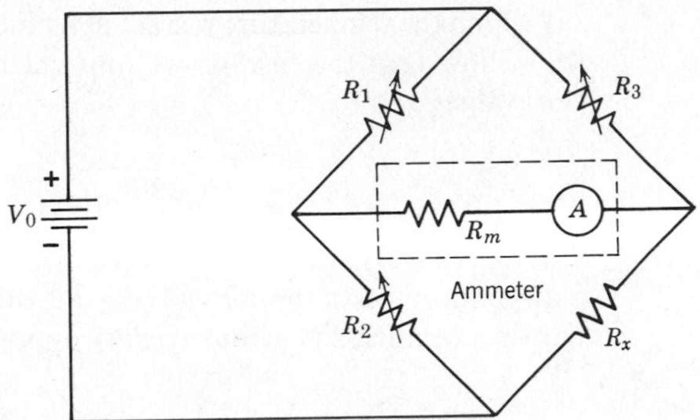

Figure P3.8
Wheatstone bridge.

conditions:

$$R_1 = R_2 = R_3 = R_x = 1\text{k}\Omega$$
$$V_0 = 12\text{ V}$$

Answer: 1 Ω.

P3.9 The fuse is a resistive circuit element that blows out (open circuits) when its current exceeds a critical value. This problem analyzes the blow-out mechanism in terms of a simple model (Fig. P3.9).

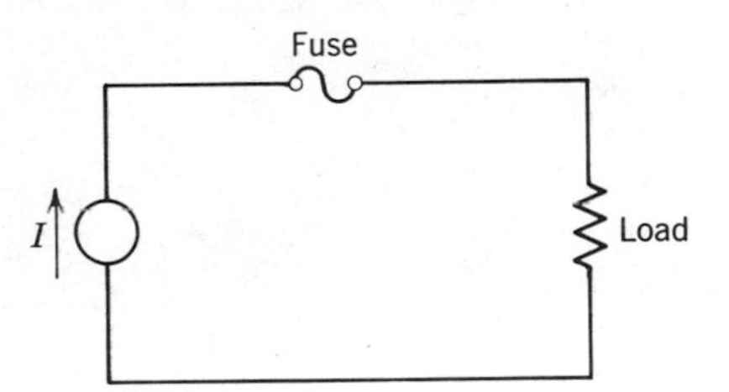

Figure P3.9

At room temperature, the fuse has a resistance R_0. When current flows, power is dissipated in the fuse causing its temperature to increase above room temperature. We model the temperature rise as being proportional to the power dissipation.

$$\Delta T = \beta P$$

where β is the proportionality constant. As the fuse heats up, its resistance increases

$$R = R_0(1 + \alpha\,\Delta T)$$

where α is another constant.

(a) Find the temperature rise ΔT as a function of the load current I.

(b) Show that the fuse blows out when the current reaches the value

$$I = \frac{1}{\sqrt{\alpha\beta R_0}}$$

P3.10 Twelve resistors, 1 Ω each, are arranged along the edges of a cube with connections at the corners (Fig. P3.10). What is the resistance R between terminals at symmetrically opposite corners of the cube?

Answer. $\frac{5}{6}$ Ω.

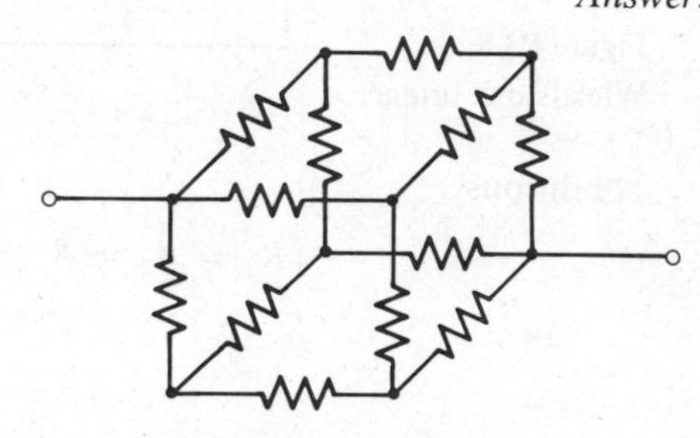

Figure P3.10
All resistors equal 1 Ω.

P3.11 A rectangular grid formed of 1 Ω resistors extends to infinity as shown in Fig. P3.11. Assume that the potential at infinity is zero volts (earth potential).
What will be the net resistance R between the nodes A and B?

Answer. $\frac{1}{2}$ Ω.

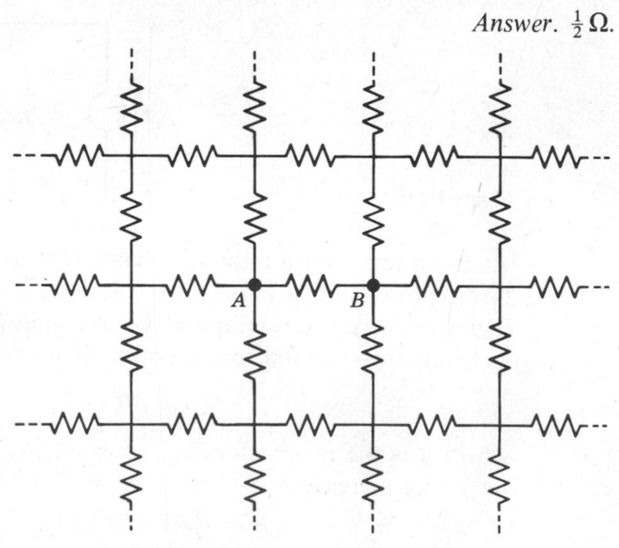

Figure P3.11

P3.12 Prove that Eq. 3.48 is correct.

P3.13 It has already been demonstrated (Section 3.2) that two resistors in series or parallel can be reduced to a single resistor of appropriate value. Similarly, equivalence relationships exist for the three-resistor networks shown in Fig. P3.13, and these relationships are useful in analyzing bridge networks and three-phase power systems. Network A is called a "Pi" or "Delta" arrangement, while Network B is variously called a "Tee," "Star," or "Wye" connection. The problem is now to find the resistance values of one network in terms of the resistance values of the other such that from the terminals the networks are equivalent.

(a) Calculate R_{T1}, R_{T2}, R_{T3} in terms of R_{P1}, R_{P2}, and R_{P3}.

Answer.

$$R_{T1} = \frac{R_{P2}R_{P3}}{R_{P1} + R_{P2} + R_{P3}}$$

$$R_{T2} = \frac{R_{P1}R_{P3}}{R_{P1} + R_{P2} + R_{P3}}$$

$$R_{T3} = \frac{R_{P1}R_{P2}}{R_{P1} + R_{P2} + R_{P3}}$$

(b) Calculate R_{P1}, R_{P2}, and R_{P3} in terms of R_{T1}, R_{T2}, and R_{T3}.

Answer.

$$R_{P1} = R_{T2} + R_{T3} + \frac{R_{T2}R_{T3}}{R_{T1}}$$

$$R_{P2} = R_{T3} + R_{T1} + \frac{R_{T3}R_{T1}}{R_{T2}}$$

$$R_{P3} = R_{T1} + R_{T2} + \frac{R_{T1}R_{T2}}{R_{T3}}$$

Note that careful labeling of the resistor values produces a pattern in the algebraic expressions for the equivalent resistances that aids in recalling the results from memory.

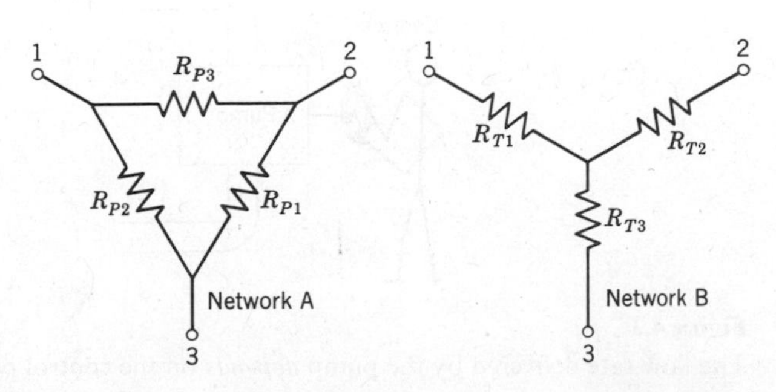

Figure P3.13

CHAPTER FOUR

DEPENDENT SOURCES

4.1 LINEAR DEPENDENT SOURCES

4.1.1 Definitions and Notation

A *dependent source* is a source element whose value depends on some other variable in the network. This concept is illustrated in Fig. 4.1 with a mechanical analog. Suppose we have a water pump that maintains a steady flow rate in a pipe (analogous to a current source). Let us presume further that the pump has a variable speed, with the speed control being adjusted by a technician. The pump can then be represented by a dependent current source because it supplies a prescribed flow rate, but the flow rate at any instant depends on the setting of the control position.

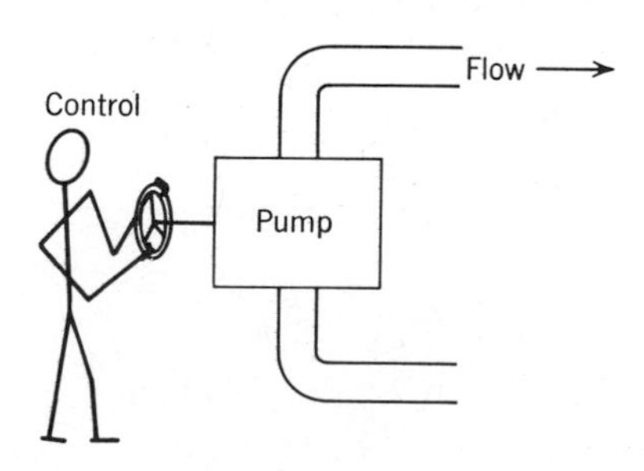

Figure 4.1
The flow rate delivered by the pump *depends* on the control position.

An electrical version of this example is shown in Fig. 4.2. A current source is shown delivering a current i_O to part of the total network. The value of i_O at any instant, however, is proportional to the voltage v_1, which appears elsewhere in the network. The proportionality constant g_m, which must have the dimensions of a conductance, is called the *transconductance*.

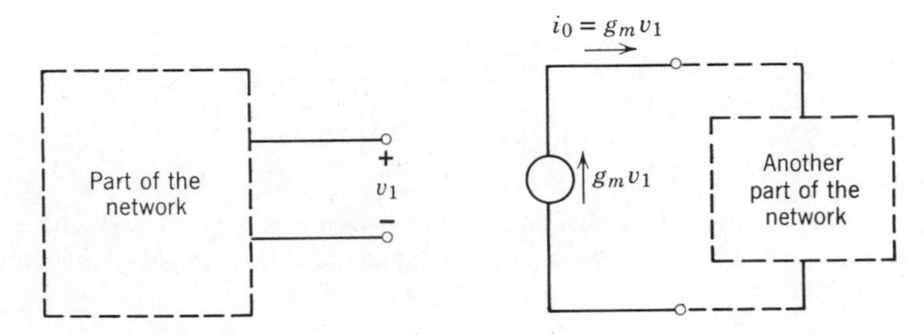

Figure 4.2
A voltage-controlled current source. The source current, i_O, *depends* on the voltage v_1. The proportionality constant g_m has the dimensions of a conductance, and is called the transconductance.

The proportionality between i_O and v_1 makes the dependent source of Fig. 4.2 a *linear dependent source*. In subsequent chapters we shall encounter dependent sources that are not linear, that is, dependent sources in which the relation between the source value and the controlling variable is not a simple proportionality. The dependent sources discussed in this chapter, however, will all be linear.

Since the "source" part of a dependent source can be either a voltage or current source, and since the control variable can be either a voltage or current, there are four kinds of linear dependent sources that one can construct. Each kind has its uses in modeling the behavior of electronic devices. The examples that follow not only illustrate different kinds of dependent sources, but also introduce some of the ways in which dependent sources can affect network behavior.

4.1.2 Examples Using Dependent Sources

Figure 4.3 shows a network in which a dependent voltage source is inserted between a Thévenin source network and a load resistor. Let us calculate the voltage v_O across the load resistor assuming the output current i_O is zero.

Writing KVL around the loop containing v_S yields

$$v_S - i_S R_S - v_1 = 0 \tag{4.1}$$

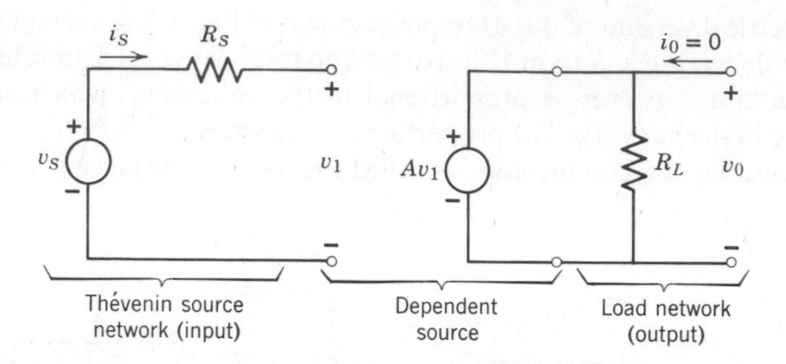

Figure 4.3
A voltage-controlled voltage source connected between a source network and a load. If the magnitude of A is greater than one, the input signal v_S is *amplified* (increased in size) by the dependent source.

Since the v_1 terminal pair is an open circuit, the current i_S must be zero. Therefore, from Eq. 4.1,

$$v_1 = v_S \tag{4.2}$$

Consider now the loop containing R_L. KVL for this loop yields

$$Av_1 - v_O = 0 \tag{4.3}$$

Substituting for v_1 from Eq. 4.2 and solving we obtain

$$v_O = Av_S \tag{4.4}$$

Thus we see that v_O is proportional to v_S. If A is greater than unity, v_O is also larger in magnitude than v_S. In this particular case, therefore, the dependent source is functioning as a *linear amplifier*, increasing the magnitude of the signal v_S by a constant factor.

Figure 4.4 shows a current-controlled current source connected between a Thévenin equivalent source network and a Thévenin equivalent load network. Notice that the control current i_B is defined as the current flowing in a specific network path (or branch), just as the control voltage in previous examples was defined as the voltage between a specific pair of nodes in the network. The choice of symbols, such as i_B for the control current, β_F for the constant of proportionality, and V_{CC} for the battery in the load network, is made in anticipation of the conventional use of these symbols in device models. We shall discover in Chapter 10, for example, that the circuit of Fig. 4.4 represents a transistor amplifier over part of the transistor's range of operation. Let us determine the relation between the source voltage v_S and the output voltage v_O assuming that the output current i_O is zero.

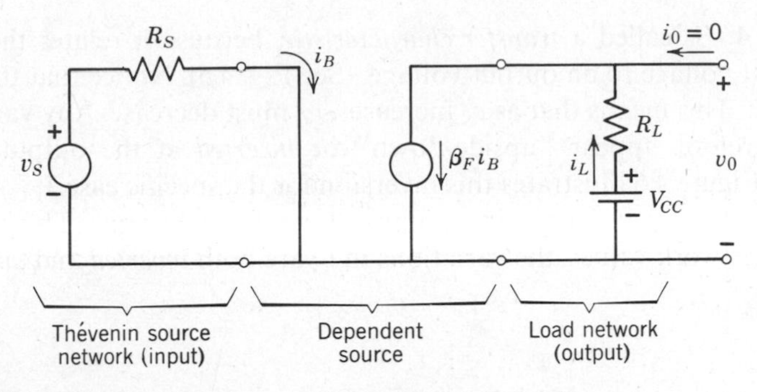

Figure 4.4
A current-controlled current source. Two of the dependent-source terminals are connected together, resulting in a three-terminal dependent source.

Since the path for i_B is a short circuit, the current flowing in the source network loop must be the short-circuit Thévenin current. That is,

$$i_B = \frac{v_S}{R_S} \tag{4.5}$$

Since we have assumed that the output current i_O is zero, the current labeled i_L is equal to $\beta_F i_B$. Therefore, we can write KVL around the load network loop as follows:

$$V_{CC} - \beta_F i_B R_L - v_O = 0 \tag{4.6}$$

Substituting for i_B and solving we obtain

$$v_O = V_{CC} - \left(\frac{\beta_F R_L}{R_S}\right) v_S \tag{4.7}$$

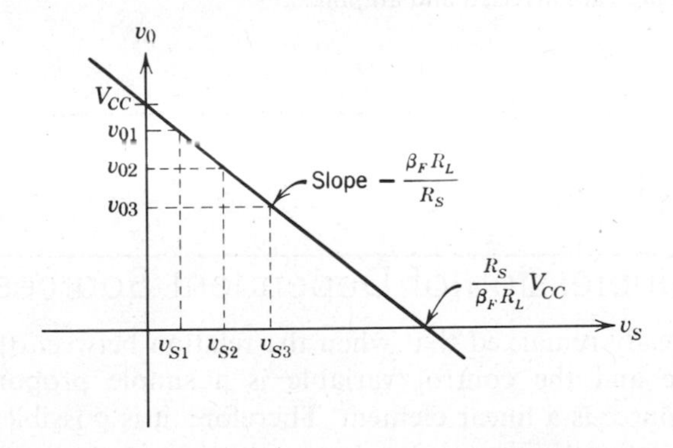

Figure 4.5
Transfer characteristic for the network of Fig. 4.4. As v_S increases, v_O decreases. Thus the network *inverts* signal variations.

Equation 4.7 is called a *transfer characteristic*, because it relates the value of an input voltage to an output voltage. (See Fig. 4.5). Notice that the slope is negative. This means that as v_S increases, v_O must decrease. Any variations in v_S, therefore, appear "upside-down" or *inverted* at the output of the network. Figure 4.6 illustrates this inversion for the specific case $V_{CC} = 12$ V. When v_S jumps up by one volt, v_O jumps down by eight volts. Thus for this choice of network values, the variations in v_S are both *inverted* and *amplified*.

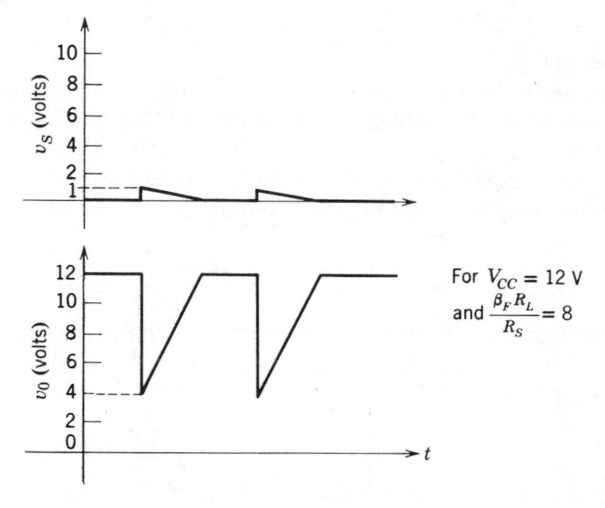

Figure 4.6
An input waveform v_S consisting of periodic 1 volt ramps is applied to the network of Fig. 4.4. The resulting output is plotted for the specific case $V_{CC} = 12$ V, $\beta_F R_L/R_S = 8$. The variations in v_S are inverted and amplified.

4.2 SUPERPOSITION AND DEPENDENT SOURCES

4.2.1 Suppression of Dependent Sources

We have already remarked that when the relation between the dependent-source value and the control variable is a simple proportionality, the dependent source is a linear element. Therefore, it is possible to use superposition in networks containing linear dependent sources, and we can use Thévenin and Norton equivalents to represent the terminal characteristics of networks containing these elements.

Recalling that superposition involves finding the response of a network to each independent source, one at a time, then summing results, a question now arises about what to do with the dependent sources when one is suppressing independent sources. The answer is very simple: *never suppress a dependent source in a superposition problem.* The reason for this categorical statement about dependent sources is that they are used to represent the behavior of resistive elements such as transistors. To suppress a dependent source, therefore, amounts to removing some elements from the network. The following example illustrates that a dependent source must not be suppressed.

4.2.2 An Example of How NOT to Use Superposition

Figure 4.7 illustrates what happens when a dependent source is incorrectly suppressed in attempting a superposition solution of a network problem. The complete network, taken from Fig. 4.3, is shown in Fig. 4.7*a* along with the correct solution found in Section 4.1.2. The independent source, v_S, can certainly be suppressed as in Fig. 4.7*b*. This yields the trivial result that v_{01} is zero; it is the same result one obtains when v_S is set to zero in the correct solution. The attempt to suppress the dependent source is shown in Fig. 4.7*c*. When the dependent voltage source is replaced by a short circuit, the output voltage v_{02} is forced to zero. The sum of the two partial solutions yields a total that is zero, *and* incorrect.

The next example illustrates how superposition can be correctly used in a network that contains a dependent source.

4.2.3 An Example of How to Use Superposition Correctly

The network of Fig. 4.4 contains two *independent* sources. Therefore, because the network is linear, we expect any response, such as the output voltage, to consist of two terms, each term proportional to one independent source. Indeed, the expression for the output voltage in Eq. 4.7 is of this form, but let us rederive Eq. 4.7 using superposition methods.

Figure 4.8 illustrates the correct superposition solution of the example of Fig. 4.4. The complete network and the previously obtained solution are shown in Fig. 4.8*a*, while Figs. 4.8*b* and *c* illustrate the two steps in the superposition solution. When v_S is suppressed, the control current i_B becomes

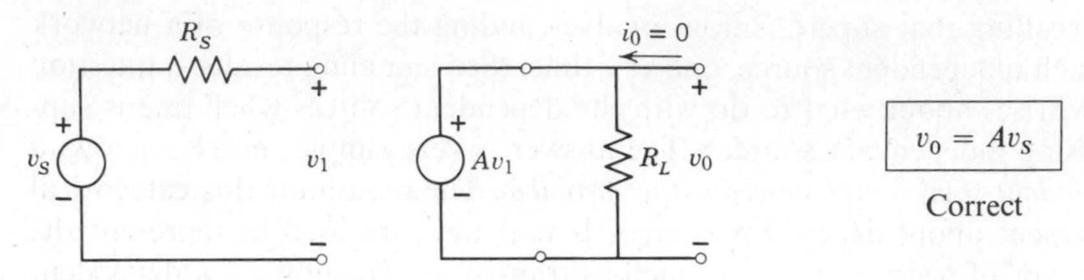

(*a*) Network of Fig. 4.3 with correct solution.

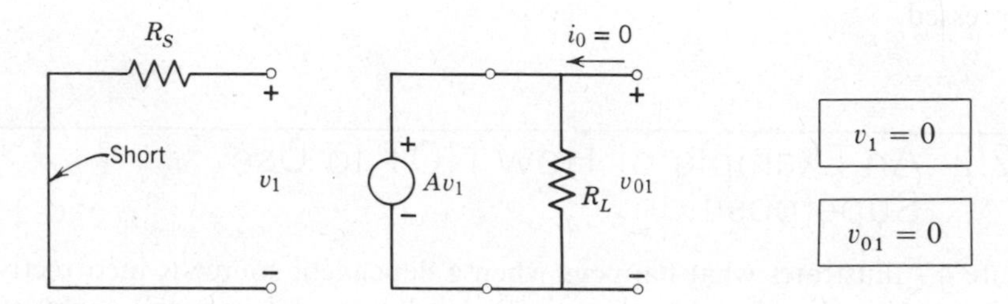

(*b*) Attempt at a superposition solution—v_S suppressed.

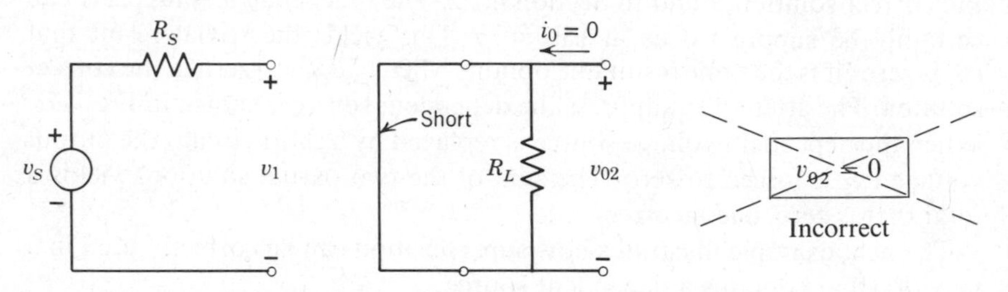

(*c*) Attempt at a superposition solution—dependent source suppressed.

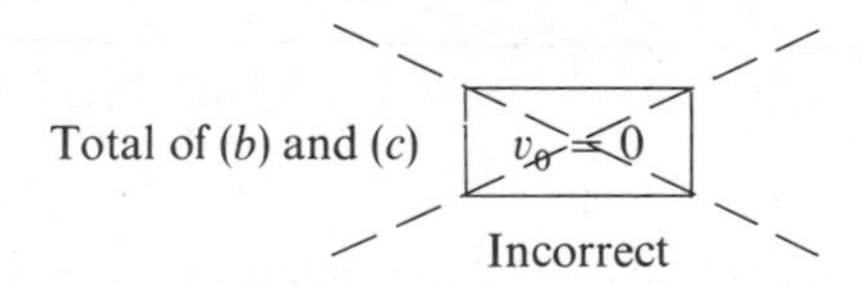

Figure 4.7
Effect of incorrectly suppressing a dependent source.

zero. Therefore, the dependent source current, $\beta_F i_B$, is also zero. Thus no current flows in R_L, yielding the result of Fig. 4.8*b* that

$$v_{O1} = V_{CC} \tag{4.8}$$

In Fig. 4.8*c*, in which V_{CC} is suppressed, the input current is v_S/R_S. The current through R_L therefore is not zero, with the result that

$$v_{02} = -\frac{\beta_F R_L}{R_S} v_S \tag{4.9}$$

Summing the two partial solutions yields the total superposition solution

$$v_O = v_{01} + v_{02} = V_{CC} - \frac{\beta_F R_L}{R_S} v_S \tag{4.10}$$

Equation 4.10 is identical to Eq. 4.7.

Notice that the dependent source is never suppressed. It is included in both of the networks involved in the superposition solution.

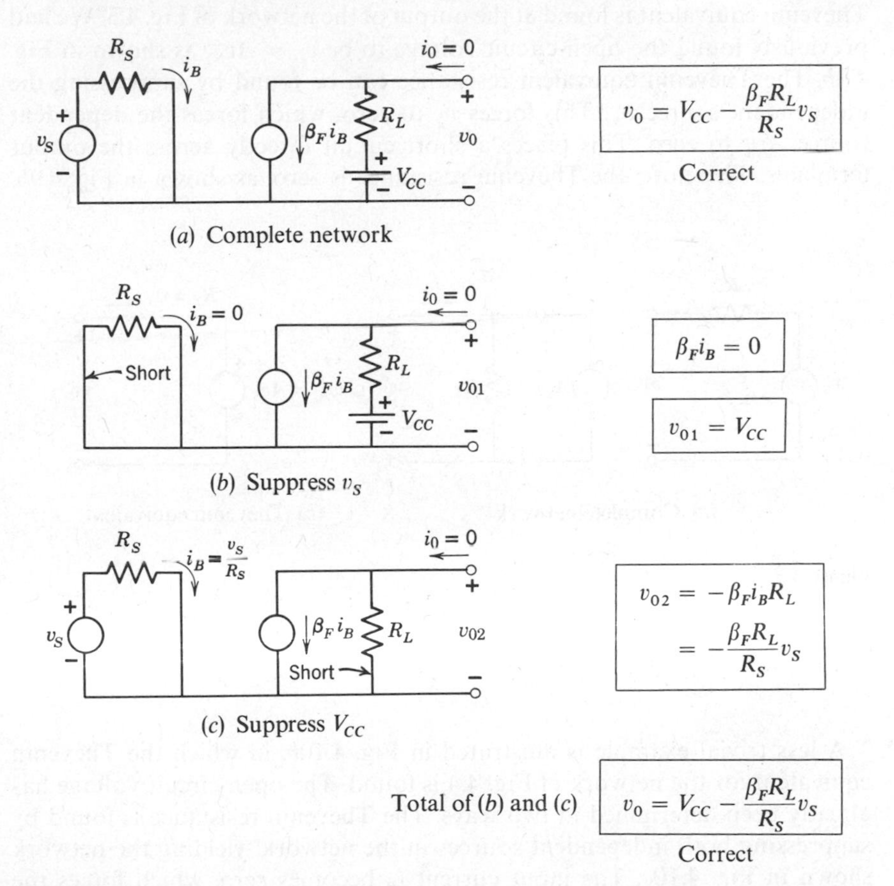

Figure 4.8
Correct use of superposition in a dependent-source network. The dependent source is never suppressed.

4.2.4 Thévenin and Norton Equivalents with Dependent Sources

Whenever superposition is valid, that is, in networks containing independent sources, resistors, and linear dependent sources, it is possible to represent the terminal characteristics of any network by a Thévenin or Norton equivalent. Dependent sources introduce several complications into the problem of finding Thévenin and Norton equivalents, but they do not alter the validity of the concept.

In the two networks we have used thus far as examples, the Thévenin equivalent issue is relatively straightforward. Figure 4.9 shows how the Thévenin equivalent is found at the output of the network of Fig. 4.3. We had previously found the open-circuit voltage to be $v_O = Av_S$, as shown in Fig. 4.9*b*. The Thévenin equivalent resistance can be found by suppressing the independent source v_S. This forces v_1 to zero, which forces the dependent source Av_1 to zero. This places a short circuit directly across the output terminals. Therefore, the Thévenin resistance is zero, as shown in Fig. 4.9*b*.

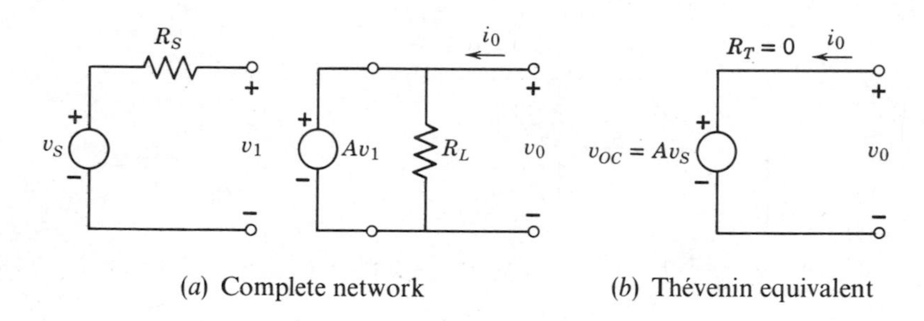

Figure 4.9
Thévenin equivalent for the example of Fig. 4.3.

A less trivial example is illustrated in Fig. 4.10*b*, in which the Thévenin equivalent for the network of Fig. 4.4 is found. The open-circuit voltage has already been determined in two ways. The Thévenin resistance is found by suppressing both independent sources in the network, yielding the network shown in Fig. 4.10*c*. The input current i_B becomes zero, which forces the current source $\beta_F i_B$ to zero. A current source that is zero is equivalent to an open-circuit, so only R_L contributes to the Thévenin resistance.

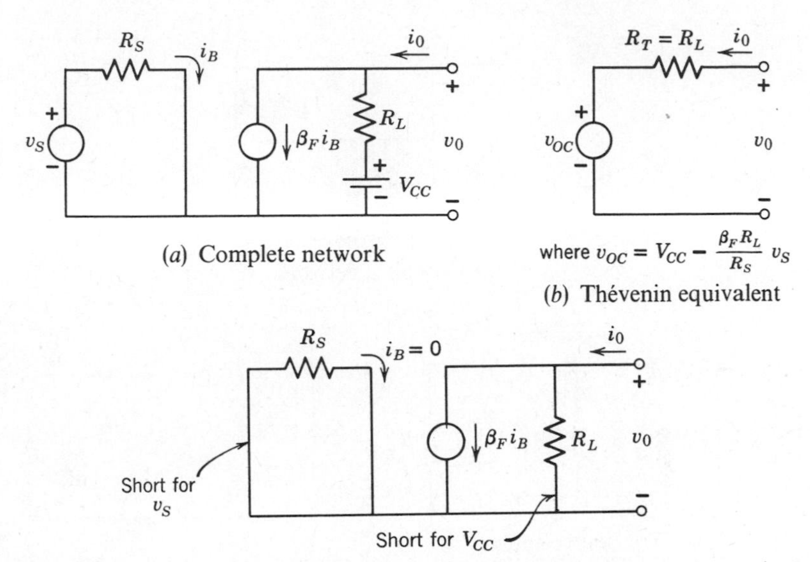

(c) Both independent sources are suppressed to find the Thévenin equivalent resistance

Figure 4.10
Thévenin equivalent for the example of Fig. 4.4.

4.2.5 Feedback and Thévenin Equivalents

Whenever there is any connection from a dependent source back to the control input for that source, that is, whenever there is any kind of *feedback* in the network, the dependent source can have enormous, and sometimes surprising effects on the terminal characteristics of a network. Two examples follow.

In Fig. 4.11, the output current from a dependent source and the control current i_B have a common current path through R_E. Therefore, the output current produces a voltage across R_E that influences the amount of control current that flows in response to an applied v_I. This is an example of feedback. Since the network contains no independent sources, there can be no open-circuit voltage or short-circuit current. If, for example, i_B is zero (open circuit), then no current can flow in R_E from either path, with the result that v_I is zero. The Thévenin equivalent of the network of Fig. 4.11*a* must therefore consist only of a Thévenin resistance.

When networks involve dependent sources and feedback, one cannot use series and parallel combination rules for resistors to find the Thévenin resistance. Instead, one must return to Ohm's law, which states that the voltage and current in a resistor are proportional. If we pass a known current

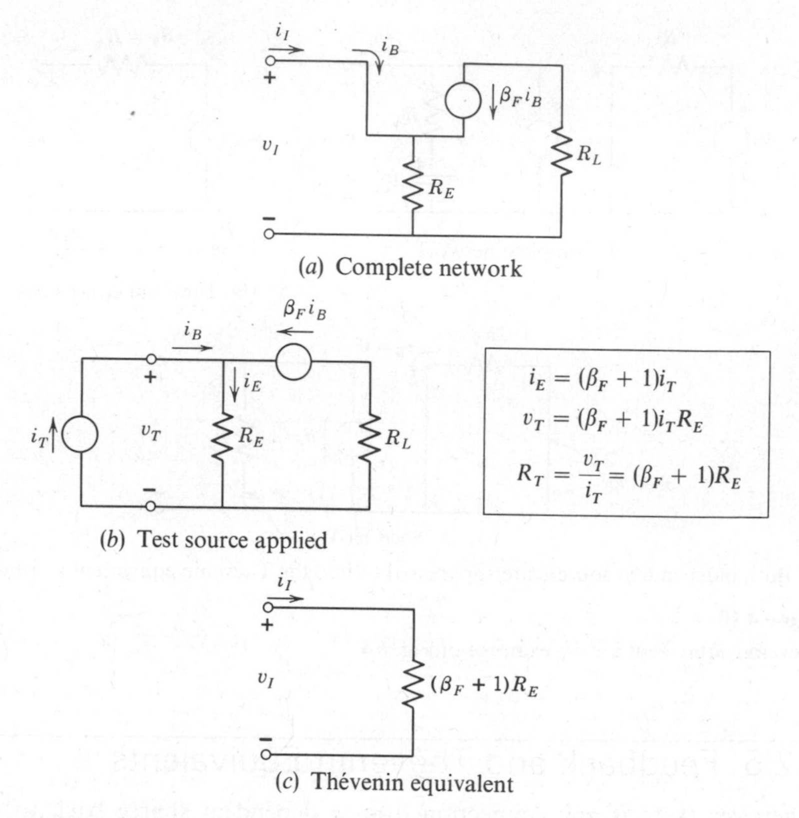

Figure 4.11
The Thévenin equivalent of a network containing a dependent source but no independent sources is simply a resistance. The network in (*b*) illustrates how the application of a test current source simplifies the determination of R_T.

through a resistor and measure the resulting voltage drop across the resistor, we can determine the value of the resistance. Similarly, in a feedback network containing a dependent source, if we apply a current source to the input and measure or calculate the resulting voltage, the Thévenin resistance can be determined.

Figure 4.11*b* illustrates the use of a test source to determine the Thévenin resistance. Examination of the network reveals that if i_B is known, the value of the current in the dependent source is also known. Therefore, we apply a test current source, i_T, which fixes i_B equal to i_T. The current through R_E is simply the sum of i_B and $\beta_F i_B$ (KCL). The voltage drop across R_E, therefore, is

$$v_T = (\beta_F i_B + i_B)R_E = (\beta_F + 1)R_E i_T \tag{4.11}$$

Notice that the source i_T has produced a response v_T which is proportional to the value of the source. The constant of proportionality is simply the equivalent resistance between the terminals. Therefore,

$$R_T = \frac{v_T}{i_T} = (\beta_F + 1)R_E \tag{4.12}$$

The complete Thévenin equivalent is shown in Fig. 4.11*c*.

Consider next the network of Fig. 4.12*a*. As in the previous example, there is no open-circuit voltage or short-circuit current. The Thévenin equivalent is simply a resistance. In this case, however, examination of the network reveals that in order to determine the value of the dependent source, it is necessary to know the voltage drop across R_S. By applying a test voltage source as in Fig. 4.12*b*, the problem is simplified. As the equations in the figure show, once v_T is specified, both i_B and $\beta_F i_B$ are known, so that i_T can be found from KCL. The resulting Thévenin resistance is shown in Fig. 4.12*c*.

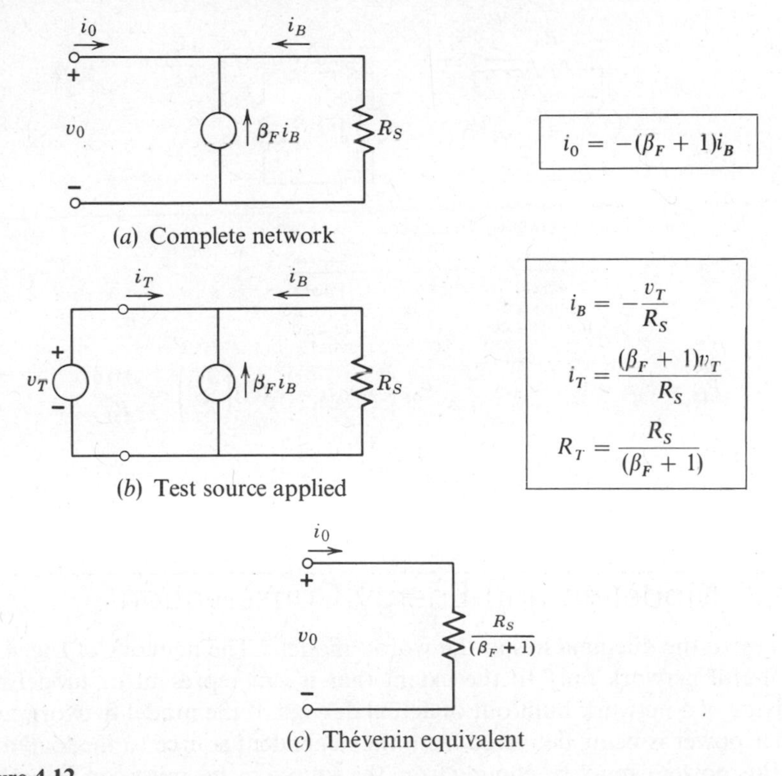

Figure 4.12
An example of the application of a test voltage source to determine a Thévenin resistance.

4.3 POWER IN DEPENDENT-SOURCE NETWORKS

4.3.1 An Example and a Dilemma

Let us return for the last time to the network of Fig. 4.3, but let us now analyze the power transfer. Figure 4.13 shows the calculation. Because the current in the input loop is zero, there is no power transferred from the source network into the dependent source. However, the power delivered from the dependent source to the load R_L is definitely not zero. Where is the extra power coming from? Clearly, it must come from the dependent source. We have been asserting that dependent sources are used to model the behavior of devices like transistors, not independent power sources. How is this dilemma to be resolved?

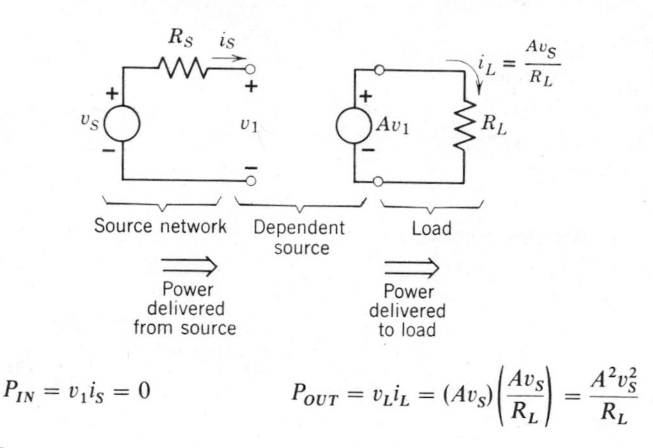

$$P_{IN} = v_1 i_S = 0 \qquad P_{OUT} = v_L i_L = (Av_S)\left(\frac{Av_S}{R_L}\right) = \frac{A^2 v_S^2}{R_L}$$

Figure 4.13
Power transfer in the network of Fig. 4.3.

4.3.2 Modeling and Energy Conservation

The key to the dilemma lies in the word "model." The network of Fig. 4.13 is a useful network only to the extent that it can represent or model the behavior of a network built out of actual devices. If the model network tells us that power is being delivered from the dependent source to the load and that this power cannot be coming from the source v_S, we must conclude that the device or combination of devices being modeled by the dependent source must include an energy source somewhere. Indeed, every electronic device that is modeled with a dependent source requires some energy source, or power supply, to operate, and we shall discover that the dependent source of Fig. 4.13 is actually representing the combined behavior of an electronic device *and* its associated power supply. This concept will be explored in the next chapter.

QUESTIONS

Q4.1 An *active* element (or network) is defined as one capable of delivering more electrical power to a load than it absorbs from appropriate source networks. A *passive* element (or network) is one that cannot provide this power gain. Using these definitions classify each element below as active or passive, and explain your reasoning:

(a) Resistor
(b) Light bulb
(c) Battery
(d) Ideal current source
(d) Linear dependent source

Q4.2 Describe mechanical analogs for each of the four kinds of linear dependent sources.

EXERCISES

E4.1 Find the Thévenin equivalent circuit of the networks in Fig. E4.1 at the indicated terminal pairs.

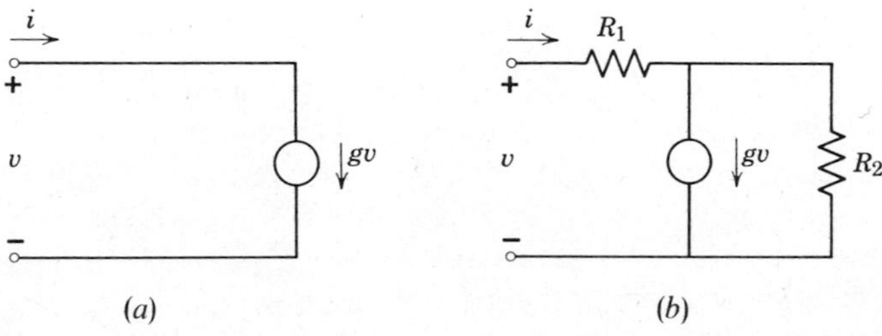

Figure E4.1

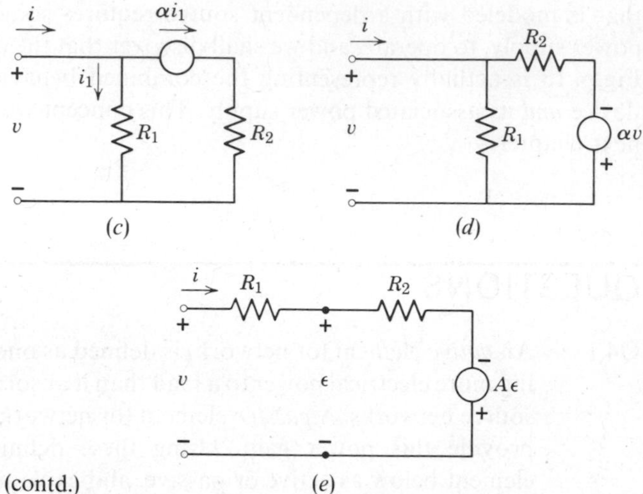

Figure E4.1 (contd.)

E4.2 Find the Thévenin equivalent of the circuit in Fig. E4.2, which is a low-frequency model for a transistor amplifier. For $i = 0$, show that the voltage gain v/v_S for the circuit is $-g_m R_1 R_L/(R_1 + R_S)$.

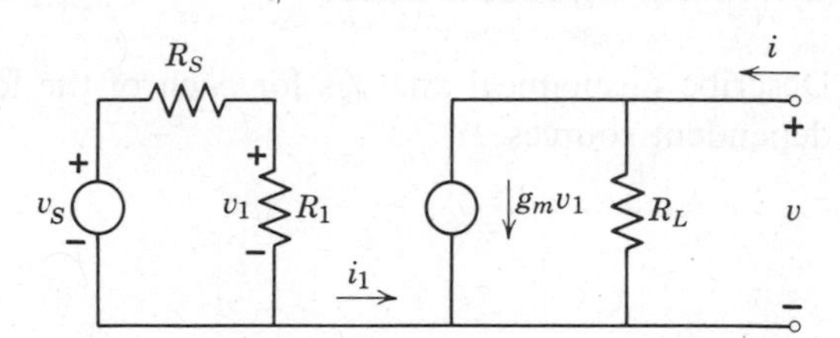

Figure E4.2

E4.3 Use KCL to *prove* that $i_1 = 0$ in the circuit of Fig. E4.2.

E4.4 Find the transfer ratios v_0/v_S indicated in Fig. E4.4.

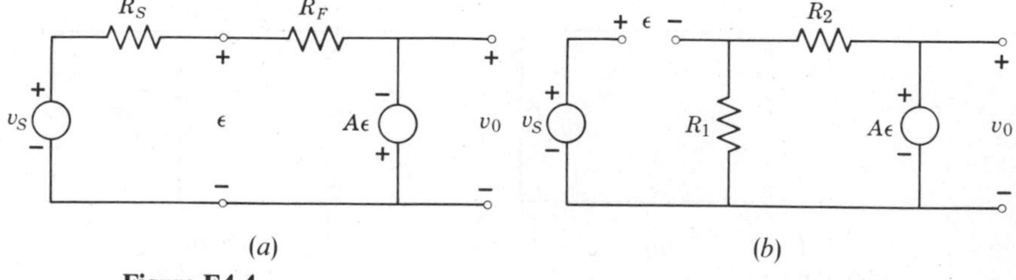

Figure E4.4

E4.5 For each circuit in Fig. E4.5, plot v_O as a function of v_1 when $i_O = 0$, and determine the Thévenin equivalent resistance between the v_O terminals.

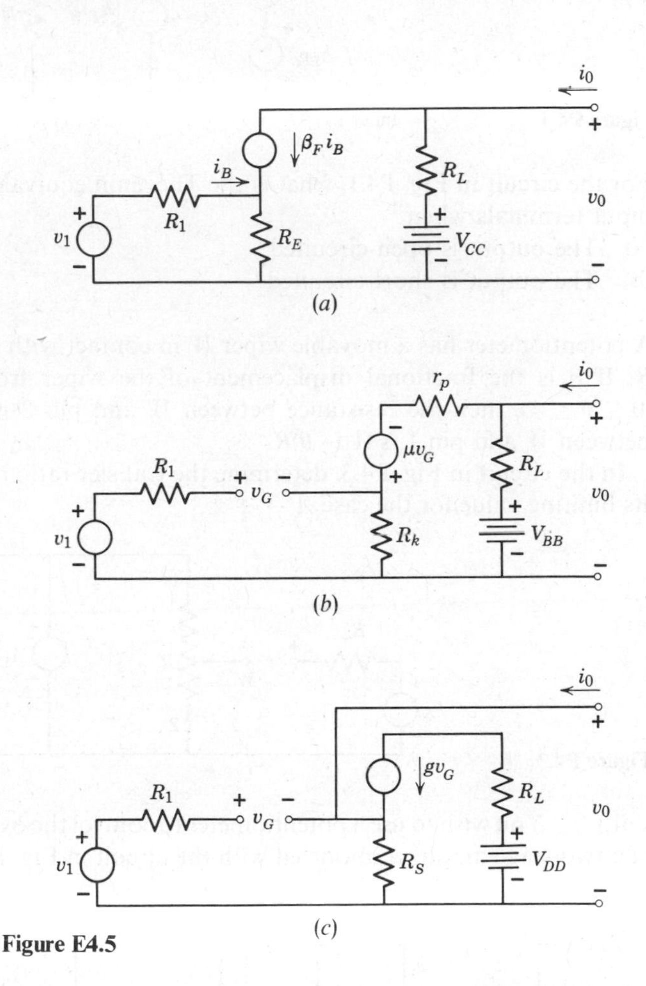

Figure E4.5

PROBLEMS

P4.1 For the circuit shown in Fig. P4.1, what is the Thévenin equivalent at the output terminals when

(a) The input is open-circuited?

(b) The input is short-circuited?

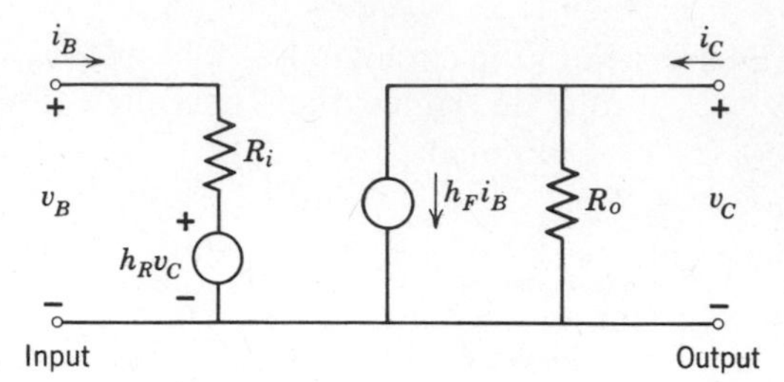

Figure P4.1

P4.2 For the circuit in Fig. P4.1, what is the Thévenin equivalent at the input terminals when

(a) The output is open-circuited?

(b) The output is short-circuited?

P4.3 A potentiometer has a movable wiper W in contact with a resistor R. If θ is the fractional displacement of the wiper from pin 2 $(0 \le \theta \le 1)$, then the resistance between W and pin 2 is θR and between W and pin 1 is $(1 - \theta)R$.

In the circuit in Fig. P4.3, determine the transfer ratio v_O/v_S, and its limiting value for the case $A \to +\infty$.

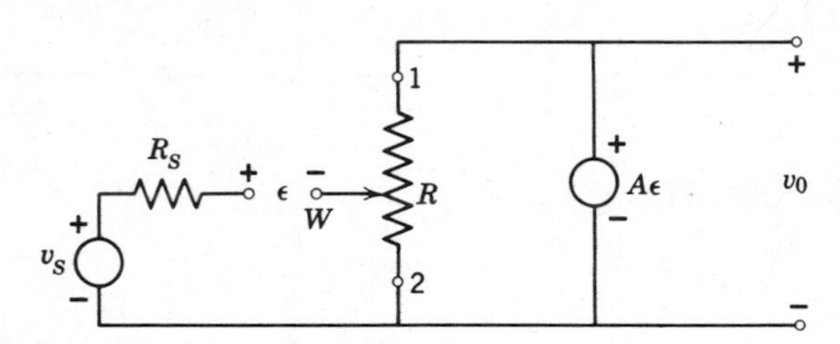

Figure P4.3

P4.4 (difficult) You wish to use a potentiometer to control the overall gain of a two-stage amplifier, modeled with the circuit in Fig. P4.4. You

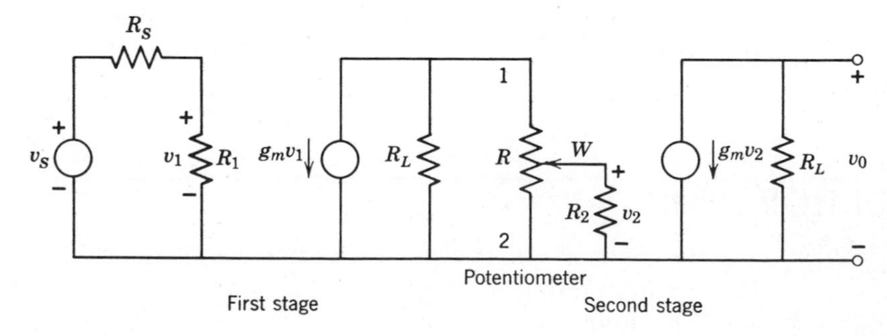

Figure P4.4

are given that $R_S = 50\,\Omega$, $R_L = 500\,\Omega$, R_1 and $R_2 = 50\,\text{k}\Omega$, and $g_m = 0.1$ s.

Choose a value for the total potentiometer resistance R such that

$$\frac{v_O}{v_S} = \theta k$$

where θ is the fractional potentiometer displacement (see P4.3) and k is a constant with magnitude in excess of 1500. It will not be possible to get perfect proportionality to θ, but your result should not have an error greater than 1% of the full-scale gain k.

CHAPTER FIVE
OPERATIONAL AMPLIFIERS

5.1 A FIRST LOOK AT OP-AMPS

5.1.1 A Versatile Building Block

The term *operational amplifier* refers to a circuit that is composed of transistors,[1] resistors, and capacitors, and that has a very large Thévenin input resistance and a very small Thévenin output resistance. It behaves, over at least a portion of its operating range, as a high-gain linear voltage amplifier. With present-day integrated-circuit technology, a high quality amplifier can be manufactured at a very modest cost. The cheapest ones cost less than $1 per unit. In this chapter, we examine how these *op-amps* can be used together with appropriately chosen feedback networks to perform a variety of signal-processing tasks in a precise, reliable, and relatively simple fashion.

Because we have not yet studied the characteristics of transistors, we will approach the op-amp from an empirical point of view. The student must accept "on faith," at least temporarily, certain facts about op-amps until we complete our discussion in later chapters. Our purpose here is simply to learn the use of these versatile components.

[1] The first operational amplifiers were made with vacuum tubes. However, with the development of integrated-circuit technology, transistor circuits with vastly superior characteristics and relatively low cost have made vacuum tube operational amplifiers obsolete.

5.1.2 Op-Amp Notation

Figure 5.1 shows the basic op-amp circuit symbol. Five terminals are shown, these comprising the minimum set needed to describe the op-amp in ideal terms. Most op-amps have several terminals in addition to the ones shown in the illustration. The additional terminals permit the connection of external networks to remove or compensate for nonideal behavior. These additional terminals and their uses are discussed in Section 5.4.

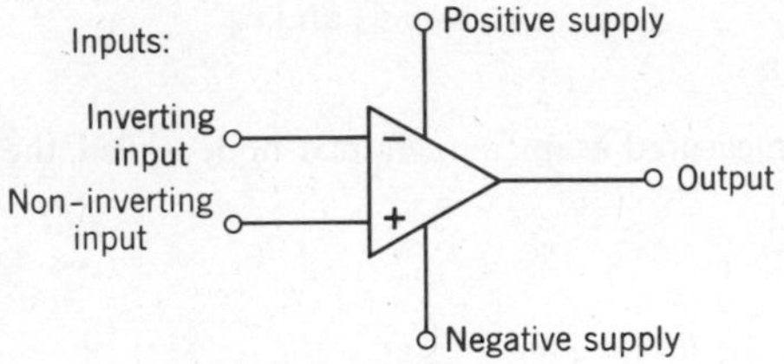

Inputs:

Inverting input

Non-inverting input

Positive supply

Output

Negative supply

Figure 5.1
Circuit symbol for an operational amplifier showing the five principal terminals. The inverting input is labeled with a minus sign and the non-inverting input is labeled with a plus sign.

Each terminal on an op-amp has a name. The conventionally used names for the two input terminals often confuse beginning students. The so-called *inverting* input is identified with a *minus* sign, and the *non-inverting* input is identified with a *plus* sign. These signs, however, are simply the *names* of the terminals and do not constrain the polarity of an applied voltage. In addition to these inputs, there is an output terminal and a pair of terminals to which dc voltages, called *supply voltages*, are connected. Batteries or regulated dc power supplies can be used; in either case, these voltage sources supply power to the components inside the op-amp and any power delivered to an output load.

Because an op-amp has several terminals, we must define voltage and current variables in a way that avoids confusion and aids in the visualization of network behavior. The use of a *reference node* greatly simplifies the notational problems (Fig. 5.2). If we wish to compare the height of two persons, we stand them up on a flat surface (the reference level) and see which one is taller. Similarly, if we wish to determine the potential difference between two points, we can measure the potential difference from each point to a reference node, then subtract one from the other.

Several symbols are conventionally used for the reference node; these are the "ground" symbols in Fig. 5.3*a*. Strictly defined, a *ground* is a point that is connected to earth potential either by a stake driven into the earth or by

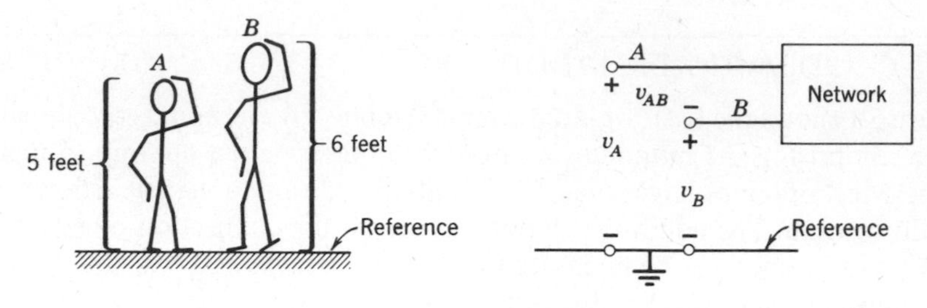

(*a*) Height difference. *B* is 1 foot taller than *A*

(*b*) Potential difference. The voltage v_{AB} is the difference between the node voltages v_A and v_B

Figure 5.2
Node voltages are measured from a reference node called the circuit common or ground.

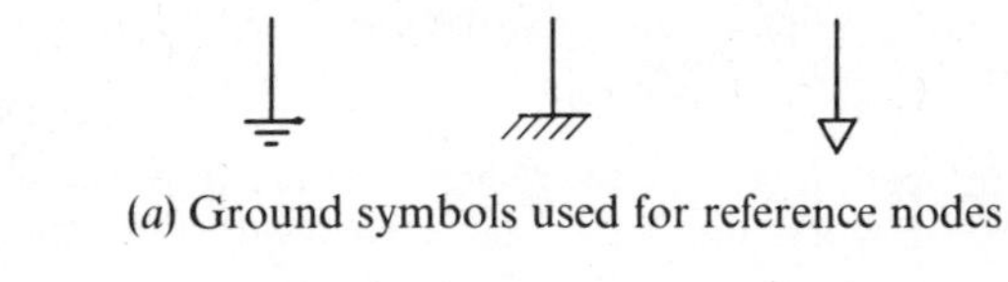

(*a*) Ground symbols used for reference nodes

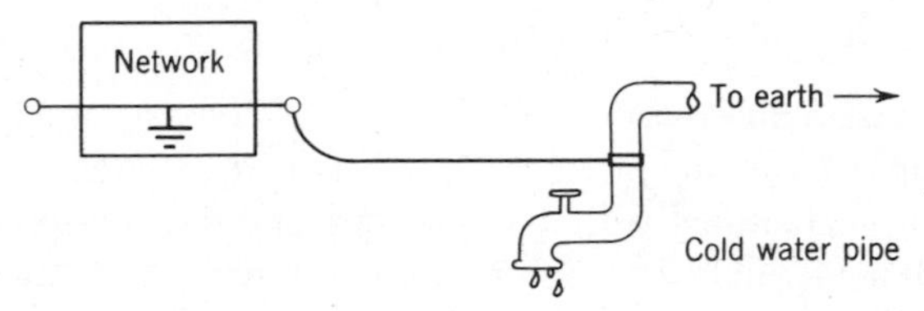

(*b*) A true ground is connected to earth potential

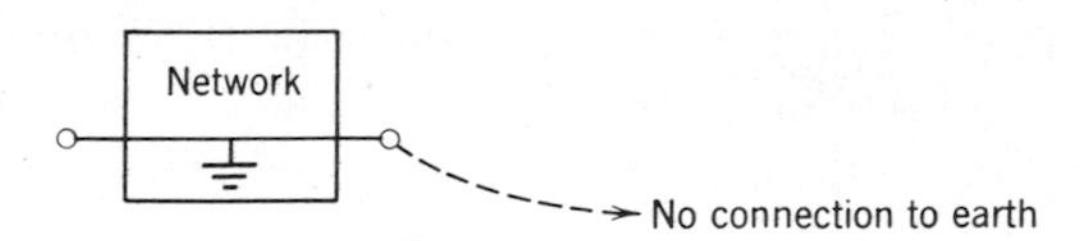

(*c*) In practice, ground symbols are used even when a network "common" or "reference" is *not* connected to earth potential

Figure 5.3
Grounds and reference nodes.

a cold water pipe. In practice, however, ground symbols are used for network reference nodes or "common" nodes, even when the reference node is not connected to earth potential.

The complete symbol for the op-amp with the reference node and both supply batteries in place is shown in Fig. 5.4*a*. The symbol V_{CC} is often used for the supply voltage, a carry-over from conventional transitor-circuit notation. The positive and negative supplies are shown as having equal magnitudes. Although this is the most common arrangement, it is not a necessary arrangement. Most op-amps operate with a wide range of supply voltages, and these voltages need not be symmetrical about the ground. Information about the choice of supply voltages and the corresponding limitations on inputs and outputs is usually provided on the op-amp data sheet. In Fig. 5.4*a* the input voltages v_- and v_+ are ordinary network variables and, as such, might be either positive or negative. The subscript is associated with the name of the input terminal, not with the algebraic sign of the voltage.

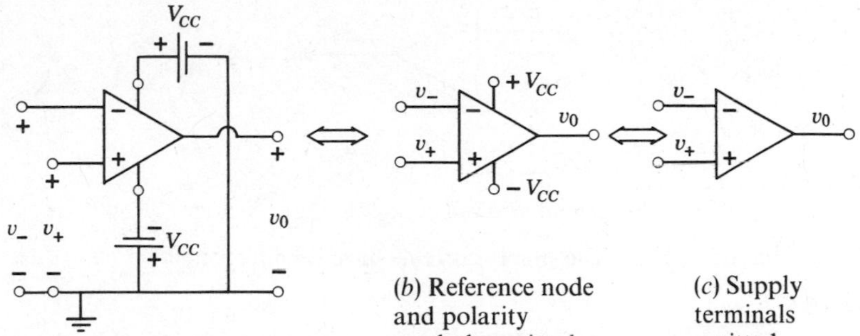

(*a*) Reference node, polarity symbols, and dc supplies shown explicitly

(*b*) Reference node and polarity symbols omitted. Connection to dc supplies indicated by labels on supply nodes

(*c*) Supply terminals omitted

Figure 5.4
Node voltage notation for operational amplifiers. The notations in (*a*), (*b*), and (*c*) are equivalent to one another.

Two shorthand notations for op-amps are shown in Figs. 5.4*b* and 5.4*c*. In Fig. 5.4*b*, the reference node is presumed; proper connections to supplies are presumed; and inputs and outputs are labeled with the appropriate node-to-reference voltages. Often the supply terminals are left off, as in Fig. 5.4*c*. However, this streamlined notation can be confusing, as explained below.

Figure 5.5*a* shows the op-amp symbol with the various terminal currents labeled. Since *all* terminals are accounted for, the op-amp itself can be treated as a big node, and we can write KCL:

$$i_- + i_+ + i_{C+} + i_{C-} + i_0 = 0 \tag{5.1}$$

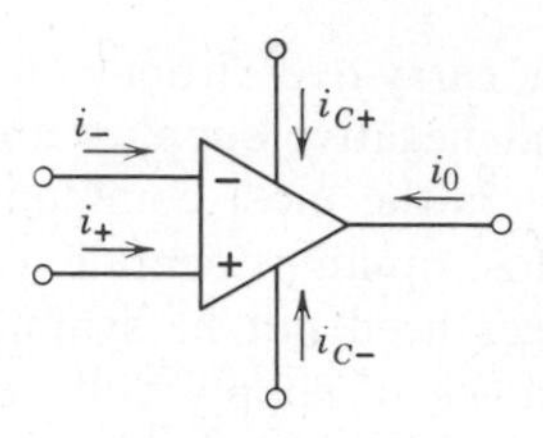

Correct: $\boxed{\text{KCL:} \quad i_- + i_+ + i_{C+} + i_{C-} + i_0 = 0}$

(*a*) KCL for an op-amp. Usually i_- and i_+ are very small. The output current i_0 flows to ground through one of the supply terminals.

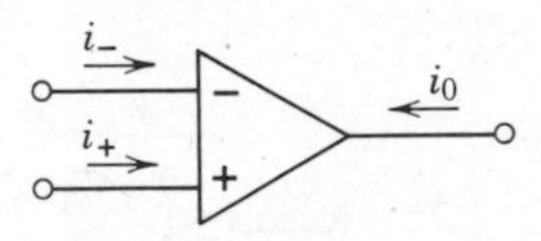

Incorrect: $\boxed{i_- + i_+ + i_0 = 0}$ (crossed out)

(*b*) Incorrect use of KCL. The supply currents have been forgotten.

Figure 5.5
Caution. The simplified op-amp notation in (*b*) obscures the role played by the supply currents.

Normally i_- and i_+ are very small compared to the other currents, so that KCL reduces approximately to

$$i_{C+} + i_{C-} = -i_0 \tag{5.2}$$

This equation tells us that the output current i_0 flows to ground through the supply terminals. If we leave the supply terminals off, as in Fig. 5.5*b* and naively write KCL, we are led to the *incorrect* conclusion that because i_- and i_+ are very small, i_0 must also be small. Our practice will be to use the simplified symbol of Fig. 5.5*b* with only inputs and outputs shown. Remember, however, that the op-amp is connected to two dc supplies.

5.1.3 Ideal Characteristics

The network behavior of op-amps will be described in two stages. Here, a rather idealized version of op-amp behavior is given. In Section 5.4, the differences between actual op-amps and the ideal model presented here are discussed.

The ideal op-amp has an infinite input resistance, a zero output resistance, and a voltage transfer characteristic of the kind shown in Fig. 5.6. The infinite input resistance means that the input currents, i_+ and i_-, are both zero; the zero output resistance means that the op-amp transfer characteristic can be modeled simply with a dependent voltage source (see Fig. 5.7). The transfer characteristic of Fig. 5.6 shows three regions of op-amp operation: a linear region and two saturation regions. Let us examine each region.

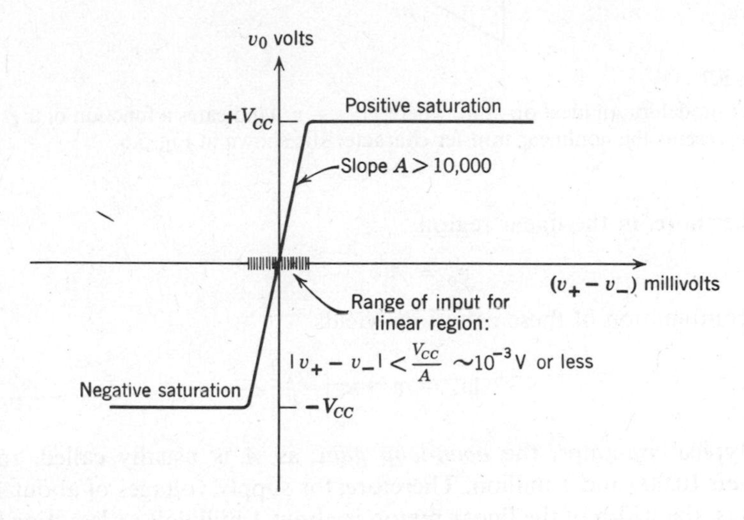

Figure 5.6
Typical op-amp transfer characteristic. The output saturates at the supply voltage. The range of input voltage for linear operation is very small.

In the linear region, the output voltage v_O lies somewhere between the two supply voltages. The functional relation between v_O and the difference in input voltages, $(v_+ - v_-)$, is a linear one, with a voltage gain A usually in excess of 10,000. In the linear region, therefore, we should be able to model the op-amp as a *linear* voltage-controlled voltage source of the kind we analyzed in Chapter 4 (Fig. 4.3).

Because the gain in the linear region is so large, and because the output voltage range is finite, the range of input voltage difference in the linear region is very small. For linear-region operation, we must have

$$|v_O| < |V_{CC}| \tag{5.3}$$

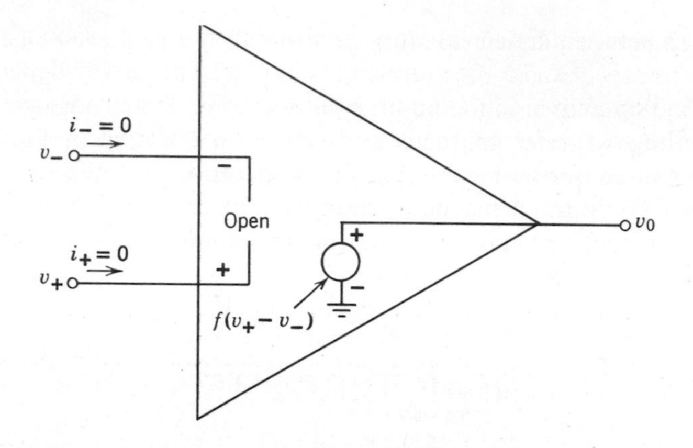

Figure 5.7
Circuit model for an ideal op-amp, where $f(v_+ - v_-)$ indicates a function of $(v_+ - v_-)$, and represents the nonlinear transfer characteristic shown in Fig. 5.6.

Furthermore, in the linear region

$$v_O = A(v_+ - v_-) \tag{5.4}$$

The combination of these equations yields

$$|v_+ - v_-| < \left|\frac{V_{CC}}{A}\right| \tag{5.5}$$

For typical op-amps, the *open-loop gain*, as A is usually called, ranges between 10,000 and 1 million. Therefore, for supply voltages of about 10 or 15 volts, the width of the linear region is about 1 millivolt or less. For high-gain op-amps, with open-loop gains of 10^6, the input voltage difference in the linear region is a few microvolts. These facts suggest a simple way of characterizing linear operation:

LINEAR OPERATION: 1. Input currents are zero.
2. Input voltages, v_+ and v_-, are *equal* (to within 1 millivolt or less.)

Operation in either saturation region is characterized by *unequal* input voltages, that is, by differences between v_+ and v_- larger than 1 millivolt. Thus, if $v_+ > v_-$ (by more than 1 millivolt) we can assume that v_O is saturated at $+V_{CC}$. Similarly, if $v_- > v_+$ by more than 1 millivolt, we can assume that v_O is saturated at $-V_{CC}$. Notice that a v_+ more positive than v_- drives the output positive. Thus v_+ is called the *non-inverting input*. Since v_- greater than v_+ drives the output negative, v_- is called the *inverting input*.

The remaining question is: How do we analyze a given network to decide whether to expect linear or saturation behavior? The answer depends on the kind of feedback used in the op-amp network. Feedback from the output to the inverting (−) terminal tends to produce operation in the linear region, while the absence of feedback (open-loop operation) or feedback to the non-inverting (+) terminal results in saturation at one of the supply voltages. These points are discussed in the following sections.

5.2 BASIC LINEAR CIRCUITS

5.2.1 Inverting Amplifiers

Figure 5.8 shows a circuit in which a feedback element R_F is connected from the output to the − terminal. Let us examine how the feedback produces operation in the linear region by (1) examining the behavior of this circuit assuming linear operation, and (2) testing the range of validity of the assumption.

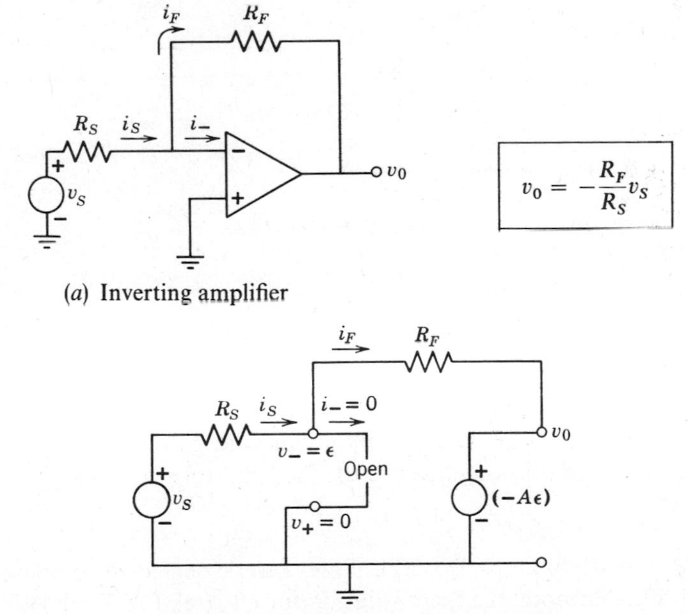

Figure 5.8
Inverting amplifier.

If we assume that the op-amp is in the linear region, we can use the equivalent circuit of Fig. 5.8*b*. The voltage on the + terminal, v_+, is zero because the + terminal is grounded. As v_S goes positive, it pulls v_- positive. When v_-'is driven positive, even by a very small amount, v_O is driven strongly negative. Because of the feedback path through R_F, a decrease in v_O pulls v_- back toward zero. Thus, a balance is achieved, in which v_- is driven just *positive* enough to drive v_O just *negative* enough to keep v_- from rising further. If the open-loop gain A is sufficiently large, this balance point occurs with v_- very near zero, even for a rather large v_S. The following algebraic analysis shows how this works.

Let us denote the equilibrium or balance value of v_- by ϵ (as in Fig. 5.8*b*) and assume that ϵ is small compared both to v_S and v_O (an assumption that will be explicitly checked at the end of the calculation). The current i_S is

$$i_S = \frac{v_S - \overset{\text{Small}}{\epsilon}}{R_S} \cong \frac{v_S}{R_S} \tag{5.6}$$

since $|\epsilon| \ll |v_S|$. The input current i_- is zero. Thus, by KCL:

$$i_F = i_S \tag{5.7}$$

Finally, by KVL,

$$v_O = \overset{\text{Small}}{\epsilon} - i_F R_F \approx -i_F R_F \tag{5.8}$$

Substituting i_S for i_F, we obtain

$$v_O = -i_S R_F \tag{5.9}$$

or

$$v_O = -\left(\frac{R_F}{R_S}\right) v_S \tag{5.10}$$

Thus, if our assumptions of linear operation and small ϵ are valid, the circuit of Fig. 5.8*a* behaves as an inverting amplifier with gain $-(R_F/R_S)$.

To test the assumptions, let us use the fact that for $v_+ = 0$ and $v_- = \epsilon$,

$$v_O = -A\epsilon \tag{5.11}$$

Since $A \gg 1$ in all op-amps, Eq. 5.11 allows ϵ to be neglected compared to v_O in Eq. 5.8. Furthermore, the exact solution for ϵ/v_S (see Fig. E4.4*a*) is

$$\frac{\epsilon}{v_S} = \left(\frac{R_F}{AR_S}\right)\left(\frac{1}{1 + 1/A + R_F/AR_S}\right) \tag{5.12}$$

Therefore, as long as

$$A \gg \frac{R_F}{R_S} \tag{5.13}$$

we can neglect ϵ compared to v_S in Eq. 5.6. The quantity R_F/R_S is called the *closed-loop gain*, the gain when feedback is applied through R_F. Equation 5.13 tells us that as long as the closed-loop gain is much less than the open-loop gain A, the input voltage ϵ is very small and can be approximated by zero when compared with v_S.

The range of v_0 for linear operation is still restricted by the requirement that $|v_0| < |V_{CC}|$. This in turn places a restriction on the magnitudes of R_F, R_S, and v_S. Thus, from Eq. 5.10, we must satisfy the following inequality for linear operation:

$$|v_S| < \frac{|V_{CC}|}{(R_F/R_S)} \tag{5.14}$$

We recall that under open-loop conditions, an input equal to the supply voltage divided by the open-loop gain (V_{CC}/A) was sufficient to saturate the amplifier. Now Eq. 5.14 reveals that under closed-loop conditions, an input equal to the supply voltage divided by the closed-loop gain will saturate the amplifier. Thus the effect of the feedback has been (1) to reduce the overall gain, (2) to permit correspondingly larger input voltage v_S without saturation, and (3) to produce an inverting closed-loop gain (R_F/R_S) that depends only on the passive, external components R_F and R_S, and not on the open-loop gain A.

The open-loop gain of an op-amp is strongly temperature dependent and varies widely from unit to unit. Therefore, the elimination of A in the closed-loop gain expression is extremely valuable. Simple choices of resistor values enable the designer to specify the gain of a linear amplifier without worrying about the detailed characteristics of the op-amp itself. The only constraints are that the open-loop gain must be large compared to the closed-loop gain and that the amplifier must not be driven into saturation.

5.2.2 Non-inverting Amplifiers and Followers

Figure 5.9*a* shows a circuit in which the input is applied to the + terminal, but the feedback is applied to the − terminal. Let us analyze this circuit assuming linear operation and assuming that the closed-loop gain will be small compared to the open-loop gain.

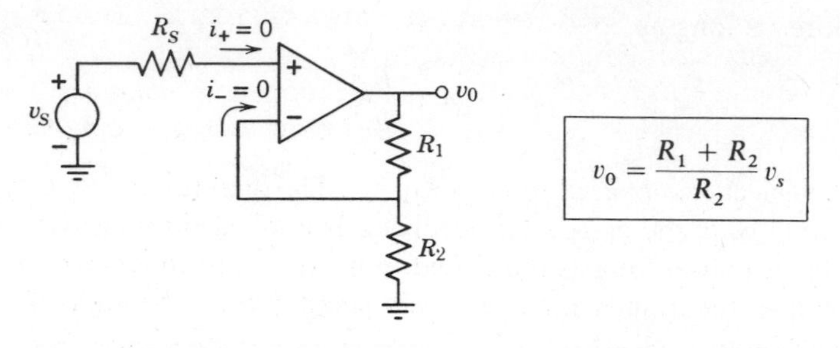

(*a*) Non-inverting amplifier

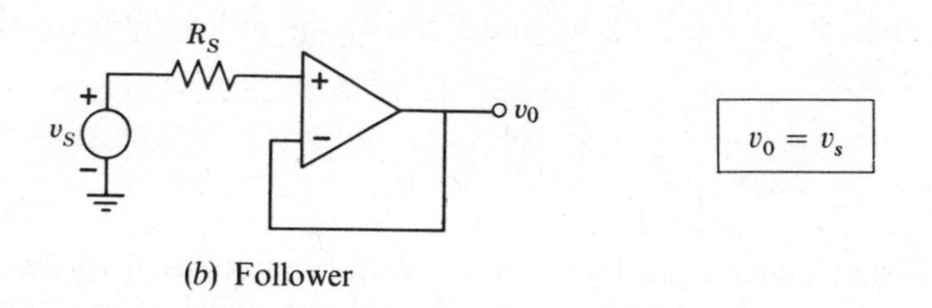

(*b*) Follower

Figure 5.9
Non-inverting amplifier and follower.

Since the input current i_+ is zero, we know that

$$v_+ = v_S \tag{5.15}$$

Furthermore, since i_- is zero, v_- must be related to v_O through a simple voltage divider.

$$v_- = \frac{R_2}{R_1 + R_2} v_O \tag{5.16}$$

Finally, we use the fact that the difference between v_+ and v_- under closed-loop conditions is essentially zero. That is $v_+ = v_-$, or

$$v_S = \frac{R_2}{R_1 + R_2} v_O \tag{5.17}$$

Solving for v_O yields

$$v_O = \left(\frac{R_1 + R_2}{R_2}\right) v_S \tag{5.18}$$

Thus the circuit of Fig. 5.9*a* is a non-inverting amplifier with gain controlled by the choice of R_1 and R_2

If we allow the resistor R_2 to get very large, the gain of the non-inverting amplifier approaches unity. In fact, if we let $R_2 \to \infty$ and $R_1 \to 0$, the circuit of Fig. 5.9*b* results. The output v_O equals the input v_S, but while the + input draws no current from the source v_S, the output v_O can supply substantial currents to external loads. This connection, called a *follower*, senses the input voltage and drives external loads with a voltage that "follows" the input.

5.2.3 Analog Addition and Subtraction

Figure 5.10 indicates how op-amps can be used to add or subtract analog waveforms. If we consider addition first, the presence of feedback to the inverting terminal through R_F assures linear operation. Therefore, $v_+ = 0$ (because v_+ is grounded) and $v_- = v_+ = 0$. The two input currents are given by

$$i_1 = \frac{v_1 - v_-}{R_1} \cong \frac{v_1}{R_1}$$
$$i_2 = \frac{v_2 - v_-}{R_2} \cong \frac{v_2}{R_2} \tag{5.19}$$

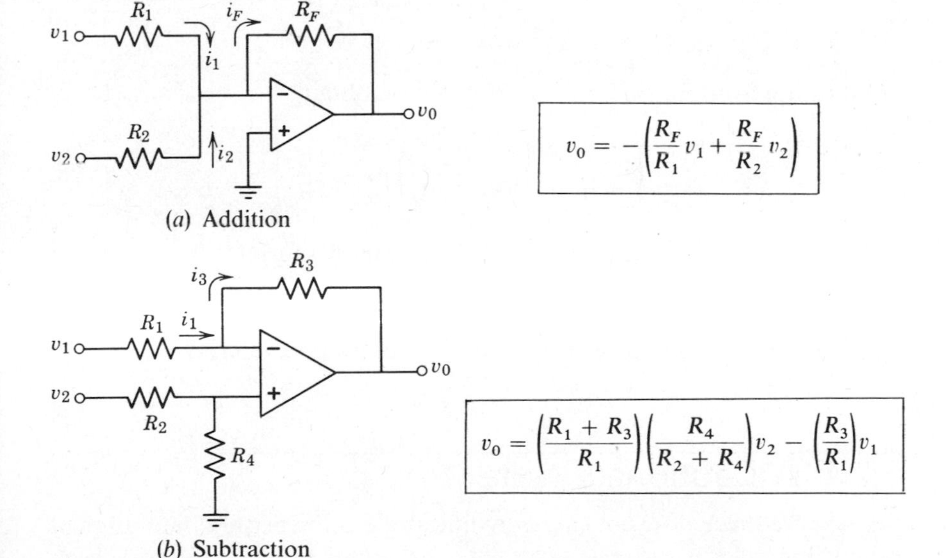

$$v_O = -\left(\frac{R_F}{R_1} v_1 + \frac{R_F}{R_2} v_2\right)$$

$$v_O = \left(\frac{R_1 + R_3}{R_1}\right)\left(\frac{R_4}{R_2 + R_4}\right) v_2 - \left(\frac{R_3}{R_1}\right) v_1$$

Figure 5.10
Analog addition and subtraction.

Because the input current i_- is zero, we must, by KCL, have

$$i_F = i_1 + i_2 \tag{5.20}$$

Therefore

$$v_O = -\left(\frac{R_F}{R_1}v_1 + \frac{R_F}{R_2}v_2\right) \tag{5.21}$$

Thus, v_O is an inverted, scaled sum of v_1 and v_2, the coefficients being determined by the choice of resistor values.

Figure 5.10*b* shows a subtraction circuit. Once again, the feedback to the negative terminal through R_3 assures linear operation. The relation between v_+ and v_2 is a simple voltage divider.

$$v_+ = \frac{R_4}{R_2 + R_4}v_2 \tag{5.22}$$

Because of the linear operation, $v_- = v_+$. Therefore, we can find i_1 to be

$$i_1 = \frac{v_1 - v_-}{R_1} = \frac{v_1 - v_+}{R_1} \tag{5.23}$$

Since the input current i_- is zero, i_3 must equal i_1, yielding

$$v_O = v_+ - \left(\frac{v_1 - v_+}{R_1}\right)R_3 \tag{5.24}$$

Substituting from Eq. 5.22 and simplifying, we obtain

$$v_O = \left(\frac{R_1 + R_3}{R_1}\right)\left(\frac{R_4}{R_2 + R_4}\right)v_2 - \left(\frac{R_3}{R_1}\right)v_1 \tag{5.25}$$

If R_1, R_2, R_3, and R_4 are all equal, Eq. 5.25 simplifies further to

$$v_O = v_2 - v_1 \tag{5.26}$$

Thus, the circuit of Fig. 5.10*b* can be used to subtract waveforms.

5.2.4 A Cautionary Note

Negative feedback does not *guarantee* linear operation because saturation of a linear amplifier is always a possibility. Therefore, it is necessary to check numerically that $|v_O| < |V_{CC}|$ in every linear circuit application. See Exercises E5.1–E5.7.

5.3 BASIC NONLINEAR CIRCUITS

5.3.1 The Comparator

When we introduced the open-loop op-amp transfer characteristic we stated that if v_+ and v_- differ by more than about 1 millivolt, the output is saturated at one of the supply voltages; the positive supply for $v_+ > v_-$, the negative supply for $v_+ < v_-$. This ability to discriminate between unequal voltages is very useful. Op-amps that are specifically designed for this application are called *comparators*. They are characterized by very high open-loop gain and by the ability to make very rapid transitions between positive and negative saturation.

One of the uses of comparators is to detect zero crossings in a waveform (Fig. 5.11). Whenever the input passes through zero, the output swings between $+V_{CC}$ and $-V_{CC}$. Thus, the output waveform is a digital signal, and the time of the transition in the output waveform carries the zero-crossing information.

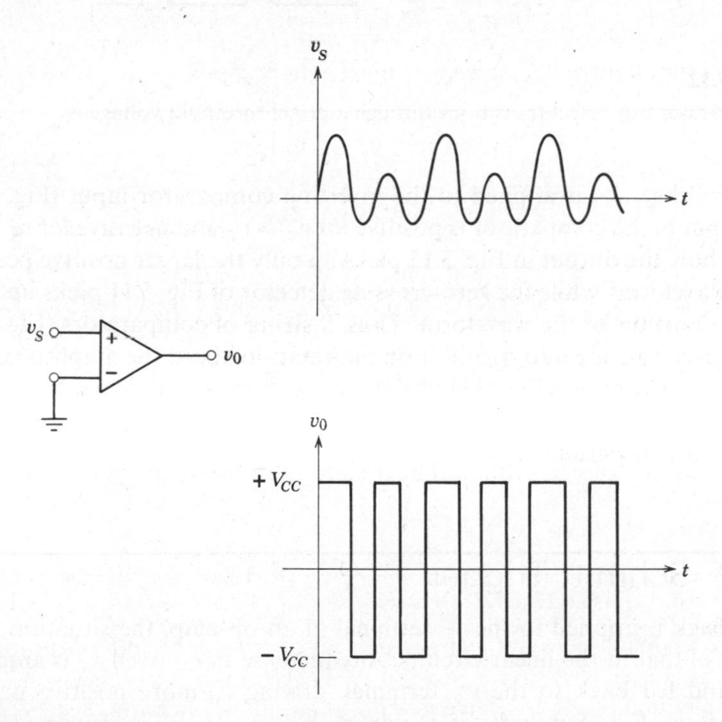

Figure 5.11
A comparator can detect zero-crossings.

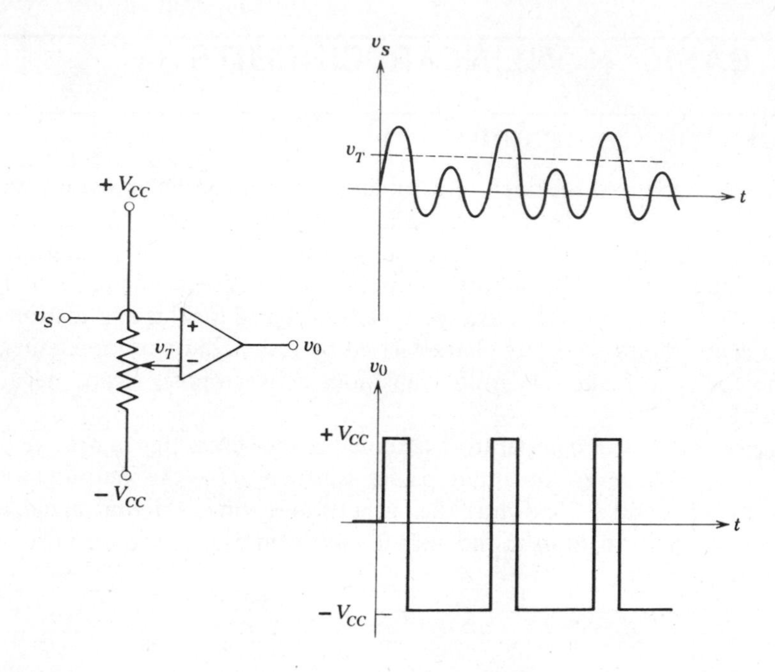

Figure 5.12
A comparator can detect transitions through a preset threshold voltage v_T.

If a voltage, v_T, is applied to the inverting comparator input (Fig. 5.12), the output of the comparator is positive for $v_S > v_T$ and negative for $v_S < v_T$. Notice how the output in Fig. 5.12 picks up only the larger positive peaks in the v_S waveforms while the zero-crossing detector of Fig. 5.11 picks up every positive portion of the waveform. Thus, a string of comparators (Fig. 5.13) can be used to code into digital form information about the amplitude of an analog waveform as a function of time. If n comparators are used, a total of $n + 1$ ranges is defined. Each pattern of outputs ($+V_{CC}$ or $-V_{CC}$) corresponds to a specific amplitude range.

5.3.2 Schmitt Trigger

If feedback is applied to the + terminal of an op-amp, the situation is the reverse of that in the linear circuits. Any increase in v_+ over v_- is amplified by A and fed back to the v_+ terminal, driving v_0 more positive until v_0 saturates at V_{CC}. Similarly, if v_+ drops below v_-, v_0 is driven strongly negative which makes v_+ even more negative until v_0 saturates at $-V_{CC}$.

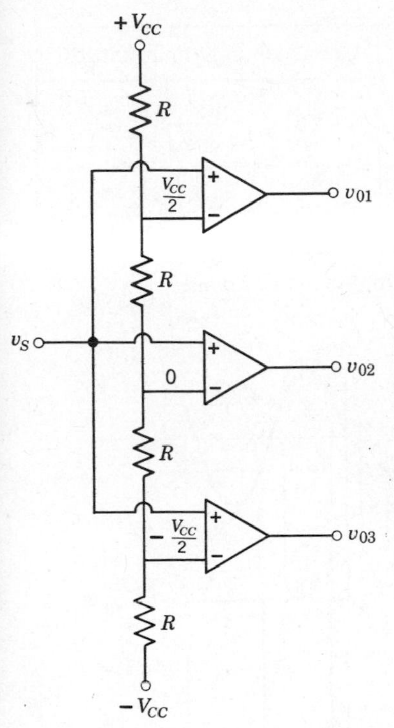

Range for v_S	Output Pattern		
	v_{01}	v_{02}	v_{03}
$v_S > \frac{V_{CC}}{2}$	+	+	+
$\frac{V_{CC}}{2} > v_S > 0$	−	+	+
$0 > v_S > -\frac{V_{CC}}{2}$	−	−	+
$\frac{-V_{CC}}{2} > v_S$	−	−	−

Figure 5.13
A simple form of digital coding of an analog amplitude. Each pattern of outputs ($+V_{CC}$ or $-V_{CC}$) corresponds to a specific amplitude range.

Thus, whenever feedback to the + terminal is present (without at least compensating feedback to the − terminal), the op-amp is in one of the two saturation states.

The Schmitt trigger circuit of Fig. 5.14 illustrates what can happen in a circuit with positive feedback. Since the feedback is to the + terminal, either v_O equals $+V_{CC}$ or it equals $-V_{CC}$. The table in Fig. 5.14*b* shows which set of input ranges correspond to each of the two possible outputs. The positive output is obtained for $v_S < R_2V_{CC}/(R_1 + R_2)$ while the negative output is obtained for $v_S > -R_2V_{CC}/(R_1 + R_2)$. In attempting to plot these ranges in Fig. 5.14*c*, we notice that the ranges overlap. Therefore, this circuit has a memory. If v_S lies between $R_2V_{CC}/(R_1 + R_2)$ and $-R_2V_{CC}/(R_1 + R_2)$, the output state depends on which way the magnitude of the input last exceeded $|R_2V_{CC}/(R_1 + R_2)|$. If, for example, the last excursion was $v_S > R_2V_{CC}/(R_1 + R_2)$, then v_O is negative and remains negative until v_S becomes sufficiently negative to drive the circuit into the positive output state. The existence of

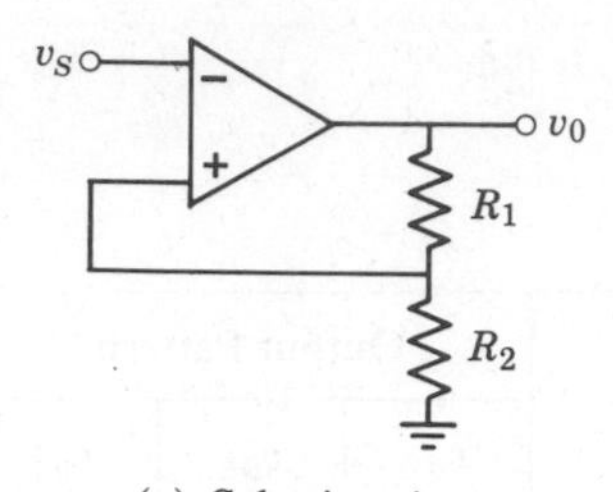

(*a*) Schmitt trigger

Output State	Corresponding Input Range
$+V_{CC}$	$v_S < \frac{R_2}{R_1 + R_2} V_{CC}$
$-V_{CC}$	$v_S > -\frac{R_2}{R_1 + R_2} V_{CC}$

(*b*) Analysis of range

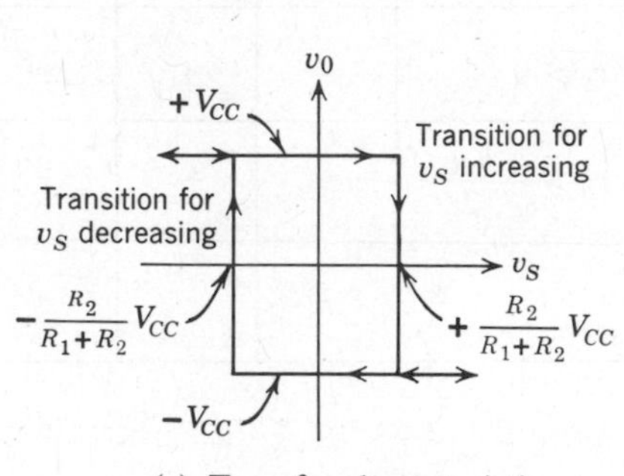

(*c*) Transfer characteristic

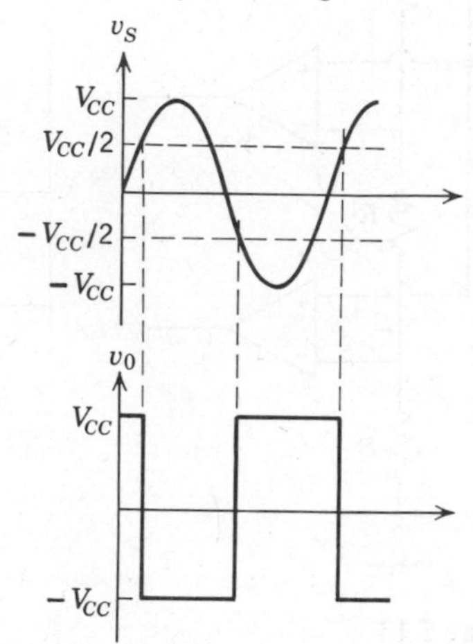

(*d*) Waveform example for the special case $R_1 = R_2$.

Figure 5.14
A Schmitt trigger has hysteresis and memory.

two stable states, depending on the past history of the circuit, is called *hysteresis.* Many circuits with positive feedback exhibit hysteresis.

An example of how a Schmitt trigger can be used is shown in Fig. 5.14*d*. This example assumes R_1 equal to R_2, so that the transition occurs at $+V_{CC}/2$ for v_S increasing and at $-V_{CC}/2$ for v_S decreasing. A sine wave with amplitude V_{CC} emerges as a square wave, delayed in time because of the hysteresis, but still with amplitude V_{CC}. Schmitt triggers are also used for the same kind of amplitude discrimination illustrated with comparators in the previous section. Because the positive feedback drives the transition between output states very hard, Schmitt triggers can be made to switch very quickly. In addition, because of the hysteresis property, the Schmitt trigger is less likely than the comparator to bounce back and forth between states if a waveform with many small ripples on it gradually passes through the transition level.

5.4 A SECOND LOOK AT OP-AMPS

5.4.1 The Point of View

Thus far we have presumed that all op-amps are ideal, in the sense that they all behave like the prototype discussed in Section 5.1.3. Actual op-amps fail to match this ideal in several ways. Many op-amps have extra terminals for the partial correction of non-idealities with external components. While a detailed account of the various imperfections in op-amps and the way in which each imperfection affects circuit performance is not necessary here, beginning students should be aware of the issues, so that they can begin to use op-amps immediately without making design errors, and so that they can understand the more detailed material and interpret the relevant ideas and terminology. In the following sections we define various op-amp imperfections, estimate magnitudes and indicate how circuit performance can be affected. It is precisely in the specification of these imperfections that op-amps differ from one another. Usually, the more ideal the behavior, the higher the price. Since most projects involve finite budgets, the student should understand when, in his own application, he should pay a higher price for a higher quality component.

5.4.2 Offset Voltage and Drift

When a short circuit is applied between the inputs of an *ideal* op-amp, the output voltage is zero. In *actual* op-amps, however, imperfect matching of components within the op-amp lead to a nonzero output voltage when the inputs are shorted together. The magnitude of this output voltage usually is directly proportional to the closed-loop gain of the circuit. It is conventional therefore, to specify this imperfection in terms of an equivalent voltage at the input, called the *input offset voltage*. If, for example, an op-amp has an input offset voltage specified at ± 1 mV maximum, and the op-amp is used in a circuit with closed-loop gain of 100, the output voltage with zero input will be somewhere between $+0.1$ V and -0.1 V.

Most op-amps have external terminals to which appropriate variable resistors can be attached for the purpose of zeroing, or nulling out this offset voltage. Op-amp data sheets give explicit instructions on what components to use, allowing the effective voltage offset to be reduced substantially. However, all parameters of op-amp circuits depend on temperature, and also tend to drift with time. It is the magnitude of the *drift* in the offset voltage, called the *offset drift*, that ultimately determines how well one can expect to null out the offset voltage. Offset drift is usually specified in terms of a number of μV/°C over a given range of temperature. For example, an offset drift might be specified as 10 μV/°C for operation between 0 and 70°C. The best way to interpret this quantity is to view the product of the offset drift and the temperature range (700 μV in the above example) as defining the maximum range within which the offset voltage will vary under any operating conditions.

The effect of whatever offset voltage remains after nulling produces a dc error in the output. Therefore, one cannot expect to amplify millivolt-sized dc signals accurately with an amplifier that has a 10 mV input offset voltage. At the same time, if the signal level is several volts, the presence of a 10 mV input offset is not likely to produce a problem.

5.4.3 Bias and Offset Currents

The input terminals of ideal op-amps draw no current. Actual op-amps, however, require small but nonzero dc currents at both inputs. The average of the two input currents is called the *bias current*, while the difference between the two input currents is called the *input offset current*. Usually, the two input currents are nearly equal, so that the offset current has a magnitude much less than the bias current.

The effect of nonzero input currents is to add small voltage drops across either the source or feedback resistors. These voltage drops can then produce output offset errors. Since bias currents flow through resistive paths, it is generally true that the larger the magnitude of the source and feedback resistors, the more likely one is to observe an error due to bias currents.

Bias and offset currents are among the most widely varying op-amp characteristics. The cheapest op-amps have bias currents in the microamp range, while expensive electrometer op-amps have bias currents in the femtoamp (10^{-15} amp) range.

In most op-amp circuits external elements may be added to null out the effect of bias and offset currents. In Fig. 5.15, a resistor R_B has been inserted between the + terminal and ground. If we assume for the moment that equal bias currents i_B flow into both input leads, the voltage on the + terminal is

$$v_+ = -i_B R_B \tag{5.28}$$

Because of the negative feedback, we can assume $v_- = v_+$. Therefore

$$i_S = \frac{v_S + i_B R_B}{R_S} \tag{5.29}$$

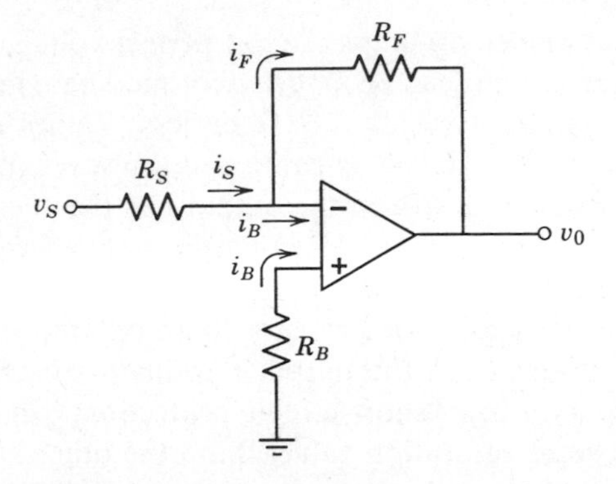

Figure 5.15
To null out the effect of equal bias currents in an inverting amplifier choose $R_B = R_S \| R_F$.

But because of i_B flowing in the − terminal, the current through R_F is

$$i_F = i_S - i_B \tag{5.30}$$

Writing KVL, we find

$$v_O = v_- - i_F R_F = -\frac{R_F}{R_S} v_S + \left[R_F - \left(\frac{R_F + R_S}{R_S} \right) R_B \right] i_B \tag{5.31}$$

Examination of Eq. 5.31 shows that if we choose

$$R_B = R_S \| R_F = \frac{R_F R_S}{R_F + R_S} \tag{5.32}$$

the bias current produces no output error.

The above example was worked out when both input currents are equal (zero offset current). With different input currents, it is still possible to find a choice of R_B that will null out the error. The critical issue then becomes the stability of bias and offset currents with temperature and time. Usually, the temperature variation and drift is specified in terms of the variation of the offset current in units of current/°C over a specified temperature range.

5.4.4 Saturation Voltage and Current, and Output Resistance

Most op-amps saturate at voltages one or two volts away from the supply voltage. Op-amp data sheets specify how close the output can go to the supply voltage, and how this saturation voltage might vary from unit to unit.

Unlike our ideal model, op-amps are not perfect voltage sources at their outputs. The output circuits can be modeled for moderate load currents with finite output resistances, typically 100 Ω or less. Values are included on op-amp data sheets. The student is cautioned, however, that the presence of the dependent voltage source in the output has the effect of making the Thévenin output resistance under closed-loop negative feedback conditions very much smaller than the specified output resistance.

If one connects too small a load resistor to an op-amp (i.e., if one tries to draw too much current from the output terminal) one of two things will happen. If the op-amp has "short circuit protection," the output current will be limited to some saturation value. If, on the other hand, the op-amp is not short-circuit protected, the power rating of the device might be exceeded, destroying the device. The data sheet for the op-amp in question will contain the relevant information.

5.4.5 Slew Rate and Compensation

Op-amps cannot change their output voltage in zero time. The *slew rate* specifies how rapidly the voltage can swing under stated closed-loop gain conditions when the input is driven with a signal. Slew rate is expressed in volts/unit time. Thus an op-amp with 2.5 V/μs slew rate can swing from −15 to +15 V in 12 μs.

The slew rate is related, although indirectly, to the frequency response of the op-amp. If the speed of variation of the output voltage is limited, a sine wave that requires too rapid a variation in output cannot be amplified by the op-amp. (See Chapter 14 for a discussion of amplifier frequency response.)

The slew rate does permit discussion of another important issue: op-amp stability. Because of the finite response time of an op-amp, the output variation lags behind the input. Since it is the output that determines the feedback, the feedback reaches the inverting terminal after the input signal. If the feedback is large enough or the delay long enough, an instability can result in which the output oscillates continuously between positive and negative values, each time overshooting the value of output that should ideally be present. Since "more feedback" corresponds to the lower closed-loop gain, it is at lower closed-loop gains that the stability can be the most serious problem. Some op-amps are *internally compensated* against this oscillation condition. They have capacitors built into them that prevent oscillation even at unity gain. Others have external terminals to which a *compensation capacitor* is to be attached, following instructions on the data sheet. As a rule, the choice of capacitor value depends on closed-loop gain, with the larger capacitance values being required for operation at a smaller closed-loop gain.

5.4.6 Common-Mode Rejection and Input Resistance

Throughout this chapter we have explicitly assumed that the inputs of an op-amp are not connected to ground through any internal path, that between these inputs there is an open circuit, and that the output responds only to the difference between v_+ and v_-, regardless of the average value of v_+ and v_-. The ways in which actual op-amps fail to meet these idealizations are manifold, and complex. In this concluding section let us look at some of these issues more carefully.

The first issue is the permissible range of input voltage. If either input voltage (or both together) gets too close to the supply voltage, the device will fail. Data sheets specify permitted maximum values. Usually, these values are within a few volts of the supply voltages.

A more critical issue is the response of the op-amp to the average of v_+ and v_-, even if we assume that both voltages lie within the permitted range. The average of v_+ and v_- is called the *common-mode signal*, while the difference between v_+ and v_- is called the *difference-mode signal*. The primary response of an op-amp is to the difference-mode signal, but there is also nonzero response to the common-mode signal.

Figure 5.16 illustrates the conventional way in which this common-mode response is expressed. The difference-mode gain A_D is the ordinary open-loop gain we have been using throughout this chapter. The common-mode gain, A_C, is the response when both + and − inputs are driven with the same

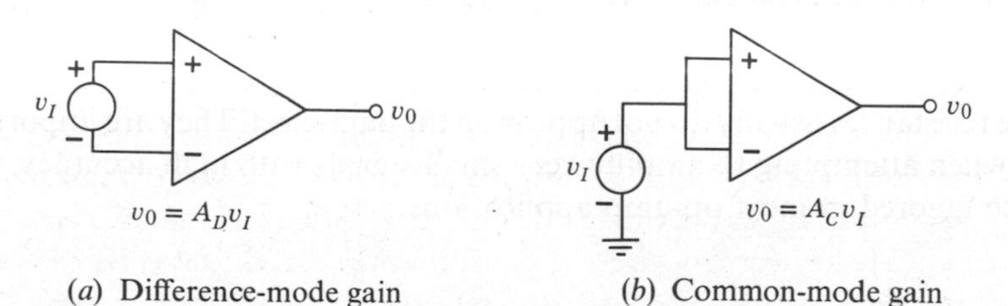

(*a*) Difference-mode gain (*b*) Common-mode gain

Expressed as a fraction: $\text{CMRR} = A_D/A_C$

Expressed in decibels: $\text{CMRR} = 20 \log (A_D/A_C)\,dB$

Figure 5.16
The common-mode rejection ratio.

signal with respect to ground. The *common-mode rejection ratio*, abbreviated either CMRR, or occasionally as CMR, is the ratio of A_D to A_C. CMRR is most often expressed in terms of *decibels* (dB), defined as follows:

$$\text{CMRR} = 20 \log \frac{A_D}{A_C} \text{ dB} \tag{5.33}$$

For example, if the difference-mode gain is 10^4, and the common-mode gain is unity, the CMRR would be 80 dB, a number typical of moderate quality op-amps.

Common-mode rejection is most important when one attempts to amplify small difference-mode signals in the presence of large common-mode signals. A major source of common-mode signals is pickup from the 60 Hz power line. It is not uncommon to find a 10 V, 60 Hz signal between two nominal "ground" points within the same laboratory. If a source of a small difference-mode signal, such as a thermocouple, is "grounded" to a point with a potential 10 V different from the op-amp ground, the common-mode signal appearing at the output can interfere significantly with the amplified difference-mode signal. Quantitative discussion of this problem is reserved for the problems, and for Chapters 17 and 18.

Finally, let us address the question of input resistance. Even after the effects of dc bias currents have been nulled out, a nonzero input current flows in response to an input voltage. This effect can be modeled by inserting an equivalent *input resistance* between the two input terminals of the op-amp, replacing the open circuit we have been using thus far. Data sheets list this input resistance, with typical values being on the order of 100 kΩ to 1 MΩ for bipolar transistor op-amps and 100 MΩ for op-amps with field-effect transistors as the input elements. Be careful, however, because in a circuit with negative feedback, the Thévenin input resistance can be quite different from the number appearing on the data sheet. This phenomenon is examined quantitatively in Problem P5.1.

There are, in addition to the input resistance between the + and − terminals, large but finite resistances between each input terminal and ground. These resistances usually do not appear on the data sheet. They are important only when attempting to amplify very small signals with high accuracy, and can be ignored in most op-amp applications.

REFERENCES

R5.1 Arpad Barna, *Operational Amplifiers* (New York: Wiley-Interscience, 1971).

R5.2 Michael Kahn, *The Versatile Op-Amp* (New York: Holt, Rinehart and Winston, 1970).

R5.3 Jerald G. Graeme and Gene E. Tobey, eds., *Operational Amplifiers: Design and Applications* (New York: McGraw-Hill, 1971).

R5.4 Bruce D. Wedlock and James K. Roberge, *Electronic Components and Measurements* (Englewood Cliffs, N.J.: Prentice-Hall, 1969), Chapter 16.

R5.5 Data sheets and applications notes available from major manufacturers (Motorola, Fairchild, RCA, National Semiconductor, Analog Devices, Philbrick, Burr-Brown, and others).

QUESTIONS

Q5.1 Is the use of a linear dependent source to model an op-amp a violation of the conservation of energy? Where does the power come from?

Q5.2 An ordinary toggle switch contains springs that retain the switch position when you let go. Consider the force from your finger as an input, and the switch position as an output. Model the toggle switch with a positive-feedback Schmitt-trigger circuit. Does the switch have hysteresis? Memory?

Q5.3 Is the use of node voltages valid when the potential difference between the reference node (circuit common) and a true ground (cold water pipe) is nonzero? Explain.

Q5.4 Is the op-amp a linear element?

Q5.5 Suppose you are driving a car, but the car you are driving has a delay between the time you turn the steering wheel and the time at which the front wheels actually begin to turn. Examine the problems of steering such a car down a nearly straight road and down a winding road. (This example may be useful in illustrating the concept of op-amp stability discussed in Section 5.4.5).

EXERCISES

E5.1 Assume linear operation and express v_O in Fig. E5.1 as a superposition of the responses to v_1 and v_2.

Answer. $v_O = \left(1 + \frac{R_F}{R_S}\right) v_2 - \frac{R_F}{R_S} v_1.$

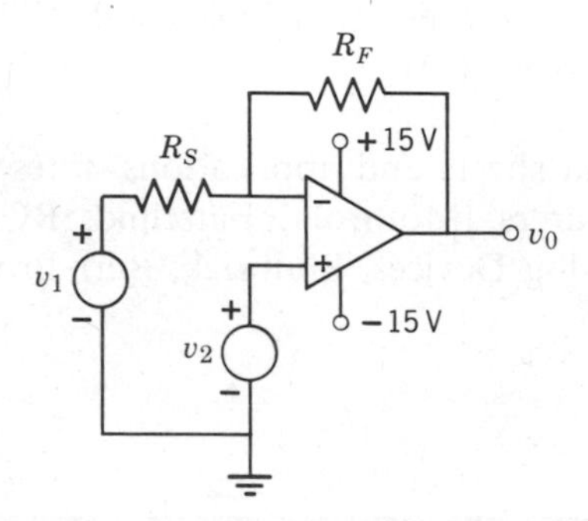

Figure E5.1

E5.2 What restrictions must be placed on v_1, v_2, R_S, and R_F to make the use of superposition valid in E5.1?

E5.3 If in E5.1, $R_F = 50\ \text{k}\Omega$, $R_S = 10\ \text{k}\Omega$, v_1 is a 100 Hz cosine wave with amplitude 0.4 V (i.e., $v_1 = 0.4 \cos 200\pi t$), and $v_2 = 0$, plot v_O as a function of time.

E5.4 Repeat E5.3 for $R_S = 1\ \text{k}\Omega$.

E5.5 Repeat E5.3 and E5.4 for $v_2 = 0.4$ V. To what extent can the superposition solution of E5.1 be used?

E5.6 For $v_1 = 8 \sin 10\pi t$, plot v_O in Fig. E5.6 as a function of time, assuming the switch between points 1 and 2 is *open*.

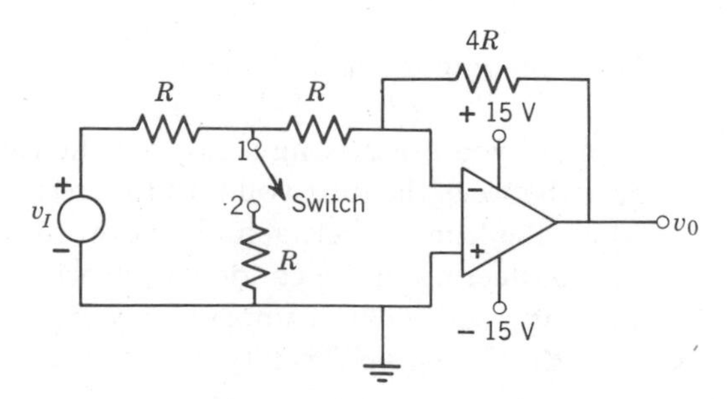

Figure E5.6

E5.7 Repeat E5.6, assuming the switch is *closed* (i.e., a short circuit between points 1 and 2).

E5.8 (a) Show that the network in Fig. E5.8 behaves like a current source of magnitude $I_0 = 1$ milliamp for a limited range of load resistors R_L.

(b) Determine the range of R_L for which I_0 is constant, and explain what happens when R_L falls outside this range.

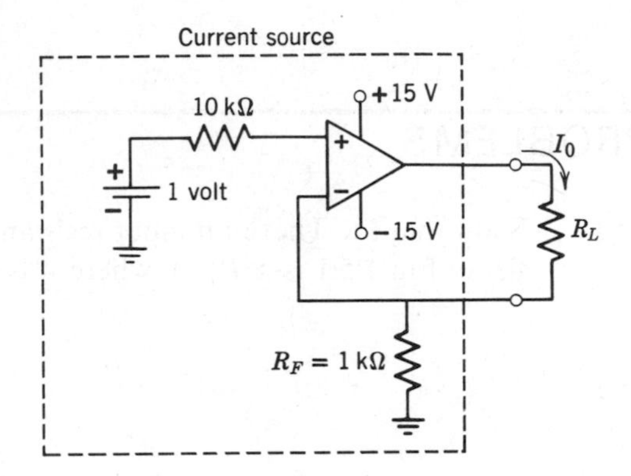

Figure E5.8

E5.9 For the circuit shown in Fig. E5.9, sketch the large-signal voltage transfer characteristic v_O versus v_I for both positive and negative changes in v_I.

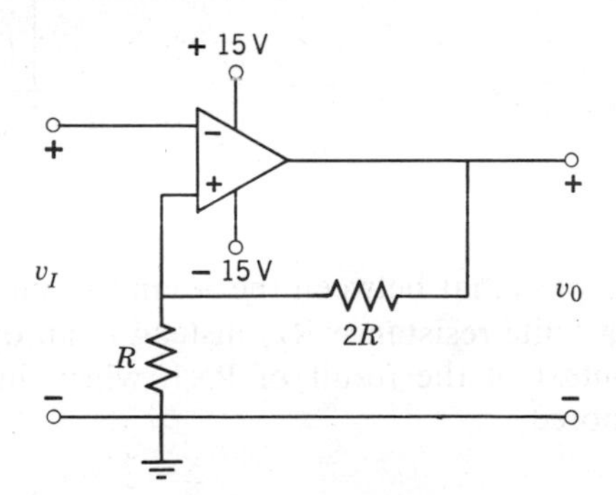

Figure E5.9

E5.10 In the network of Fig. 5.13, plot the output voltages v_{O1}, v_{O2}, and v_{O3} as functions of time when v_S is the waveform shown in Fig. E5.10.

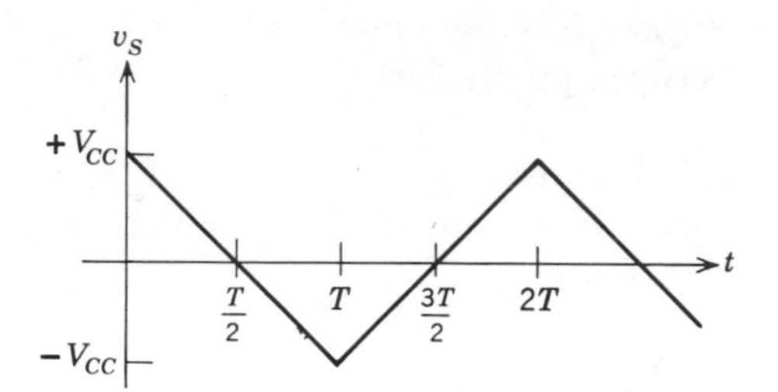

Figure E5.10

PROBLEMS

P5.1 Show that the Thévenin input resistance R_T of the inverting amplifier of Fig. P5.1 is $\approx R_F/A$, where A is the open-loop op-amp gain.

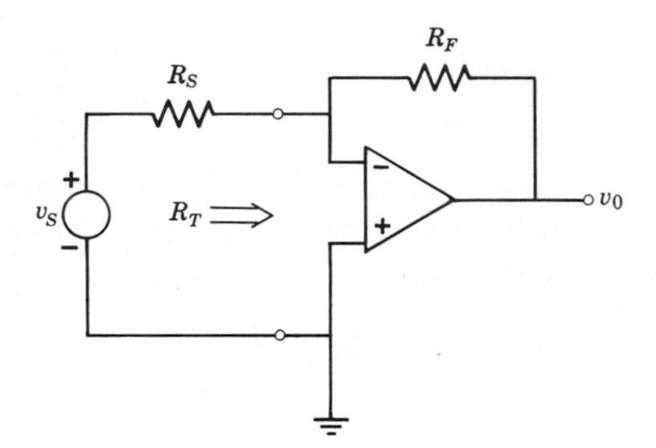

Figure P5.1

P5.2 Suppose that between the − and + inputs, an op-amp has a large but finite resistance R_{IN} instead of an open circuit. Discuss, in the context of the result of P5.1, when this resistance can be safely ignored.

P5.3 Show that the Thévenin input resistance R_T of the non-inverting amplifier of Fig. P5.3 is approximately R_{IN} times the ratio of the open-loop gain to the closed-loop gain.

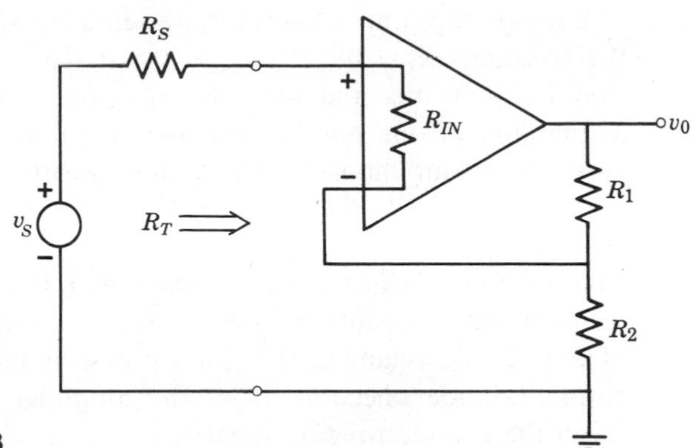

Figure P5.3

P5.4 Find the Thévenin output resistance in Fig. P5.4 assuming that the op-amp has a nonzero output resistance R_{OUT}. Use your result to justify the neglect of R_{OUT} under normal circumstances.

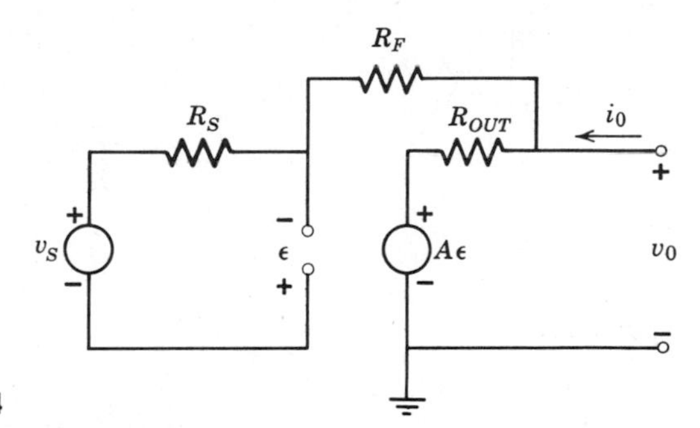

Figure P5.4

P5.5[2] A threshold logic unit (TLU) is a circuit that produces a digital output ("HIGH" or "LOW" voltage level) in response to n analog (continuous) voltage inputs according to the following condition:

The digital output will be "HIGH" whenever

$$A_0 + A_1v_1 + A_2v_2 + \ldots + A_nv_n > 0$$

where $v_1 \ldots v_n$ are the n analog voltages, and $A_0 \ldots A_n$ are constants. (A_0 has the dimension volts and $A_1 \ldots A_n$ are dimensionless.) Otherwise, the digital output will be "LOW." Thus, the TLU can make a decision based on the values of the inputs $v_1 \ldots v_n$.

[2] Throughout the book, design problems are identified by boldface problem numbers.

Use op-amps and resistors to design a two-input TLU that has the constants $A_0 = 1$ volt, $A_1 = 1$, and $A_2 = 2$. You may assume that $|v_1| < 6$ volts and $|v_2| < 8$ volts, and you may assume the availability of one positive and one negative power supply with voltages of your choosing. The digital "HIGH" output should be at the positive supply; the "LOW" at the negative supply.

P5.6 Figure P5.6*a* shows a source network with both a difference-mode source v_S and a common-mode source v_C. Find v_O in the amplifier of Fig. P5.6*b*, assuming that the op-amp is ideal and has perfect common-mode rejection. Does the amplifier circuit as a whole reject the common-mode signal v_C?

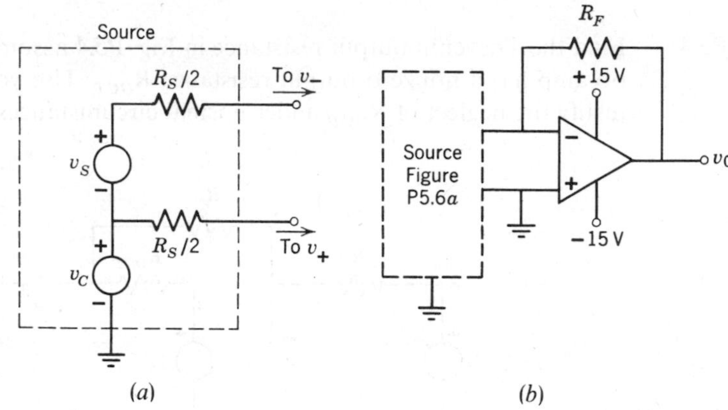

Figure P5.6

P5.7 Show that for the proper choice of R_+ (i.e., $R_+ = R_F$), the output v_O in Fig. P5.7 does not depend on v_C. Such an amplifier circuit is called a *differential amplifier*.

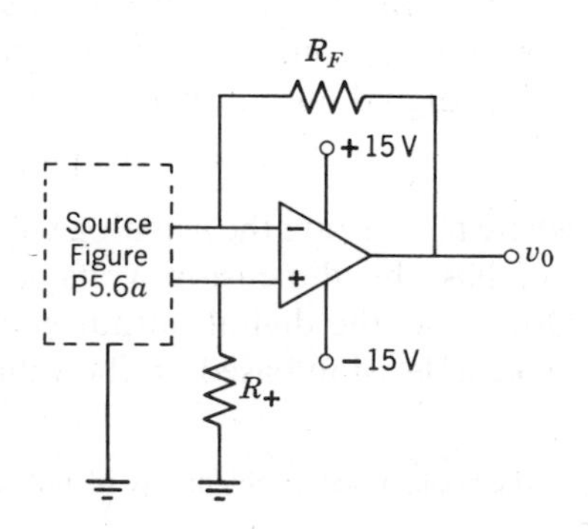

Figure P5.7

P5.8 Assuming $V_1 > V_2$ in Fig. P5.8, fill in the table below with the values of v_O for each range of v_I values.

v_I	v_O
$v_I > V_1$	
$V_1 > v_I > V_2$	
$V_2 > v_I$	

Notice that v_O is greater than zero only when $v_I > V_2$ *and* $v_I < V_1$ Thus, this circuit performs a simple logic function, giving a positive output only when a particular set of input conditions is satisfied.

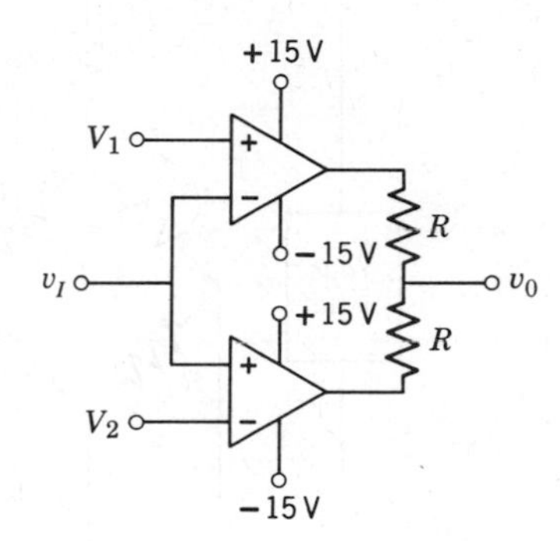

Figure P5.8

P5.9 Given an input waveform $v_{IN}(t)$ and a square wave $v_{SQ}(t)$ with a 1 volt peak amplitude, design a circuit (Fig. P5.9) to give an output of $+V_{CC}$ whenever

$$0 < v_{IN} < 2 \text{ volts} \qquad \textit{and} \qquad v_{SQ} > 0$$

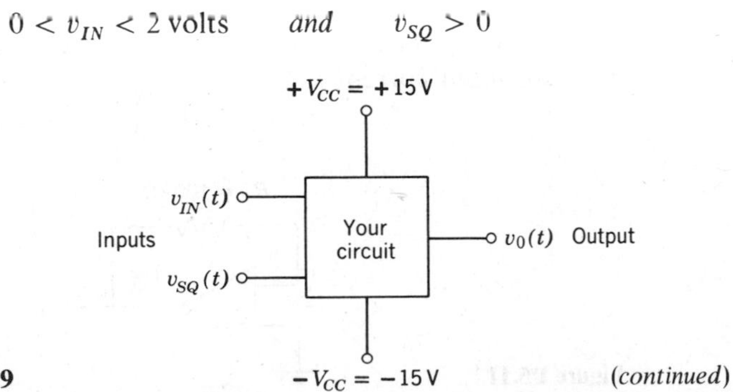

Figure P5.9

(continued)

and an output of less than $+V_{CC}/2$ (i.e., it could be zero, or even negative) for all other combinations of v_{IN} and v_{SQ}. Use resistors and comparators in your design (see P5.8). A follower may also be useful in some designs. You may assume the availability of ± 15 V power supplies.

P5.10 Figure P5.10 shows a resistance bridge, one leg of which might be a transducer for measurement of temperature or displacement (strain). Assuming R_1, R_2, R_3, and R_4 might vary between 10 and 500 Ω, design a circuit that will supply a *constant* current $I_0 = 0.5$ mA to the bridge (see E5.8).

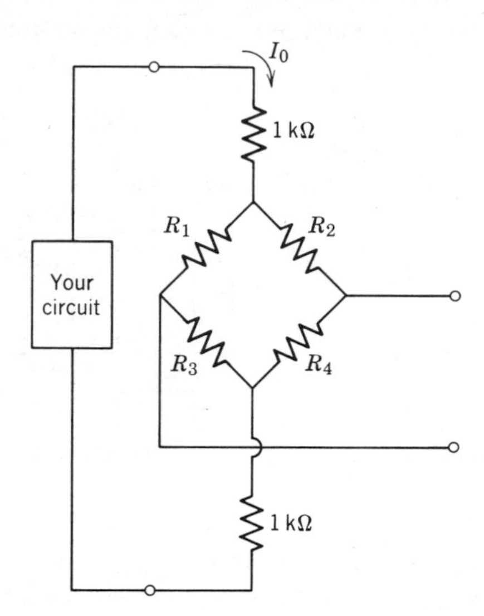

Figure P5.10

P5.11 Show how the circuit of Fig. P5.11 can be used to measure the nonzero input current i_-.

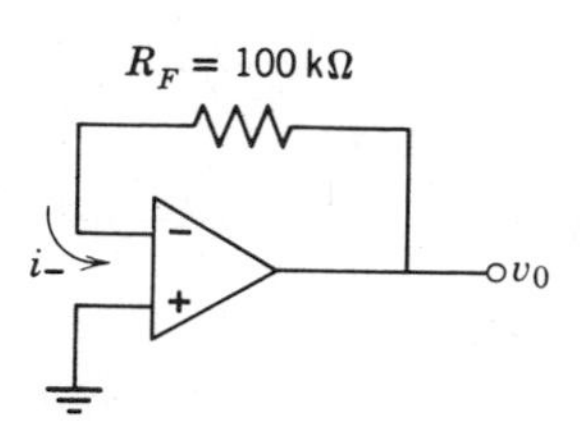

Figure P5.11

CHAPTER SIX
CAPACITANCE AND INDUCTANCE

6.0 INTRODUCTION

Up to now we have dealt exclusively with resistive networks. Most of these networks had responses to excitations that depended only on the instantaneous value of the excitation.[1] This chapter initiates the discussion of a much broader class of networks in which responses to a time-varying source depend on the past behavior of the source as well as on its instantaneous value. Two new kinds of network elements, capacitance and inductance, are needed to model the behavior of these networks. We begin the discussion with these elements.

6.1 Capacitance

6.1.1 The Ideal Capacitance

The *ideal capacitance* is an element in which the voltage is proportional to charge rather than to current. Figure 6.1 shows both the circuit symbol for a capacitance and a sketch of the parallel-plate structure discussed in Section 2.1.2, which we can now identify as a capacitance element. The electric

[1] The lone exception was the Schmitt trigger of Section 5.3.2. It was pointed out there that the responses of positive-feedback circuits often depend on the past history of an excitation.

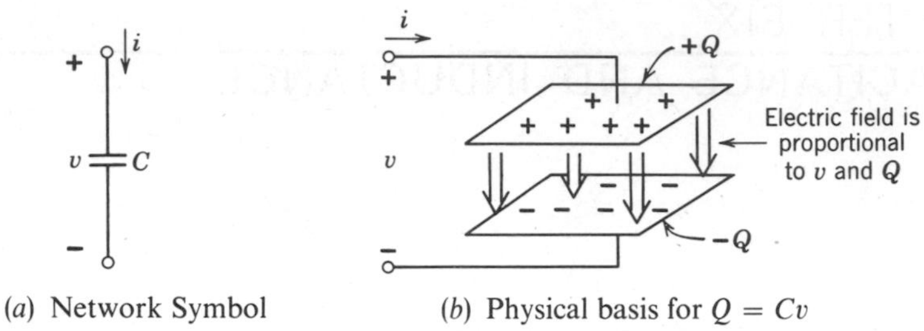

(a) Network Symbol

(b) Physical basis for $Q = Cv$

Figure 6.1
The ideal capacitance. The charge on the plates is proportional to the voltage.

field between the plates is proportional both to the voltage v and the total charge Q. The resulting proportionality between Q and v can be written as

$$Q = Cv \tag{6.1}$$

in which the constant of proportionality C is called the *capacitance* of the element, and depends only on the geometrical arrangement of conductors and insulators. The unit of capacitance is the *farad* (abbreviated F).

To find the v-i characteristic of an ideal capacitance, let us calculate the time rate of change of Eq. 6.1, assuming that C is a constant.[2]

$$\frac{dQ}{dt} = C\frac{dv}{dt} \tag{6.2}$$

But dQ/dt is simply the current i flowing into the + terminal. Therefore, the v-i characteristic can be written

$$i = C\frac{dv}{dt} \tag{6.3}$$

Since current flows only when v is changing, we conclude that *the ideal capacitance is an open circuit for dc current*, a result that nicely matches our intuitive view that the insulating gap between the plates should not permit a steady flow of charge.

How, in fact, can we understand the flow of any nonzero current "through" a capacitance, given that there is an insulating gap? When a current i is flowing, charge is entering the + terminal at the rate of Q coulombs/second (Fig. 6.2). At the same time, charge is leaving the − terminal at the rate of Q coulombs/second. After one second, there is a net charge of $+Q$ on the

[2] If C is made to vary with time, Eq. 6.2 contains an additional term $v(dC/dt)$ on the right-hand side. Time-varying capacitances are useful in frequency modulation and in parametric amplification, two subjects that fall beyond the scope of this text.

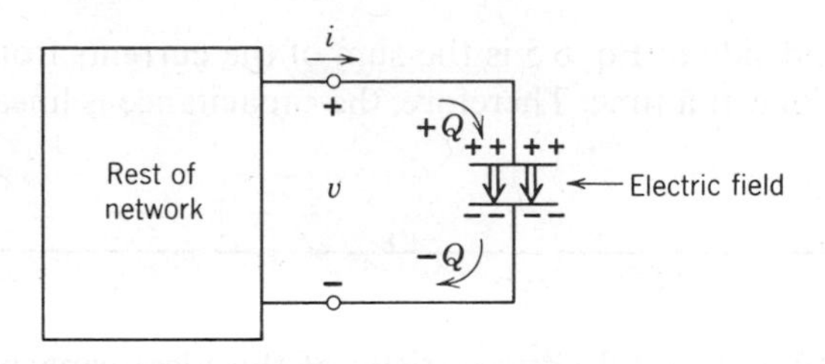

Figure 6.2
Time-varying currents appear to flow "through" the capacitance, but no charge actually crosses the gap. If a charge $+Q$ flows to one plate, an equal charge leaves the other plate.

+ plate, and, there is a net charge of $-Q$ on the − plate (after the charge of $+Q$ leaves). Thus, the overall charge of the capacitance is still zero, but there has been a *charge transfer* through the rest of the network from the − plate to the + plate. Since nonzero currents change the charge, and since voltage and charge are proportional, current can flow only when the voltage is changing.

6.1.2 Linearity

Ideal capacitances are *linear* elements as the following example shows. In Fig. 6.3 a capacitance is connected to the series combination of two voltage sources. The total voltage on the capacitance is obtained from KVL.

$$v_C = v_1 + v_2 \tag{6.4}$$

The current in the circuit is found with the aid of Eq. 6.3.

$$i_C = C\frac{dv_C}{dt} = C\frac{d}{dt}(v_1 + v_2) = C\frac{dv_1}{dt} + C\frac{dv_2}{dt} \tag{6.5}$$

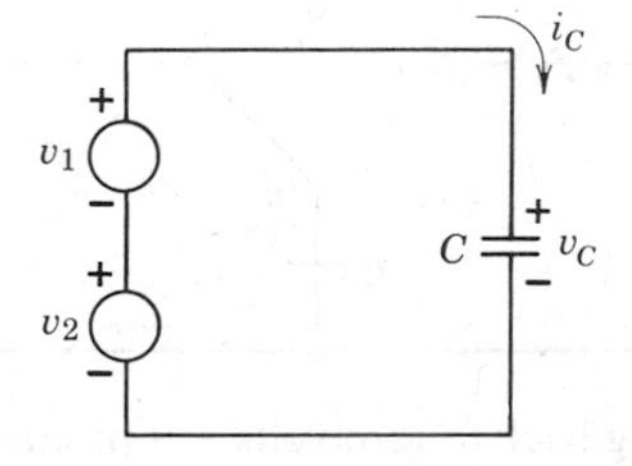

Figure 6.3
The linearity of capacitance.

The right-hand side of Eq. 6.5 is the sum of the currents from each voltage source taken one at a time. Therefore, the capacitance is linear.

6.1.3 Energy Storage

An interesting and useful characteristic of the ideal capacitance is that it *stores* electrical energy. If we refer to Eq. 6.3, the power absorbed by a capacitance is

$$P = v_C i_C = v_C C \frac{dv_C}{dt} = \frac{d}{dt}(\tfrac{1}{2}Cv_C^2) \tag{6.6}$$

Since power absorption is identical to the *rate* of energy absorption, the quantity $\frac{1}{2}Cv_C^2$ must represent the total electrical energy absorbed by the capacitance when it charges from 0 to v_C volts. This absorbed energy is actually stored in the electric field between the capacitance plates, and is called *electric stored energy.*

To show that this electric energy is actually stored in recoverable form let us examine the network of Fig. 6.4. The capacitance is connected to a

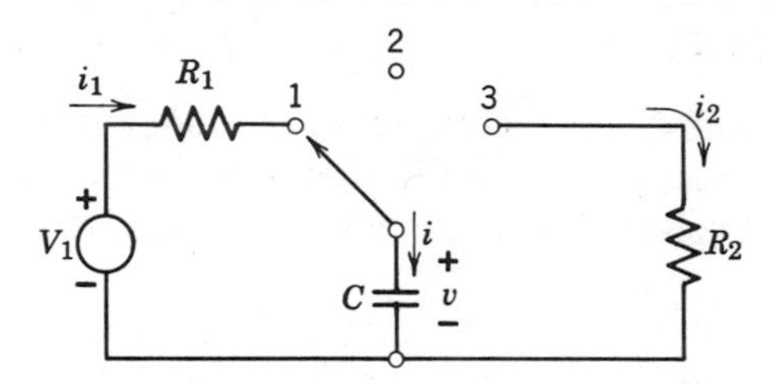

(*a*) Both i and v are positive when charging from zero to V_1 volts.

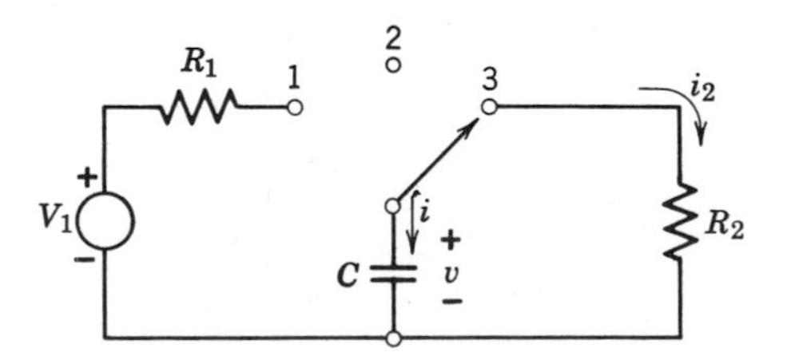

(*b*) When discharging back to zero volts, v is positive, but i is negative.

Figure 6.4
The capacitance stores energy in recoverable form.

three-position switch. Assume that we begin in position 2 (open circuit) with the capacitance discharged ($v = 0$ volts). When the switch is moved to position 1, current will flow through R_1 until v charges up to V_1 volts.[3] Equation 6.6 tells us that the power absorbed during this charging process is positive, and that the total energy absorbed is $\frac{1}{2}CV_1^2$.

When the switch is moved back to position 2, no current can flow into or out of the capacitance. Its voltage, therefore, must remain exactly constant. If the switch is then moved to position 3, a conducting path appears between the capacitance terminals. There is now a positive voltage of V_1 volts across the resistance R_2. By Ohm's law, therefore, positive current must flow in the direction of the i_2 reference arrow.[3] The terminal current i for the capacitance must therefore be negative. Since the voltage is still positive, the power is negative. Thus the capacitance is delivering power to the load resistance R_2. The total amount of energy delivered to this resistance as the voltage decreases from V_1 down to 0 is equal to

$$\underbrace{\tfrac{1}{2}CV_1^2}_{\text{Energy before discharge}} - \underbrace{0}_{\text{Energy after discharge}} = \tfrac{1}{2}CV_1^2 \qquad (6.7)$$

Thus the total energy delivered to the resistance during discharging is the same as the total energy absorbed by the capacitance during the charging. No energy is lost in the capacitance itself. We say, therefore, that an ideal capacitance is a *lossless energy storage element.*

6.1.4 Capacitors

Actual capacitance elements are called *capacitors*. They are made in many shapes and sizes, but they always consist of two conductors separated by some kind of insulator. Among the insulating materials used are mica, paper, or simply an air gap. The capacitance values for actual capacitors are usually much less than 1 farad. Physically convenient units of capacitance are the microfarad ($\mu F = 10^{-6}$ farad) and the picofarad ($pF = 10^{-12}$ farad).

The insulators in actual capacitors are very good open circuits, but they are not perfect. At sufficiently high voltages, the insulators can suffer electrical breakdown, and a sudden discharge can take place. A less dramatic imperfection is sometimes important even at low voltages. Tiny amounts of current called the *leakage current* can flow through the insulator or through

[3] The precise time-variation of this current is taken up in Section 6.3.

oily or moist films on the surface of the capacitor. A charged capacitor, therefore, will discharge spontaneously. This discharge is usually very slow, with discharge times measured in minutes or even days.

6.1.5 Parallel and Series Combinations

Figure 6.5*a* shows two capacitors C_1 and C_2 in parallel. Since the capacitors are in parallel, the voltage v is the same for both capacitors. The total charge, therefore, is

$$Q = Q_1 + Q_2 = C_1 v + C_2 v = (C_1 + C_2)v \tag{6.8}$$

Thus the parallel combination behaves like an equivalent capacitor of value $C = C_1 + C_2$. Capacitances in parallel add.

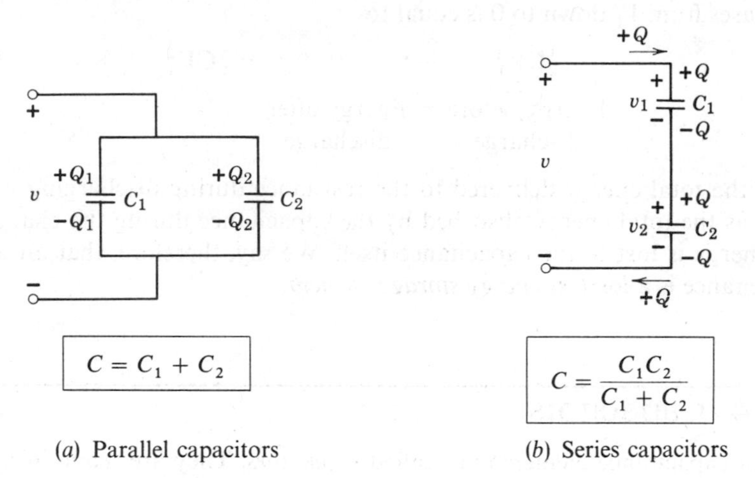

Figure 6.5
Parallel and series capacitors.

Figure 6.5*b* shows two capacitors in series. If a charge $+Q$ flows into the $+$terminal of C_1, an equal charge must leave the $-$ terminal of C_1 and enter the $+$ terminal of C_2, causing an equal charge $+Q$ to leave the $-$ terminal of C_2. Both capacitors thus have a charge of magnitude Q. The total voltage therefore is

$$v = v_1 + v_2 = \frac{Q}{C_1} + \frac{Q}{C_2} = Q\left(\frac{C_1 + C_2}{C_1 C_2}\right) \tag{6.9}$$

Equation 6.9 is similar in form to Eq. 6.1, but with an equivalent capacitance $C_1C_2/(C_1 + C_2)$ for the series combination of C_1 and C_2.

6.2 INDUCTANCE

6.2.1 The Ideal Inductance

The *ideal inductance* is an element with a *v-i* characteristic complementary to that of the ideal capacitance. Its coil-like circuit symbol is shown in Fig. 6.6*a*. In an ideal inductance, the voltage is proportional to the rate of change of the current:

$$v = L\frac{di}{dt} \tag{6.10}$$

The constant of proportionality, L, is called the *inductance*. The unit of inductance is the *henry* (abbreviated H).

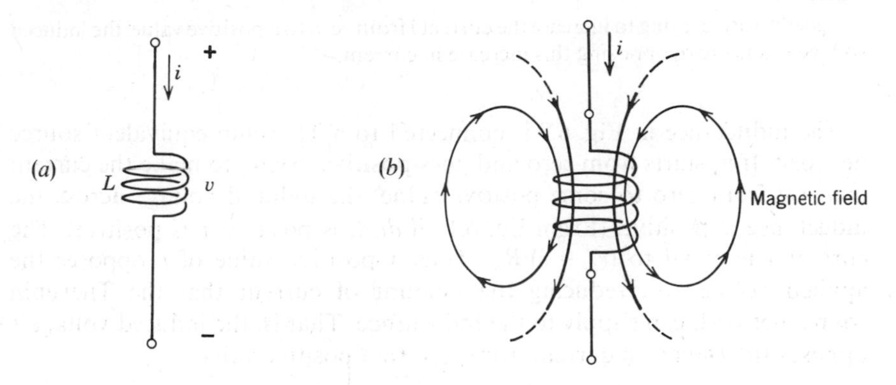

Figure 6.6
The ideal inductance.

The physical basis of inductive behavior can be seen in two ways. In simplest terms, inductance is associated with the momentum of the electric current. Just as it takes mechanical energy to start or stop the flow of water in a frictionless pipe, it takes electrical energy to start or stop the current in a perfect conductor. Referring to the definition, Eq. 6.10, we note that when the current is constant, di/dt is zero, and v is zero. Thus for dc currents, the ideal inductance behaves like a perfect conductor, a *short circuit*.

In the language of electromagnetic fields, the basis for inductive behavior is Faraday's law of induction. A current in a wire sets up a magnetic field around the wire, the strength of the field being proportional to the current. In certain configurations of conductors, particularly coils, the magnetic fields produced by each little segment of current add up to produce a relatively large total magnetic field in the center of the coil (Fig. 6.6*b*). If the current i

changes, the magnetic field must change proportionately. Faraday's law of induction says that when a circuit encloses a region containing a magnetic field, a voltage is *induced* in the circuit whenever the magnetic field linking the circuit changes. The polarity of the induced voltage always *opposes* the original current change. Figure 6.7 contains an example.

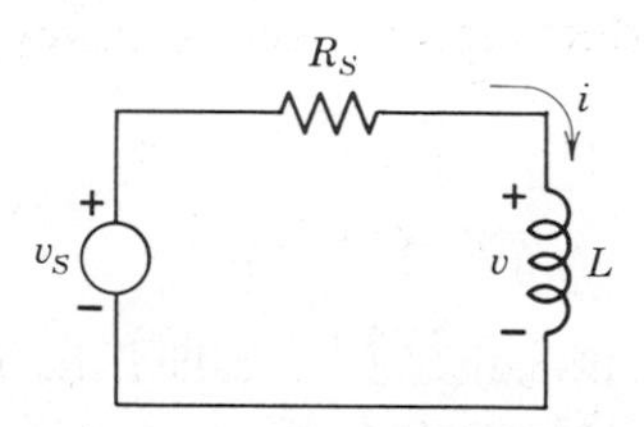

Figure 6.7
If v_S goes positive trying to increase the current i from zero to a positive value, the induced voltage v is positive opposing this increase in current.

The inductance in Fig. 6.7 is connected to a Thévenin equivalent source network. If v_S starts from zero and goes positive, trying to make the current increase from zero to some positive value, the induced voltage across the inductance is positive. (From Eq. 6.10 if di/dt is positive, v is positive). The current i is equal to $(v_S - v)/R_S$. Thus a positive value of v opposes the applied voltage v_S, reducing the amount of current that the Thévenin source network can supply to the inductance. That is, the induced voltage v opposes the *change* in current from zero to a positive value.

6.2.2 Linearity

Like the ideal capacitance, the ideal inductance is linear. Figure 6.8 shows an inductance connected to two parallel current sources. Suppressing the i_2 source and finding the voltage due to i_1,

$$v_{(from\, i_1)} = L\frac{di_1}{dt} \tag{6.11}$$

Similarly suppressing the i_1 source,

$$v_{(from\, i_2)} = L\frac{di_2}{dt} \tag{6.12}$$

For the complete network,

$$i_L = (i_1 + i_2) \tag{6.13}$$

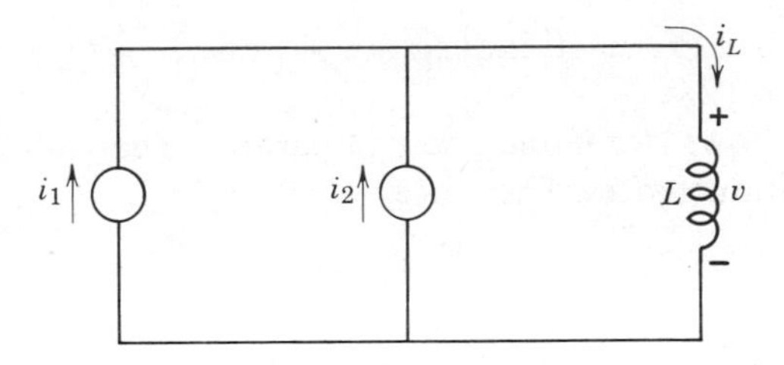

Figure 6.8
The linearity of an inductance.

Inserting Eq. 6.13 into the definition of inductance, Eq. 6.10, we see that

$$v_{total} = L\frac{d}{dt}(i_1 + i_2) = L\frac{di_1}{dt} + L\frac{di_2}{dt} \tag{6.14}$$

Comparison of this result with the sum of Eqs. 6.11 and 6.12 shows that the inductance is linear.

6.2.3 Energy Storage

Inductances store energy just as capacitances do, but since the mechanism involves a magnetic field instead of an electric field, we call the energy in an inductance *magnetic stored energy*. If we refer to Eq. 6.10, the power absorbed by the inductance is

$$P = vi = \left(L\frac{di}{dt}\right)i = \frac{d}{dt}(\tfrac{1}{2}Li^2) \tag{6.15}$$

Since power is rate of change of energy, $\frac{1}{2}Li^2$ must be the total magnetic energy stored in the inductance when it is carrying a current i.

Just as in the case of the capacitance, this energy is all stored in recoverable form. The ideal inductance, therefore, is also a *lossless energy storage element*

6.2.4 Series and Parallel Inductances

Figure 6.9*a* shows two inductances L_1 and L_2 in series. The total voltage is

$$v = v_1 + v_2 \tag{6.16}$$

But since both inductances carry the same current i,

$$v = L_1\frac{di}{dt} + L_2\frac{di}{dt} = (L_1 + L_2)\frac{di}{dt} \tag{6.17}$$

which is equivalent to a single inductance of value $L = L_1 + L_2$. Inductances in series add.

Figure 6.9*b* shows two inductances in parallel. The voltage v must be the same for both inductances. Therefore,

$$v = L_1 \frac{di_1}{dt} = L_2 \frac{di_2}{dt} \tag{6.18}$$

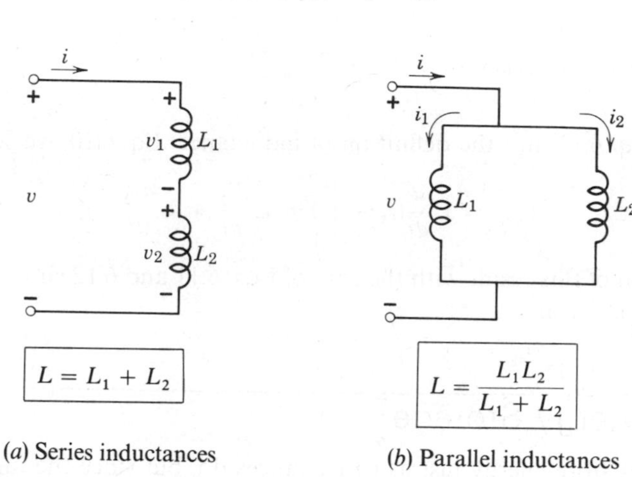

(*a*) Series inductances (*b*) Parallel inductances

Figure 6.9
Series and parallel inductances.

The rate of change of total current i, therefore, is

$$\frac{di}{dt} = \frac{di_1}{dt} + \frac{di_2}{dt} = \left(\frac{1}{L_1} + \frac{1}{L_2}\right) v \tag{6.19}$$

which has the form of Eq. 6.10, except that the total inductance of the parallel combination is $L = L_1 L_2/(L_1 + L_2)$.

6.2.5 Inductors

Actual inductance elements are called *inductors*. They usually consist of a coil. Often this coil is wound on a core of magnetic material (iron or ferrite) that enhances the strength of the magnetic field in the coil.

The behavior of actual inductors departs from the ideal more than the corresponding behavior of actual capacitors. One reason is that all real coils and core materials have some resistance. This resistance is an inherent part of the actual element. (An important exception is a coil made out of superconducting wire,[4] which behaves like a perfect inductance.) Inductors, therefore, are often modeled as an ideal inductance in series with a small resistance, as shown in Fig. 6.10.

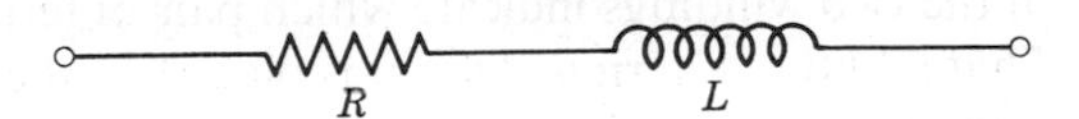

Figure 6.10
A real inductor is modeled as an ideal inductance in series with a resistance.

6.2.6 Mutual Inductance and Transformers

The phenomenon of magnetic induction, which is the physical basis of the behavior of inductances, is not limited to single coils. Whenever a circuit is linked by a changing magnetic field, a voltage is induced that can produce a current. For example, in Fig. 6.11 two coils are shown in close proximity so that the magnetic field set up by the current i_2 partially links the lower coil. When i_2 changes, a voltage is induced at the v_1 terminals. Similarly a changing current in the lower coil induces a voltage in the upper coil. This magnetic coupling between circuits is called *mutual inductance.*

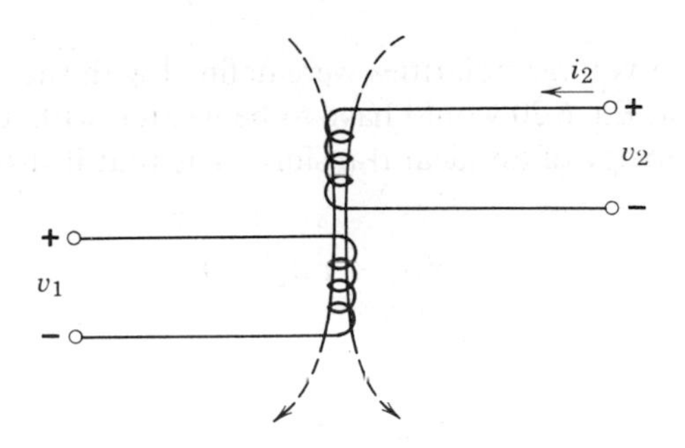

Figure 6.11
Mutual inductance.

A *transformer* consists of two or more coils, usually with differing numbers of turns, wound around a common core. There is no connection between the

[4] In some metals, such as pure lead and pure tin, the resistance vanishes completely when the metal is cooled to temperatures near that of liquid helium (4 K, or −269°C). This phenomenon is called superconductivity.

coils, but for time-varying waveforms, the mutual inductance between the coils enables currents in one of the coils to produce voltages in the other coil.

The *ideal transformer* is a four-terminal network element with the symbol shown in Fig. 6.12. The *turns ratio* between the two windings is shown as $1:n$. The dots on the two windings indicate which pair of terminals have the same polarity. That is, if the $+$ terminal for v_1 is placed on the end with a dot, and if the $+$ terminal for v_2 is also placed on the end with the dot (as in Fig. 6.12) the relation between v_1 and v_2 for an ideal transformer is written

$$v_2 = nv_1 \tag{6.20}$$

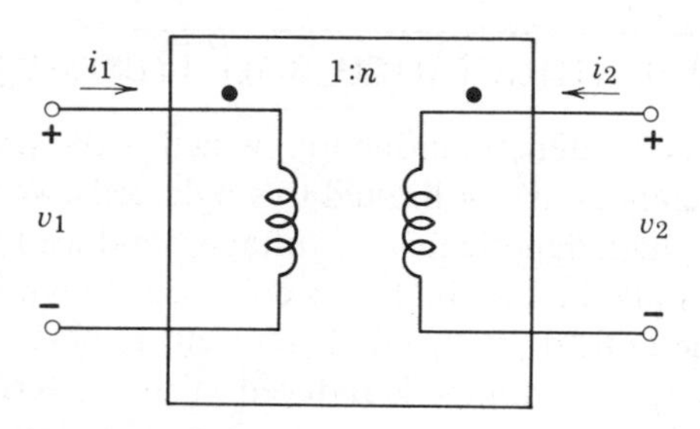

Figure 6.12
Ideal transformer.

If one of the two voltage polarities were defined with the $-$ terminal at the end with the dot, Eq. 6.20 would have to be written with a minus sign.

A second property of an ideal transformer is that it dissipates no power. That is,

$$v_1 i_1 + v_2 i_2 = 0 \tag{6.21}$$

which yields

$$\frac{i_1}{i_2} = -\frac{v_2}{v_1} = -n \tag{6.22}$$

If we assume that n is greater than unity and that a source is connected to the v_1 terminals and a load connected to the v_2 terminals, then the voltage on the load is larger than the voltage from the source, but the current in the load is correspondingly smaller than the current from the source. This is called the *step-up* configuration because the voltage from the source is stepped up by a factor of n. One could equally well connect a source network to the v_2 terminals and a load to the v_1 terminals. Here the load voltage is less than the source voltage, but the load current is correspondingly larger than the source current. This is called the *step-down* configuration because the voltage from the source is stepped down by a factor of n. In either configuration, the winding connected to the source is called the *primary*, while the winding connected to the load is called the *secondary*.

Let us analyze the terminal characteristics of the network of Fig. 6.13, in which a resistance has been connected across the v_2 winding of an ideal transformer. There are no independent sources in this network, so the open circuit voltage is zero. Therefore, the Thévenin equivalent must be a simple resistance. To find the value of that resistance, assume that a test current i_T is flowing in the + terminal at the v_1 pair. That is, assume

$$i_1 = i_T \tag{6.23}$$

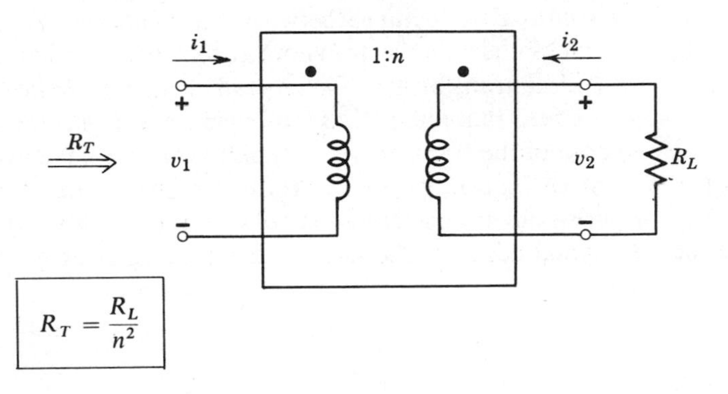

Figure 6.13
An ideal transformer can change the equivalent resistance of a network without itself dissipating any power.

The transformer equation, Eq. 6.22, gives

$$i_2 = -\frac{i_T}{n} \tag{6.24}$$

By Ohm's law

$$v_2 = -i_2 R_L = +\frac{i_T R_L}{n} \tag{6.25}$$

Equation 6.22 again gives

$$v_1 = \frac{v_2}{n} = \frac{i_T R_L}{n^2} \tag{6.26}$$

Thus the Thévenin resistance at the v_1 terminal pair is then

$$R_T = \frac{v_1}{i_T} = \frac{R_L}{n^2} \tag{6.27}$$

The resistance in the secondary circuit appears in the primary circuit as a resistance $1/n^2$ times as large.

Actual transformers differ from the ideal transformer in several ways. First, the ideal transformer equation (Eq. 6.22) has no provision for rejecting dc currents, even though real transformers have no dc coupling between primary and secondary. Second, the ideal transformer equation assumes perfect magnetic coupling between the two coils, that is, that all the magnetic field set up by the primary links the secondary, and vice versa. In practice, there is always some magnetic field that links only one of the coils, and that therefore contributes to a simple inductance of one winding or the other rather than to the mutual inductance between the windings. Figure 6.14 shows a simple circuit model for a real transformer that adds two inductances, L_e and L_m, to an ideal transformer. The so-called *leakage inductance* L_e models the magnetic field that fails to link both windings. It has the effect of reducing the response of the transformer to rapidly varying waveforms. The *magnetizing inductance* L_m is in parallel with the primary of the ideal transformer, and thus shorts out any dc component of i_1. The effect of L_m, therefore, is to prevent dc components of the current from being coupled between primary and secondary.

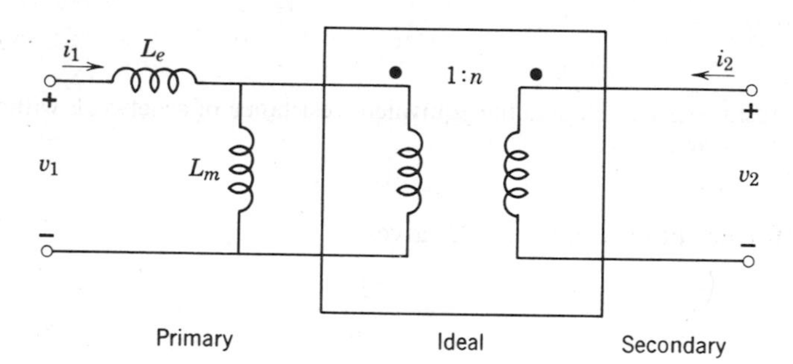

Figure 6.14
A simple circuit model for a real transformer.

The windings in transformers have resistances that should be included in the model whenever these resistances are comparable to the Thévenin resistance of the remaining network. There is, in addition, a capacitance between the two windings that permits the coupling of common-mode signals from the primary to the secondary. This last issue will become important in Chapter 17 when we consider grounding techniques in analog instrumentation systems.

6.3 STEP RESPONSE IN *RC* AND *LR* NETWORKS

6.3.1 Initial and Final States

This section deals with the determination of the response of *RC* and *LR* networks to step sources (see Fig. 6.15*a*). The step function of unit height at $t = 0$ is written $u_{-1}(t)$, the amplitude factor A indicating the height of the step in v_S.

To determine a response to v_S (e.g., the time-dependence of the capacitor voltage v_C) it is useful to divide the total time axis into three intervals: (1) an initial time interval before the step occurs during which v_S can be considered a dc source, (2) a time interval just after the step occurs during which the response to the sudden change in v_S takes place, and (3) a final time interval beginning long enough after the step so that one can once again think of v_S as a dc source. The division between the initial and middle intervals obviously occurs at the instant of the step, at $t = 0$ in the example of Fig. 6.15. The division between the middle and final time intervals is less clear

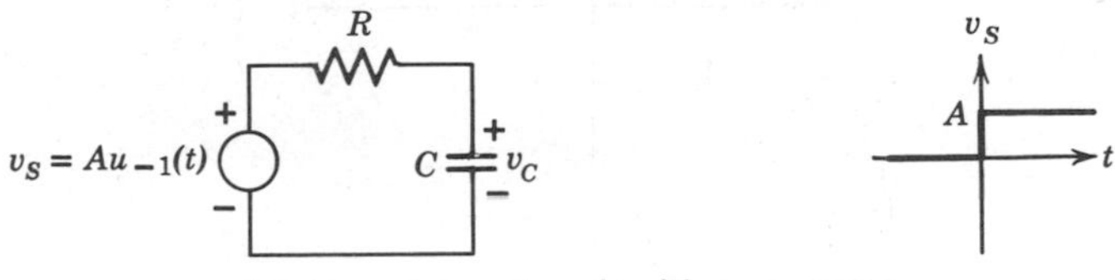

(*a*) Complete network with step source

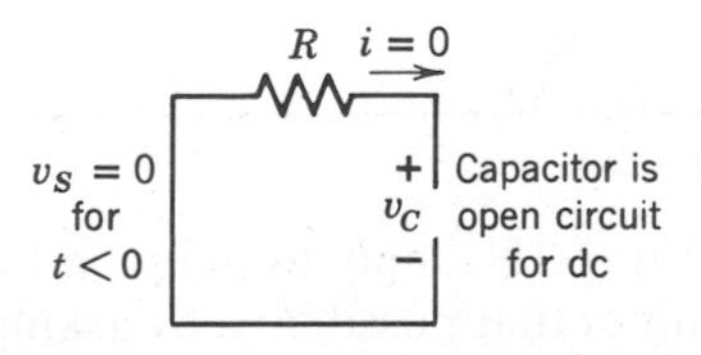

(*b*) Network for finding initial state, $t < 0$, $v_C(\text{initial}) = 0$

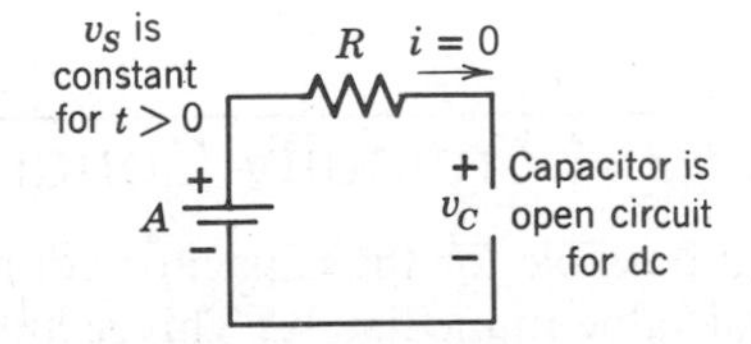

(*c*) Network for finding final state, $t \to \infty$, $v_C(\text{final}) = A$

Figure 6.15
The concept of initial and final states.

at the moment, and we shall delay being more precise until Section 6.3.3. For now let us concentrate on the initial and final intervals, the times during which the network has settled down to its appropriate dc response.

Our goal in considering the initial and final time intervals is to determine as simply as possible the initial and final values of the response of interest. In Fig. 6.15, we seek the voltage on the capacitor. We recall that the current in a capacitor is proportional to the rate of change of capacitor voltage. Under dc conditions, therefore, when all voltages are constant, capacitor currents must be zero. In order to find the initial and final capacitor voltages in Fig. 6.15, we can use the networks of Figs. 6.15*b* and *c*, in which v_S has been replaced by the dc value appropriate to the time interval, and the capacitor has been replaced by an open circuit, a legitimate replacement when considering responses to dc sources. For dc sources, there will be no current flowing in the resistor R. By KVL, therefore, the capacitor voltage is simply equal to the open-circuit voltage of the source network. Thus, v_C(initial) is equal to zero, and v_C(final) is equal to A. Figure 6.16 shows a partial graph of $v_C(t)$, with the initial and final values of the capacitor voltage indicated. The only question remaining now is how to find the time variation of v_C during the middle interval, when the capacitor voltage is changing from its initial to its final value.

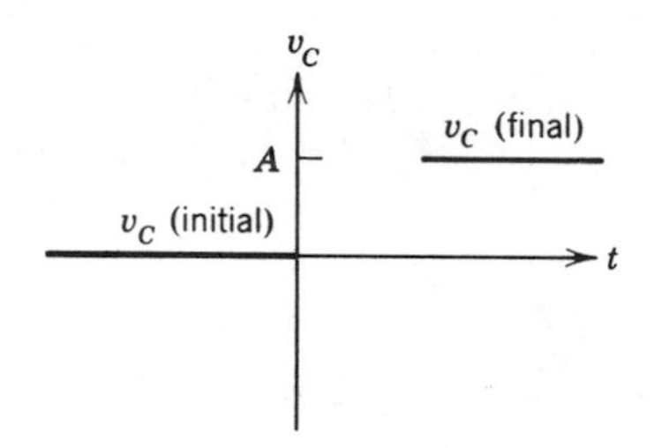

Figure 6.16
How does the transition between initial and final states take place?

6.3.2 Continuity Conditions

Is it possible for the capacitor voltage in Fig. 6.15 simply to jump up to its final value immediately? This section examines that possibility, by asking in detail what happens when we try to make a sudden change either in the voltage across a capacitor or in the current through an inductor.

Figure 6.17*a* shows a circuit that has been idealized by assuming that the source network has no source resistance. A sequence of possible source and response waveforms is shown in Figs. 6.17*b* and *c*. In Fig. 6.17*b* the source changes from 0 up to A volts at three different rates. The first takes the longest time to make the change, and changes at the corresponding slow rate of A/t_1 V/s. The other source waveforms change more rapidly and therefore require less time to accomplish the total change from 0 to A.

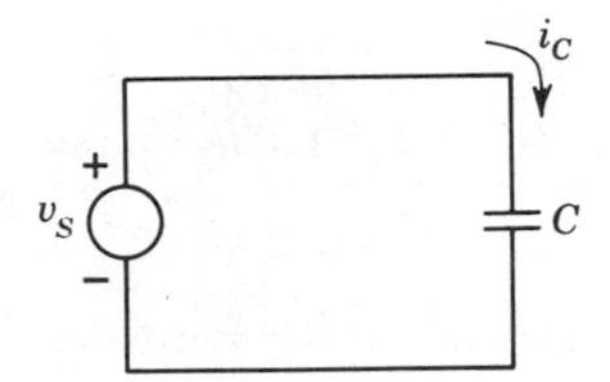

(*a*) Idealized network with no source resistance

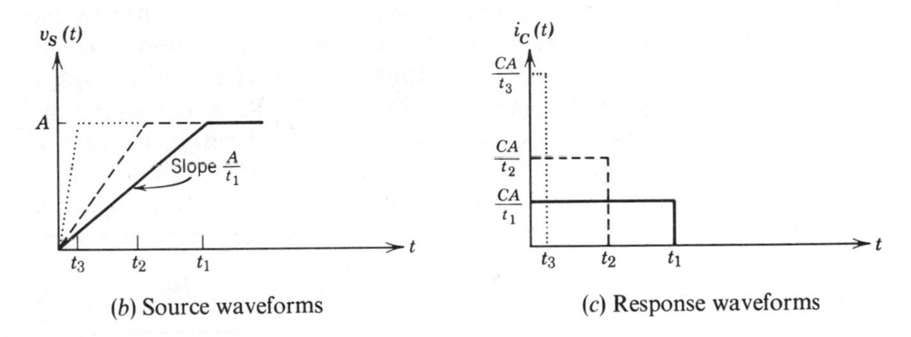

(*b*) Source waveforms

(*c*) Response waveforms

Figure 6.17
Approaching the step response.

The response waveforms are shown in Fig. 6.17*c*. The current $i_C(t)$ is proportional to the rate of change of the source waveform. As the source change gets more rapid, the response grows in amplitude, and becomes correspondingly shorter in duration so that the area under the response waveform (i.e., the total charge on the capacitor) remains equal to CA. In the limiting case of the instantaneous step change in the source (Fig. 6.18*a*) the response is infinite at just one instant in time, the total "area" under this infinite response still being equal to CA. This idealized limiting case is often useful. It is given a special name, the *impulse function* and is represented by the heavy arrow at $t = 0$ in Fig. 6.18*b*. The impulse function is labeled by its total area CA instead of by its amplitude. The unit impulse is given the symbol $u_0(t)$.

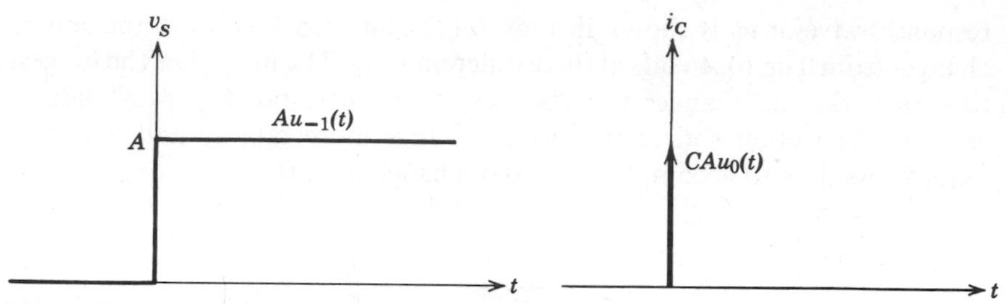

(a) Step function as a limiting case of Fig. 6.17b (b) Impulse function as a limiting case of Fig. 6.17c

Figure 6.18
Step and impulse function for the network of Fig. 6.17.

In practice, real source networks, such as the Thévenin network shown in Fig. 6.19, have nonzero source resistances and finite risetimes, the risetime being a measure of the time necessary to make a step transition.[5] During the transition, the rate of change of v_S is approximately A/τ_r, where τ_r is the risetime. Thus, if a capacitor were to be able to change its voltage as rapidly as the source v_S, a current of magnitude CA/τ_r would result. Because of the finite source resistance, however, the source network cannot supply more than a maximum current of A/R, and this only when the capacitor voltage is zero. Unless this maximum current is much greater than CA/τ_r,

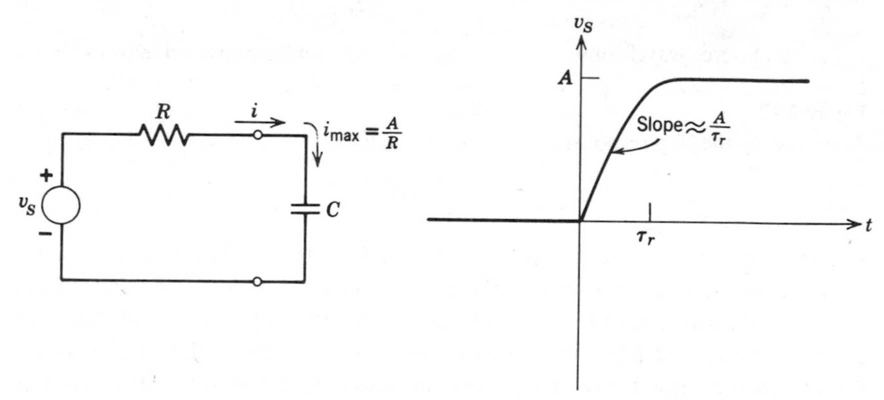

Figure 6.19
A real step source has a finite risetime τ_r and a nonzero source resistance R. The maximum current the source network can deliver is limited to A/R.

[5] Strictly speaking, the *risetime* is defined as the time for a step waveform to go from 10 to 90% of its final value. Our less strict usage in this section differs slightly from the normal technical definition.

it is impossible for the capacitor to charge up quickly enough to follow the source waveform. This can be stated algebraically by noting that if

$$\frac{A}{R} \ll \frac{CA}{\tau_r} \tag{6.28}$$

the capacitor voltage cannot hope to follow the step. Therefore, when

$$RC \gg \tau_r \tag{6.29}$$

we can assume that the capacitor voltage will lag behind the step, the amount of time lag being related to the size of the product RC.[6] For an *ideal* step source, τ_r is exactly zero and, therefore, every network will exhibit some kind of time delay related to the RC product. Said another way, we can model any real step source having a short enough risetime with an ideal step source, provided we remember that "short enough," as defined by Eq. 6.29, depends on the network in which the step source is used.

It is possible to make a similar analysis for a network containing an inductor. The result analogous to Eq. 6.29 is that when

$$\frac{L}{R} \gg \tau_r \tag{6.30}$$

the step source can be modeled by an ideal step and the response will lag behind the step in a way related to the L/R quotient.

Let us now restrict our attention to those networks where it is legitimate to represent the source network as an ideal step source together with an appropriate Thévenin or Norton resistance. This class of networks includes almost all cases in which step response is of interest. Because the source networks are composed entirely of finite-sized elements, they cannot supply an infinite impulse of current to a capacitor to change its voltage instantaneously, nor can they supply infinite impulses of voltage to an inductor to change its current instantaneously. The voltage across a capacitor just after a step in a source waveform must be the same as the voltage just before the step. Similarly, the current in an inductor immediately after a step must be the same as the current just before the step. These relations are called *continuity conditions*, and can be expressed algebraically as follows:

If a step change occurs in a source waveform at $t = 0$, and we use $t = 0^-$ and $t = 0^+$ to distinguish the network just before the step change from the network just after the step change, then for a capacitor

$$v_C(0^-) = v_C(0^+) \tag{6.31}$$

[6] A more detailed and less approximate analysis shows that there is *always* a time lag of order RC in the response to a ramp of finite slope. If $RC \ll \tau_r$, however, this delay is so small as to be negligible on the time scale fixed by τ_r.

and for an inductor

$$i_L(0^-) = i_L(0^+) \tag{6.32}$$

These continuity conditions tell us, among other things, that the transition between the initial and final states in the example of Fig. 6.16 cannot simply be a sudden jump in capacitor voltage. In the following section, we shall see that the change in stored energy that must take place between such initial and final states occurs with a characteristic time dependence, called the natural response.

6.3.3 The Natural Response

Networks that contain energy storage elements (C's and L's) differ in one very important way from purely resistive networks. In the resistive networks, whenever the independent sources are set to zero, all voltages and currents are zero. By contrast, in networks with energy storage elements, there can be nonzero voltages and currents after the independent sources are set to zero. These voltages and currents constitute the *natural response* of the network. The natural response persists until all of the stored energy is dissipated in the resistances. We have already described this phenomenon in a qualitative way in Section 6.1.3. A capacitance was charged up by a voltage source and was then switched across a resistance. Current flowed in the resistance until the capacitance discharged to zero volts. Let us now develop the exact time dependence of this natural response.

Figure 6.20 shows a basic RC network. It has no independent source. Using KVL, we obtain

$$v + iR = 0 \tag{6.33}$$

By the definition of the capacitance (Eq. 6.3)

$$i = C\frac{dv}{dt}$$

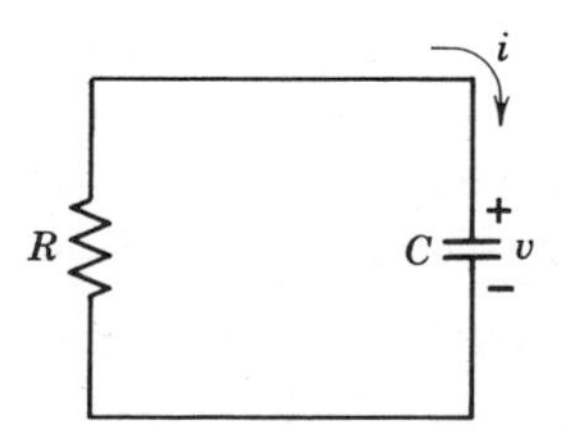

Figure 6.20
An RC network with no independent sources. Energy stored in the capacitor at $t = 0$ can produce an exponentially decaying current for $t > 0$.

Therefore

$$RC\frac{dv}{dt} + v = 0 \tag{6.34}$$

This equation is called a *differential equation*. It relates the rate of change of a quantity to the instantaneous value of that quantity. Both KVL and KCL will yield differential equations whenever a network contains C's or L's. There are many elegant and powerful mathematical methods for solving such equations. Fortunately, one can develop a thorough understanding of the properties of linear networks[7] without an extensive mathematical background. Once we have examined Eq. 6.34 and its solution, we shall develop methods for finding network responses without having to write down KVL and KCL in differential equation form.

Returning to Eq. 6.34, we rewrite it as

$$\frac{dv}{dt} = \left(-\frac{1}{RC}\right) \times (v) \tag{6.35}$$

$$\text{Slope} = (\text{constant}) \times (\text{function})$$

The solution of this equation must be a function with the property that its slope at any instant must be proportional to the value of the function at that instant. The *exponential function* has this property. If a is a constant and x is a general variable, the exponential function e^{ax} is defined by the relations

$$e^{a \cdot 0} = 1 \tag{6.36}$$

and

$$\frac{d}{dx}(e^{ax}) = a(e^{ax}) \tag{6.37}$$

where e is the base of natural logarithms, equal to 2.718.

Examining Eq. 6.35 above, we see that by identifying the general variable x with time, and by choosing

$$a = -\frac{1}{RC} \tag{6.38}$$

we can solve Eq. 6.35 with an exponential function $e^{-t/RC}$. The function $e^{-t/RC}$ happens to have value unity at $t = 0$ (from Eq. 6.36). To construct an exponential function with value A at $t = 0$, we simply write

$$v = A\,e^{-t/RC} \tag{6.39}$$

[7] The same cannot always be said for nonlinear networks.

Now since

$$\frac{dv}{dt} = A\frac{d}{dt}(e^{-t/RC}) = -\frac{1}{RC}(A\,e^{-t/RC}) \tag{6.40}$$

this more general form of v is also a solution of Eq. 6.35. It represents initial stored energy at $t = 0$ decreasing exponentially to zero because of current flow in the resistor of Fig. 6.20. The magnitude of this current is

$$i = -\frac{v}{R} = -\frac{A}{R}e^{-t/RC} \tag{6.41}$$

The value of the constant A depends on the state of the network just after the independent sources were either set to zero or switched out of the network. Assume for the moment that $A = 1$, and let us examine the exponential function itself.

Figure 6.21 shows the exponential function plotted in standard form. The quantity τ is called the *time constant*. For the previous example, we had

$$\tau = RC \tag{6.42}$$

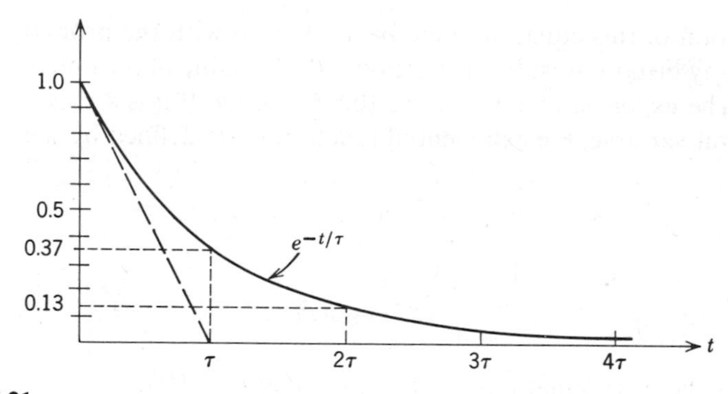

Figure 6.21
An exponential decay. τ is called the time constant of the decay.

The value of the exponential function is 1 at $t = 0$. For this reason, we often use the network conditions at $t = 0$ to find the value of constants like A in Eq. 6.39. Furthermore, the initial slope, shown by the dashed line, intersects the axis at $t = \tau$. When this point in time is reached, the exponential function has decayed to

$$e^{-\tau/\tau} = e^{-1} = 0.37 \tag{6.43}$$

After two time constants ($t = 2\tau$) the function has decayed further to $e^{-2\tau/\tau} = e^{-2} = (0.37)^2 = 0.13$, and so on. During each succeeding interval τ, the function gets smaller by a factor of e^{-1}. After five time constants ($t = 5\tau$), the exponential is 99% gone, and is usually considered to be negligible thereafter.

The exponential $e^{-t/\tau}$ can also be used to represent the natural response when the final value is nonzero (see Fig. 6.22). In this case the final value is included in the total response, the characteristic function being of the form $(1 - e^{-t/\tau})$.

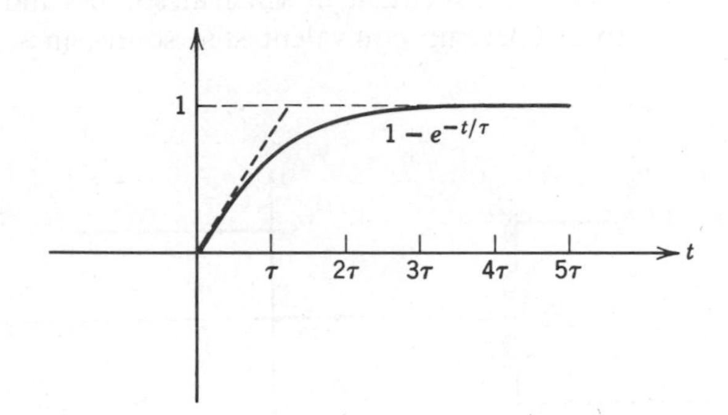

Figure 6.22
Exponential decay to a nonzero final value.

This exponential behavior of the form $e^{-t/\tau}$ is the characteristic natural response in a network containing either a single capacitor or a single inductor whenever a source undergoes a change that demands a change in the stored energy. Thus for a step source the voltage on a capacitor or the current through an inductor will go from its intial to its final value along an $e^{-t/\tau}$ kind of curve, this being the time-dependence characteristic of a change in stored energy. The following section illustrates how the exponential natural response can be combined with the continuity conditions of the previous section to allow us to solve by inspection any step response problem in *LR* and *RC* circuits.

6.3.4 Single-Time-Constant Circuits

Networks which contain either one capacitor or one inductor, along with an arbitrary combination of sources and resistors, are called single-time-constant networks, since the natural response always has an $e^{-t/\tau}$ behavior.

The time constant τ can always be found by direct inspection of the network. In an *RC* circuit, the time constant is simply the product of the capacitance C with the Thévenin resistance of the network viewed from the capacitor's terminals. In an LR network, the time constant is the inductance L divided by the Thévenin resistance seen from the inductor's terminals. To find any response to a step source in such a network, first determine the initial and final states, then use the continuity conditions to determine what value to assign to the capacitor voltage or the inductor current at the instant of the step, and finally connect this value to the final value with an $e^{-t/\tau}$ waveform. The following examples illustrate the method.

Figure 6.23 shows a general LR circuit, in which all sources and resistances have been reduced to a Thévenin equivalent step source in series with a

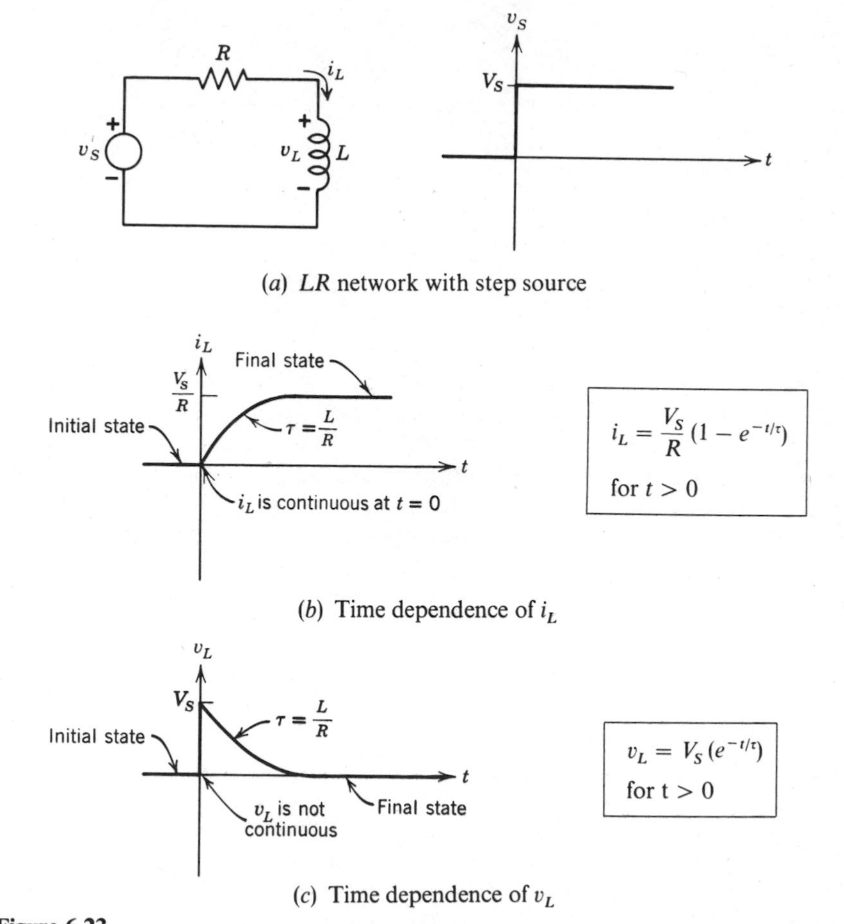

Figure 6.23
Step response in an LR circuit. The quantity $(1 - e^{-t/\tau})$ characterizes the exponential variation of i_L to its final state.

Thévenin resistance. The time constant for this circuit is L/R. The initial state, with $v_S = 0$, is found by replacing the inductor with a short circuit. In this case both v_L and i_L are zero, v_L because the inductor behaves like a short circuit, and i_L because the intial value of the source is zero. In the final state, long after the natural response has died out, the voltage v_L is once again zero, while the current has increased to V_S/R (remember the inductor is a short circuit for dc). We can now apply the continuity conditions to find out what the value of i_L must be at $t = 0^+$, just *after* the step occurs. Since the current in an inductor must be continuous (Eq. 6.32), $i_L(0^+)$ must be zero. The characteristic natural response, which takes the current from this value to its final value, is of the form $1 - e^{-t/\tau}$, where $\tau = L/R$. Thus the total waveform for i_L is

$$i_L = u_{-1}(t)\frac{V_S}{R}(1 - e^{-Rt/L}) \tag{6.44}$$

There is no continuity condition on the voltage across the inductor. Indeed, if the current i_L cannot change instantaneously, we expect the voltage on the inductor to rise instantaneously to V_S to prevent current flow through R. This voltage dies away exponentially as the current i_L builds up. The total voltage waveform is

$$v_L = u_{-1}(t)V_S\, e^{-Rt/L} \tag{6.45}$$

Figure 6.24 shows an example with a capacitor. In Fig. 6.24*a* a step source with two resistors and a capacitor is shown. In Fig. 6.24*b*, the network has been reduced to Thévenin equivalent form. From this second circuit the time constant is

$$\tau = R_T C = (R_1 \| R_2)C \tag{6.46}$$

The initial values of both i_C and v_C are zero, the current because the capacitor is an open circuit for dc, and the voltage because the initial value of the source is zero. The final values are zero for the current, and $R_2 V_S/(R_1 + R_2)$ for the voltage, this being the amplitude of the step in the Thévenin equivalent network.

Since the voltage across a capacitor cannot change instantaneously we know that $v_C(0^+)$ must be zero. The decay to the final state is described by a $(1 - e^{-t/\tau})$ time dependence. Furthermore, since the voltage across the capacitor is zero at $t = 0^+$, the current at $t = 0^+$ must be equal to the short-circuit current of the Thévenin equivalent circuit. That is, the current in the capacitor jumps from zero at $t = 0^-$ to v_{OC}/R_T at $t = 0^+$, this current then decaying to zero as the capacitor charges up to its final value. The

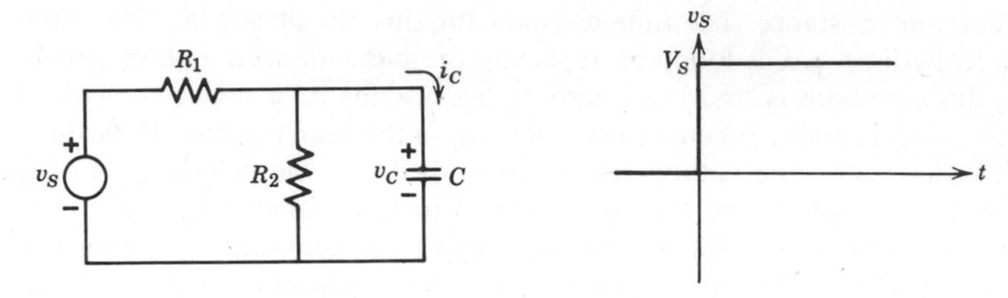

(*a*) *RC* network with step source

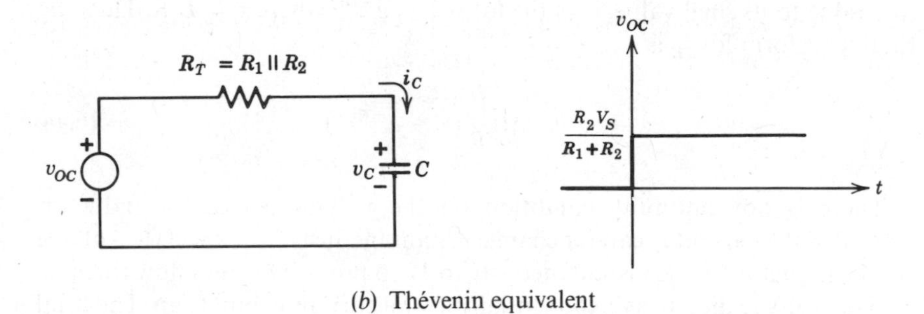

(*b*) Thévenin equivalent

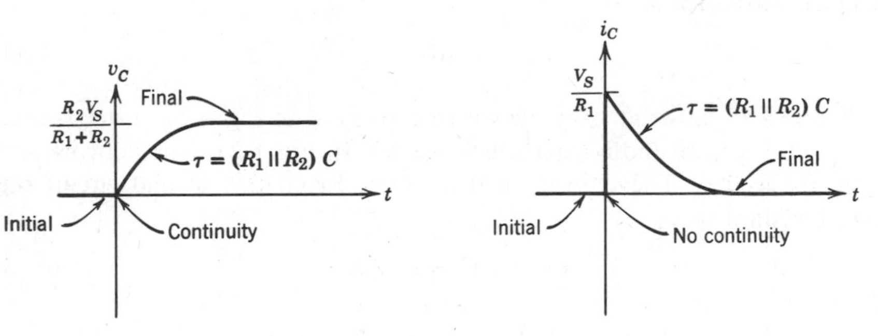

(*c*) Current and voltage responses

Figure 6.24
Step response in an *RC* circuit.

full form of the voltage and current waveforms are

$$v_C = u_{-1}(t)\frac{R_2 V_S}{R_1 + R_2}(1 - e^{-t/(R_1 \| R_2)C}) \tag{6.47}$$

$$i_C = u_{-1}(t)\frac{V_S}{R_1} e^{-t/(R_1 \| R_2)C} \tag{6.48}$$

The above examples are perfectly general, except that they both had sources with initial values of zero. Examples with nonzero initial values are included in the exercises at the end of the chapter.

6.4 CIRCUIT APPLICATIONS

6.4.1 Analog Integration and Differentiation

Having now introduced the inductor and capacitor and examined how to treat step responses in simple networks, we can inquire into some new circuits that include energy storage elements.

Figure 6.25 shows how capacitors can be used in combination with op-amps and resistors to perform the mathematical operations of integration and differentiation. In the case of the integrator (Fig. 6.25*a*), because of the feedback to the negative terminal, one can presume linear operation. As a result, v_- will equal v_+ (which is zero) so the input current i_S is v_S/R_S. Since ideally no current enters the $-$ terminal ($i_- = 0$), the current i_F must also equal v_S/R_S. The total charge on the capacitor is proportional to the integral of this current over all time. Thus the output voltage is

$$v_O = -\frac{1}{RC}\int_{-\infty}^{t} v_S\,dt \tag{6.49}$$

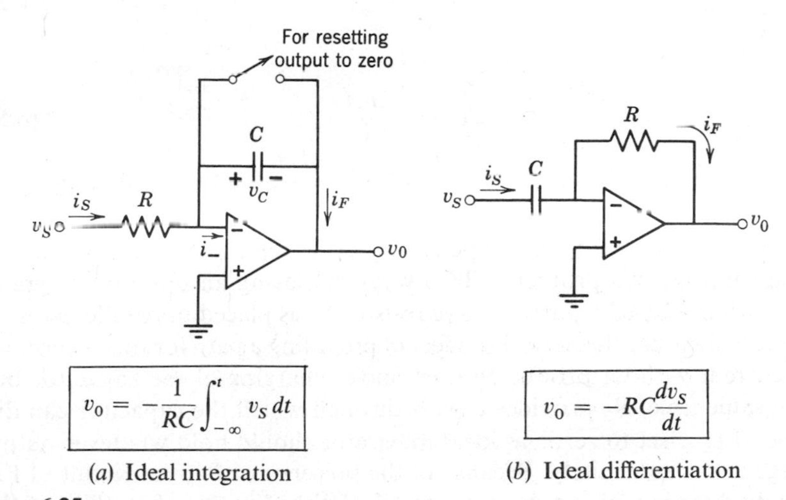

(*a*) Ideal integration (*b*) Ideal differentiation

Figure 6.25
Analog integration and differentiation.

Often, a switch is added to permit the output to be set to zero prior to starting the calculation of an integral.

If the capacitor and resistor are exchanged, as in Fig. 6.25*b*, the circuit becomes a differentiator. The input current is $C(dv_S/dt)$. The output voltage, therefore, is simply

$$v_O = -RC\frac{dv_S}{dt} \tag{6.50}$$

In both of the above cases, the op-amps were presumed ideal. Non-idealities of the kind introduced in Section 5.4 will affect the operation of these circuits. The most significant non-ideality, however, and the one that cannot be ignored, is the bias current for the integrator. We recall from Section 5.4.3 that all op-amps require small dc bias currents at both inputs. In the circuit of Fig. 6.25*a*, the bias current can enter the + terminal from ground, but there is no path for the bias current to enter the − terminal. At first glance, it appears that this current could come from v_S, but unless v_S has precisely the correct average dc value, the dc component of the current i_S will not match the required bias current. Therefore, the only place from which to get this bias current is the capacitor. Of course, if one draws a steady current from a capacitor, it gradually charges up. In particular, if v_S were zero, then i_F would equal the negative of the − terminal bias current. Denoting this current by i_-, the voltage across the capacitor follows the time dependence.

$$v_C = -\frac{i_- t}{C} \tag{6.51}$$

or

$$v_O = +\frac{i_- t}{C} \tag{6.52}$$

until v_O reaches V_{CC}, at which point the op-amp saturates making the integrator inoperable.

There are many ways of compensating for the bias current, each method introducing its own problems. Two ways of biasing an op-amp integrator are shown in Fig. 6.26. First, a large resistor R_2 is placed in parallel with the capacitor. This has the desirable effect of providing a path for the − terminal bias current without producing continuous charging of the capacitor but, at the same time, R_2 provides a path through which the capacitor can discharge. If v_S is set to zero, an ideal integrator should hold whatever output voltage it has indefinitely. Because of the presence of R_2, the circuit of Fig. 6.26*a* discharges with a time constant R_2C, thus limiting the utility of this circuit to time scales much shorter than R_2C.

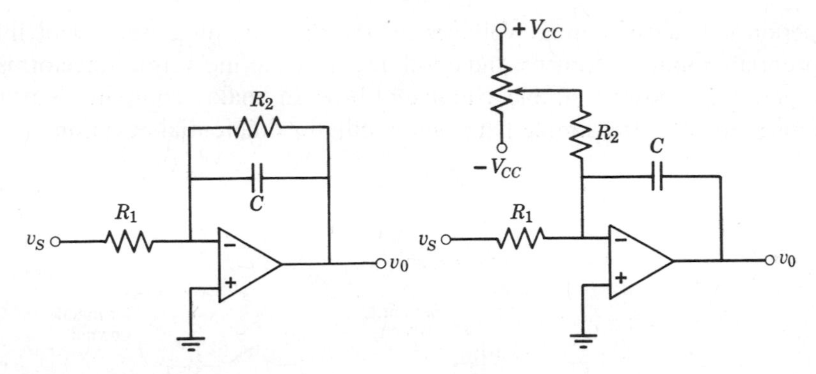

(*a*) R_2 provides a dc path for the bias current, but allows the capacitor to discharge with time constant R_2C.

(*b*) The potentiometer and R_2 provide the bias current without discharging the capacitor, but they cannot correct for drifts in bias current.

Figure 6.26
Two ways of biasing op-amp integrators.

The circuit of Fig. 6.26*b* has the advantage of providing exact biasing without establishing such a discharge path, but the bias must be readjusted as the bias current drifts. If one were to add a resistor R_1 in series with the + terminal, as suggested in Section 5.4.3, then R_2 need supply only the relatively smaller input offset current, and requires adjustment only when the offset current drifts.

As a general rule, the more precisely one wishes to calculate an integral, the shorter must be the time scale over which to expect this accuracy. Drifts ultimately limit the precision of all integrators.

6.4.2 The Analog Computer

We learned in Chapter 5 how to use op-amps and resistors to perform scaled additions and subtractions of waveforms. Now with the inclusion of integrators, it becomes possible to set up circuits that perform remarkably complex computational functions, including the solution of differential equations. An *analog computer* is a collection of op-amps, relays, potentiometers, and display devices that permits flexible assembly of analog computation circuits with the addition of resistors, capacitors, or even nonlinear elements as input and feedback elements. The "programming" of such a computer involves connecting the op-amps so that the correct computation is performed, setting initial voltages on the various integrators, applying appropriate input waveforms, and observing the resulting output waveforms.

Figure 6.27 shows a simple example of how an analog computer is programmed to solve the simple-harmonic-oscillator differential equation

$$\frac{d^2x}{dt^2} = -Kx \tag{6.53}$$

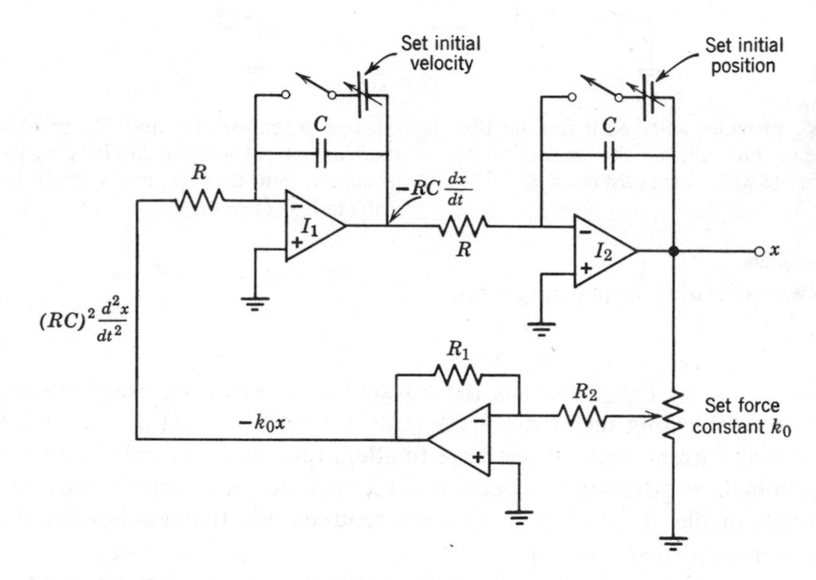

Differential equation: $\frac{d^2x}{dt^2} = -\left[\frac{k_0}{(RC)^2}\right]x$

Figure 6.27
An example of how an analog computer would be used to solve a differential equation. When the switches are opened, the time dependence of the position x, velocity dx/dt and acceleration d^2x/dt^2 corresponding to the set initial conditions could be monitored with an oscilloscope.

If the output is called x, the input to integrator I_2 is $-RC(dx/dt)$, and the input to integrator I_1 must then be $(RC)^2(d^2x/dt^2)$. The voltage on the input to I_1 is obtained by feedback from the x output, first by a potentiometer, then by an inverting amplifier. Thus the specific differential equation represented here is

$$\frac{d^2x}{dt^2} = -\left[\frac{k_0}{(RC)^2}\right]x \tag{6.54}$$

Two switches are shown that force the voltages on the capacitors (hence the outputs of the integrators) to preset values, permitting any combination of initial position (x) and velocity (dx/dt) to be set into the computer. When the switches are opened, both integrators work simultaneously, the time evolution of the various waveforms simply being the solutions to the differential equation. For the example shown, as we know from studies of the harmonic oscillator in elementary physics, the output is simply a sinusoidal oscillation with frequency $\sqrt{k_0}/2\pi RC$, the amplitude depending on the initial choices of position and velocity.

6.4.3 A Square-Wave Oscillator

Figure 6.28 shows how a Schmitt trigger (Section 5.3.2) can be combined with time-delayed feedback to the negative terminal to produce a square-

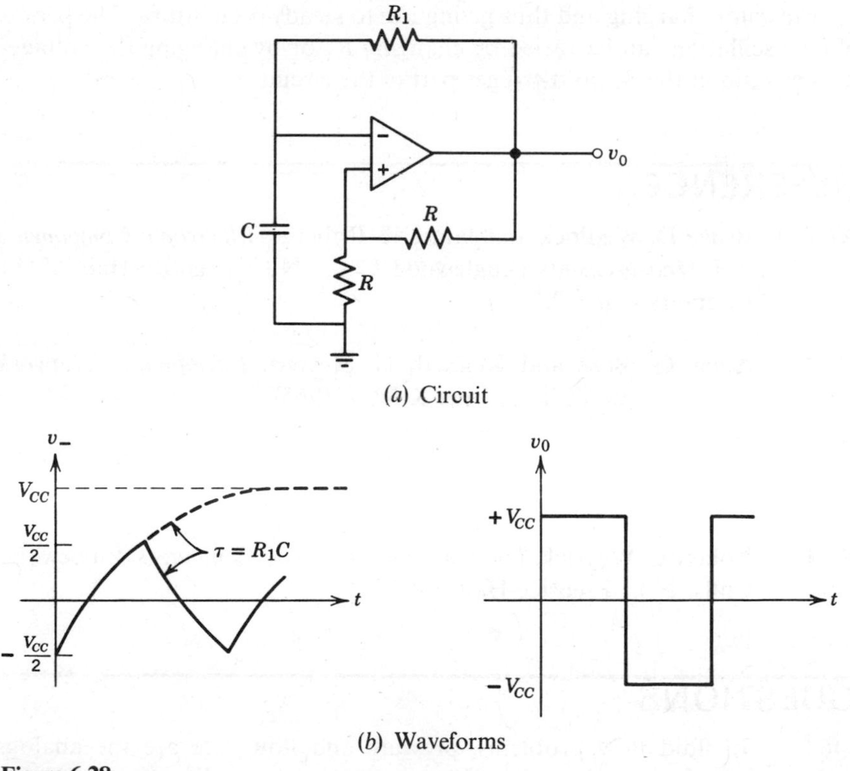

Figure 6.28
A square-wave oscillator

wave oscillator. Because of the capacitor connected between the − terminal and ground, the voltage on the − input cannot change instantaneously. Instead, current must flow through R_1, charging the capacitor toward the instantaneous value of v_0. Because of the time delay inherent in this charging process, the positive feedback Schmitt trigger dominates the circuit operation. The output voltage can be expected to be either at $+V_{CC}$ or at $-V_{CC}$. The feedback to the − terminal charges the capacitor toward v_0 but before this charging process can go to completion, the voltage difference $v_+ - v_-$ changes sign, causing v_0 to switch to the other supply voltage. Waveforms are shown in the illustration.

Suppose at $t = 0$ that the output has just switched to $+V_{CC}$. One can presume, therefore, that since v_+ was at $-V_{CC}/2$ just *before* $t = 0$, v_- was also at $-V_{CC}/2$ just before $t = 0$. Furthermore, since the voltage across a capacitor cannot change instantaneously, the voltage v_- is still at $-V_{CC}/2$ just after the op-amp switches state. The capacitor now charges toward $+V_{CC}$. When the voltage reaches $V_{CC}/2$, the op-amp switches state again, reversing the capacitor charging and thus giving rise to steady oscillations. The period of the oscillation can be varied by changing R_1 or by changing the voltage-divider ratio in the Schmitt-trigger part of the circuit.

REFERENCES

R6.1 Bruce D. Wedlock and James K. Roberge, *Electronic Components and Measurements* (Englewood Cliffs, N.J.: Prentice-Hall, 1969), Chapters 8 and 9.

R6.2 Amar G. Bose and Kenneth N. Stevens, *Introductory Network Theory* (New York: Harper & Row, 1965).

R6.3 Hugh H. Skilling, *Electrical Engineering Circuits* (New York: John Wiley, 1957).

R6.4 Robert C. Weyrick, *Fundamentals of Analog Computers* (Englewood Cliffs, N.J.: Prentice-Hall, 1969).

QUESTIONS

Q6.1 In fluid flow problems, pressure and flow rate are the analogs of voltage and current. What are the analogs of resistance, charge, capacitance, and inductance?

Q6.2 An active element (as opposed to a passive element) is capable of power gain, that is, of delivering more power to a network than it absorbs from a source network. Classify capacitors, inductors, and transformers as either active or passive.

Q6.3 Can the insertion of a transformer in a network change the time constant of that network? Give examples to justify your answer.

Q6.4 In practice, it is almost always preferable to design analog circuits so they employ integrators rather than differentiators. Can you think of any reasons why this might be true? (*Hint.* Consider the possibility that there will be superimposed on the signals being processed small random variations called noise.)

EXERCISES

E6.1 Find the equivalent capacitance C in each of the networks in Fig. E6.1.

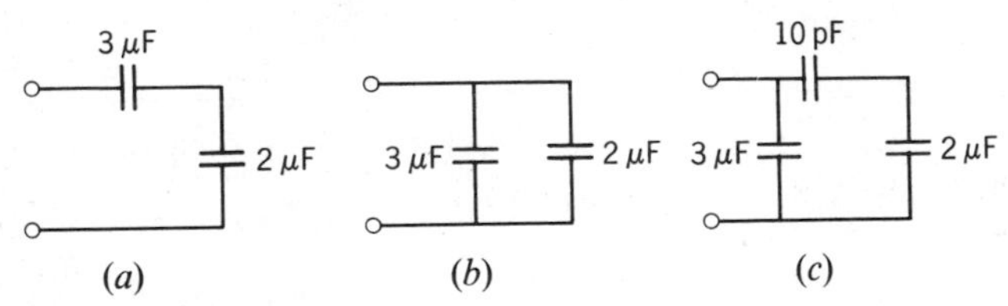

Figure E6.1

E6.2 Find the equivalent inductance L in each of the networks in Fig. E6.2.

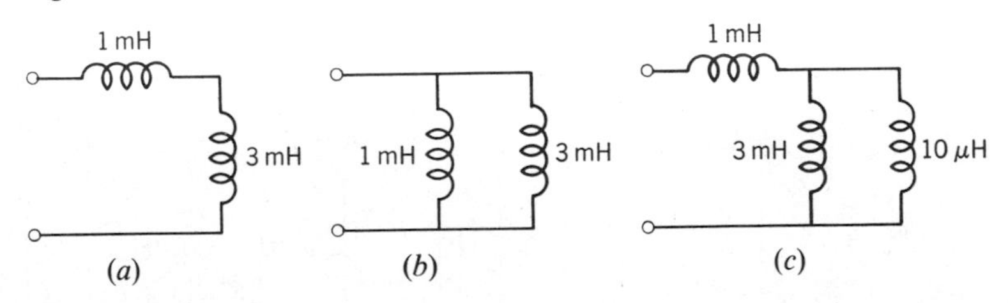

Figure E6.2

E6.3 Find the equivalent element in Fig. E6.3 when N is (a) a resistor R; (b) a capacitor C; (c) an inductor L.

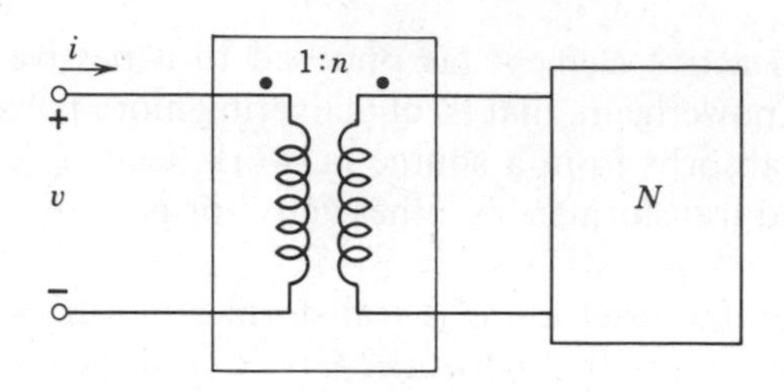

Figure E6.3

E6.4 Find the Thévenin equivalent of the networks in Fig. E6.4. Observe the polarities carefully.

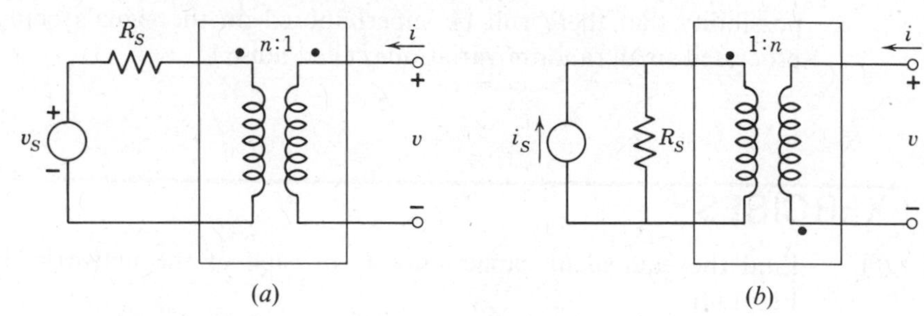

Figure E6.4

E6.5 Assume $v_S(t) = Au_{-1}(t)$. Find the indicated response in each of the networks in Fig. E6.5.

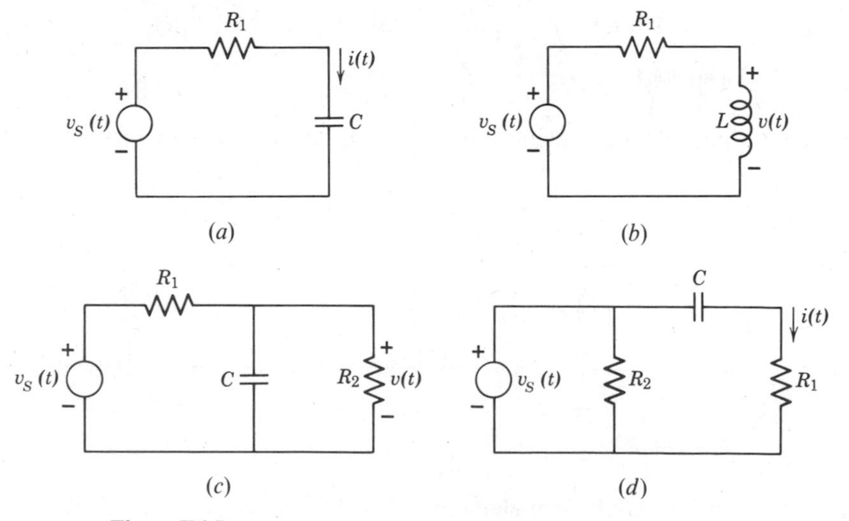

Figure E6.5

E6.6 Assume $i_S = Bu_{-1}(t)$. Find the indicated response in each of the networks in Fig. E6.6.

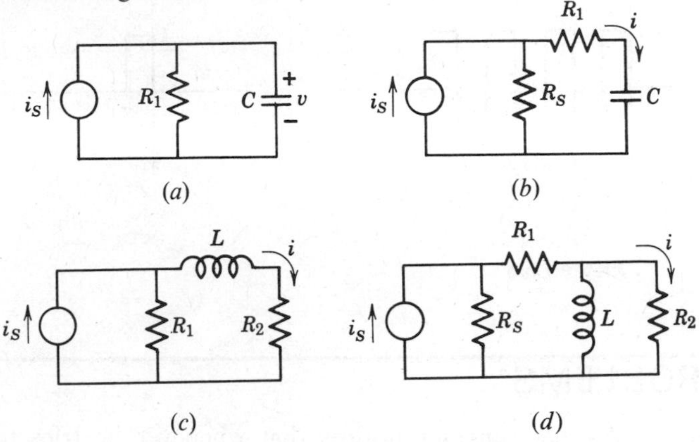

Figure E6.6

E6.7 The networks in Fig. E6.7 involve nonzero initial conditions. Find the indicated responses.

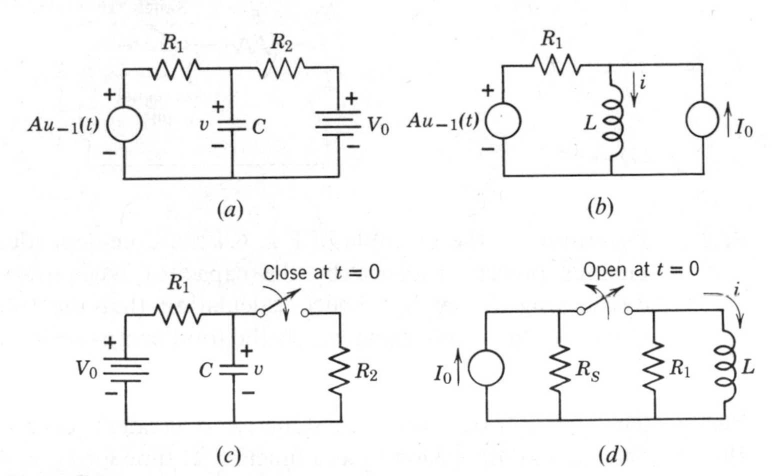

Figure E6.7

E6.8 For the ideal integrator of Fig. 6.25*a*, sketch the waveforms for v_O assuming $v_O = 0$ at $t = 0$ and assuming v_S is given by the following periodic waveforms.

(a) $v_S = u_{-1}(t) 3 \cos \omega t$

(b) Figure E6.8*a*.

(c) Figure E6.8*b*.

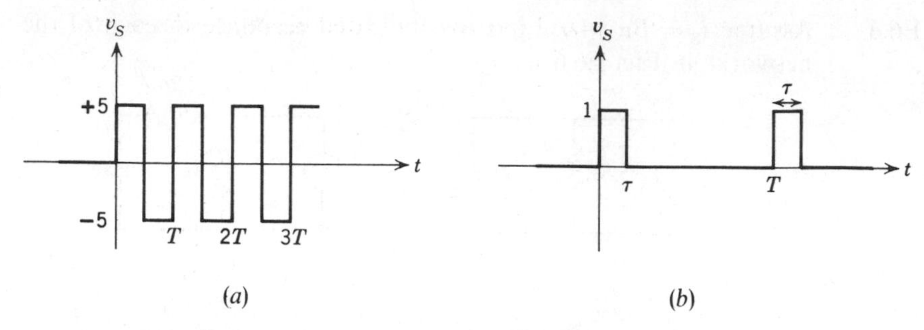

Figure E6.8

PROBLEMS

P6.1 An experimenter notices that whenever he tries to turn off his electromagnet, he gets a big spark at the switch.
(a) Why does he get the spark?
(b) Suggest a way to suppress it.

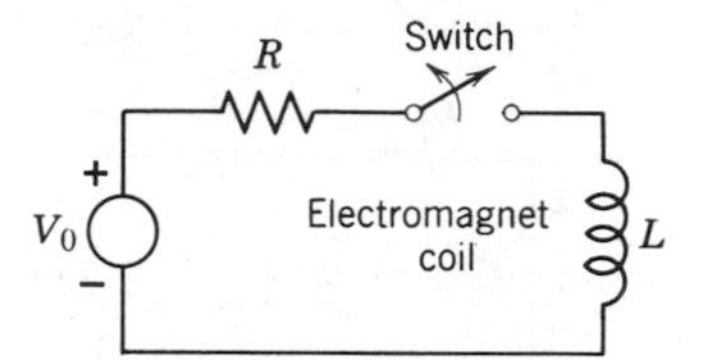

Figure P6.1

P6.2 Determine in the example of Fig. 6.4 the time-dependent instantaneous power absorbed by the capacitor when charging and discharging. Show by explicit calculation that the total energy absorbed during charging equals the total energy delivered during discharging.

P6.3 The "mysterious box" contains a dependent current source. Sketch and dimension v_2 as a function of time for $v_S = Au_{-1}(t)$.

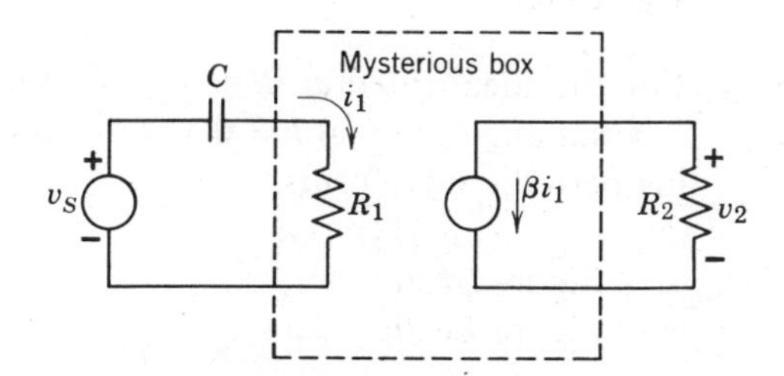

Figure P6.3

P6.4 Find v_2 as a function of time in Fig. P6.4 using the transformer model of Fig. 6.14 but with L_e replaced by a short circuit.

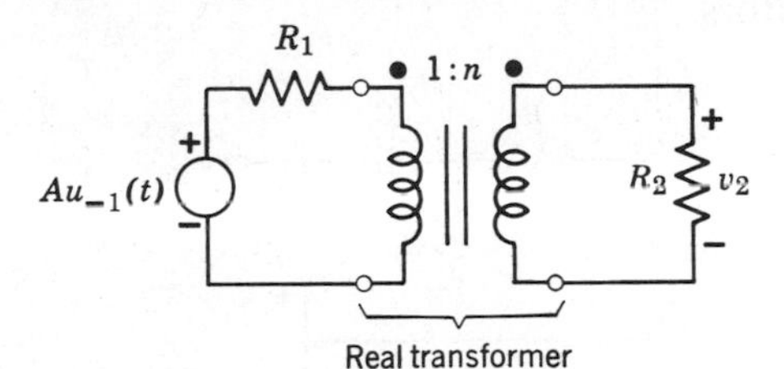

Figure P6.4

P6.5 (a) Find the current i_2 in Fig. P6.5*a* for $v_S = 3u_{-1}(t)$. Sketch your result.

(b) Find the output voltage v_0 in the circuit in Fig. P6.5*b*, assuming $v_S = 3u_{-1}(t)$. Sketch your result.

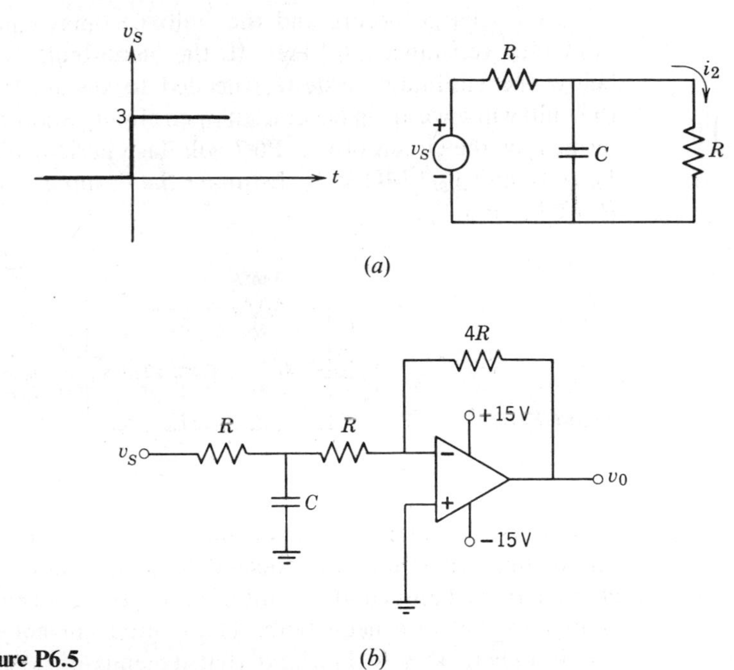

Figure P6.5

P6.6 Let us examine the effect of finite open-loop gain on the operation of an integrator.

The circuit of Fig. P6.6*a* can be represented by the equivalent circuit in Fig. 6.6*b*. Find R_{eq} and C_{eq}. When v_1 is a step, how does

i_1 differ from the case when A is infinite? Discuss how this effect might (or might not) interfere with the operation of actual integrators.

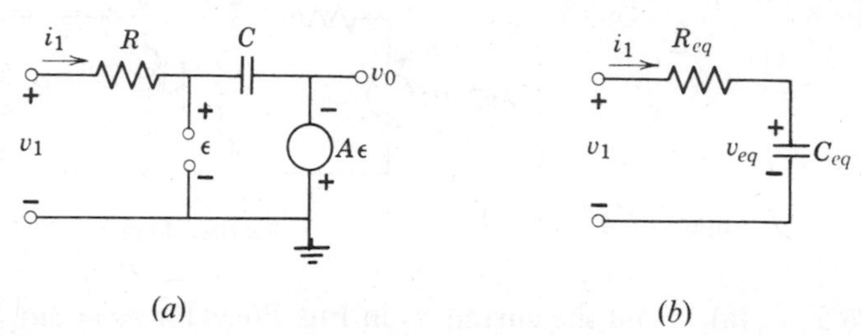

Figure P6.6

P6.7 A neon bulb has the property that for $v < V_T$, where V_T is a threshhold voltage, the bulb is an open circuit. Once v reaches V_T, however, a discharge occurs and the bulb becomes equivalent to a moderate resistance ($\approx 1\ \text{k}\Omega$). If the neon-bulb current drops below the minimum value (I_S) needed to sustain the discharge, the bulb will once again become an open circuit. Show in qualitative terms how the circuit of Fig. P6.7 will flash periodically as long as $V_B > V_T$ and $V_B/1\ \text{M}\Omega < I_S$. Estimate the flashing rate in terms of R, C, V_B, and V_T.

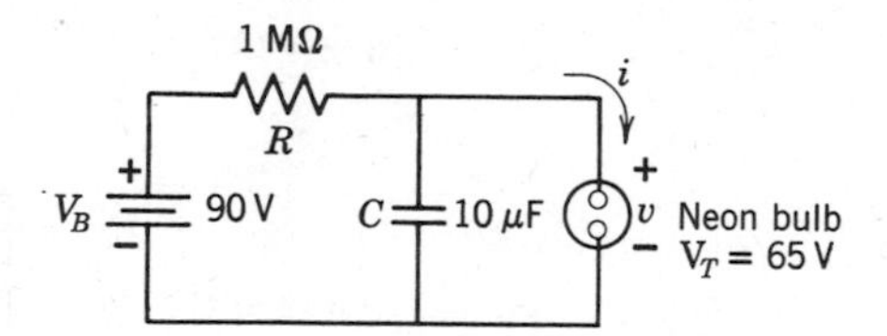

Figure P6.7

P6.8 Figure P6.8 is identical to Fig. P6.7 except that a unijunction transistor has replaced the neon bulb. When the base 2 lead of a unijunction transistor is connected to a dc supply, then the v-i characteristic between the emitter and base 1 terminals is very similar to that of a neon bulb. The emitter current i is zero for $v < V_P$ (where V_P is a threshold that depends on V_B), but when v reaches V_P, the emitter-base 1 equivalent can be represented by a small resistance until the current i drops below a minimum value, I_S, which depends in detail on the structure of the transistor. Show that this circuit will generate a periodic ramp waveform at v_C if $V_B/R < I_S$. Estimate the period of this ramp in terms of R, C, and

V_B for the case $V_P = 0.6\ V_B$ assuming rapid complete discharge of the capacitor once V_P is exceeded.

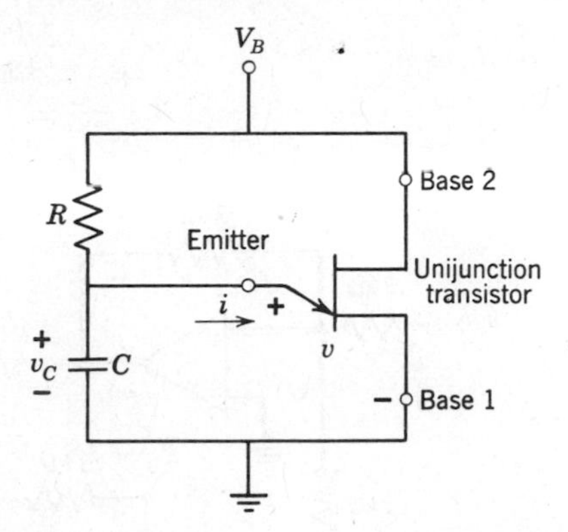

Figure P6.8

P6.9 You are given a 1 V square wave with a 10 ms period (Fig. P6.9*a*). You need the waveform in Fig. P6.9*b*. Design a circuit using op-amps, resistors, and capacitors that will produce v_2 from v_1. Assume $\pm$ 15 V power supplies. (*Hint.* First integrate, then amplify into saturation to clip off the top of the triangles, then attenuate to the right voltage scale).

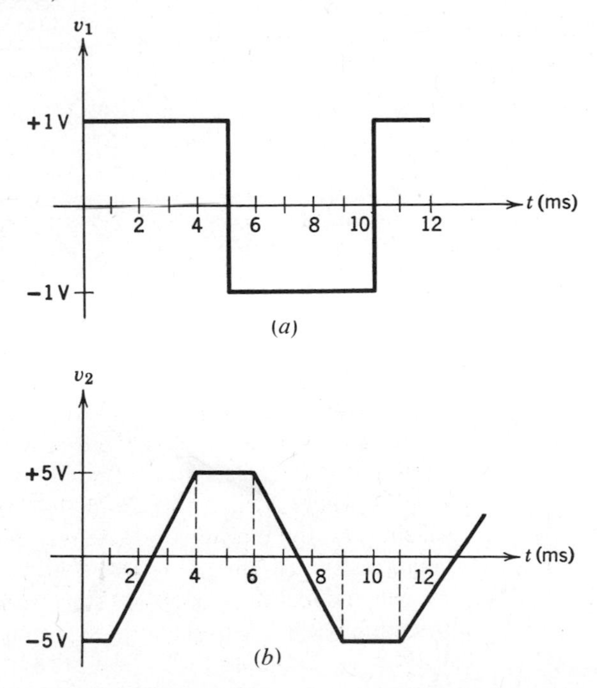

Figure P6.9

P6.10 What is the differential equation relating v_O to v_S in the network of Fig. P6.10?

Answer: $0.1 \frac{dv_O}{dt} + 0.3v_O = v_S$.

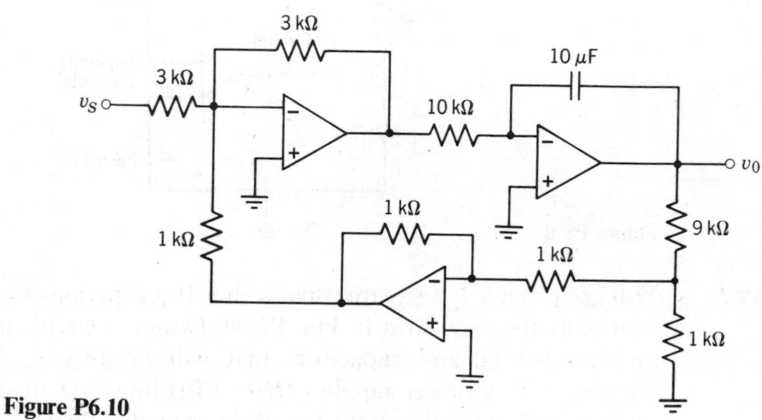

Figure P6.10

PART B

ELECTRONIC DEVICES

CHAPTER SEVEN
ELECTRICAL CONDUCTION PROCESSES

7.1 LOOKING BACK AND LOOKING FORWARD

The reader who has digested a reasonable fraction of the first six chapters can no longer be considered a novice. Basic concepts in network theory, device modeling, and signal processing have been introduced, along with five specific kinds of components: resistors, op-amps, capacitors, inductors, and transformers. The approach taken thus far has of necessity been somewhat heuristic, particularly in regard to the op-amp. In the absence of fundamental background ideas, we presented the terminal characteristics of the op-amp as experimental facts, promising to return with a slightly more fundamental treatment at a later point. Since the intelligent use of electronic devices requires at least a qualitative understanding of device operation, it is now appropriate to begin to develop that understanding.

We begin with a descriptive discussion of the conduction processes that take place in electronic devices. Then, as we introduce successively more complex device structures in the chapters to follow, our descriptive view of device operation will help us to understand not only the normal network properties of the device, but also the limitations of device usage in each potential application. As in the first six chapters, we shall continue to ask with each new device: What signal-processing tasks can be performed with this new device that could not be performed before? In this fashion we hope to keep the user's point of view before us at all times.

7.2 CONDUCTION IN HOMOGENEOUS MATERIALS

7.2.1 Metals

Metals such as copper, silver, tin, and aluminum, which are considered to be good conductors of electricity, are formed by the mutual bonding of the parent atoms into a regular, geometrical array or *lattice* by forces acting at the microscopic atomic scale. This bonding process is such that the outer or valence electrons from each parent atom are released and are free to move throughout the solid in response to applied electric fields. The remaining electrons and the nucleus of each parent atom are not mobile but rather remain bound in a relatively fixed position determined by the crystal lattice.

Since each atom was electrically neutral to start, the solid metal is also electrically neutral. Furthermore, when an atom gives up its valence electrons, the remaining electrons and the nucleus of the atom bound in the crystal lattice can be represented by a bound positive ion, with charge equal and opposite to that given up by the atom through the loss of the valence electrons. Finally, since the bound positive ions are uniformly distributed throughout the solid, the mobile electrons must also be uniformly distributed, because if some region should become devoid of free electrons, the resulting net positive charge would attract sufficient mobile electrons to the region to restore electrical neutrality.

On the basis of this qualitative description we can construct the following model for a metal: a uniform distribution of bound positively charged metal ions uniformly surrounded by a "gas" of mobile electrons. Such a model is represented in a two-dimensional schematic fashion in Fig. 7.1 where it has been assumed that each atom gives up one free electron to the mobile electron gas. From this model we can summarize several important facts governing electrical conduction in metals.

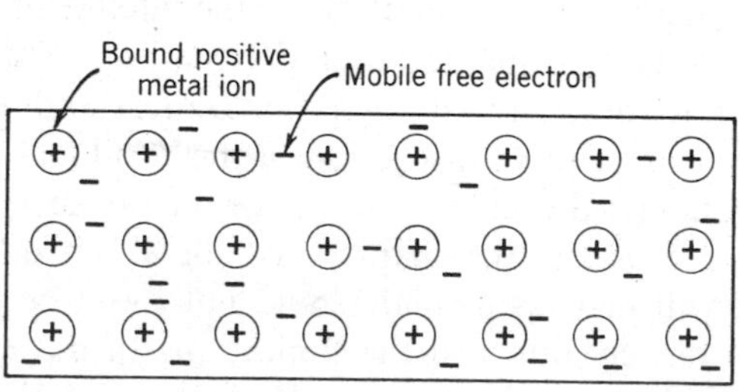

Figure 7.1

Two-dimensional schematic representation of the electron gas model of a metal. Each atom is assumed to give up one free electron and to produce a bound metal ion of charge $+q$.

1. Although the positive ions play an important role in maintaining electrical neutrality, they cannot contribute to the conduction process since they are immobile. Thus in a metal, current is carried by a single type of mobile charge—the free electron.[1]
2. Since there are approximately 10^{23} atoms per cubic centimeter in a metal, there will be approximately 10^{23} free electrons per cubic centimeter available to contribute to the conduction process. It is this relatively large density of mobile charge carriers that make metals such good electrical conductors.
3. The charge-carrier density of free electrons available to contribute to the conduction process depends primarily on the number of valence electrons of the parent atom. Thus for a given metal, the carrier density is fixed.
4. Over distances large compared to the interatomic spacing a metal is everywhere electrically neutral.

7.2.2 Pure Semiconductors

The conduction process in *semiconductors* differs from that in a metal in several respects. The most important differences are that semiconductors conduct current through two distinct and independent modes of electron motion, and that the relative importance of these two modes can be controlled over a wide range through the addition of minute quantities of appropriate elements to the basic semiconductor material. Although one mode of electron motion in semiconductors can be described in terms of a free electron gas, analogous to the situation in a metal, the other mode must be described in terms of the flow of *positive* charge. It is the existence of and the ability to control precisely these two independent charge-transport mechanisms that makes possible all junction semiconductor devices, including the diode and bipolar transistor.

To gain some understanding of conduction in semiconductors, we begin by considering the case of an absolutely pure or *intrinsic* semiconductor crystal. As with the metal, a semiconductor is composed of a regular geometric lattice of parent atoms. Most of the valence electrons, however, are

[1] The modes of electron motion within this "free electron gas" can be quite complex, and can be accurately described only in the language of quantum mechanics. See the References cited at the end of this chapter.

not free to move throughout the crystal. Instead, they are involved in holding the atoms in their fixed positions in the crystal lattice through the process of *covalent bonding*. At 0 K (absolute zero) every valence electron will be involved in the bonding process; no charge carriers will be available for conduction. As the temperature is increased, however, a fraction of the valence electrons break free from the covalent bonds and produce a mobile or free electron gas capable of supporting an electric current. When a bond is broken, a net charge deficiency appears in the total bond network. This deficiency behaves as an apparent mobile positive charge, also capable of carrying an electric current. These apparent positive charges are called *holes*, since they arise from the lack of an electron in a covalent bond. In the intrinsic semiconductor it follows that the number of free electrons must equal the number of holes since an electron escaping from a covalent bond simultaneously produces both a free electron and a hole. The crystal remains electrically neutral, since the number of positive charges equal the number of negative charges. See Fig. 7.2 for a schematic representation of an intrinsic semiconductor.

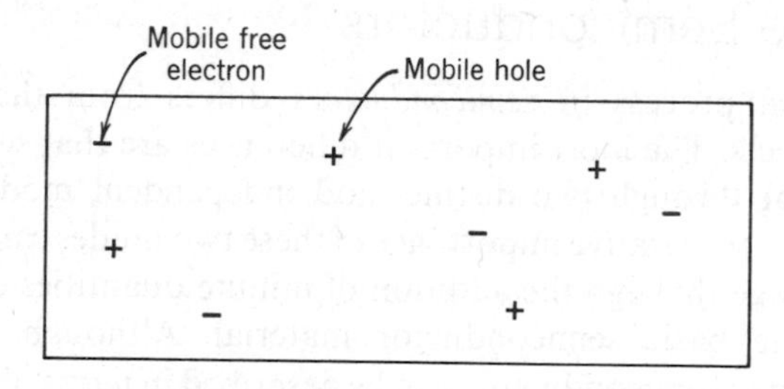

Figure 7.2
Two-dimensional schematic representation of an intrinsic semiconductor.

The most common semiconductor materials for electronic devices are germanium and silicon. In their intrinsic state at normal room temperature (300 K) the number of free electrons and holes are 10^{13} carriers per cubic centimeter in germanium and 10^{10} in silicon. Since the number of atoms per cubic centimeter is still about 10^{23}, the number of charge carriers is only a tiny fraction of the number that would be present if these materials behaved as metals. Because of the much smaller number of available charge carriers, their conduction properties are much poorer than those of metals. They do, however, conduct sufficient current so that they cannot be classified as insulators; hence, the term *semiconductor*.

7.2.3 Doped Semiconductors

If semiconductors were available only in their intrinsic state, their usefulness would be severely limited. The construction of devices such as diodes and transistors depends not only on the fact that there are two types of charge carriers available—electrons and holes—but also on the ability to control precisely the relative concentrations of electrons and holes.

We have just seen that an electron escaping from a covalent bond becomes a free electron and simultaneously produces a hole. If we can somehow produce a free electron *without breaking a covalent bond*, then no hole will be formed and the concentration of free electrons can be made to exceed the concentration of holes. Similarly, if a broken covalent bond can be produced without the release of a free electron, then the hole concentration can be made larger than that of the free electrons. These situations can be achieved by *doping* the basic semiconductor material through the addition of minute quantities of foreign elements called *impurities*.

The parent atoms of germanium and silicon each have four valence electrons, and the covalent bonding process involves each of these four electrons. If in the semiconductor crystal we substitute an impurity atom with five valence electrons for one of the parent atoms, then the impurity atom will have one more electron than is required to complete the bonding process. This extra electron can now become a mobile free electron *without* breaking a covalent bond and, consequently, no hole is produced. When the fifth electron becomes free, it leaves behind a positive, singly-charged ion since the impurity atom acts, in essence, like an atom in a metal, giving up one free electron for the conduction process with an immobile positive ion remaining fixed in the crystal structure. Because this type of impurity atom produces a free electron, it is called a *donor*. The elements containing five valence electrons are found in Group V of the periodic table. Phosphorous, arsenic, and antimony are used as donor impurities in germanium and silicon.

In addition to the free electrons produced by the donor impurities, there will still be some broken covalent bonds, each producing a free electron and a hole. Thus, both types of mobile charge carriers remain, but the number of free electrons now exceeds the number of holes. In this case the free electrons are called the *majority carrier* and the holes are called the *minority carrier*. A semiconductor in which the majority carrier is negatively charged is termed *n-type*. A schematic representation of an *n*-type semiconductor, showing the mobile electrons, mobile holes, and immobile donor ions, is shown in Fig. 7.3*a*.

If an element having only three valence electrons is substituted as an impurity atom for one of the parent atoms in a semiconductor crystal, then a broken covalent bond is produced without an accompanying free

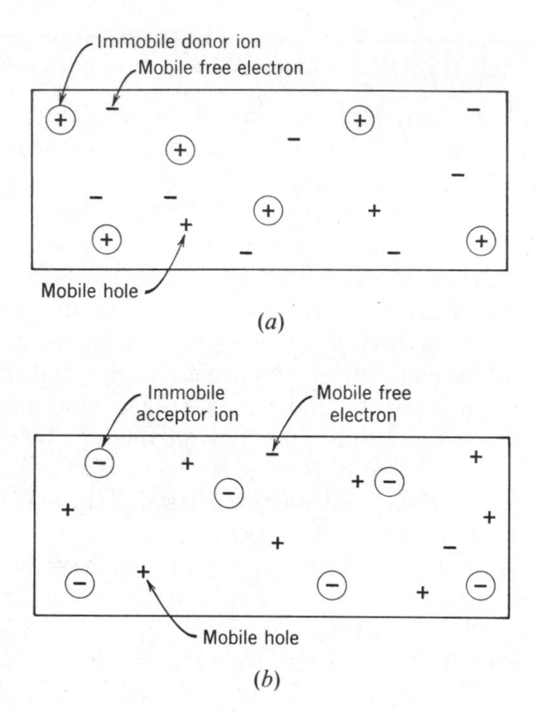

Figure 7.3
Two-dimensional schematic representation of doped semiconductor materials. (*a*) *n*-type, (*b*) *p*-type. Note in each case that the crystal is electrically neutral.

electron. In this way it is possible to increase the number of holes over the number of free electrons. Since this impurity atom is neutral when it has three valence electrons, the motion of the hole away from the impurity atom —which in reality is an electron filling the broken covalent bond formed by the missing electron of the impurity atom—leaves behind a negatively charged immobile ion fixed at the site of the impurity atom. Since this type of impurity atom takes on an electron to complete the bonding process, it is called an *acceptor*. As in *n*-type material there will still be some broken covalent bonds that produce both free electrons and holes; thus, both types of mobile charge carriers will be present. However, in this case the hole will be the majority carrier and the electron the minority carrier. Because the majority carrier is positively charged, the semiconductor is called *p*-type. See Fig. 7.3*b* for a schematic representation showing the three types of charges in *p*-type material. Elements with three valence electrons are found in Group III of the periodic table. Boron, aluminum, gallium, and indium are used as acceptor impurities in germanium and silicon.

Before leaving the topic of impurities it is worthwhile to consider the quantity of an impurity atom required to change the concentration of electrons or holes. Suppose, for example, we desire n-type silicon containing 10^{15} free electrons per cubic centimeter. Since at room temperature there are only 10^{10} free electrons per cubic centimeter in intrinsic silicon, the additional electrons will have to come from donor impurities. Thus, the addition of 10^{15} donor atoms to each cubic centimeter of silicon is required. Since each cubic centimeter of silicon contains 5×10^{22} atoms, we are adding only *one impurity atom for every* 50 *million silicon atoms.* If we use phosphorous, whose atomic weight is nearly equal to that of silicon, then we must add only 20 *micrograms* of phosphorous to a one-kilogram melt. Yet this tiny addition of impurity increases the number of free electrons by five orders of magnitude! This sensitivity of the mobile carrier concentrations to such minute quantities of impurities indicates why extreme care must be taken in refining and purifying silicon and germanium for use in semiconductor devices.

The addition of impurity atoms to a semiconductor also affects the concentration of minority carriers. Although a derivation of this point is beyond the scope of this discussion, the result is quite simple. If the addition of impurities increases one carrier concentration over its value in the intrinsic semiconductor by a given factor, then the opposite carrier's concentration is reduced from its value in the intrinsic material by the same factor. In the example discussed above, the electron concentration in silicon was increased from 10^{10} to 10^{15} cm^{-3}, a factor of 10^5, through the addition of phosphorous. The hole concentration is then correspondingly reduced by a factor of 10^5, from 10^{10} to 10^5 cm^{-3}.

On the basis of our quantitative description and models of the charge carriers, here is a summary of several of the important facts governing electrical conduction in semiconductors.

1. A semiconductor has two independent species of mobile charge carriers: the positively charged hole and the negatively charged free electron

2. The relative concentrations of free electrons and holes in a semiconductor may be adjusted by the addition of minute quantities of appropriate impurity elements. Materials in which holes are the majority carrier are called p-type, while those in which free electrons are the majority carrier are called n-type.

3. The difference in charge density between free electrons and holes in a doped semiconductor is exactly balanced by the charged immobile impurity ions. Thus, over distances large compared to the interatomic spacing, a semiconductor is electrically neutral.

4. The concentration of majority carriers in semiconductor materials used in diodes and transistors is typically six or more orders of magnitude smaller than in a metal. Thus, compared with metals, semiconductors are relatively poor electrical conductors.

7.2.4 The *v-i* Characteristic of the Homogeneous Bar

The next step in understanding conduction processes is to show the relationship between the motion of the mobile charge carriers and the resulting electric currents to the voltages applied at the terminals of a simple device structure.

The simplest possible device structure is the homogeneous bar (Fig. 7.4*a*). A *homogeneous* structure is characterized by complete uniformity throughout its entire volume. To give a specific example, we have chosen a *p*-type semiconductor, although the following discussion could apply equally well to an *n*-type semiconductor, an intrinsic semiconductor, or a metal. The fact that the *p*-type bar is homogeneous means that the hole, free-electron, and acceptor-ion concentrations do not vary with position within the bar.

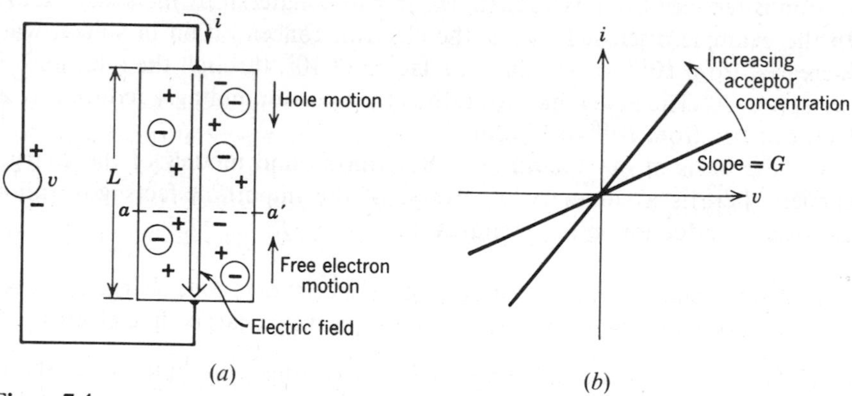

Figure 7.4
Volt-ampere characteristic of a homogeneous *p*-type semiconductor bar.

The bar is everywhere electrically neutral, with a portion of the required negative charge being provided by the immobile acceptor ions that cannot contribute to the electric conduction process.

The application of a voltage between the ends of the bar results in an electric field that forces the holes to move or *drift* in the direction of the electric field and the free electrons to move or *drift* in the opposite direction. Alternatively, one may view this charge motion as the attraction of the positive holes toward the negative battery terminal and the attraction of the negative free electrons toward the positive battery terminal. In any event, the result of this charge carrier motion is the net transport of charge through a cross-sectional plane, such as *a-a'* in Fig. 7.4*a*, with a corresponding electric current. Notice that while the flow of holes and the flow of electrons across *a-a'* are in opposite directions, holes and electrons also carry opposite charges. Thus the net charge transported by holes and the net charge transported by electrons add algebraically to produce the total terminal current i.

So long as the applied voltage is not too large, the v-i characteristic of a homogeneous semiconductor bar is linear, obeying Ohm's law,

$$i = Gv \tag{7.1}$$

where G is the conductance of the bar, measured between its terminals. The conductance of any bar of material is given by

$$G = \sigma\left(\frac{A}{L}\right) \tag{7.2}$$

where A is the cross-sectional area of the bar, L is the bar's length, and σ is the *conductivity* of the material. If the area A and the length L are expressed in meters, the appropriate unit for the conductivity is $(\Omega\text{-m})^{-1}$. More often, the area and length are expressed in centimeters. In these mixed units, the conductivity is expressed in $(\Omega\text{-cm})^{-1}$.

In pure metals and intrinsic semiconductors, the conductivity depends only on the temperature of the sample. Typical values of σ at room temperature are 10^4 to 10^6 $(\Omega\text{-cm})^{-1}$ in metals, and 10^{-5} to 10^{-6} $(\Omega\text{-cm})^{-1}$ in pure silicon. Doping a semiconductor increases its conductivity. If a semiconductor has n free electrons/cm^3 and p holes/cm^3, the conductivity can be expressed as

$$\sigma = q(\mu_e n + \mu_h p) \tag{7.3}$$

where q is the electronic charge (1.6×10^{-19} coulombs), and μ_e and μ_h are constants called the *mobilities* of free electrons and holes in the semiconductor material. For σ in $(\Omega\text{-cm})^{-1}$, the mobilities have units $\text{cm}^2/\text{V-s}$. The magnitudes of electron and hole mobilities in silicon and germanium are on the order of 1000 $\text{cm}^2/\text{V-s}$ at room temperature. The mobility of a charge carrier relates the magnitude of the average drift velocity of the carrier to the strength of an applied electric field. Holes drift along the direction of the field; electrons drift opposite to the direction of the field.

If a bar is doped p-type, then $p \gg n$ and Eq. 7.3 simplifies to

$$\sigma = q\mu_h p \qquad (p\text{-type}) \tag{7.4}$$

and the v-i relationship (Eq. 7.1) becomes

$$i = q\mu_h p\left(\frac{A}{L}\right)v \tag{7.5}$$

If two bars of identical size but differing acceptor concentrations are compared, then Eq. 7.5 shows that the bar with the largest acceptor concentration, having more holes, will conduct a larger current for a given voltage (see Fig. 7.4*b*).

The substitution of an *n*-type bar for the *p*-type bar in Fig. 7.4*a* also results in a linear v-i characteristic in accordance with Ohm's law. In this case, since $n \gg p$, Eq. 7.3 may be simplified to

$$\sigma = q\mu_e n \qquad (n\text{-type}) \tag{7.6}$$

so that the v-i relationship becomes

$$i = q\mu_e n\left(\frac{A}{L}\right)v \tag{7.7}$$

7.3 JUNCTIONS AND CONTACTS

7.3.1 The *p-n* Junction

A semiconductor structure more complex than the homogeneous bar is formed when *p*-type and *n*-type bars are metallurgically joined to form a *p-n junction* (Fig. 7.5). One of the important features of this structure is the asymmetrical nature of the mobile hole and electron concentrations in the neighborhood of the junction. For instance, in the *p*-type region holes are the majority carrier and may have a concentration on the order of 10^{15} cm^{-3}. On the other hand, in the *n*-type region holes are the minority

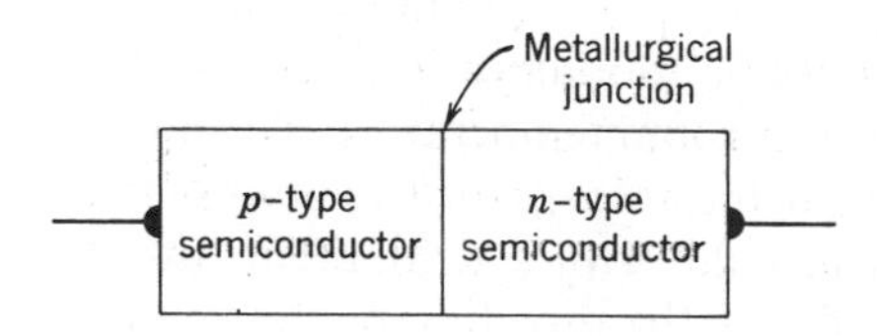

Figure 7.5
p-n junction structure.

carrier, with a concentration that may be on the order of $10^5\ \text{cm}^{-3}$. Thus, in the neighborhood of the junction there must be a substantial change in hole concentration as one moves from the p to the n region. A similar argument applied to the electrons shows that the electron concentration must also undergo a substantial change with position in the neighborhood of the p-n junction. These large variations in carrier concentration with position, called *carrier concentration gradients*, give rise to a new mode of net charge transport and produce a highly asymmetrical v-i characteristic for this structure.

7.3.2 Diffusion Current

An electric current is defined as the net transport of charge through a cross-sectional plane. We have discussed in Sections 2.1 and 7.2 how an electric field, exerting a force on the individual charge carriers, produces a net charge transport and a corresponding electric current. In this section we examine briefly a second mechanism by which charge may be transported, and then qualitatively describe how this mechanism influences the v-i characteristic of the p-n junction.

In discussing the motion of charge carriers up to this point, we have made the tacit assumption that in the absence of an electric field the carriers are motionless. In fact, nothing could be further from the truth! The same thermal energy which produces broken covalent bonds also imparts a kinetic energy to the charge carriers such that at room temperature mobile carriers are moving in random directions, but with an average velocity of 10^7 cm/s.

This random motion of charge carriers in a crystal is analogous in many ways to the random motion of individual gas molecules in the atmosphere, the main difference being that the gas molecules are uncharged. Let us first consider the simple case of gas molecules to see how a net transport of carriers can occur. Figure 7.6 shows a closed box with a removable partition.

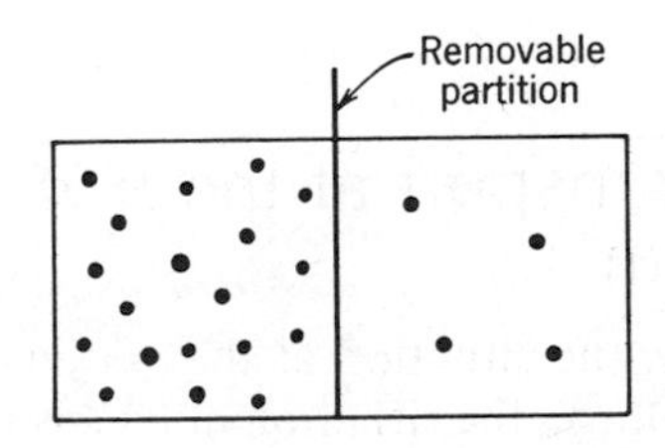

Figure 7.6
Gas molecule experiment to illustrate transport by diffusion.

In the left half there is a large concentration of gas molecules, while on the right the gas molecules exist in a very small concentration. In each half of the box the molecules are moving with their thermal velocities in a random fashion. However, since there are more molecules on the left than on the right, during a given time interval more molecules will strike and rebound off the partition from the left than from the right. If we remove the partition, those molecules that previously struck the partition will now move to the opposite half, with the result that there will be a net flow of molecules from left to right. This net flow will continue until the concentrations of molecules in each half become equal and the entire volume becomes homogeneous. At this point the *net* transfer is zero even though there is still motion of individual molecules in both directions.

This mechanism of net transport through random thermal motion in the presence of a concentration gradient is called *diffusion.* Notice that no external force, such as results from an electric field acting on a charged particle, is involved. The collisions between particles, which serve to insure that their velocities maintain a random distribution, is the only source of force involved.

A second example of net transport through diffusion is the familiar case of the spreading of a drop of colored ink placed in a container of water. Initially, the ink molecules have a high concentration in the location of the droplet, and zero concentration outside this region, producing a concentration gradient. In time the random thermal velocities of the ink molecules cause the molecules to spread throughout the container until a homogeneous mixture exists.

The net flow of gas molecules or of ink molecules constitute a diffusion current, but not an electric current. A corresponding diffusion current of charge carriers produces a net charge transport and thus an electric current. In a homogeneous material, however, where there are no gradients of carrier concentration, the random thermal motion of the charge carriers produces zero net charge transport and makes zero contribution to the conduction process. In the junction structure of Fig. 7.5, the gradients are large, and diffusion must be taken into account.

7.3.3 Charge Transport at the *p-n* Junction in Equilibrium

Let us consider, initially, the situation at the *p-n* junction of Fig. 7.5 in the absence of an applied voltage, the situation often termed *thermal equilibrium.*

As already shown, in the neighborhood of the junction plane there will exist large concentration gradients of both electrons and holes. These gradients, through the mechanism of diffusion, will produce a flow of holes from the *p*-type to the *n*-type region and a flow of electrons from the *n*-type to the *p*-type region. Recalling the experiment with the gas molecules (Fig. 7.6), one might expect this diffusion of electrons and holes to continue until their respective concentrations are equalized throughout the entire structure. This, however, is not the case. When holes diffuse from the *p*-region, they leave behind an equal number of immobile, negatively charged acceptor ions. Similarly, electrons diffusing from the *n*-region leave behind positively charged, immobile donor ions. Thus in the vicinity of the junction the diffusion of holes and electrons results in a region with excess, immobile negative charge in the *p*-type material, and a region with excess, immobile positive charge in the *n*-type material. These regions of excess, immobile charge adjacent to the junction comprise what is called a *space-charge layer* (Fig. 7.7). The space-charge regions to each side of the junctions are charged with polarity opposite to that of the mobile carriers that have diffused out of the respective regions. As the diffusion process continues and the charged regions increase in size, they exert an increasing attractive force on the majority carriers. This force opposes the diffusive flow. Thus the initial process of diffusion is seen to be self-limiting, continuing to the point where the attractive force of the unneutralized impurity ions in the space-charge layer just cancels the diffusive flow, producing zero net charge transport and zero current.

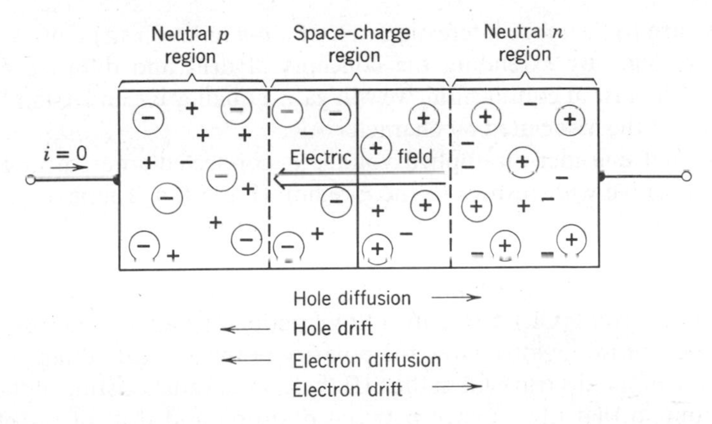

Figure 7.7
p-n junction in equilibrium.

One view of the space-charge layer is to recognize that the regions of immobile charge density produce an electric field whose magnitude is proportional to the size of the charged regions and whose direction produces a drift of electrons or holes opposite to the diffusive flow. On this basis we may view the net transport of a given carrier type as the difference between the net transport due to diffusion and the net transport due to drift. In equilibrium the drift and diffusion components for electrons and for holes exactly cancel each other, producing zero terminal current. Figure 7.7 shows the *p-n* junction in equilibrium along with the respective current components.

Because of the two-mode transport of carriers in the neighborhood of the junction, resulting from the charged nature of electrons and holes, the diffusion of electrons and holes does not spread their concentrations uniformly throughout the structure. Instead, the diffusion and drift are localized in the neighborhood of the junction. Far from the junction the *p* and *n* regions are neutral and homogeneous, unaffected by the presence of the junction. The transition from the region of space charge near the junction to regions of neutrality is not completely abrupt; it is, however, pictorially convenient to assume an abrupt transition. Thus in Fig. 7.7 the dashed lines indicate a boundary between the space-charge region and the neutral *p* and *n* regions.

7.3.4 The *v-i* Characteristic of the *p-n* junction

We now turn to the more interesting case of a *p-n* junction structure with an applied voltage. By extending the concepts of drift and diffusive carrier flow from the case at equilibrium, we will gain a qualitative understanding of the nature of the associated *v-i* characteristic.

Let us first consider an applied voltage v, connected so as to make the *p*-region positive with respect to the *n*-region (Fig. 7.8*a*). The first question is how this voltage divides between the two neutral regions and the space-charge region. Although a detailed analysis of this question is beyond our qualitative description, the applied voltage appears almost entirely across the space-charge layer (SCL), and only a tiny fraction of v appears across the *p*- and *n*-type neutral regions. Now the polarity of the applied voltage is such that it *reduces* the electric field in the SCL below the value existing in thermal equilibrium so that the balance between diffusion and drift of the charge carriers is destroyed. Specifically, since the field and the corresponding drift currents are reduced, the diffusion components dominate the charge-transport process, with a net flow of holes from the *p*- to the *n*-type region and a corresponding flow of electrons from the *n*- to the *p*-type region. As in the homogeneous bar, the net transport of holes and electrons in opposite

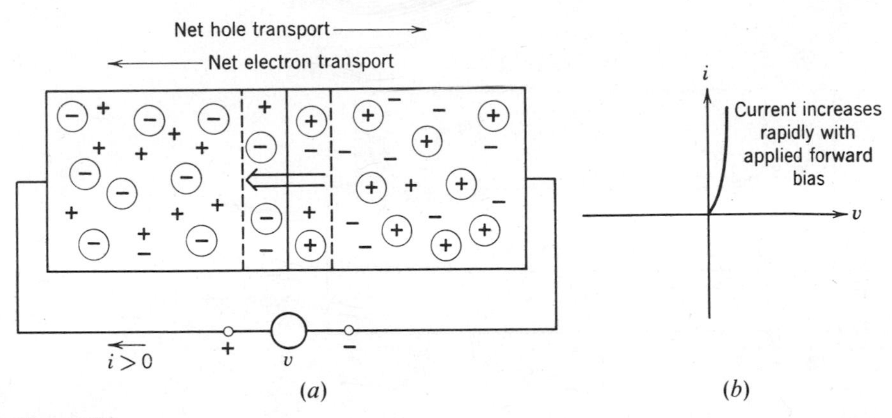

Figure 7.8
p-n junction with forward bias ($v > 0$).

directions algebraically add to establish a net positive terminal current through the structure. Since the concentration gradients are so large, it requires only small values of v, less than one volt, to produce a substantial current. The polarity of applied voltage that produces these relatively large currents is called a *forward voltage* or *forward bias*.

A forward-bias voltage reduces the field in the space-charge layer, allowing diffusion of majority carriers across the junction to the side where they are in the minority. Thus, under forward-bias conditions, the concentrations of *minority* carriers near the junction increase substantially. This increase in minority carrier concentration near a forward-biased junction, by diffusion across the junction, is called *injection of minority carriers.* (We shall see in Chapter 9 that this injection process is central to the operation of a junction transistor.)

We next consider the situation when a voltage of opposite polarity, a *reverse voltage*, is applied to the *p-n* junction (Fig. 7.9*a*). As with forward bias, the applied voltage appears almost exclusively across the SCL and changes the electric field from its equilibrium value. However, in this case the polarity is such that the applied voltage increases the electric field in the SCL. As the field increases, it opposes the diffusion of majority carriers so strongly that the diffusive components of charge transport are virtually stopped. That is, the field is directed so as to hold majority carriers in their respective neutral regions and to prevent their diffusion across the SCL. Also, the field direction is such that it attracts *minority* carriers from their respective neutral regions and moves them by drift across the SCL. The net transport of holes from the *n*- to the *p*-type regions and electrons from the *p*- to the *n*-type region now produces a negative value of i, or *reverse current*, through the structure.

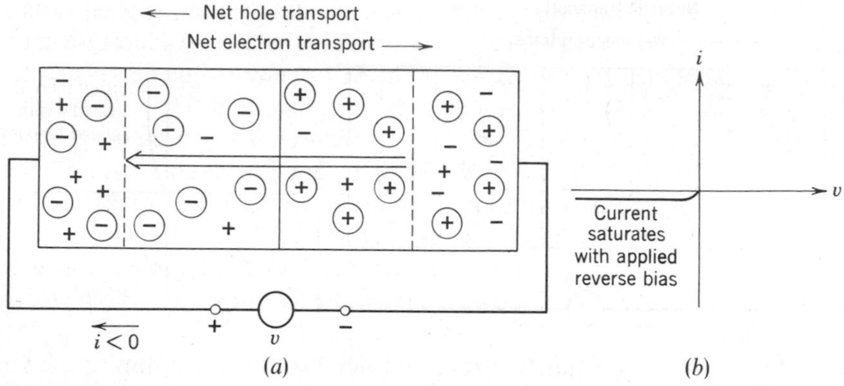

Figure 7.9

The *p-n* junction with reverse bias ($v < 0$). The increased width of the space-charge layer (much exaggerated) is an example of the depletion of carriers produced by a reverse-biased junction.

We have already seen that in neutral material the concentration of minority carriers is many orders of magnitude less than the concentration of majority carriers. Hence, the reverse current, which is proportional to the minority-carrier concentration, is very much smaller than the current obtained under forward bias. In addition, once the reverse voltage has increased to the point where the diffusion of majority carriers has ceased, typically a fraction of a volt, further increases in the reverse voltage produce little or no change in the magnitude of the reverse current (Fig. 7.9*b*). Although increasing the reverse voltage produces an accompanying increase in the electric field in the SCL, the reverse current is limited by the supply of minority carriers in the neutral regions. Thus, once the field has reached the point that it extracts all the minority carriers that the neutral regions can supply, the current becomes independent of further increases in the field strength. This constant reverse current is called the *reverse saturation current* of the *p-n* junction, since the value of current reaches a maximum or saturation value as the reverse voltage is increased.

The process of extracting minority carriers across a reverse-biased junction is called *collection.* Although the collected reverse saturation current is independent of reverse voltage, it is affected by mechanisms that cause the minority-carrier concentrations to change. Some examples of these are changes in temperature, irradiation of the semiconductor material by light or X-rays, or the injection of additional minority carriers by means of a second *p-n* junction. The last is an extremely important example, as it forms the basis of operation of the junction transistor.

Along with the changes in the electric field produced by an applied voltage v, there is a corresponding change in the width of the SCL. An increase in the field is accompanied by an increase in the SCL width, while decreasing the field decreases the SCL width (see Figs. 7.7, 7.8*a*, and 7.9*a*). Although a substantial change in width is shown in the figures, it has been exaggerated for clarity. In reality, the width of the SCL and its changes with applied voltage are usually only a small fraction of the width of the total structure. The increased width of the SCL is an example of the *depletion* of carriers that takes place near a reverse-biased junction. This depletion phenomenon is the basis of operation of the junction field-effect transistor, discussed in Chapter 9.

In summary, the *p-n* junction supports significant current flow in the forward direction (from *p*- to *n*-type material), but permits only a very small current in the reverse direction. Typical forward currents are in the range of milliamperes to well over an ampere, depending on the size of the structure and its power dissipation capability. On the other hand, reverse currents are about six orders of magnitude less, lying in the nanoampere to microampere range.

7.3.5 Ohmic Contacts and Schottky Barriers

In fabricating a semiconductor device for use in a circuit, it is necessary to attach metal contacts to the semiconductor. If the v-i characteristic of the semiconductor structure, such as a *p-n* junction, is to be faithfully retained for the device as a whole, then the junctions between the metal contacts and the semiconductor should not produce any significant modification of the v-i characteristic. Some metals produce useful contacts to semiconductors; others do not. The explanation lies in the details of the quantum mechanical energy states for the electrons in the metal, and for the electrons and holes in the semiconductor. Without attempting such an explanation here, it is possible to classify metal-semiconductor contacts into two rough categories: Ohmic contacts, and Schottky barriers.

An *Ohmic contact* is a junction with a v-i characteristic that is perfectly linear, being the electrical equivalent of a homogeneous bar (an ideal resistor). Metals that form Ohmic contacts with semiconductors are used for making contacts to semiconductor devices, because the addition of a small series resistance in the Ohmic contact does not alter the basic v-i characteristic of the device.

In some cases, depending on the specific metal and semiconductor, a space-charge layer can get set up in the semiconductor in the vicinity of a metal-semiconductor junction. The name given to this kind of junction is a *Schottky barrier*. The *v-i* characteristic of such a junction is very similar to the *v-i* characteristic of the *p-n* junction described earlier and some devices incorporating Schottky barrier junctions are now coming into use.

7.4 THE SEMICONDUCTOR DIODE

7.4.1 Circuit Symbol

A semiconductor diode is a two-terminal device containing a single *p-n* junction. The general circuit symbol for a semiconductor diode is shown in Fig. 7.10, along with its relationship to the *p-n* structure. The arrow portion of the symbol indicates the direction of the forward current. The diode terminals are labeled "anode" and "cathode," terms that are carry-overs from the era of vacuum tube diodes. Forward bias results when the anode is positive with respect to the cathode, and the diode carries a forward current. Reverse voltage requires the cathode to be more positive than the anode, and the reverse current is limited to the small saturation current.

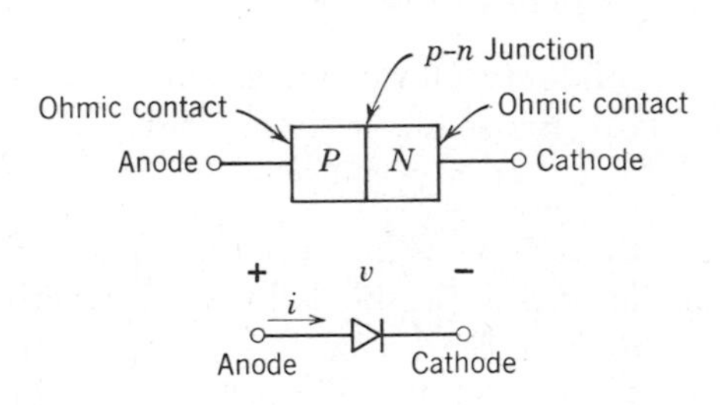

Figure 7.10
Circuit symbol for a semiconductor diode.

7.4.2 The Exponential Diode

A theoretical analysis of the *p-n* junction structure yields a single equation, shown below, which correctly describes both forward-bias and reverse-bias operation.

$$i = I_S(e^{qv/kT} - 1) \tag{7.8}$$

In the above equation, I_S is the reverse saturation current of the diode, q is the electronic charge ($q = 1.6 \times 10^{-19}$ coulombs), k is Boltzmann's constant ($k = 1.38 \times 10^{-23}$ joules/K), and T is the absolute temperature (degrees Kelvin). A convenient mnemonic is that the quantity kT/q has the dimension of a voltage, and that at room temperature, the magnitude of this voltage is

$$\frac{kT}{q} = 26\,\text{mV} \qquad \text{(at 300 K)} \tag{7.9}$$

For v much greater than 26 mV, the exponential factor $e^{qv/kT}$ must be much larger than unity; hence the "1" in Eq. 7.8 may be neglected. Thus, for forward bias with $v \gg 26$ mV, Eq. 7.8 becomes

$$\text{Forward bias:} \qquad i = I_S e^{qv/kT} \gg I_S \tag{7.10}$$

Equation 7.10 is a simple exponential relationship between i and v that agrees with our physical reasoning that the forward current is much larger in magnitude than the reverse current I_S, and that the forward current increases very rapidly with small increases in v. For example, at room temperature, with $kT/q = 26$ mV, the forward-bias current increases by a factor of 10 for every 60 mV increase in v. In reverse bias, however, with $v < -26$ mV, the exponential term in Eq. 7.8 becomes small compared with the "1" and the current reduces to

$$\text{Reverse bias:} \qquad i = -I_S \tag{7.11}$$

It is seen thus that Eq. 7.8 successfully combines a reverse saturation current $-I_S$ with a large forward-bias current.

Although the derivation of the exponential relation in Eq. 7.8 depends on some physical idealizations, the result is in good experimental agreement with actual junction diodes, at least over the major portion of their operating region (see Section 7.4.3, below). Because of the assumed idealizations, the v-i relationship in Eq. 7.8 is often called the "ideal" semiconductor diode. We wish to reserve the term "ideal" for another use, and shall therefore use the term *exponential diode* to refer to Eq. 7.8.

Figure 7.11 shows a plot of the v-i characteristic of an exponential diode for the specific choice $I_S = 10$ pA (1 pA $= 10^{-12}$ A). The circuit symbol in Fig. 7.11 is written with an I_S alongside to remind us that we are assuming the diode to be an exponential diode. Because of the linear voltage and current scales in Fig. 7.11, and because the current axis was chosen to display the characteristic all the way to a maximum current of 100 mA, the characteristic appears to have a very sharp nonlinearity. The current is zero until a particular voltage is reached, at which point i increases rapidly with increasing v. The voltage at which the current appears to depart significantly from zero

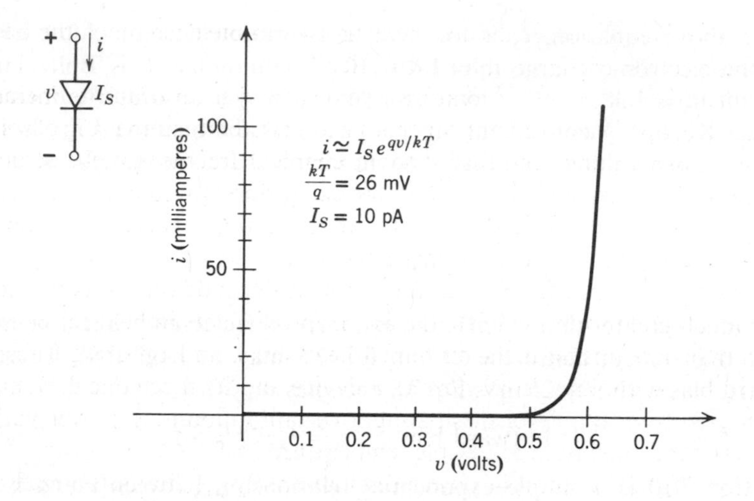

Figure 7.11

Circuit symbol and v-i characteristic of the exponential diode.

is often called the *threshold voltage* of the diode, and is typically on the order of 0.2 to 0.3 V in germanium devices and 0.6 to 0.7 V in silicon structures. Figure 7.11 illustrates the forward characteristic of a silicon diode, whose rated forward current is 100 mA and whose reverse saturation current, I_S, is 10 pA. Notice that the current appears to depart from zero at about +0.5 V. This is simply a result of plotting the exponential on a linear scale of 10 mA per division. Between reverse bias and +0.5 V forward bias the current increases from −10 pA to about +1 mA, a tremendous relative change. On the linear current scale chosen for Fig. 7.11, however, this change is not apparent.

7.4.3 Maximum Power Dissipation

In any practical application of a device care must be taken to insure that excessive voltage or current is not applied, lest the device be destroyed. In this and the following section we discuss some of the factors that establish the maximum voltage and current limits for the diode.

The arch-enemy of any electrical component is excessive heating. In resistive elements, the power dissipated in the element is converted to heat, which raises the temperature of the element above its surroundings or *ambient*. The maximum temperature that a device can withstand coupled with its ability to transfer the generated heat to the ambient sets a limit on the maximum power dissipation for the device.

The maximum device temperature is governed by several factors, including changes in the semiconductor properties with temperature, melting of solders used to fabricate the structure, and mechanical fracture of the structure due to unequal coefficients of thermal expansion. In silicon devices, the maximum temperature is limited to about 200°C, whereas germanium structures are seldom operated above 100°C.

The ability to transfer heat to the ambient depends on the construction and mechanical mounting of the device. Improvements in heat rejection are obtained by mounting units on finned heat sinks, and by using forced air or even liquid cooling. In any event, a device and its mechanical mounting arrangement is capable of dissipating a certain amount of power without exceeding the maximum allowable temperature.

The maximum allowable power dissipation $P_{D,\max}$ limits the maximum product of voltage and current in the device

$$vi \leq P_{D,\max} \tag{7.12}$$

If we plot Eq. 7.12 in the v-i plane, as in Fig. 7.12, we obtain a hyperbola in the first and third quadrants indicating the boundary for allowable power dissipation in the device. If the operating point of a diode is permitted to cross this boundary, leaving the safe operating region, thermal destruction through excessive heating will result.

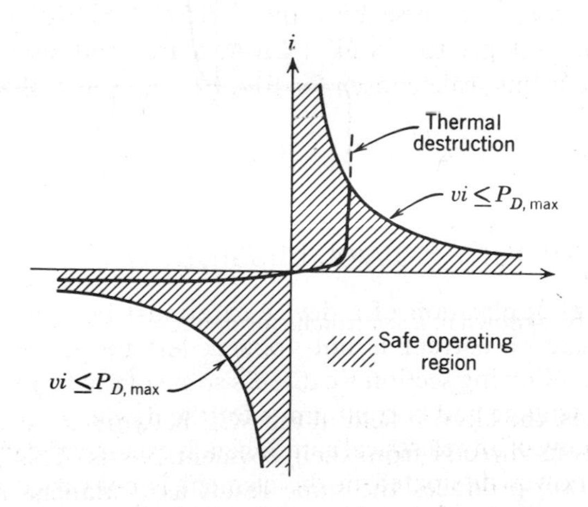

Figure 7.12
Boundaries of maximum allowable power dissipation in the v-i plane.

7.4.4 Diode Voltage Limitations

Although the maximum power dissipation sets an absolute limit on device operation, beyond which irreversible destruction takes place, there are other effects, not necessarily destructive, that will cause a diode characteristic to depart significantly from the behavior described in Section 7.3. One such effect, called *reverse voltage breakdown*, sets a limit on the reverse voltage a diode can withstand before a substantial reverse current flows.

In Section 7.3 we found that increases in the reverse voltage of a diode was accompanied by a small reverse saturation current and an increasing electric field in the space-charge layer (SCL). As the electric field increases, so does the velocity of the reverse-current mobile carriers crossing the SCL. At some point, these carriers attain sufficient speed so that through collisions they knock additional electrons from the covalent bonds in the SCL, producing both a free electron and a hole. These new carriers add to the reverse current and, through additional collisions, may themselves produce still more mobile electrons and holes. This process, called *avalanche multiplication*, produces a very rapid increase in the reverse current (Fig. 7.13). The reverse-bias voltage that the diode will withstand before reverse breakdown occurs is called the *maximum reverse blocking voltage*, V_R, of the diode.

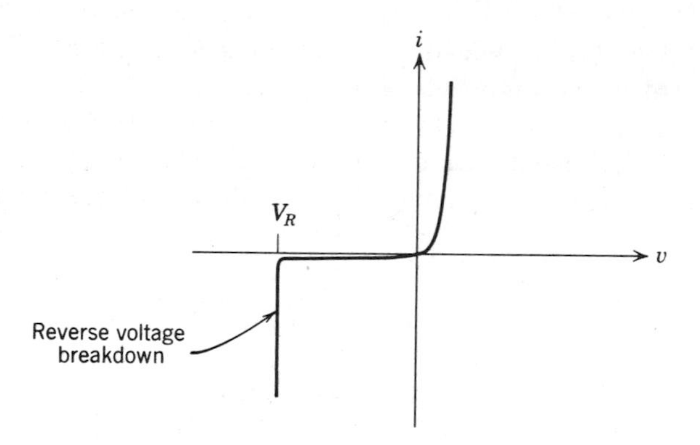

Figure 7.13
Reverse voltage breakdown in a semiconductor diode.

In some cases, the electric field in the SCL becomes so strong that it can dislodge electrons directly from their covalent bonds. This process, called *Zener breakdown*, produces the same result as avalanche multiplication; a rapid increase in reverse current once the reverse blocking voltage is exceeded. Generally, Zener breakdown is dominant in diodes that break down below six volts reverse voltage, while avalanche multiplication is dominant in diodes that break down above six volts.

Clearly, reverse breakdown drastically alters the basic v-i characteristic of a diode. Thus, if one requires a diode that blocks reverse current, one must select a unit with a reverse blocking voltage larger than any reverse voltage the diode will encounter in the circuit.

Although the terminology "reverse breakdown" implies a destructive process, this is not the case. Diodes may be operated in and out of the breakdown region with no irreversible changes, provided that the maximum power dissipation limits are not exceeded.

7.4.5 Zener Diodes

In the reverse breakdown region, a diode has the property that the voltage is nearly independent of the current. A simple linear model for a diode in reverse breakdown consists of nothing more than a battery whose voltage equals the diode breakdown voltage. Thus, if a constant voltage at some point in a circuit is required, one can employ a diode operating in the reverse breakdown region. Diodes intended for this mode of operation are called *voltage reference diodes*, or sometimes *Zener diodes*, although the mechanism of breakdown may be either Zener breakdown or avalanche multiplication. Similarly, the voltage at which breakdown occurs is often called the *Zener voltage*. Commercial voltage reference diodes are available with breakdown voltages from about 2.4 to 200 volts, providing the user with a wide range of voltages to choose from. The circuit symbol for a voltage reference diode is shown in Fig. 7.14. Frequently, the breakdown voltage is written beside the symbol.

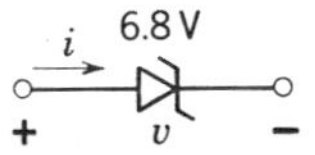

Figure 7.14
Circuit symbol for 6.8 V voltage reference diode (Zener diode).

Manufacturers often specify a minimum reverse current at which the voltage reference diode is to be operated to insure that the breakdown mechanism is well established. The maximum current limit is set by the maximum permissible power dissipation.

REFERENCES

R7.1 Paul E. Gray and Campbell L. Searle, *Electronic Principles: Physics, Models, and Circuits* (New York: John Wiley, 1969), Chapters 2 to 4.

R7.2 Paul D. Ankrum, *Semiconductor Electronics* (Englewood Cliffs, N.J.: Prentice-Hall, 1971).

R7.3 Charles L. Alley and Kenneth W. Atwood, *Semiconductor Devices and Circuits* (New York: John Wiley, 1971).

R7.4 Ben G. Streetman, *Solid State Electronic Devices* (Englewood Cliffs, N.J.: Prentice-Hall, 1972).

QUESTIONS

Q7.1 An insulator is a material in which all electrons are occupied in bonding, and none are available to conduct electric currents. Does an intrinsic semiconductor more closely resemble a metal or an insulator? What about a doped semiconductor?

Q7.2 Would you expect group IV metals such as Sn and Pb, which have four valence electrons, to be useful in doping germanium and silicon?

Q7.3 A given metal-to-semiconductor contact can supply or accept any amount of mobile charge to or from the semiconductor, as needed. Would this contact be classified as an Ohmic contact or as a Schottky barrier?

EXERCISES

E7.1 A No. 24 copper wire has a diameter of 0.02 in. What length of wire is needed to make a resistance of 25 Ω, assuming a conductivity of 6×10^5 $(\Omega\text{-cm})^{-1}$? (*Answer.* 1000 ft.)

E7.2 What length of a 0.02 in square cross-section bar of silicon, doped n-type to 10^{15} electron/cm^3, is needed to make a resistance of

25 Ω? Assume an electron mobility of 1000 cm^2/V-s. (*Answer*. 0.01 cm.)

E7.3 Plot the exponential diode relation on two different current scales: (a) 10 μA maximum; and (b) 10 mA maximum. Use $I_S = 10$ pA and $kT/q = 26$ mV. Notice that the threshold effect at ~0.6 V is pronounced in (*b*) but not visible in (*a*).

CHAPTER EIGHT
DIODE CIRCUITS AND APPLICATIONS

8.1 THE DIODE AS A NETWORK ELEMENT

The v-i characteristic of a typical semiconductor diode is shown in Fig. 8.1. The physical origin of the nonlinearities in this v-i characteristic have been discussed in the preceding chapter. In this chapter we examine, first, how to analyze networks containing nonlinear elements such as diodes, and, second, a variety of specific circuit applications involving diodes together with other elements.

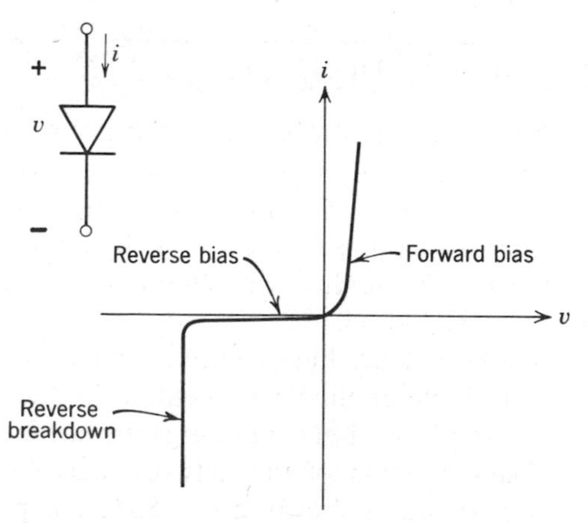

Figure 8.1
Typical diode characteristic illustrating three regions of operation.

Generally, diode applications can be classified according to which of the three regions of diode operation (see Fig. 8.1) are used. Switching and rectifying applications involve transitions between the reverse-bias and forward-bias regions. In such applications, care must be taken to choose a diode with a reverse breakdown voltage sufficiently large to prevent undesired reverse breakdown. The reverse breakdown region is employed primarily in voltage reference applications. There, the diode is chosen for the specific value of reverse voltage at which reverse breakdown occurs. We will assume that reverse breakdown can be safely ignored except in those circuits specifically employing the reverse breakdown part of the characteristics.

8.2 DIRECT METHODS OF ANALYSIS

The diode is the first network element we encounter that is strikingly nonlinear in the middle of its normal operating range. (The op-amp, of course, is nonlinear, but only near the extremes of the normal operating region.) It is now appropriate, therefore, to develop methods of network analysis that one can employ with networks containing nonlinear elements. Both KVL and KCL can be used, since their validity does not depend on the linearity or nonlinearity of the network elements. However, we must exercise caution in the use of superposition, Thévenin equivalents, and Norton equivalents, because these methods are explicitly restricted to linear networks.

8.2.1 Graphical Methods: The Load Line

One generally useful approach is to separate the linear network from the nonlinear elements and carry out a graphical solution for the voltage and current in the nonlinear elements. This method has the distinct advantage of not requiring any approximations in dealing with the nonlinear elements, because one can measure the nonlinear *v-i* characteristics in the laboratory with whatever accuracy is desired.

To illustrate the method of graphical solution, let us consider the network shown in Fig. 8.2*a*. A single nonlinear element, a diode, is connected to a network of arbitrary complexity, but containing only linear resistive elements and sources. The *linear* portion of the network may be replaced by its Thévenin equivalent network as shown in Fig. 8.2*b*. The problem now is to determine the voltage v_1 and current i_1. Once this voltage and current are known, the solution for the remainder of the voltages and currents within the

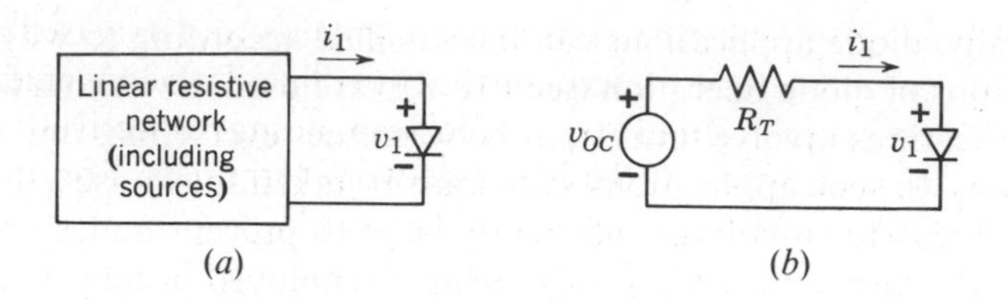

Figure 8.2
(*a*) General nonlinear network. (*b*) Linear network replaced by Thévenin equivalent circuit.

linear network involve only linear equations and can be carried out in the usual way.

The first step in the solution is to separate the linear network from the nonlinear element, as shown in Fig. 8.3*a*. Next, we determine the relationship between v_D and i_D, the *v-i* characteristic of the nonlinear element, by experimental measurement or from another source such as a manufacturer's data sheet, and plot this relationship as shown in Fig. 8.3*b*. The third step in the solution is to find the relationship between v_L and i_L, the *v-i* characteristic for the linear network. From Fig. 8.3*a*, we have

$$v_L = v_{OC} - i_L R_T \tag{8.1}$$

where v_{OC} and R_T are the Thévenin equivalent voltage and resistance of the linear network. Equation 8.1 is then plotted on the graph containing the *v-i* characteristic of the nonlinear element. Since Eq. 8.1 is a linear relationship

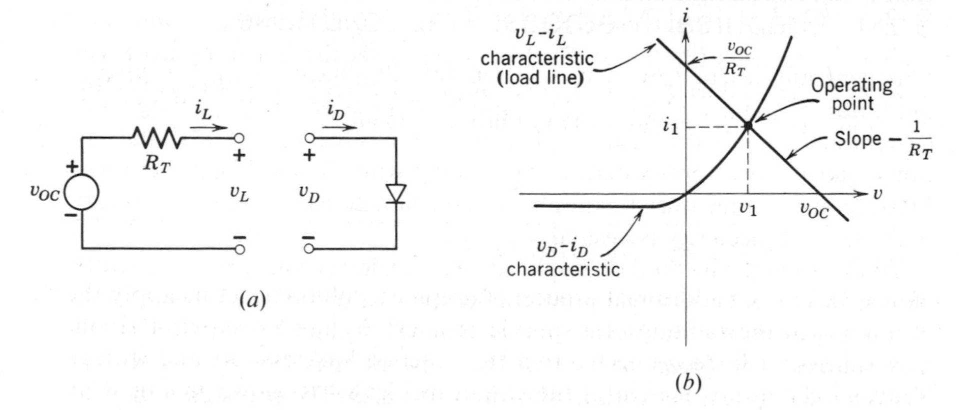

Figure 8.3
Graphical solution to nonlinear networks.

between v_L and i_L, the equation plots as a straight line, and only two points on the line need be calculated to determine the entire line. Two convenient points are the intercepts at $v_L = 0$ and at $i_L = 0$ corresponding respectively to a short circuit and an open circuit at the linear network terminals. For $v_L = 0$, Eq. 8.1 yields

$$\begin{aligned} i_L &= \frac{v_{OC}}{R_T} \\ v_L &= 0 \end{aligned} \tag{8.2}$$

whereas for $i_L = 0$ we obtain

$$\begin{aligned} v_L &= v_{OC} \\ i_L &= 0 \end{aligned} \tag{8.3}$$

These points, which are the *x*- and *y*-axis intercepts of the line in question, are indicated in Fig. 8.3*b* along with the resulting line. The slope of the line is seen to be $-1/R_T$, so that for small values of R_T the line approaches the vertical and for large values of R_T the line becomes horizontal.

If the nonlinear element is connected to the linear network, as in Fig. 8.2*b*, then we have the following circuit constraints imposed by the connection:

$$\begin{aligned} v_L &= v_D = v_1 \\ \text{and} \qquad i_L &= i_D = i_1 \end{aligned} \tag{8.4}$$

From Fig. 8.3*b* there is only one point where v_L equals v_D *and* i_L equals i_D: the intersection of the two *v*-*i* characteristic curves. Thus the required values of v_1 and i_1 must be the values of voltage and current at this intersection point, and can be read directly off the graph.

The linear *v*-*i* characteristic plotted in Fig. 8.3*b* is known as a *load line*, since it represents the locus of all possible loads the linear network can present to the nonlinear element. Also, the intersection v_1-i_1 is often called the *operating point* or *Q*-point of the nonlinear element.

8.2.2 Example Calculation

Before discussing additional aspects of graphical solutions, let us apply the technique to the solution of a specific problem. Figure 8.4 shows a circuit consisting of a diode connected to a linear network of resistors and voltage sources. The linear portion of the circuit in Fig. 8.4 is shown in Fig. 8.5*a*, and our first step is to find the Thévenin equivalent circuit for the linear network.

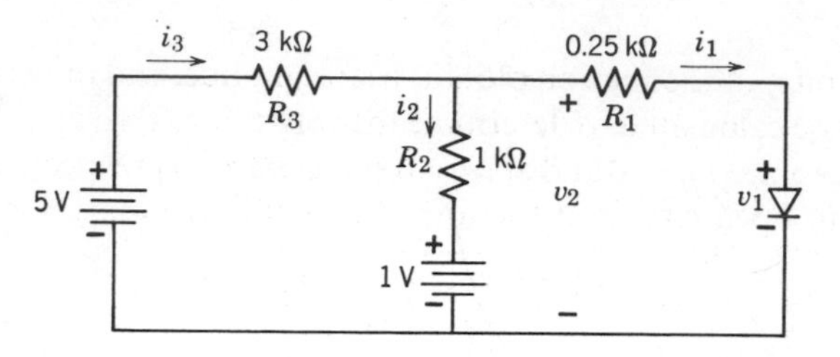

Figure 8.4
Example of a nonlinear network.

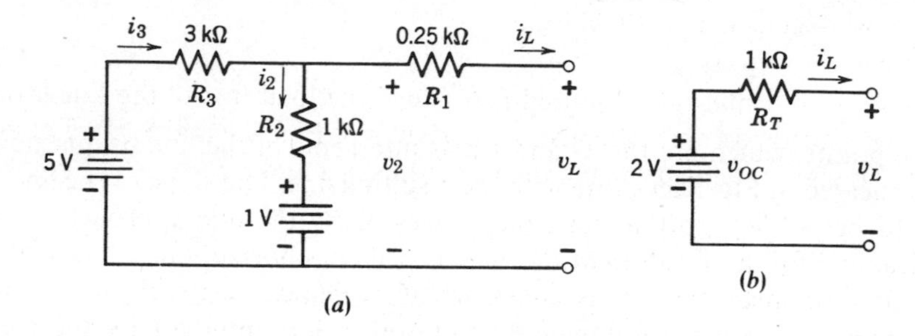

Figure 8.5
(*a*) Linear portion of network of Fig. 8.4. (*b*) Thévenin equivalent circuit for network in (*a*).

The Thévenin equivalent voltage source is equal to the open-circuit voltage at the terminals; that is, v_L under the condition $i_L = 0$. If $i_L = 0$, then $v_L = v_2$, since there can be no voltage drop across R_1. Also, with $i_L = 0$,

$$i_2 = \frac{5 - 1}{R_2 + R_3} \tag{8.5}$$

$$v_{OC} = v_2 = 1 + i_2 R_2 = 1 + \left(\frac{R_2}{R_2 + R_3}\right)(5 - 1) \tag{8.6}$$

Substitution of numerical values for R_2 and R_3 yields $v_{OC} = 2$ volts. The Thévenin equivalent resistance of the linear network is found by calculating the resistance at the network terminals with the voltage sources replaced by short circuits. Thus

$$R_T = R_1 + (R_2 \| R_3) = 1\ \text{k}\Omega \tag{8.7}$$

The complete Thévenin equivalent circuit is shown in Fig. 8.5*b*.

The graphical solution for v_1 and i_1 is shown in Fig. 8.6. The nonlinear *v-i* characteristic of the diode, obtained from experimental measurements or

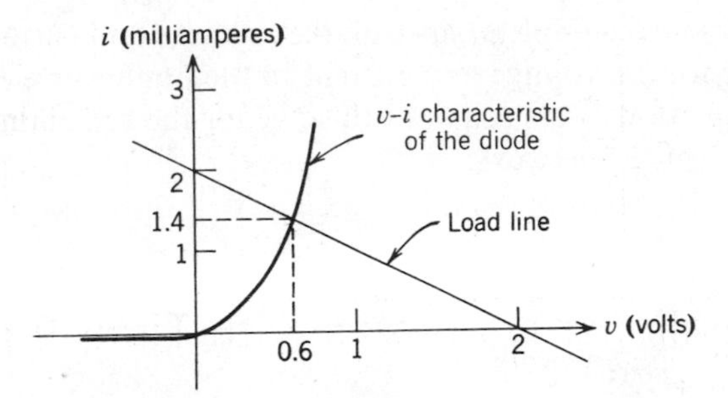

Figure 8.6
Graphical solution to network in Fig. 8.4.

data sheet information, is plotted first. Next, the load line of the linear network is plotted. In this case, using the results of Eqs. 8.6 and 8.7, we obtain two points on the line through the use of Eqs. 8.2 and 8.3:

$$v_L = 0$$
$$i_L = \frac{v_{OC}}{R_T} = \frac{2\text{ V}}{1\text{ k}\Omega} = 2\text{ mA} \tag{8.8}$$

and

$$i_L = 0$$
$$v_L = v_{OC} = 2\text{ V} \tag{8.9}$$

The load line is then drawn through these two points. Finally, the solution for v_1 and i_1 is obtained by noting the values of v and i that correspond to the intersection of the load line and the nonlinear characteristic. From Fig. 8.6 these values are seen to be

$$v_1 = 0.6\text{ V}$$
$$i_1 = 1.4\text{ mA} \tag{8.10}$$

The solution for the remainder of the voltages and currents in the network of Fig. 8.4 is now straightforward. From the graphical solution we have the values for v_1 and i_1, given by Eq. 8.10. For v_2 we write

$$v_2 = v_1 + i_1 R_1 = 0.6 + 1.4 \times 0.25 = 0.95\text{ V} \tag{8.11}$$

The currents i_2 and i_3 can now be obtained:

$$i_2 = \frac{v_2 - 1}{R_2} = -0.05\text{ mA} \tag{8.12}$$

$$i_3 = \frac{5 - v_2}{R_3} = 1.35\text{ mA} \tag{8.13}$$

The solution is now complete, since all the voltages and currents are known. Notice that once the voltage and current in the nonlinear element is found, only *linear* equations were required to solve for the remaining voltages and currents in the linear network.

8.2.3 Nonlinear Networks with Time-Varying Sources

In the preceding discussion of graphical solutions to nonlinear networks, it was tacitly assumed that the linear network contained only dc sources. Although graphical solutions are used most frequently for dc problems, the technique can also be applied to nonlinear resistive networks with time-varying sources.

If the linear network contains time-varying sources, then the Thévenin equivalent voltage source in Fig. 8.2*a* will be a function of time. However, the Thévenin resistance of the linear network remains a constant, since it is independent of the amplitude of the sources. Thus, the load line of the linear network has a constant slope, $-1/R_T$, but moves parallel to itself in the v-i plane such that the voltage axis intercept is always equal to the instantaneous Thévenin equivalent voltage. Since the load line is moving in the v-i plane, its intersection with the nonlinear v-i characteristic also moves, producing the corresponding time variation of voltage and current in the nonlinear element.

Figure 8.7*a* illustrates a time-varying Thévenin equivalent voltage v_{OC}. The axes are rotated from the normal position to facilitate the subsequent graphical constructions. At a particular instant of time, t_1, the voltage is v_1. Thus, a load line is drawn in the v-i plane with slope $-1/R_T$ and v-axis intercept v_1 (Fig. 8.7*b*). The voltage and current through the nonlinear element at time t_1 are then given by the intersection of the load line and the v-i characteristic of the nonlinear element. Figure 8.7*c* shows a plot of the current i_1 through the nonlinear element as a function of time. At time, t_1, the value of current is transferred from the intersection of the nonlinear characteristic and the load line corresponding to v_1 in the v-i plane to the current versus time plot in Fig. 8.7*c*.

This construction is repeated as often as necessary to establish the required plot of i_1 as a function of time. In Fig. 8.7, three additional times are illustrated. Notice that the slope of the load line remains constant, but the load line moves parallel to itself according to the variation of v_{OC}. Notice also that the load line appears in the third quadrant for negative values of v_{OC}, and produces a negative value of i_1 in this example.

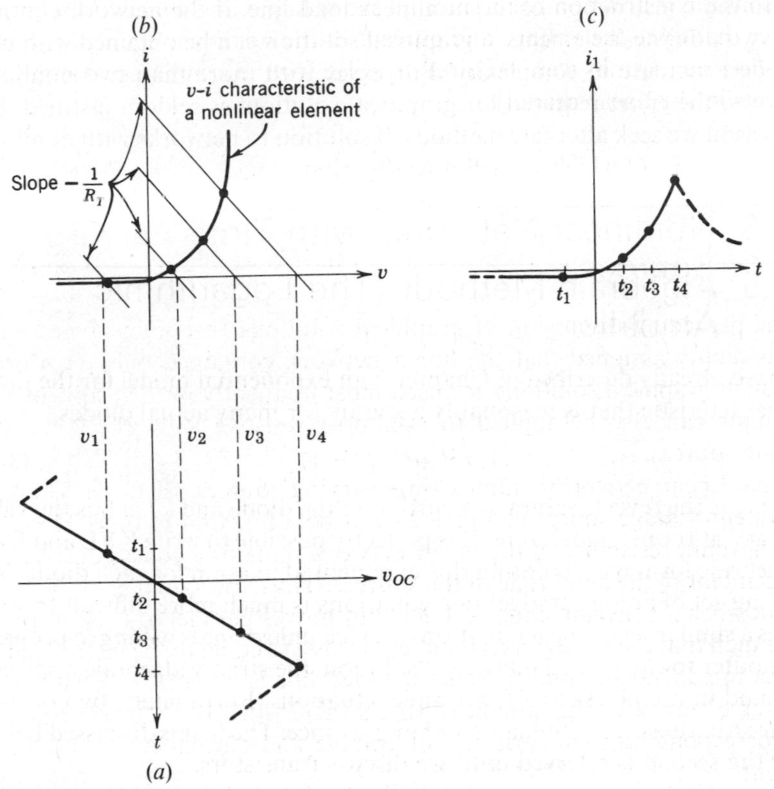

Figure 8.7
Graphical solution of a nonlinear network containing a time-varying source. (*a*) Time-varying Thévenin voltage v_{OC}. (*b*) Graphical solution with load line. (*c*) Current in nonlinear element i_1. Note the distortion of the current waveform compared to the voltage waveform.

8.2.4 Networks with More than One Nonlinear Element

If a circuit contains more than one nonlinear element, graphical solutions, while possible, become more complex. First, it is not possible to separate the circuit into a linear part and a single nonlinear element. Thus the network facing a given nonlinear element contains at least one nonlinear element, and the *v-i* characteristic at its terminals will produce a nonlinear load line. Of course, the intersection of the nonlinear load line with the remaining nonlinear characteristic produces the desired operating point; the difficulty is with the construction of the nonlinear load line. If the network contains only two nonlinear elements, a graphical solution can be obtained with only a modest increase in complexity. For cases with more than two nonlinear elements, the effort required for graphical solutions is seldom justified. For this reason we seek alternate methods of solution to networks with nonlinear elements, some of which we will now describe.

8.2.5 Algebraic Methods: The Logarithmic Amplifier

We have already described in Chapter 7 an exponential model for the diode v-i characteristic that is reasonably accurate for many actual diodes.

$$i = I_S(e^{qv/kT} - 1) \tag{8.14}$$

where I_S is the reverse saturation current of the diode and kT/q has the value of 26 mV at room temperature. It is perfectly possible to write KVL and KCL in algebraic form incorporating this exponential relation for each diode. The resulting set of nonlinear algebraic equations is much more difficult to solve than is a similar set of linear equations. In fact, unless one is willing to program a computer to carry out a numerical solution, the strictly algebraic approach is limited in usefulness to a few simple situations. Fortunately, two of these simple situations are of fundamental importance. The first is discussed below, while the second is reserved until we discuss transistors.

Figure 8.8 shows an exponential diode (I_S placed next to the diode symbol indicates an exponential diode) used as a negative feedback element around an op-amp. Because of the negative feedback, we presume initially that the op-amp is in the linear region. Since v_+ is zero, we can take v_- as zero also. Therefore the current i_1, which equals the current i_D, is given by

$$i_1 = i_D = \frac{v_1}{R} \tag{8.15}$$

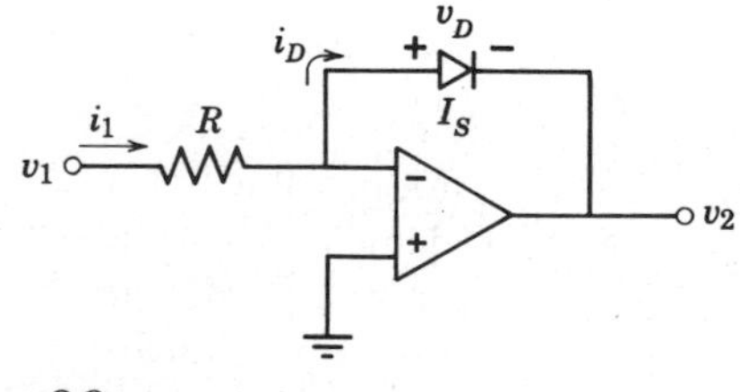

$$v_2 = -\left(\frac{kT}{q}\right)\ln\left(\frac{v_1}{I_S R}\right)$$

for $\frac{v_1}{R} \gg I_S$

Figure 8.8
A logarithmic amplifier.

If this current represents a forward bias current for the diode, that is, if i_1 is positive and much greater than the saturation current I_S, we can approximate (see Section 7.4.2) the complete exponential relation by

$$i_D \approx I_S\, e^{qv_D/kT} \tag{8.16}$$

Furthermore, writing KVL from the v_- node to the v_2 node, we obtain

$$v_2 = -v_D \tag{8.17}$$

Finally, substituting for i_D and v_D in Eq. 8.16 yields

$$\frac{v_1}{R} = I_S\, e^{-qv_2/kT} \tag{8.18}$$

which can be solved to obtain a logarithmic transfer characteristic

$$v_2 = -\frac{kT}{q}\ln\left(\frac{v_1}{I_S R}\right) \tag{8.19}$$

Thus, as long as v_1 is sufficiently positive to supply a current through R much larger than the saturation current I_S, the diode is forward biased into the purely exponential region, permitting the construction of an analog circuit that computes the logarithm of an input voltage. Such circuits are widely used in instrumentation.

Note that the logarithmic amplifier of Fig. 8.8 works only for v_1 positive. If v_1 becomes negative, the current i_1 attempts to become negative, but is limited by the small reverse saturation current of the diode. The diode becomes strongly back biased, the v_- node voltage becomes negative, and the amplifier output v_2 saturates at the $+V_{CC}$ supply.

8.3 PIECEWISE LINEAR MODELS OF DIODE BEHAVIOR

The graphical and algebraic methods of analysis presented in the previous section, while extremely accurate in their representations of the nonlinear elements, are much too cumbersome for general use. In networks with several nonlinear elements, both methods become very complex. Furthermore, in a network containing an energy storage element, the graphical technique cannot handle adequately the dynamic responses to time-varying inputs.

An alternate and very useful approach is to represent the nonlinear v-i characteristic in pieces with a set of approximate but linear v-i characteristics, each one valid over a suitably restricted range of voltage and current. We have already used this approach in discussing op-amps. The complete op-amp transfer characteristic was represented in three linear segments: a linear gain region and two saturation regions. As long as we were careful to check that the voltages and currents were in the proper ranges, it was possible to represent the nonlinear op-amp with a piecewise linear model.

In this section, we shall develop some piecewise linear models for the semiconductor diode. These models will include resistors, sources, and a new idealized nonlinear element called the ideal diode.

8.3.1 Ideal Diode

The *p-n* junction diode permits large forward currents to flow with only a small forward voltage, and supports large reverse voltages with only a tiny reverse saturation current. If we use a diode in a circuit where the voltages are large compared to the diode forward voltage and the currents are large compared to the diode reverse saturation current, then an adequate diode circuit model can be derived by neglecting both the diode forward voltage and diode reverse current, that is, by assuming they are zero. The resulting circuit element is called the *ideal diode*, and is characterized by the following set of equations:

$$v = 0 \quad \text{for} \quad i > 0 \tag{8.20a}$$

$$i = 0 \quad \text{for} \quad v < 0 \tag{8.20b}$$

The circuit symbol for the ideal diode is shown in Fig. 8.9, along with its v-i characteristic. Since the ideal diode permits no voltage when the current is positive, it reduces to a short circuit for positive current. Similarly, for negative voltage the ideal diode reduces to an open circuit. Thus, in the ideal

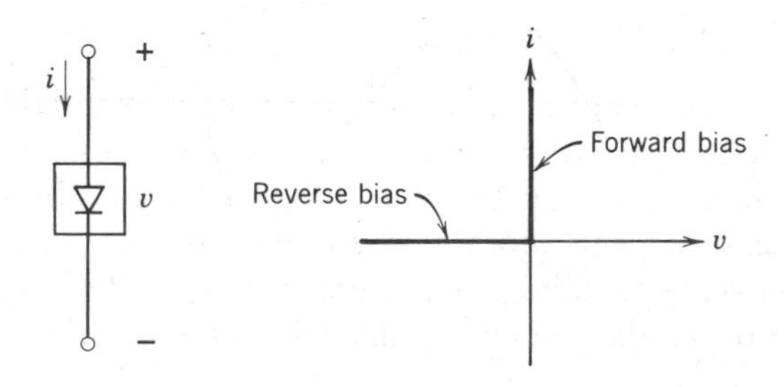

Figure 8.9
(*a*) Circuit symbol for the ideal diode. (*b*) v-i characteristic of the ideal diode.

diode model the continuous curvature of the *p-n* junction diode characteristic is eliminated. Instead, whenever the diode has a forward currrent, it is replaced by a short circuit, and whenever it has a reverse voltage, it is replaced by an open circuit.

8.3.2 Examples: Rectifier Circuits

To illustrate the use of the ideal diode model and to develop some familiarity with its application, we next consider two examples of diode rectifier circuits and employ the ideal diode model in their solution.

The key to solving any network containing ideal diodes is to determine the ranges of input variable for which the diode is an open circuit or a short circuit. Once these ranges have been determined, the diodes may be replaced by an open or short circuit, resulting in a completely linear network. One must remember, however, to use the correct linear network as the input variable changes from one range to another.

The first example is the *half-wave rectifier* circuit shown in Fig. 8.10*a*, where we have substituted the ideal diode model for the junction diode used

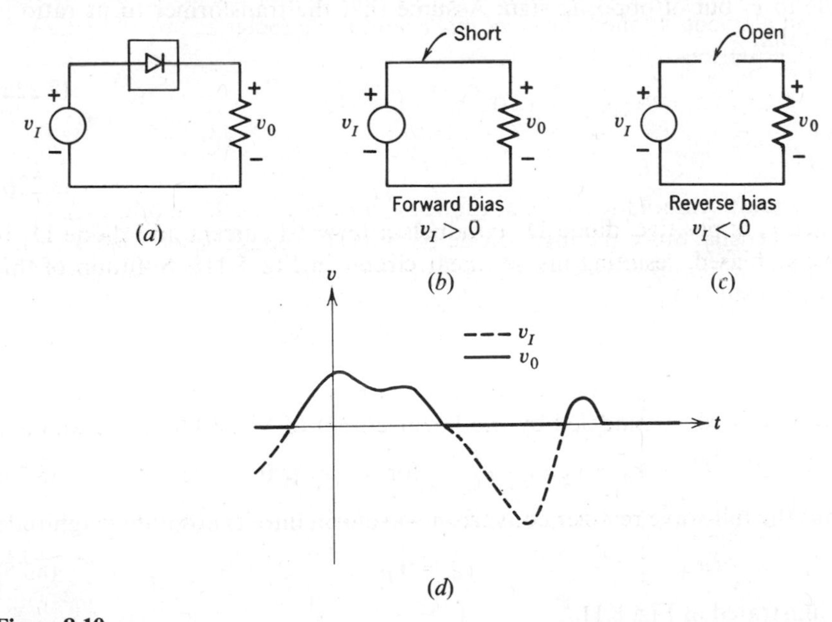

Figure 8.10
Half-wave rectifier. (*a*) Circuit. (*b*) and (*c*) Linear circuit models for restricted ranges of v_I. (*d*) Input and output waveforms.

in the actual circuit realization. We see that if v_I is positive, a forward current will flow through the ideal diode and, hence, the ideal diode may be replaced by a short circuit. On the other hand, if v_I is negative, the diode will be reverse-biased and, hence, the ideal diode may be replaced by an open circuit. Thus, we may construct two linear circuit models, one valid for $v_I > 0$ and the other valid for $v_I < 0$, to represent the behavior of the half-wave rectifier (see Figs. 8.10*b* and 8.10*c*). The problem is now reduced to the solution of two linear networks, a skill that we have already developed. From Fig. 8.10*b* we have

$$v_O = v_I \qquad \text{for} \qquad v_I > 0 \tag{8.21a}$$

and from Fig. 8.10*c*

$$v_O = 0 \qquad \text{for} \qquad v_I < 0 \tag{8.21b}$$

Thus the half-wave rectifier has the property that the output voltage equals the input voltage when the input is positive, and the output voltage is zero when the input is negative. This behavior is illustrated in the waveforms of Fig. 8.10*d*.

The second example is the *full-wave rectifier* circuit (Fig. 8.11*a*). Here, a center-tapped transformer is used to generate two voltages equal in magnitude to v_I but of opposite sign. Assume that the transformer turns ratio is such that

$$v_1 = v_I \tag{8.22a}$$

and

$$v_2 = -v_I \tag{8.22b}$$

When v_I is positive, diode D_1 conducts a forward current and diode D_2 is reverse biased, resulting in the linear circuit in Fig. 8.11*b*. Solution of this circuit results in

$$v_O = v_1 = v_I \qquad \text{for} \qquad v_I > 0 \tag{8.23}$$

When v_I is negative, diode D_2 conducts a forward current and diode D_1 is reverse biased, as modeled by the linear circuit of Fig. 8.11*c*. Here we have

$$v_O = v_2 = -v_I \qquad \text{for} \qquad v_I < 0 \tag{8.24}$$

Thus the full-wave rectifier converts a waveform into its absolute magnitude

$$v_O = |v_I| \tag{8.25}$$

as illustrated in Fig. 8.11*d*.

Although the input voltage to a half-wave or full-wave rectifier might contain no dc component (the waveform has zero time average value), the

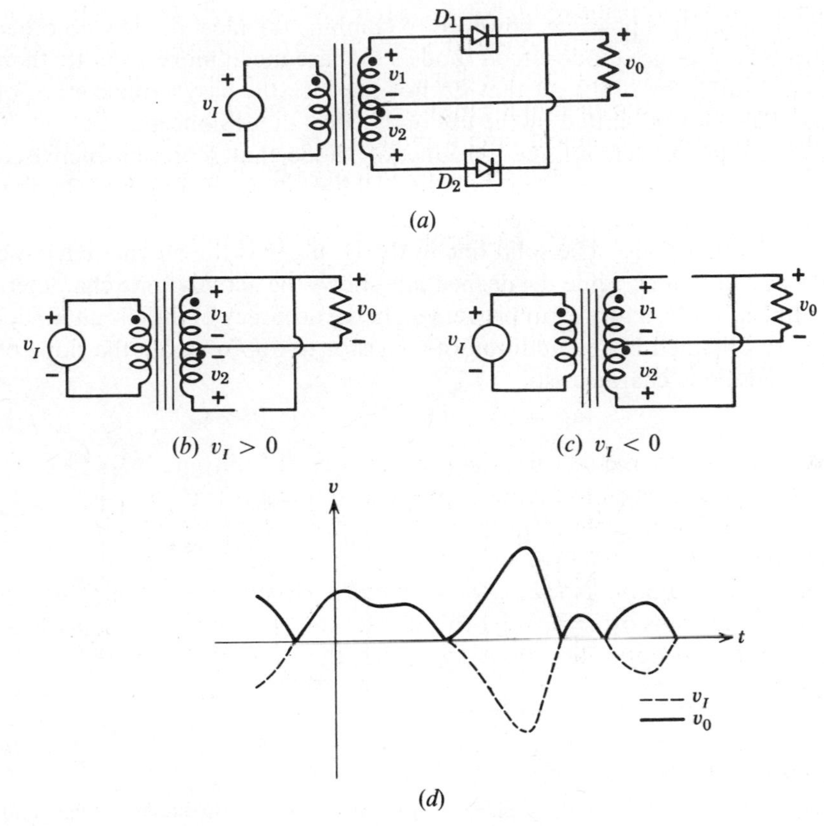

Figure 8.11
Full-wave rectifier. (*a*) Circuit. (*b*) and (*c*) Linear circuit models for restricted ranges of v_I. (*d*) Input and output waveforms.

output voltages, since they are always of positive polarity, do have a dc component. These circuits find wide applications in power supplies, where they convert the 60.Hz ac input to a dc output for powering amplifiers and similar circuits, and in signal processing when it is desired to convert an ac signal to a dc signal.

8.3.3 Improved Models for Forward Bias

In many circuits, the source voltages present are *not* large compared to the diode threshold voltage (≈ 0.6 V in silicon). In such circuits, the representation of a forward-biased diode as a perfect short circuit is too drastic a simplification. It is possible, however, to combine the ideal diode with other elements to produce diode circuit models that are much more accurate than the ideal diode by itself, yet that do not sacrifice the easy visualization of circuit behavior permitted by the use of the ideal diode concept.

Two circuit models for the *p-n* junction diode that represent increased accuracy over the simple ideal diode model are shown in Fig. 8.12. In Fig. 8.12*a*, a voltage source has been combined with the ideal diode to represent the threshold voltage. The solid line in the *v-i* plane is the *v-i* characteristic of the circuit model, while the dashed line shows the actual diode characteristic. In Fig. 8.12*b*, a further improvement in the accuracy of the circuit model has been obtained by the addition of a resistor to approximate the slope of the actual diode characteristic.

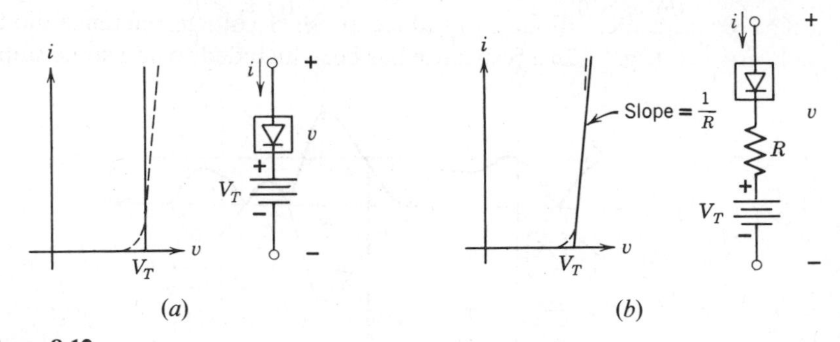

Figure 8.12

Piecewise linear models for the semiconductor diode (*a*) with threshold voltage, and (*b*) with threshold voltage and forward resistance.

In the reverse region, $v \ll -kT/q$, the current in semiconductor diodes is normally so small that it may be safely assumed to be zero in all but the most critical applications. Thus, either model of Fig. 8.12 is normally adequate for the reverse direction, since both yield zero reverse current.

It might appear initially that the relatively high degree of accuracy exhibited by the model of Fig. 8.12*b* would dictate its universal use. It is, however, impossible to decide on a model without consideration of the remainder of the circuit. For example, the piecewise linear models would fail completely to represent the exponential curve exploited in the logarithmic amplifier. On the other hand, in the case of full-wave rectifier operating with 100 volt signals, the ideal diode by itself is fully adequate. It is far better to assume the simplest, reasonable model at the outset to gain a basic understanding of a circuit function. Then the effects of successively more accurate and complex models on the circuit performance can be explored. In many cases the more accurate model will make no significant change in the circuit calculations; in some cases it will. There is no general rule that can be applied; each case must be considered separately, with experience as the best long-term guide.

8.3.4 Models for Reverse Breakdown

In the reverse breakdown region of the diode characteristic, the voltage is nearly independent of the current. Two possible piecewise linear models for the reverse breakdown region are shown in Fig. 8.13. The voltage V_Z is the magnitude of the so-called *Zener voltage*, this name being chosen for both the avalanche and Zener breakdown mechanisms. In Fig. 8.13*a* a battery is combined with an ideal diode to produce an ideal voltage reference diode characteristic. In Fig. 8.12*b* a resistance has been included to add some slope

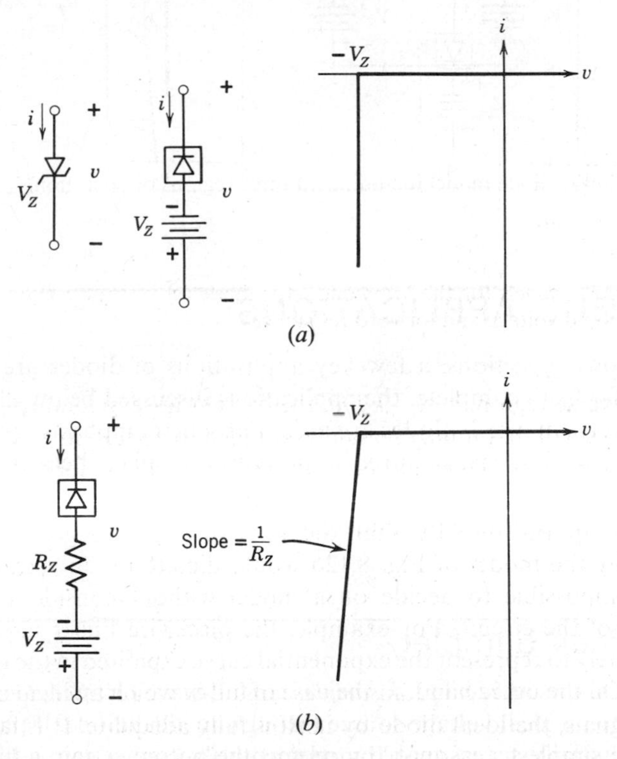

Figure 8.13
Piecewise linear models for voltage reference diode in reverse region. (*a*) Ideal reference diode. (*b*) Reference diode with Zener impedance.

to the characteristic in the breakdown region. Data sheets for voltage reference diodes often specify a value for this resistance, the *Zener impedance*, at some nominal operating point on the breakdown portion of the characteristic.

Note that both models of Fig. 8.13, although correctly approximating the reverse-breakdown and reverse-bias regions of diode operation, do not include any provision for forward bias. If a single circuit model is desired that covers all three regions of diode operation, it must contain at least two ideal diodes. See Fig. 8.14 for a simple example of such a model.

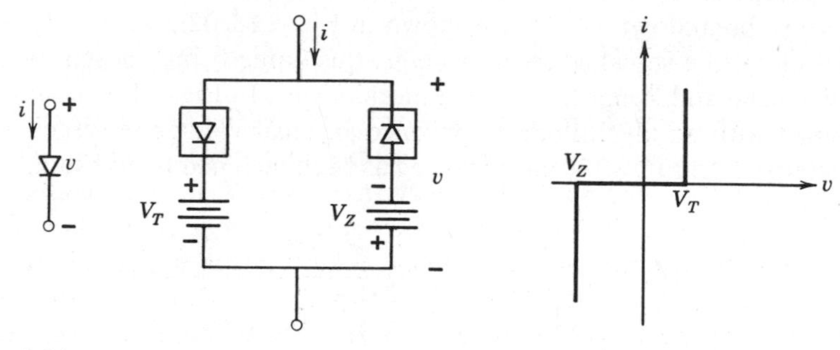

Figure 8.14
A piecewise linear diode model including all three regions of operation.

8.4 DIODE APPLICATIONS

In the following sections, a few key applications of diodes are presented. Although far from complete, the applications discussed below illustrate the variety of ways in which diodes are used. Particular emphasis is given to the behavior of rectifier circuits used in dc power supplies, because an understanding of these circuits is important to the users of these supplies, even if the supplies are purchased commercially.

8.4.1 Peak Sampler

Figure 8.15 shows an ideal *peak sampler* circuit, in which an ideal capacitor is connected to an ideal voltage source through an ideal diode. If the capacitor voltage is initially set to zero (by momentarily connecting a short circuit across its terminals), and v_I subsequently varies to a positive maximum as shown in the figure, the combined effect of the diode and capacitor is to cause

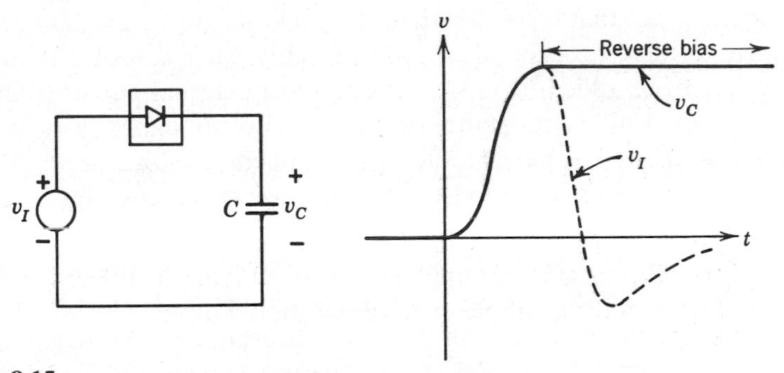

Figure 8.15
Ideal peak sampler.

the capacitor to charge up to the peak value of v_I, after which the capacitor holds this voltage indefinitely. Let us examine this process step by step.

As v_I goes positive, the diode is forward-biased. A forward-biased ideal diode is equivalent to a short circuit. Therefore the diode conducts whatever forward current is needed to keep the capacitor voltage, v_C, exactly equal to the source voltage, v_I. Once v_I reaches its maximum, however, and begins to decrease, the ideal diode becomes reverse biased because a reverse current is needed to discharge the capacitor. A reverse-biased ideal diode, however, is an open circuit. Thus, once v_I passes through its maximum, the capacitor cannot discharge, and it therefore retains its voltage indefinitely. The only way, in this idealized version of the circuit, to return the capacitor voltage to zero is to connect an external short circuit once again across the capacitor terminals.

The circuit of Fig. 8.15 is ideal. In practice, there are several ways in which actual peak samplers depart in their behavior from the ideal, as shown in Fig. 8.16. First, with an actual diode replacing the ideal diode, the capacitor

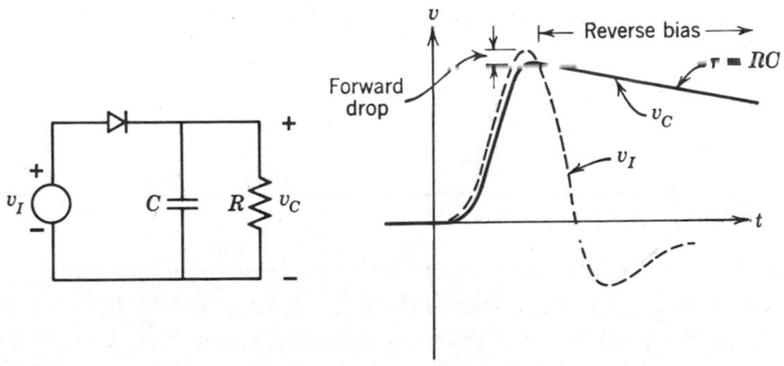

Figure 8.16
Effect of forward diode drop and load resistor on peak sampler.

charges to a voltage that is smaller than the peak of v_I by an amount equal to the forward voltage drop on the diode, roughly 0.6 V for silicon. Second, a resistor has been added to represent either an external circuit (such as a voltmeter) or the leakage resistance of the capacitor. In either case, the finite resistance shunting the capacitor allows the capacitor to discharge with time constant RC even after the diode becomes reverse-biased. Thus the peak voltage on the capacitor is less than the actual waveform peak, and it decays at a rate dependent on the current paths for discharge. (The reverse saturation current of the diode also provides a discharge path for the capacitor.)

Peak samplers in which the above errors are largely eliminated can be constructed using op-amps and feedback. They are widely used for measuring the peak amplitudes of transient waveforms. The peak sampler also illustrates very clearly how the rectifier circuit in a dc power supply works, as discussed further in the next section.

8.4.2 A Power Supply Rectifier

Figure 8.17 shows a practical full-wave rectifier circuit for converting ac power to partially filtered dc. Although the circuit looks very much like the full-wave rectifier of Fig. 8.11, the operation of this circuit is most easily understood in terms of the peak sampler. The center-tapped transformer produces two time-varying waveforms of opposite polarity, v_1 and v_2.

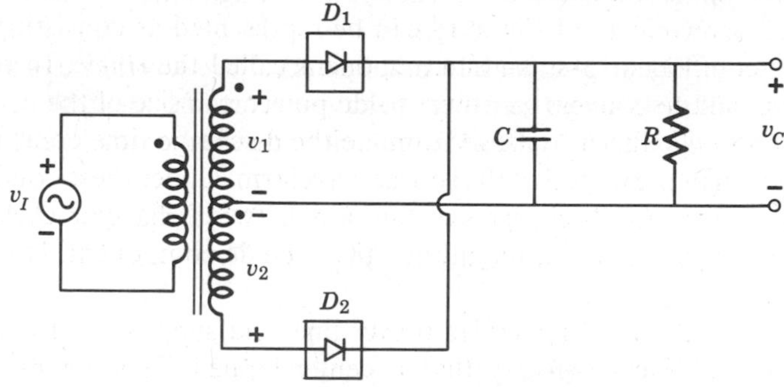

Figure 8.17
A full-wave rectifier with capacitor driven by a sine-wave source.

Whenever v_1 exceeds the capacitor voltage v_C, diode D_1 will conduct, charging up the capacitor, and whenever v_2 exceeds v_C, diode D_2 will conduct (see the waveforms of Fig. 8.18). Thus the combination of D_1, D_2, and the

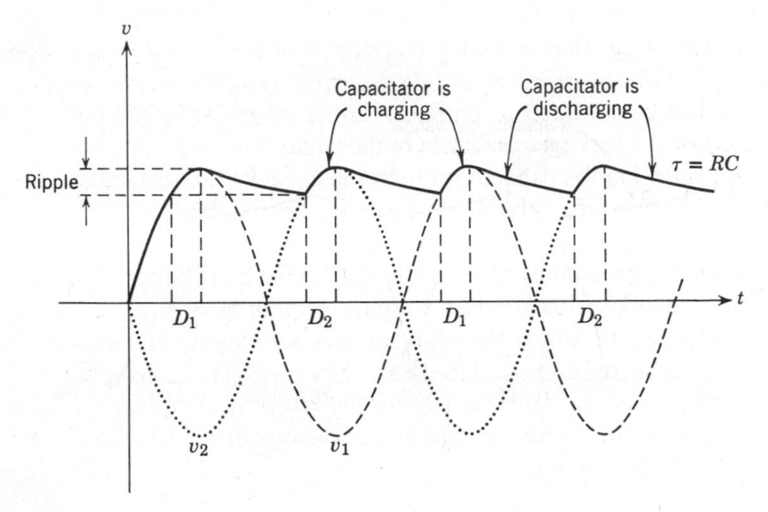

Figure 8.18
Waveforms for the full-wave rectifier.

capacitor acts like a symmetrical peak sampler, charging up the capacitor whenever the magnitude of either source voltage exceeds v_C. During the time intervals when neither v_1 nor v_2 exceed the capacitor voltage, the capacitor discharges toward zero volts through the load resistance R with a time constant RC. Thus the output waveform is a pulsating dc waveform, resulting from alternate cycles of charging through D_1, discharging, charging through D_2, and discharging.

The output waveform of Fig. 8.18 can be represented as consisting of an average dc component plus an ac component called the *ripple*. In rectifier circuits designed to convert ac power to dc power, the size of the capacitor is usually chosen sufficiently large to make the discharge time constant RC much longer than the period of the source waveform. Under these conditions, the capacitor does not discharge substantially between charging cycles, and the ripple is much smaller in magnitude than the dc component. Here is an example.

Suppose v_I is the 115 Vac, 60 Hz power line, and suppose that the transformer has a 24 Vac secondary that is center-tapped. This means that v_1 and v_2 are equal but opposite 60 Hz sinusoids with rms amplitudes of 12 Vac. Thus, since $V_{peak} = \sqrt{2}V_{rms}$,

$$v_1 = -v_2 = (\sqrt{2} \times 12)\cos(120\pi t) \tag{8.26}$$

The peak amplitude is $V_{peak} = 16.8$ V. During each charging portion of the cycle, therefore, v_C charges up to 16.8 V (see Fig. 8.19).

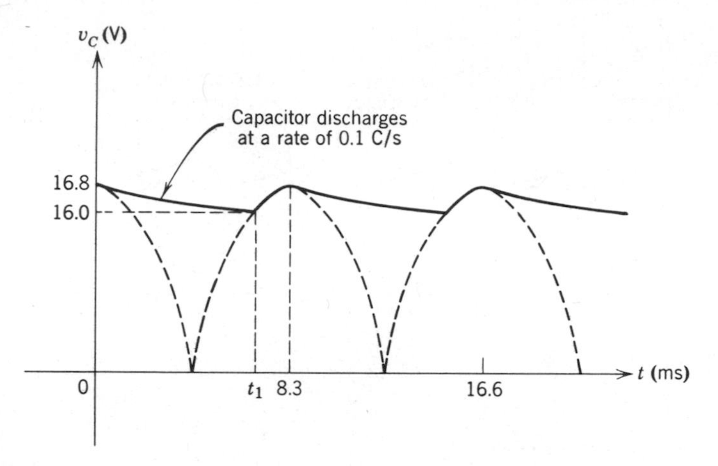

Figure 8.19
The full-wave rectifier.

Let us suppose that the rectifier is delivering an average dc load current of $I_R = 0.1$ A through the load resistor R (i.e., R is adjusted to a value near 160 Ω so that it carries 0.1 A dc). Let us also suppose $C = 1000\ \mu F$. The RC time constant is then on the order of $160 \times (1000 \times 10^{-6}) = 0.16$ s, which is about 10 times the 16.6 ms period of the 60 Hz sinusoid, and about 20 times the 8.3 ms period of the 120 Hz full-wave rectified waveform in Fig. 8.19. Therefore the discharge of the capacitor is very slight over one discharge cycle and can be represented as a constant rate of 0.1 A or 0.1 C/s. The duration of the discharge part of the cycle, represented by t_1 in Fig. 8.19, is nearly equal to the period of the 120 Hz rectified wave. Approximately, therefore, the change in the capacitor voltage during discharge is

$$
\begin{aligned}
v_{ripple}(\text{peak-to-peak}) &= \frac{I_R \cdot t_1}{C} \\
&\simeq \frac{(0.1)(8.3 \times 10^{-3})}{10^{-3}} = 0.83 \text{ V}
\end{aligned}
\tag{8.27}
$$

That is, when delivering 0.1 A to the 160 Ω load, the capacitor discharges by 0.83 V between charging cycles. The peak-to-peak ripple is then 0.83 V and the average dc voltage is 16.4 V.

If the load current were to be increased, the ripple would become larger while the average dc component of v_C would become smaller. Thus, the dc component of the rectifier output depends strongly on the size of the load. Such a supply is said to be *unregulated*.

8.4.3 Diode Clamps and Limiters

Since diodes become strongly conducting when the forward voltage exceeds the threshold, diodes can be used to *limit* or *clamp* the excursion of voltages in a network. In the circuit of Fig. 8.20, the diode will conduct a forward current whenever the input voltage v_I exceeds the battery voltage V. Thus the linear circuit of Fig. 8.20*b* is valid for $v_I > V$, and yields

$$v_O = V \qquad \text{for} \qquad v_I > V \tag{8.28}$$

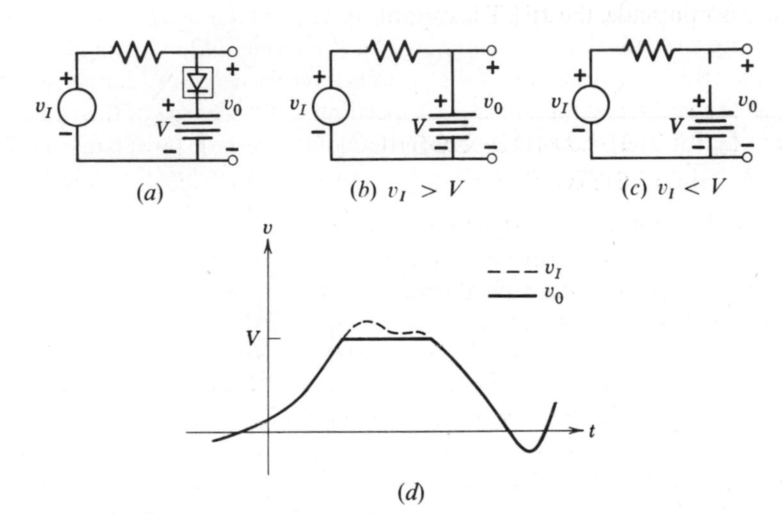

Figure 8.20
Diode limiter. (*a*) Circuit. (*b*) and (*c*) Linear circuit models for restricted ranges of v_I. (*d*) Input and output waveforms.

When the input voltage is less than the battery voltage, the diode is reverse-biased, yielding the circuit of Fig. 8.20*c*. In this case

$$v_O = v_I \qquad \text{for} \qquad v_I < V \tag{8.29}$$

Thus the function of the limiter is to prevent the output voltage from ever exceeding the battery voltage, as shown with the waveforms in Fig. 8.20*d*.

By reversing the diode, the limiter can be made to prevent the output voltage from ever becoming less than the battery voltage. Also, by reversing the polarity of the battery, limiting can be made to occur for negative values of output voltage. This circuit and its variations are used as protection circuits to prevent burnout of semiconductor devices through excessive voltage, and to produce waveforms of constant amplitude for FM and digital signal applications.

Figure 8.21 contains an example of the use of a diode as a protection device. A dc source is connected to a relay coil (modeled as an inductor). Since the diode is back-biased, all the current flows in the inductor. When the switch is opened, the current in the inductor must drop rapidly to zero, producing a strongly negative transient in v_L. If an alternate path is not provided for this current, the transient will get big enough to produce a spark across the switch (this principle is used to ignite the spark plug in an automobile engine). The diode of Fig. 8.21 provides an alternate path for the current, limiting the negative excursion of v_L to the threshold voltage of 0.6 V. Sparks across the switch are prevented, and the current in the inductor decays smoothly, dissipating the inductive stored energy in the diode. Often, a small resistor is put in series with the diode. This permits a larger negative excursion of v_L, but has the combined advantages of speeding the discharge of the inductive stored energy, and protecting the diode from excessive power dissipation. A diode used in this connection is often called a "free-wheeling" diode.

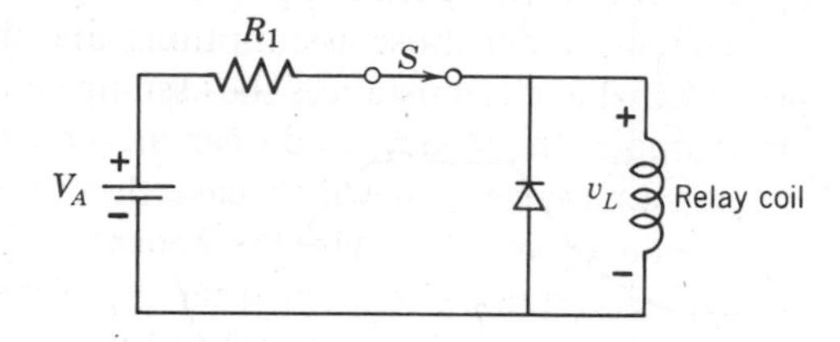

Figure 8.21
The "free-wheeling" diode limits the negative excursion of v_L when the switch S is opened.

8.4.4 A dc Power Supply Employing a Zener Diode Regulator

As a final example, let us examine the circuit of Fig. 8.22, which includes not only a full-wave rectifier circuit, but also a Zener diode used to limit or regulate the output voltage. Because this circuit contains more than one nonlinear element, extreme care must be exercised in analyzing its performance. For example, in Figure 8.19 while the peak voltage on v_C is 16.8 V regardless of the rest of the circuit, the amount of ripple *and* therefore the *average* dc component of v_C depends on the magnitude of the total rectifier load current. But in Fig. 8.22, since the circuit consisting of R_1, the Zener diode, and R_L has a nonlinear element, we cannot represent it as a simple Thévenin equivalent resistance. To understand such a circuit we must make

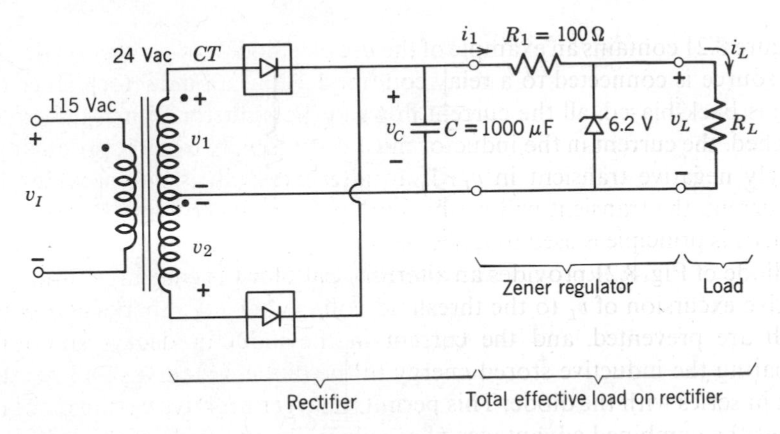

Figure 8.22
The use of a Zener diode to produce a regulated dc power supply.

reasonable assumptions about the operating points of the various nonlinear elements, analyze the circuit under those assumptions, and then check back to find whether and under what circumstances the assumptions were justified.

In Fig. 8.22, for example, the Zener diode can never become forward-biased. Either i_1 is greater than i_L, in which case the Zener is in reverse breakdown, or i_1 is equal to i_L, in which case the Zener is in reverse bias and does not conduct. Let us assume, therefore, that the Zener is in reverse breakdown, analyze the operation of the circuit, and then see what range of values of the load resistor R_L are consistent with this assumption.

Let us suppose that the Zener diode has a 6.2 V reverse breakdown voltage, and a Zener impedance of 10 Ω. Then the effective load on the rectifier can be represented by the linear network of Fig. 8.23*a*. Furthermore, because of the small Zener impedance and the relatively larger value of R_1, an *approximate* Thévenin equivalent for the circuit of Fig. 8.23*a* can be constructed, as shown in Fig. 8.23*b*. This extremely simple equivalent is accurate to about 10% so long as the value of R_L is not any smaller than about 60 Ω. By using this simple equivalent, one can construct another approximate circuit, shown in Fig. 8.23*c*, from which an estimate of the rectifier load current i_1 can be made.

We know that v_C charges up to a peak voltage of 16.8 V during the charging cycle. Therefore, just at the peak of the v_C waveform, the current i_1 has magnitude

$$i_1(\text{peak}) = \frac{16.8 - 6.2}{100} = 0.106 \text{ A} \tag{8.30}$$

This is the same magnitude of load current we analyzed in Section 8.4.2.

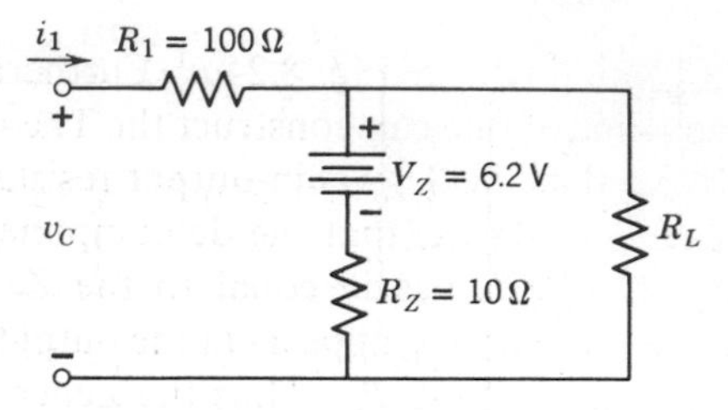

(*a*) Reverse breakdown equivalent circuit

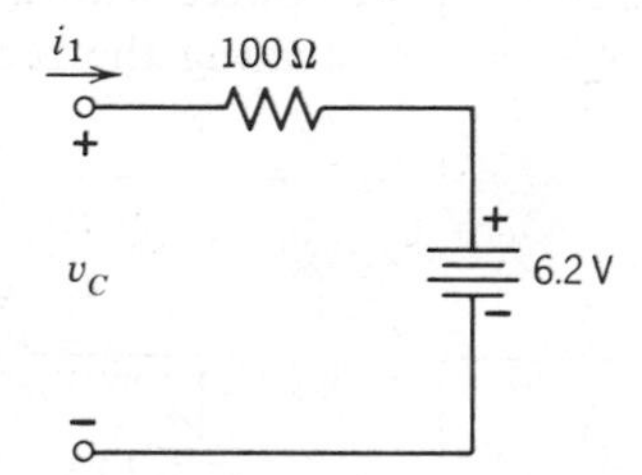

(*b*) Thévenin equivalent for (*a*) accurate to 10%

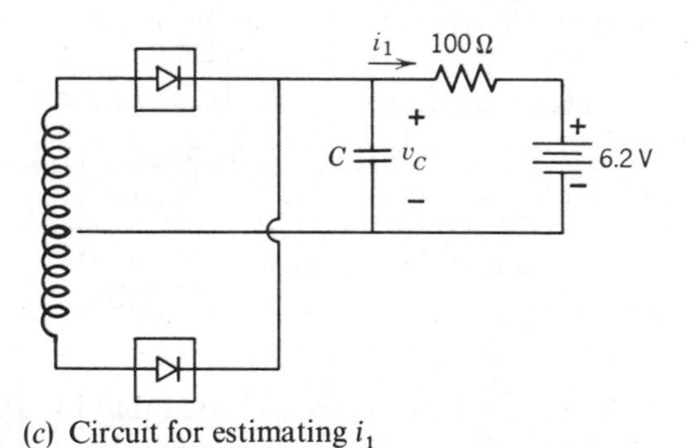

(*c*) Circuit for estimating i_1

Figure 8.23
Equivalent rectifier load assuming that the Zener diode is in reverse breakdown.

There we saw that with a load current of about 0.1 A, the capacitor discharge was very slow and could be modeled by a constant rate. Furthermore the ripple in this case was 0.83 V peak-to-peak, and the average dc component of v_C was therefore 16.4 V. Thus, by replacing the rectifier load with an approximate equivalent circuit, we have succeeded in finding an approximate operating point for the rectifier, and we know that the waveform v_C will look like that shown in Fig. 8.19.

Since we know now how v_C will behave as a function of time *under the actual loading conditions present in the circuit*, we can for analysis purposes replace the rectifier-capacitor combination by a voltage source that has the proper waveform for v_C, as shown in Fig. 8.24*a*.[1] Furthermore, since the circuit of Fig. 8.24*a* is now linear, one can construct the Thévenin equivalent shown in Fig. 8.24*b*. Notice that the Thévenin output resistance is approximately equal to the Zener impedance, that the dc component of the open-circuit Thévenin voltage (7.1 V) is nearly equal to the Zener breakdown voltage, and that the ripple present in v_C appears at the output but attenuated by the voltage divider between R_Z and R_1. Thus the Zener limiter has had the desirable effects of producing a dc source with a relatively low source resistance and, at the same time, attenuating the rectifier ripple to about 80 mV peak-to-peak.

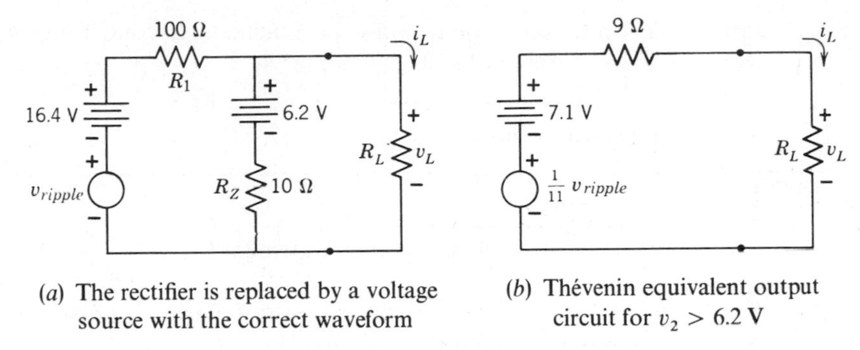

(*a*) The rectifier is replaced by a voltage source with the correct waveform

(*b*) Thévenin equivalent output circuit for $v_2 > 6.2$ V

Figure 8.24
Circuits for finding i_L and v_L.

In deriving the circuit of Fig. 8.24*b* we assumed that the Zener diode is in reverse breakdown. This is equivalent to assuming that the load voltage v_L is greater than 6.2 V. We can now translate our assumption about the Zener diode operating point into a restriction on the magnitude of R_L. If v_L must be greater than 6.2 V, the dc drop across the 9 Ω Thévenin resistance must be less than 0.9 V. Therefore the load current i_L must be less than 0.1 A. To keep i_L below 0.1 A, the load resistor must be chosen greater than $6.2/0.1 = 62\ \Omega$. If the load resistor is made smaller than 62 Ω, the current through the Zener diode drops to zero, and the Zener enters the reverse-bias region. In that case, the operation of the circuit reduces to the unregulated full-wave rectifier discussed in Section 8.4.2, but with an equivalent load resistance of $R_1 + R_L$.

[1] Note. If the loading conditions were to change substantially, for example, by replacing the 6.2 V Zener diode with a 7.5 V Zener diode, then it would be necessary to reanalyze the rectifier circuit to find a new v_C waveform.

REFERENCES

R8.1 Paul E. Gray and Campbell L. Searle, *Electronic Principles: Physics, Models, and Circuits* (New York: John Wiley, 1969), Chapter 6.

R8.2 Paul D. Ankrum, *Semiconductor Electronics* (Englewood Cliffs, N.J.: Prentice-Hall, 1971), Chapter 4.

R8.3 Jacob Millman and Herbert Taub, *Pulse, Digital, and Switching Waveforms* (New York: McGraw-Hill, 1965), Chapters 7 and 8.

R8.4 James J. Brophy, *Basic Electronics for Scientists*, Second Edition (New York: McGraw-Hill, 1972), Chapter 4.

QUESTIONS

Q8.1 Is a diode an active or passive network element?

Q8.2 Make up several circuits that illustrate the failure of the superposition method for nonlinear elements.

Q8.3 Examine any piece of commercial line-operated electronic equipment (e.g., radio, television, or phonograph). Try tracing the 60 Hz power line to the power transformer (there will be a switch and a fuse along the way), then trace the transformer output to the rectifier diodes and filter capacitor. *Caution.* Be sure the equipment is not plugged in when you explore power input connections.

Q8.4 Is the waveform distortion (or shape change) that is illustrated in Fig. 8.7 a necessary consequence of nonlinearities in one or more circuit elements?

Q8.5 Does the ideal diode consume power? What about actual diodes?

EXERCISES

E8.1 Plot an exponential diode characteristic for the case $I_S = 10$ pA (10^{-11} amps) and $kT/q = 26$ mV. Use a current scale that extends to 100 mA. *Hint*. The diode relation can be rewritten in the form

$$v = \left(\frac{kT}{q}\right) 2.3 \log \left(\frac{i + I_S}{I_S}\right)$$

At what forward voltage does the diode appear to "turn on"?

E8.2 Use your graph from E8.1 to determine graphically the current and voltage waveforms indicated in Fig. E8.2. Is the sine-wave shape modified by the nonlinearity?

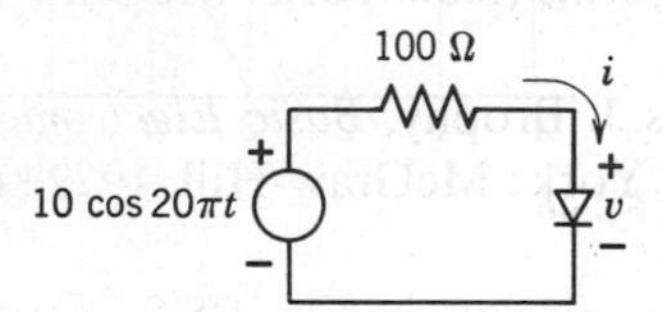

Figure E8.2

E8.3 Use your graph from E8.1 to determine choices for V_0 and R_0 in Fig. E8.3*b* to match the exponential diode characteristic in the vicinity of $i = 30$ mA.

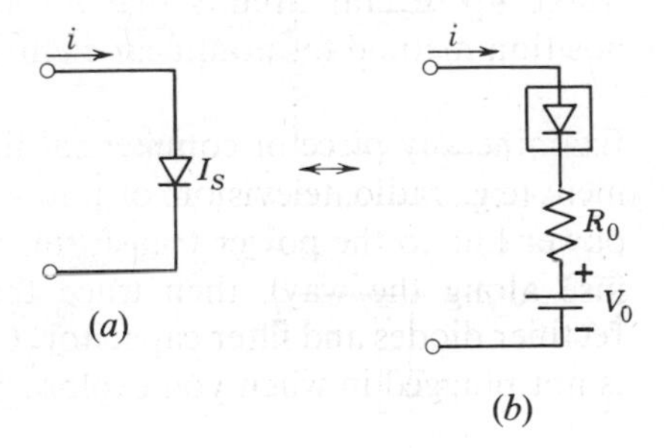

Figure E8.3

E8.4 The terminal characteristics of network *A* are plotted in Fig. E8.4*b*. Determine the operating point current, i_A, in Fig. E8.4*a* for $v_0 = 12$ V and $R_1 = R_2 = 2\ \Omega$.

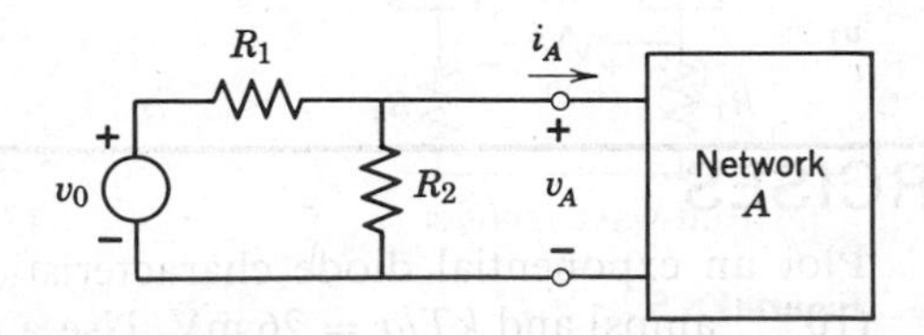

Figure E8.4*a*

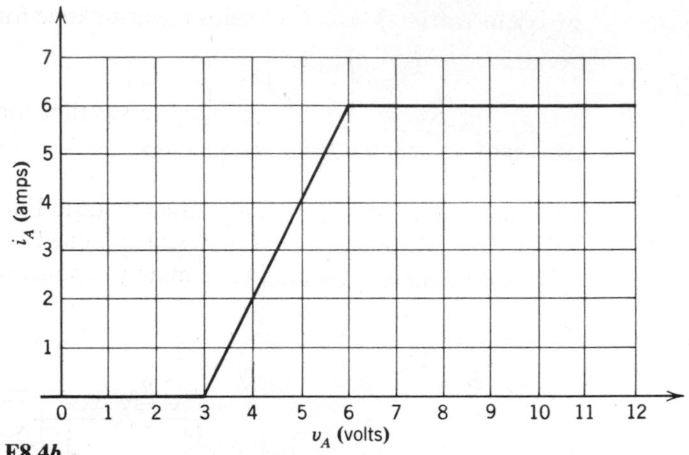

Figure E8.4*b*

E8.5 For the circuits in Fig. E8.5 (i) specify the ranges of v_1 for which each diode is ON (short circuit) or OFF (open circuit), (ii) sketch and dimension the v_1-i_1 characteristic, and (iii) sketch and dimension the v_2-v_1 transfer characteristic.

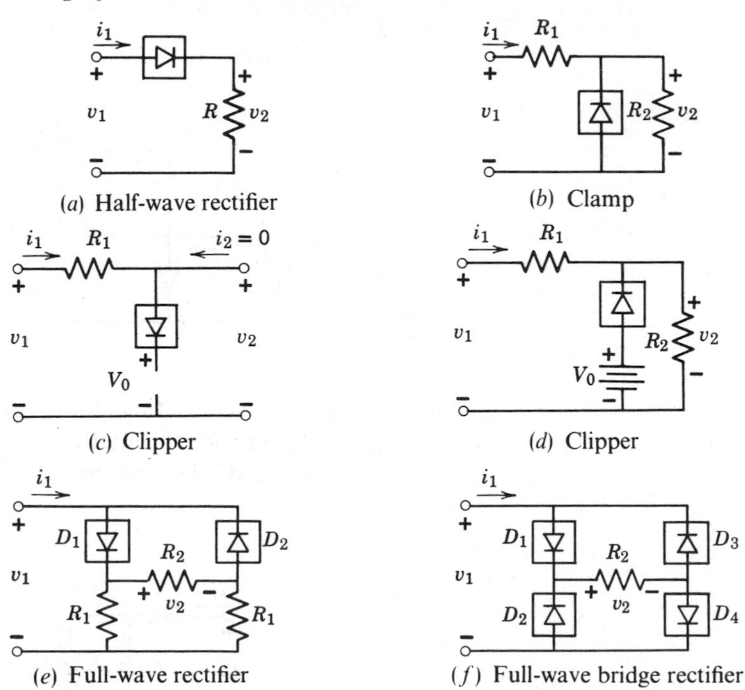

Figure E8.5

E8.6 For each circuit in Fig. E8.5, plot i_1 versus time for $v_1 = A \sin \omega t$. (Assume $A > V_0$ in *c* and *d*.)

E8.7 For each circuit in Fig. E8.5, plot v_2 versus time for $v_1 = A \cos \omega t$. (Assume $A > V_0$ in *c* and *d*.)

E8.8 Sketch and dimension the v_1-i_1 characteristic in Fig. E8.8, assuming the diode is ideal. How is your result modified if the diode is an actual junction diode (i.e., approximately exponential)?

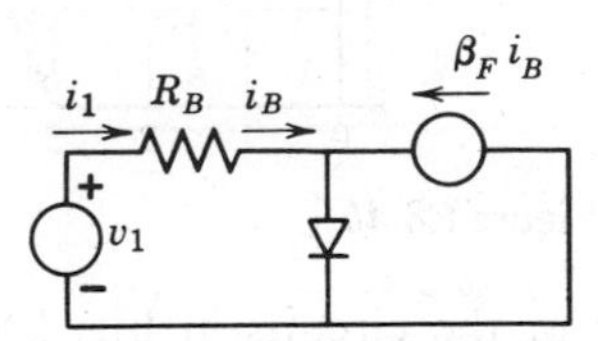

Figure E8.8
Transistor model: input

E8.9 Sketch and dimension the v_2-i_2 characteristic for Fig. E8.9 for the following choices of I_S (assume $\beta_F = 50$).

$$I_S = 0, 1, 2 \text{ mA}$$

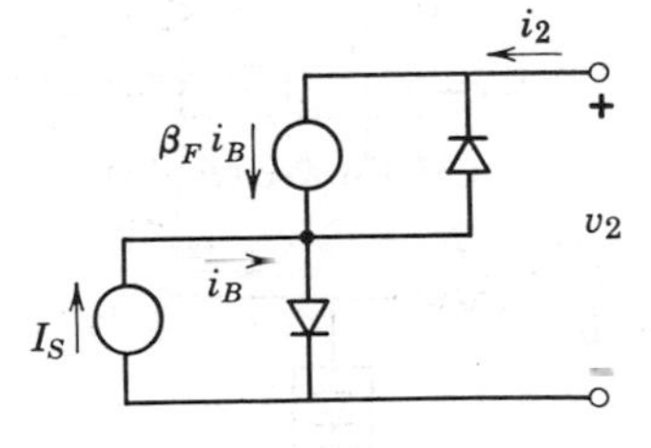

Figure E8.9
Transistor model: output

E8.10 Sketch and dimension i_1 and v_2 versus time for $v_1 = A \cos \omega t$ (Fig. E8.10). Assume the diodes are ideal, but then describe the modification to your result if the diodes are not ideal. Assume $\omega RC = 10$.

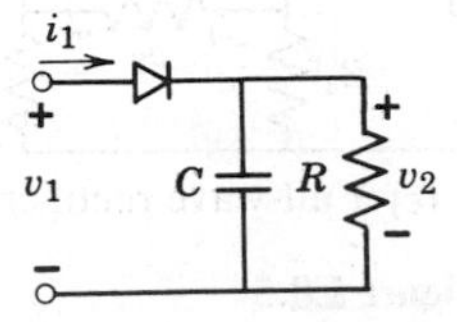

Figure E8.10
Peak sampler

E8.11 Sketch i_1 and v_2 versus time in Fig. E8.11 for when v_1 is a square wave of period T and amplitude $\pm A$. Assume $RC = 10T$. Repeat for $RC = T/10$. Discuss the difference between the two results.

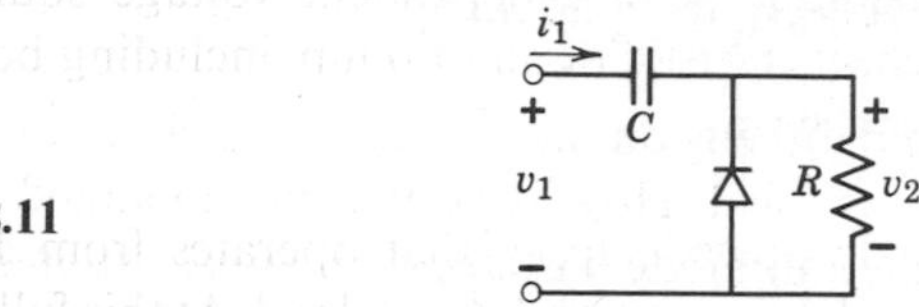

Figure E8.11
Clamp

PROBLEMS

P8.1 One way of reducing the ripple in capacitive-input rectifier circuits is to increase the size of the capacitor. Refer to Exercise E8.10, and discuss what happens to the peak current through the diode (i_1) as the size of the capacitor increases. How might this place a practical limit on the capacitor size?

P8.2 Op-amps and feedback can be used to "sharpen up" the nonidealities of diode circuits. Determine the function of each network in Fig. P8.2, and describe how the high-gain op-amps in conjunction with feedback networks reduce the effects of the forward

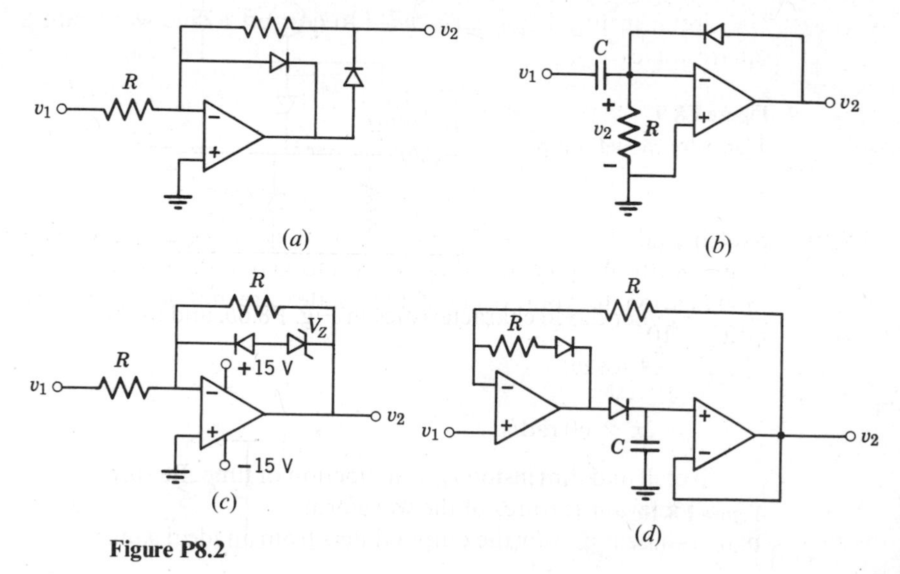

Figure P8.2

voltage drop in a diode. (*Hint.* There is one precision clipper, one precision clamp, one peak sampler, and one precision half-wave rectifier.)

P8.3 Use ideal diodes and a dependent voltage source to model the transfer characteristic of an op-amp, including both the linear and saturation regions.

P8.4 Design a dc power supply that operates from 115 Vac at 60 Hz and that will deliver up to 1 A to a load. At this full-rated current the dc voltage across the load should be 15 V, and the peak-to-peak ripple should not exceed 0.2 V. Determine peak-voltage and peak-current ratings and average power ratings for the components in your design. Assume that any Zener diodes employed in your design have Zener resistances of 5 Ω.

P8.5 Power supplies can be characterized by their response to variations in the 60 Hz line voltage (line regulation) or in the load current (load regulation). For the simple rectifier circuit of Section 8.4.2, determine:

(i) The percent change in dc voltage for a 20% change in ac input voltage.

(ii) The percent change in dc voltage for a change in load current from no load (0 mA) to full load. Assume a full load of 200 mA.

P8.6 The circuit in Fig. P8.6*a* is *intended* to convert a sine wave into a square wave.

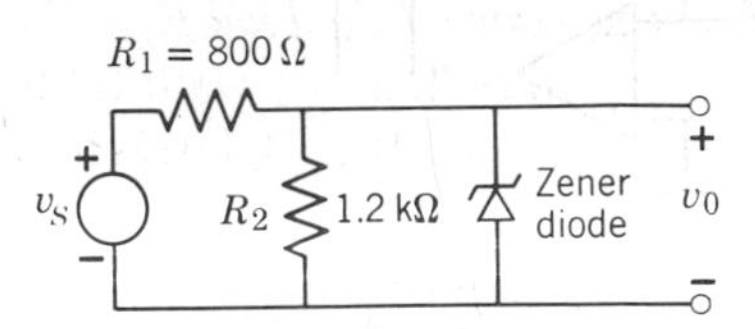

Figure P8.6*a*

Use the Zener diode characteristics in Fig. P8.6*b*, and assume

$$v_S = A \cos \omega t$$
$$A = 20\text{ V}$$
$$\omega = 2\pi \times 60\text{ rad/s}$$

(a) Sketch and dimension v_O as a function of time. Be sure to label all relevant features of the waveform.

(b) Comment on *why* the output differs from an ideal square wave.

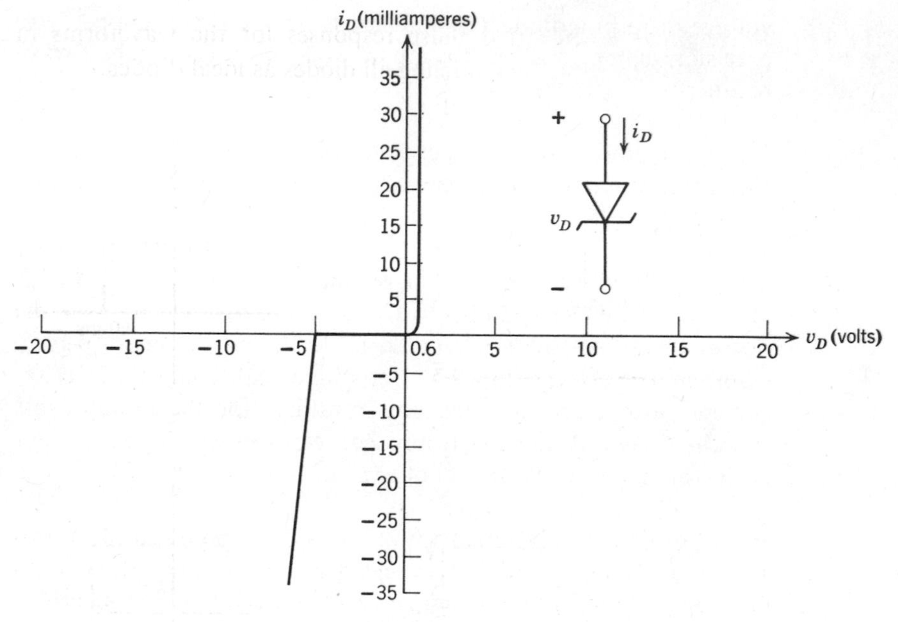

Figure P8.6*b*

P8.7 This problem examines the load and line regulation of a Zener diode power supply. For the characteristics of the Zener diode see Fig. P8.6*b*.

(a) *Assume the load R_L is not connected*

(i) Find v_O assuming $i_O = 0$ and assuming $v_S = +20$ V.

(ii) How much does v_O change if v_S changes to 18 V?

(b) *Now connect the load R_L*

(i) In real life we want to draw current from the supply into a load. If R_L (600 Ω) is connected across the output, find v_O, assuming $v_S = 20$ V. (*Hint.* Make a Thévenin equivalent of the linear part of the circuit.)

(ii) What is i_O in this case?

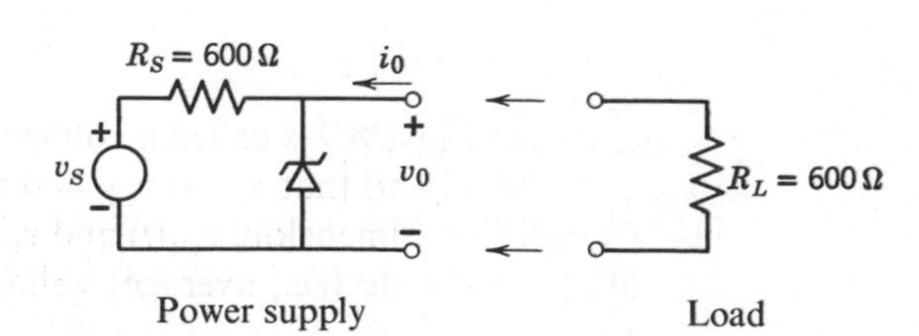

Figure P8.7 Power supply Load

P8.8 Determine the indicated pulse responses for the waveforms in Figs. P8.8*a* to P8.8*c* by modeling all diodes as ideal diodes.

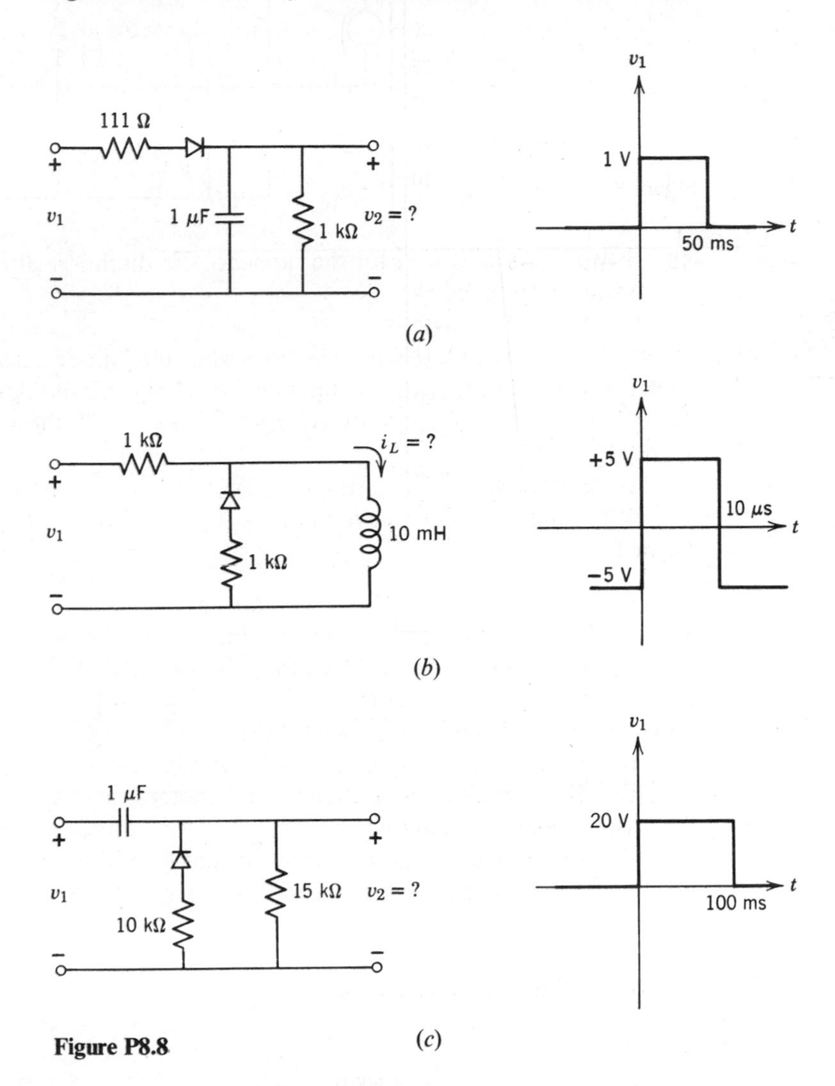

Figure P8.8

P8.9 The circuit in Fig. P8.9 is called a voltage doubler. Assume that the diodes are ideal and that $v_{C1} = v_{C2} = 0$ at $t = 0$.

(a) Sketch and dimension, $v_{C1}(t)$ and $v_{C2}(t)$ for $0 \leq t < 0.05$ s.

(b) What is the dc (i.e., average) voltage for v_{C1}? For v_{C2}? For v_o?

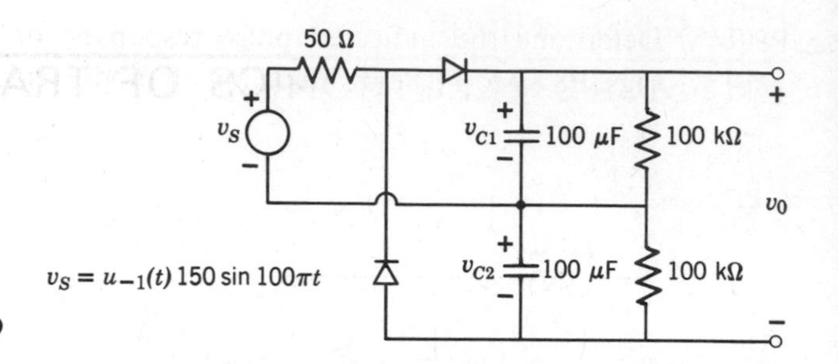

Figure P8.9

(c) How long does it take for the capacitors to discharge after the power is removed?

P8.10 Figure P8.10 illustrates the use of a "free-wheeling" diode across an inductor to prevent excessive voltages in switching circuits. Assume the switch has been closed for a long time. At $t = 0$, the switch opens.

(a) Sketch and dimension $i_L(t)$ and $v_L(t)$.

(b) What range of values of R will keep $|v_L| < 100$ V?

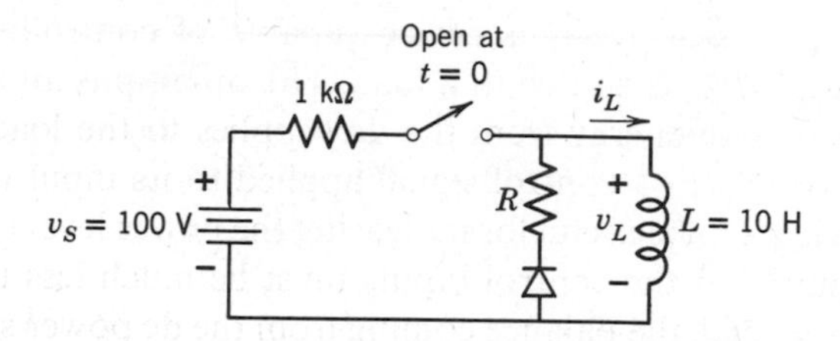

Figure P8.10

CHAPTER NINE
PHYSICAL ELECTRONICS OF TRANSISTORS

9.0 ACTIVE DEVICES AND CONTROL ELEMENTS

An *active* device is a device capable of controlling the flow of electrical energy from a source to a load. The op-amp is an active device, controlling the flow of energy from the dc supplies to the load and feedback elements according to a control signal applied to its input (see Fig. 9.1). If an active device is to be useful for increasing the power level of a signal, then the power required at the control inputs must be much less than the power delivered to the load, the balance coming from the dc power sources. The combination of the active device and its associated power source then functions as an *amplifier*, and is said to have *power gain.*

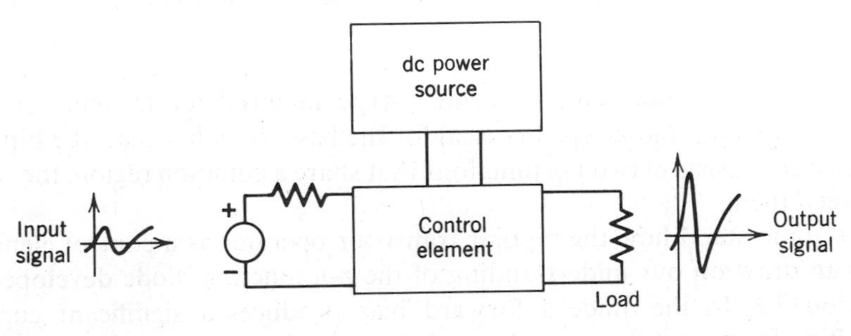

Figure 9.1
Basic amplifier configuration.

The varieties of situations requiring power gain are too numerous to list. Virtually every signal-processing or switching function involves some form of amplification or power gain. The weak signals from a radio antenna, for example, must be amplified to the point where the speech or music components of the signal can be used to drive a loudspeaker. The energy that drives the loudspeaker comes from a local power source, but is controlled by the weak signal from the antenna.

Network elements that can be used for the control of power flow are called *control elements.* Mechanical relays, vacuum triodes, and transistors all fall under this heading. To function as a control element, a device normally requires at least three electrical terminals, with the v-i characteristic at the output terminal pair being dependent on the voltage or current at the other terminals.

The transistor, in its various forms, is the most widely used active device in modern electronic circuits. Transistors may be divided into two general categories: the *bipolar* transistor, and the *unipolar* or *field-effect* transistor. In this chapter we will qualitatively explore the physical behavior of these active devices and develop their v-i characteristic curves.

9.1 BIPOLAR TRANSISTORS AS CONTROL ELEMENTS

9.1.1 The Physical Basis of Transistor Operation: Injection, Diffusion, and Collection

A typical bipolar transistor structure is shown in Fig. 9.2. It consists of a p-type central region, called the *base*, which is sandwiched between two n-type regions, called the *emitter* and the *collector*. This arrangement is known as an *n-p-n transistor*. It is also possible to construct a complementary form, the *p-n-p* transistor, by using p-type material for the emitter and collector regions and n-type material for the base. In either case, the bipolar transistor consists of two *p-n* junctions that share a common region, the base, between them.

To understand how the bipolar transistor operates as a control element, we can draw on our understanding of the *p-n* junction diode developed in Section 7.3. In the diode a forward bias produces a significant current resulting from a net flow of holes and electrons from the regions where they are majority carriers to the regions where they are minority carriers. This

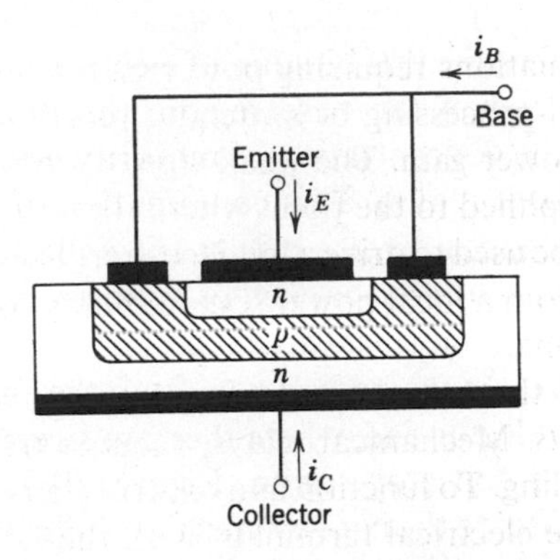

Figure 9.2
Bipolar transistor structure.

injection process is reviewed schematically in Fig. 9.3. The total terminal current i is given by the sum of the two current components arising from the hole flow and the electron flow. Under reverse bias, the *p-n* junction diode is characterized by a small reverse saturation current, arising from the *collection* of minority carriers from their respective regions by the electric field in the space-charge layer. The magnitude of this reverse saturation current depends on the available concentration of minority carriers, and is small in the diode.

The bipolar transistor works as a control element by combining *injection* at one of its *p-n* junctions with *collection* at the other *p-n* junction. In the *normal gain* or the *active gain* region of operation, the emitter-base junction is maintained in forward bias, and the collector-base junction is held in reverse bias. By doping the emitter region much more heavily than the base, most of the injection of minority carriers is made to occur into the base side of the junction. Thus, under forward-bias conditions, there is a large buildup of minority carriers (electrons in an *n-p-n* transistor) on the base side of the

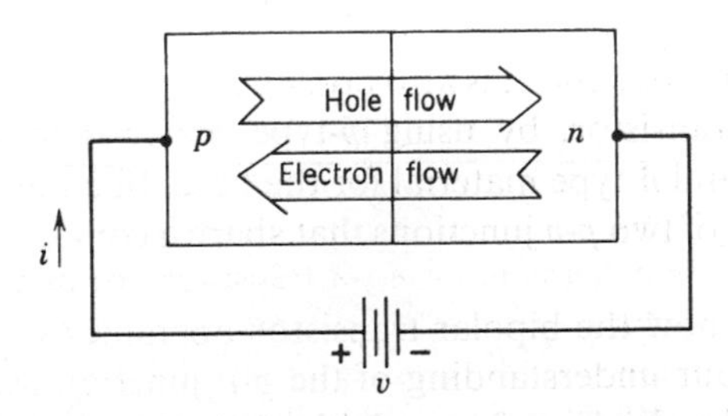

Figure 9.3
Net flow of holes and electrons in a forward-biased *p-n* junction.

emitter-base junction. While this buildup of electron concentration is taking place at the emitter end of the base region, the reverse-biased collector-base junction keeps the concentration of minority electrons very low at the collector end of the base region. This combination of a forward bias at the emitter-base junction and a reverse bias at the collector-base junction thus establishes a large *concentration gradient* of minority carriers across the base region. Normal thermal motions, therefore, produce a diffusive flow of minority electrons through the base region, from the emitter end, where they are in excess, to the collector end, where they are swept across the collector-base junction into the collector region. Figure 9.4 illustrates this flow of electrons from the emitter to the base by injection, across the base by diffusion, and into the collector by collection. Nearly all the electrons entering the base region from the emitter reach the collector. A small fraction, however, *recombine* with holes to form complete covalent bonds. Because of this recombination process, and because of the injection of holes from the base to the emitter, some holes must be supplied to the base via the base terminal. These holes cannot be supplied from the collector because they are the minority carrier there, and are few in number.

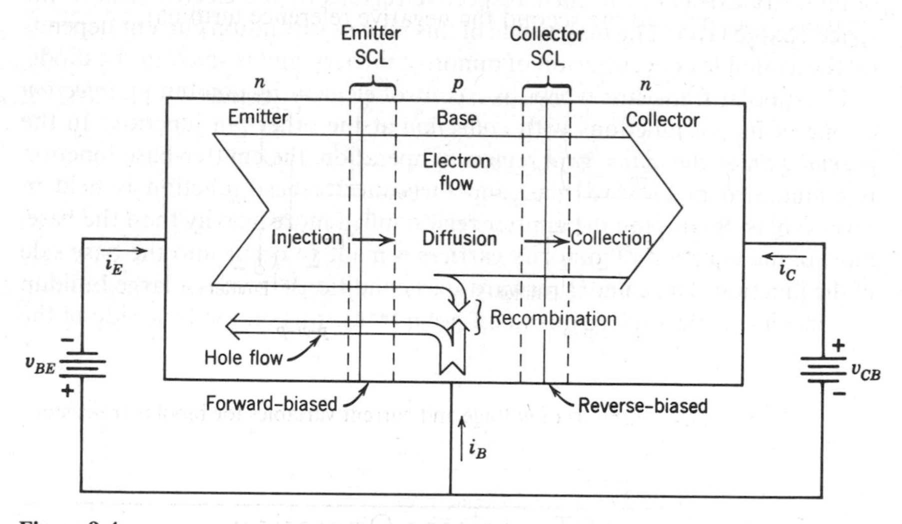

Figure 9.4
Electron and hole flow patterns in an *n-p-n* transistor operating in the active-gain region.

In summary, the collector junction behaves as a reverse-biased diode *whose saturation current is controlled by the injection of electrons at the emitter junction.* The collector current is independent of the collector-to-base voltage, provided that a reverse bias is established at the collector junction. Thus, the basic property of a control element, in this case the dependence of the output (collector) current on an input variable (emitter current or emitter-to-base voltage) has been demonstrated.

9.1.2 Circuit Symbols and Terminal Variables for Bipolar Transistors

The circuit symbols for bipolar transistors, along with their conventional terminal voltage and current designations, are shown in Fig. 9.5. The only difference between *n-p-n* and *p-n-p* device symbols is the direction of the arrow on the emitter lead, which indicates the direction of forward current in the emitter-base diode. Note that the reference directions for the terminal currents are always defined entering the device, irrespective of the direction of positive current flow. Thus at least one of the terminal currents must be algebraically negative. The voltage subscripts denote the terminals between which the voltage is measured, with the first subscript indicating the positive reference terminal and the second the negative reference terminal.

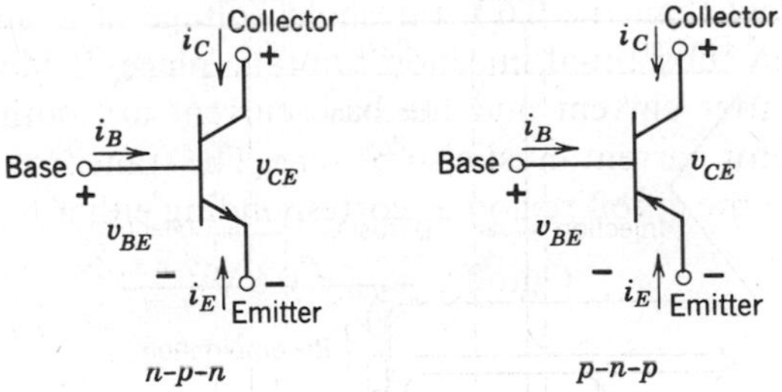

Figure 9.5
Circuit symbols and definitions of voltage and current variables for bipolar transistors.

9.1.3 Regions of Transistor Operation

Since the transistor has three terminals, there are six possible ways to connect it in a given circuit. Only three of these provide useful amplification, and of these three, only one, the *common-emitter* connection illustrated in Fig. 9.6, is ordinarily used as a vehicle for discussing the *v-i* characteristics of the device. The name common emitter derives from the fact that the emitter terminal is common to both the input and output circuits.

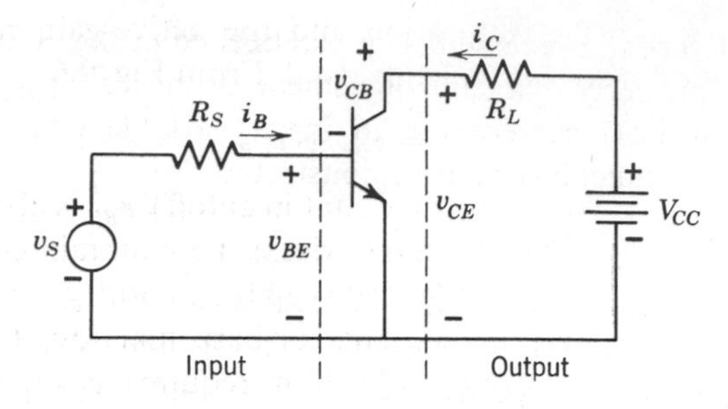

Figure 9.6
Common-emitter amplifier connection for an *n-p-n* transistor.

There are four regions of transistor operation, three of which will be discussed here, and the fourth in Section 9.1.6. The *active-gain* region has the emitter-base junction in forward bias and the collector-base junction in reverse bias. Other regions of operation occur if the emitter-base junction is not forward biased, or if the collector-base junction is not reverse-biased. The first of these is called *cutoff*, the other *saturation.*

Injection of minority carriers into the base region will only occur if the emitter-base junction is forward-biased. If, therefore, the base-to-emitter voltage v_{BE} is less than the 0.6 V threshold voltage of a (silicon) semiconductor diode, no substantial injection can take place. Under these circumstances, the emitter current and the base current are both zero,[1] and by KCL, the collector current must also be zero. The transistor is in *cutoff.* We can characterize the cutoff region as corresponding either to

$$\text{Cutoff:} \quad v_{BE} < 0.6 \text{ V} \tag{9.1}$$

or, equivalently,

$$\text{Cutoff:} \quad i_B = 0 \tag{9.2}$$

Once minority carriers are injected into the base region, their subsequent diffusion across the base and collection at the collector-base junction depends on having sufficient reverse bias on the collector-base junction. If this junction is not reverse-biased, injection into the base will take place from the collector as well as from the emitter. Under these circumstances, the transistor is *saturated.* The terminal currents in saturation are controlled by the external network and not by the transistor.

[1] "Zero" is used here because the reverse saturation currents of back-biased diodes can be considered negligible.

The boundary between saturation and the active-gain region depends on the magnitudes of the currents involved. From Fig. 9.6

$$v_{CE} = v_{BE} + v_{CB} \tag{9.3}$$

When the transistor is conducting (i.e., not in cutoff), v_{BE} is about 0.6 to 0.7 V (we shall continue to assume silicon devices). To maintain collector action, v_{CB} must be of a magnitude sufficient to keep the minority-carrier concentration low at the base side of the collector-base junction. For signal-level currents (tens of milliamperes) collection requires $v_{CB} \approx 0.3$ V, so that active-gain operation requires $v_{CE} \gtrsim 1$ V. For larger currents, on the order of amperes, it might be necessary to maintain $v_{CB} \approx 1$ V to avoid saturation, while for very small currents, it is possible to maintain collector action and avoid saturation with v_{CB} being negative by a few tenths of a volt. Because of this dependence of the saturation point on current level, transistor data sheets will list a *saturation voltage* $V_{CE,sat}$ for specific current conditions. As discussed above, the saturation voltages will usually be in the range 0.2 to 1.5 V, depending on the current conditions, and depending somewhat on the details of the transistor structure itself. The requirements for active gain operation, avoiding both cutoff and saturation conditions, therefore, can be stated as

Active-gain:

$$i_B > 0 \quad (\text{or } v_{BE} \approx 0.6 \text{ V}) \tag{9.4a}$$

and

$$v_{CE} > V_{CE,sat} \tag{9.4b}$$

9.1.4 *v-i* Characteristics

The *v-i* characteristics of a bipolar transistor can now be discussed in the context of the three regions of operation, cutoff, saturation, and active gain. Between the base and emitter terminals, the transistor looks like an ordinary semiconductor diode with one difference: the emitter and base currents have different magnitudes, with the base current the smaller of the two. Figure 9.7*a* shows a typical input characteristic, with base currents in the tens of microamperes under forward-bias conditions. The figure also shows the cutoff region corresponding to $v_{BE} < 0.6$ V.

The output characteristics of the transistor are shown in Fig. 9.7*b*. Notice that there is a family of curves, one for each value of the base current. Curves for selected values of i_B are shown in the figure. For the moment, let us concentrate on the curve labeled "cutoff," which corresponds to $i_B = 0\ \mu$A. It also corresponds in Fig. 9.7*b* to a collector current of zero, in agreement with our notion of the cutoff region. Next, we observe that the various curves

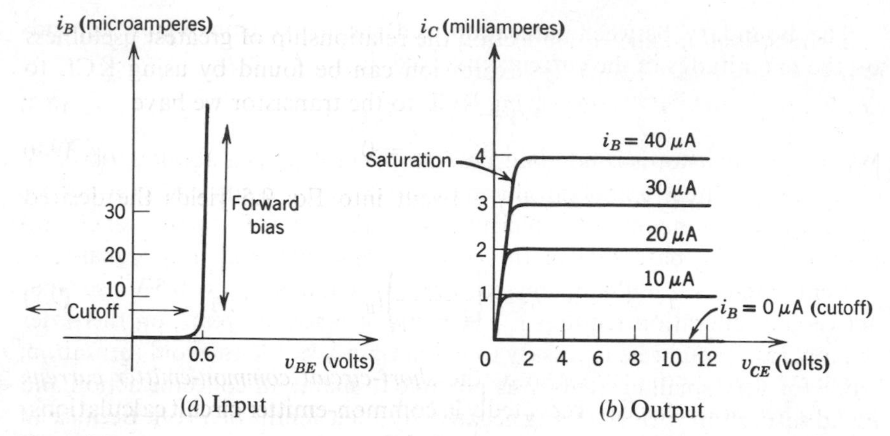

Figure 9.7
Common-emitter input and output characteristics.

for different values of i_B merge into a single curve near $v_{CE} = 0$. This is the saturation region, in which v_{CE} has become so small that reverse bias on the collector-base junction can no longer be maintained, and collector currents can flow independent of base current. The portions of the characteristics with the nearly horizontal curves depending on base current constitute the active-gain region. The reason for this multiplicity of curves is discussed in detail below.

The charge-carrier flow pattern of Fig. 9.4 shows that the collector current (opposite in direction to the electron flow) results from the substantial electron transport through the base and across the collector junction. This electron flow is proportional to but somewhat less than the emitter current. Quantitatively we can write

$$i_C = -\alpha_F i_E \tag{9.5}$$

The minus sign on the right-hand side of Eq. 9.5 arises because i_E in Fig. 9.4 is defined as positive when current flows into the emitter terminal. The proportionality constant α_F, sometimes called the *short-circuit common-base current gain*, is a positive number whose magnitude is slightly less than unity. Only that fraction of the emitter current carried by electrons produces a corresponding collector current. The fraction of i_E arising from hole flow at the emitter junction does not appear at the collector. Furthermore, not all the electrons crossing the emitter junction reach the collector, a small fraction being lost in the base region through recombination. However, these two defects can be held to such small values that for a typical transistor α_F lies in the range of 0.95 to 0.995.

In the common-emitter connection the relationship of greatest usefulness is between i_C and i_B. Such an expression can be found by using KCL to eliminate i_E from Eq. 9.5. Applying KCL to the transistor we have

$$i_E + i_B + i_C = 0 \tag{9.6}$$

Solving for i_E and substituting the result into Eq. 9.5 yields the desired expression:

$$i_C = \left(\frac{\alpha_F}{1 - \alpha_F}\right) i_B \tag{9.7}$$

The term $\alpha_F/(1 - \alpha_F)$ is known as the *short-circuit common-emitter current gain*, and, because it occurs repeatedly in common-emitter circuit calculations, it is usually denoted by a separate symbol β_F.

$$\beta_F = \left(\frac{\alpha_F}{1 - \alpha_F}\right) \tag{9.8}$$

Equation 9.7, therefore, can be simplified to yield

$$i_C = \beta_F i_B \tag{9.9}$$

Since α_F is usually quite close to unity, Eq. 9.8 shows that β_F can be quite large. The typical range of values for α_F (0.95–0.995) results in a range for β_F of 20 to 200.

Notice that in Eq. 9.9, the collector current is *independent* of v_{CE}, provided only that v_{CE} is large enough to insure active-gain operation. Therefore, a plot of i_C versus v_{CE} will be horizontal as in Fig. 9.7*b*. Furthermore, for every value of i_B, there will be a different horizontal output characteristic in the active-gain region. The practice, therefore, is to plot v_{CE}-i_C characteristics only for selected values of i_B. In Fig. 9.8*a* a graphical representation of Eq. 9.9 has been plotted for the active gain region, while for comparison purposes, the actual output characteristics of an *n-p-n* transistor are shown in Fig. 9.8*b*. The simple model of Eq. 9.9 is quite a good approximation to the actual characteristics, even including cutoff. The errors are that Eq. 9.9 does not apply to the saturation region, and that Eq. 9.9 does not adequately represent the small but nonzero upward slope of the actual characteristics.

The common-emitter output characteristics may be used to determine the parameter β_F for any transistor by calculating the ratio of i_C to i_B with data taken directly from the curves. Although it is true that the ratio of i_C to i_B will exhibit a small dependence on i_C and v_{CE} in an actual transistor, these variations are usually small enough to permit a single average quantity β_F to model the entire active-gain region. However, it is reasonable that measurements on an actual transistor should be made at or near the values of

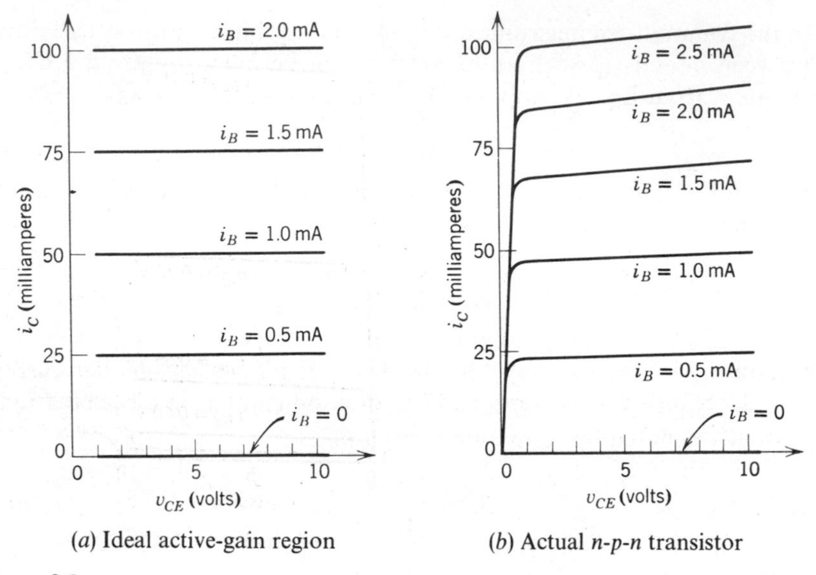

(*a*) Ideal active-gain region　　(*b*) Actual *n-p-n* transistor

Figure 9.8
Common-emitter output characteristics with i_B as the input parameter.

i_C and v_{CE} that will be encountered in the eventual circuit application, thus minimizing the effects of variations of i_C/i_B with operating point.

If the base-to-emitter voltage v_{BE} is chosen as the input control variable, the development of the corresponding *v-i* characteristic at the output terminal requires a relationship between v_{BE} and i_C. Referring to Fig. 9.4 again, we see that the emitter current will be related to the base-to-emitter voltage through the *v-i* characteristic of a forward-biased semiconductor diode

$$i_E = -I_{ES}\, e^{qv_{BE}/kT} \tag{9.10}$$

where I_{ES} represents the saturation current of the emitter-base junction of the transistor. The minus sign occurs because the defined direction of i_E is opposite to the forward current in the emitter-base diode. The substitution of Eq. 9.10 into Eq. 9.5 produces the required characteristic equation.

$$i_C = \alpha_F I_{ES}\, e^{qv_{BE}/kT} \tag{9.11}$$

As was true with base current control, operation in the active-gain region will only occur for v_{CE} greater than $V_{CE,sat}$. Figure 9.9*a* shows the resulting output characteristics based on Eq. 9.11, while Fig. 9.9*b* illustrates the measured characteristics on the same transistor used for Fig. 9.8*b*. Compare the two sets of common-emitter output characteristics, noting particularly the markedly nonlinear variation of collector current with v_{BE}.

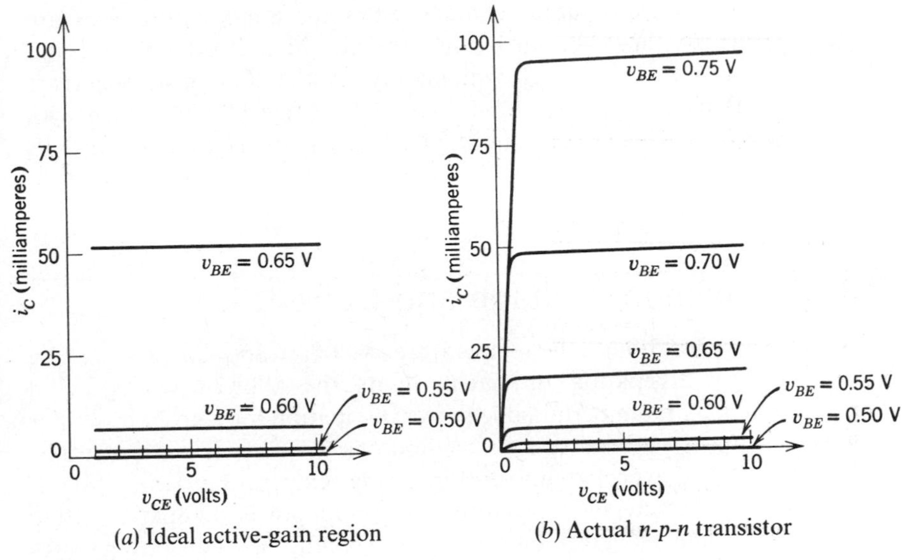

Figure 9.9
Common-emitter output characteristics with v_{BE} as the input parameter.

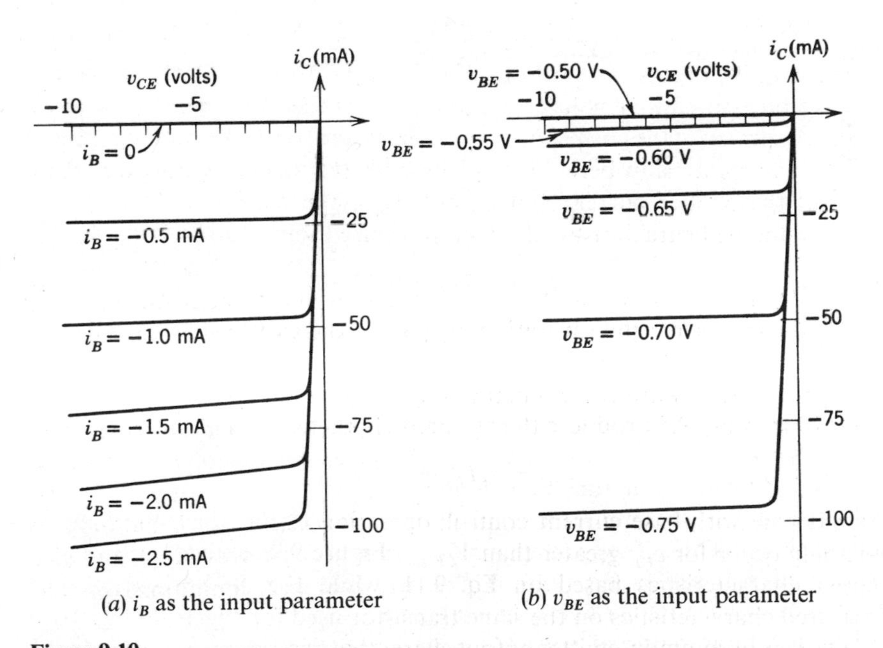

Figure 9.10
Common-emitter output characteristics of an actual *p-n-p* transistor.

The common-emitter output characteristics for a *p-n-p* transistor are complimentary to those for an *n-p-n* transistor in that all voltages and currents change sign. Figure 9.10 contains a typical set of *p-n-p* characteristic curves for both base current and base-to-emitter voltage. Compare them with the curves in Figs. 9.8*b* and 9.9*b*, noting particularly the signs of all voltages and currents.

9.1.5 Maximum Voltage and Current Limits

As was the case for diodes, the temperature rise of a transistor produced by internal power dissipation ultimately limits the allowed voltages and currents. The maximum permissible power dissipation for a given transistor will depend on its physical size, its method of construction and mounting, and the maximum expected ambient operating temperature. Evaluation of these thermal parameters will produce a maximum power dissipation rating for the device under the given set of operating conditions. This rating serves to limit the maximum internal temperature of the structure to a safe level, usually from 150 to 200°C in silicon units.

In a transistor operating in the active-gain region, virtually all the power dissipation occurs at the collector junction. Therefore, we can equate the internal power dissipation to the product of v_{CE} and i_C:

$$v_{CE}i_C = p_D \tag{9.12}$$

If we set p_D equal to the maximum permissible power dissipation, we obtain an equation for permissible values of v_{CE} and i_C.

$$v_{CE}i_C \leq P_{D,max} \tag{9.13}$$

When Eq. 9.13 is evaluated with the equal sign and plotted on the same coordinates as the common-emitter output characteristics, it produces a pair of hyperbolas, one in the first quadrant and one in the third quadrant, denoting the boundary for v_{CE} and i_C beyond which excessive power dissipation may destroy the device. Such a plot for the first quadrant characteristics is shown in Fig. 9.11 along with the safe operating regions based on the criterion of avoiding thermal destruction.

Equation 9.13 does not set an explicit limit on either v_{CE} or i_C but only on their product. Thus, if v_{CE} is kept small, i_C can become very large, and vice versa, and still satisfy Eq. 9.13. There are, however, limitations on the maximum values that v_{CE} and i_C may assume.

The maximum value of collector current that a transistor can carry is limited by the current carrying capacity of the fine wire used to connect the active regions of the device to its terminal leads. If excessive current occurs, this wire will melt, causing an open circuit at one or more of the device terminals. Manufacturer's data sheets will specify a maximum permissible collector current, denoted as $I_{C,max}$, that should not be exceeded.

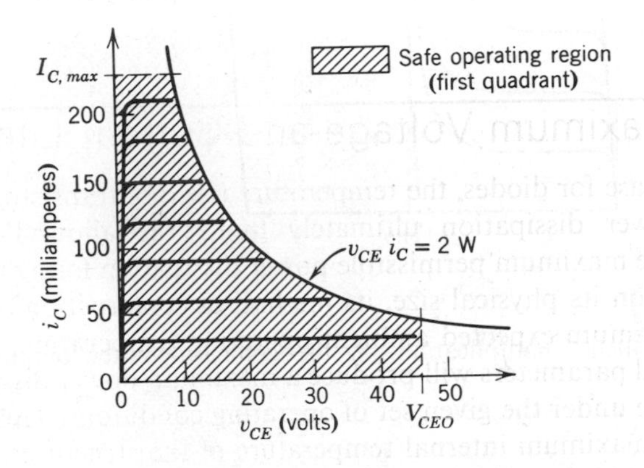

Figure 9.11
Safe operating region (first quadrant) of an *n-p-n* transistor with a maximum power dissipation of 2 W.

The maximum value of collector-to-emitter voltage that may be applied to a transistor is governed by the process of avalanche multiplication in the collector junction. Recall that in the junction diode increasing the reverse bias on the junction eventually causes the reverse current to increase sharply. In the transistor, increasing v_{CE} will also eventually cause breakdown of the collector junction. However, in the common-emitter connection, the output characteristics are very sensitive to avalanche multiplication, which causes the characteristics to bend sharply upward at a voltage one-half to one-third of the basic breakdown voltage of the collector junction. This effect is shown in Fig. 9.12, along with the breakdown characteristic of the collector-base junction by itself. Notice that the upward bending characteristics approach a maximum value of v_{CE} asymptotically. This asymptotic voltage is called the *sustaining voltage*, since substantial values of i_C result for zero base current. It is denoted on manufacturers' data sheets by V_{CEO} or BV_{CEO}. Although under certain circumstances the transistor may be operated above V_{CEO}, extreme care must be taken. For most applications one should limit v_{CE} to a value less than V_{CEO}.

These maximum values of i_C and v_{CE} are shown dotted in Fig. 9.11, and they, along with the locus of $P_{D,max}$, form the boundary of the safe first-

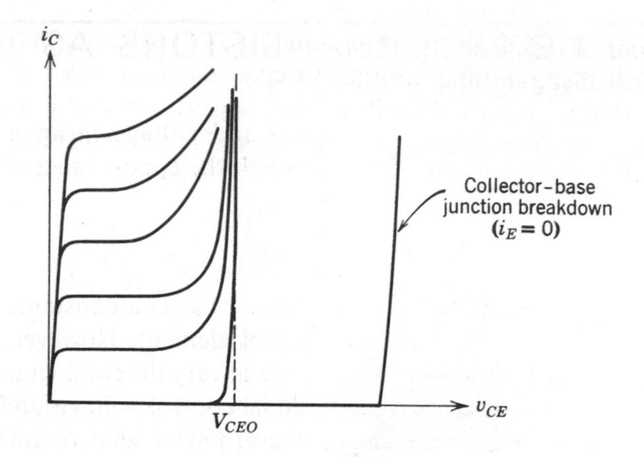

Figure 9.12
Effect of avalanche multiplication on the common-emitter output characteristic curves.

quadrant operating region for the transistor. In selecting a transistor for a specific application, one should first determine the locus of v_{CE} and i_C for the circuit, and then check to see that it lies within the safe operating region of the device.

9.1.6 Transistor Operation in the Reverse Region

The bipolar transistor is a somewhat symmetrical device, since a region of one type of semiconductor is placed between two regions of an oppositely doped semiconductor. Thus it is perfectly possible to interchange the roles of collector and emitter and operate the transitor in reverse with the emitter-base junction reverse-biased and the collector-base junction forward-biased. Normally, however, the emitter and collector are fabricated very differently, with the result that superior amplifying characteristics are obtained in the normal connection, and very poor amplifying characteristics are obtained in the reverse region. Nevertheless, transistor characteristics *do* continue into the third quadrant (the reverse region), and a current gain β_R can be defined there analogous to β_F in the active-gain region. In this book, we generally ignore reverse-region operation, as it is encountered relatively rarely in signal-processing circuits. (See Section 16.2.3 for one important application to digital circuits.)

9.2 FIELD-EFFECT TRANSISTORS AS CONTROL ELEMENTS

In Chapter 7 we found that the current and voltage in an n-type semiconducting bar were linearly related through the conductance of the bar

$$G = q\mu_e n\left(\frac{A}{L}\right) \tag{9.14}$$

Since the conductance G in the two-terminal bar is a constant, there is no possibility of utilizing the bar as a control element. However, through a more complex structure, it becomes possible to vary the conductance between two terminals in response to a third-terminal control voltage, and this forms the basic mechanism of operation for the group of semiconductor control elements known as *field-effect transistors* (FETs).

Examination of Eq. 9.14 shows that if the conductance of a bar is to be controlled, there are two possible ways in which it may be accomplished. Either the carrier concentration n or the geometry (A/L) must be made responsive to a control voltage. The electronic charge q is a fundamental constant not subject to control, and the mobility μ_e is also an independent constant in the first-order analysis of FETs.

9.2.1 Junction Field-Effect Transistors

Figure 9.13 is a schematic diagram of an FET that operates on the basis of varying the cross-sectional area of a semiconductor bar. The basic bar is shown as n-type, and this portion of the device is called the *channel*. The terminals at the ends of the bar are labeled *source* and *drain*. Along one side of the bar is a *p-type* region, which forms a p-n junction along the length of the bar. The p-region is termed the *gate* and is connected to the third or

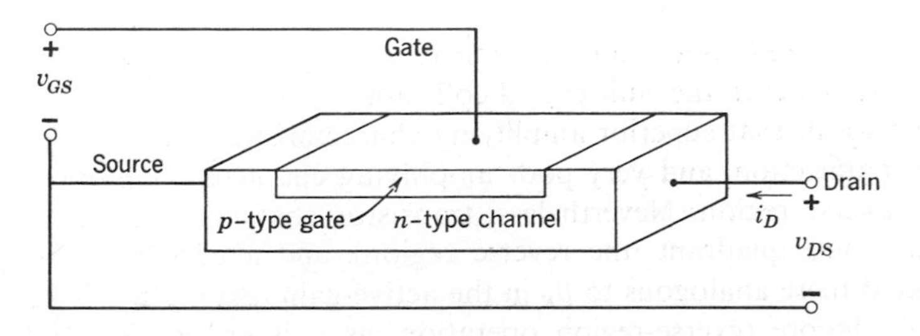

Figure 9.13
Structure of an n-channel junction field-effect transistor (JFET).

control terminal of the structure. This general form of FET is known as the *junction field-effect transistor* (JFET).

In our analysis of the *p-n* junction under reverse bias, we found that the space-charge layer (or *depletion region*), which contained only fixed ionized impurities and virtually no mobile charge carriers, increased in width with increasing reverse bias. If the gate of the JFET is reverse-biased, the resulting depletion region will penetrate into the *n*-type channel, reducing the portion of the total cross-sectional area that contains mobile carriers. Although the actual physical dimensions of the structure do not change, the cross-sectional area that can carry current is varied by the applied gate voltage. Thus, by changing the reverse bias at the gate, the conductance between the drain and source terminals can be controlled.

The maximum conductance occurs for $v_{GS} = 0$, since this results in minimum penetration of the depletion region and, hence, maximum cross-sectional conduction area. For some value of reverse bias on the gate, the depletion region penetrates the entire cross section, pinching off the channel, and the conductance is reduced to zero (open circuit). The value of v_{GS} that produces this condition is called the *pinch-off voltage* and is denoted by V_P. For an *n*-channel JFET V_P is a negative voltage. When v_{GS} becomes more negative than V_P, the conductance between drain and source remains zero. This behavior of the depletion layer and the corresponding *v-i* characteristics for small values of v_{DS} is summarized in Fig. 9.14 for several values of v_{GS}. Notice that we have again achieved the property of a control element, a variation of the output characteristics in response to an input voltage.

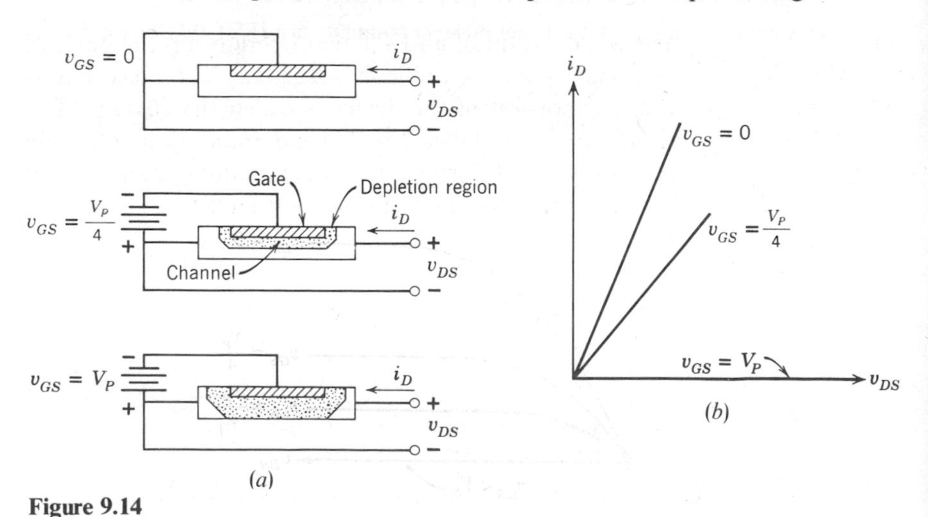

Figure 9.14
(*a*) Effect of gate voltage on the depletion layer in an *n*-channel JFET. (*b*) Corresponding output characteristics for small values of v_{DS}.

If v_{DS} becomes comparable to or larger than the magnitude of V_P, the characteristic curves depart from the linear behavior shown in Fig. 9.14. This is caused by the drain-to-source voltage, which produces a larger reverse bias (and hence a wider depletion layer) at the drain end of the channel than at the source end. Thus, increasing v_{DS} further reduces the conducting cross section and produces a characteristic curve that departs from the extension of the linear curve near the origin. When v_{DS} reaches the point where pinch-off occurs at the drain end of the channel, the drain current cannot increase in response to increases in v_{DS} but, instead, remains constant (see Fig. 9.15).

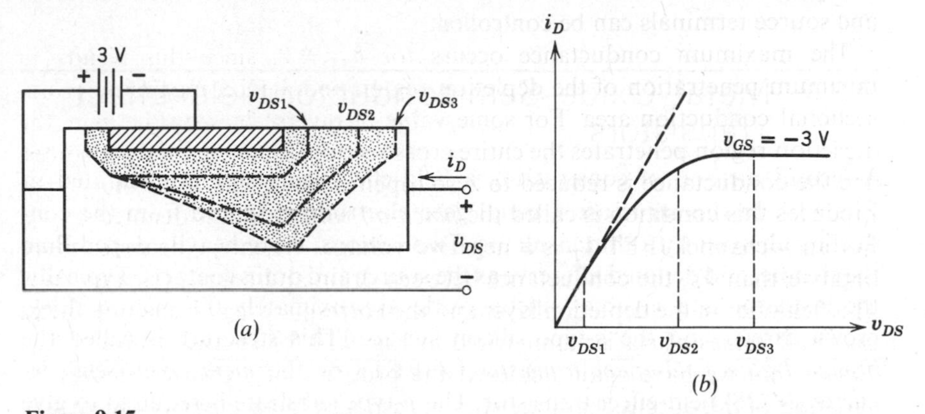

Figure 9.15
(*a*) Effect of increasing v_{DS} on the depletion layer boundary in a JFET. (*b*) Corresponding output characteristic curve.

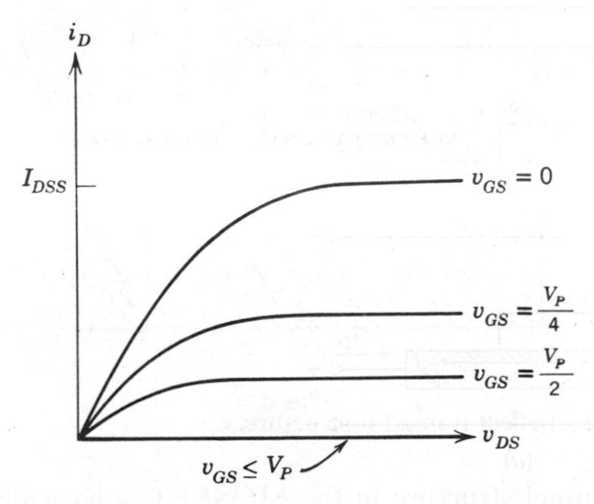

Figure 9.16
Complete output characteristic curves for an *n*-channel JFET. I_{DSS} is defined as the limiting drain current when $v_{GS} = 0$.

Even though the drain current is independent of the drain-to-source voltage once pinch-off occurs, control of i_D is retained by v_{GS}. A complete set of drain characteristics is shown in Fig. 9.16 for a typical *n*-channel JFET. Notice that the range of control voltage is $V_P \leq v_{GS} \leq 0$. Positive values of v_{GS} produce little control since the depletion-layer changes for forward bias are very small. Also, if the gate junction is forward-biased, a substantial gate current must be supplied. However, under reverse bias, only the reverse-saturation current of the gate junction is required, and is so small that it may be neglected in most applications. Since under these conditions the input power is miniscule, the JFET is capable of large power gain.

9.2.2 Metal-Oxide-Semiconductor Field-Effect Transistors

A second method of constructing a field-effect transistor is illustrated in Fig. 9.17. In this structure a metal gate electrode is spaced from the conducting channel by a thin insulator. Two regions, more heavily doped than the channel and denoted n^+, serve as the source and drain contacts. Typically, the insulator is a layer of silicon oxide approximately 0.1 micron thick, grown directly on the *n*-type silicon surface. This structure is called the *insulated-gate field-effect transistor* (IGFET), or the *metal-oxide-semiconductor* (MOS) field-effect transistor. The *p*-type substrate is required to give the assembly sufficient physical size and strength for handling and does not play a direct role in the electrical performance.

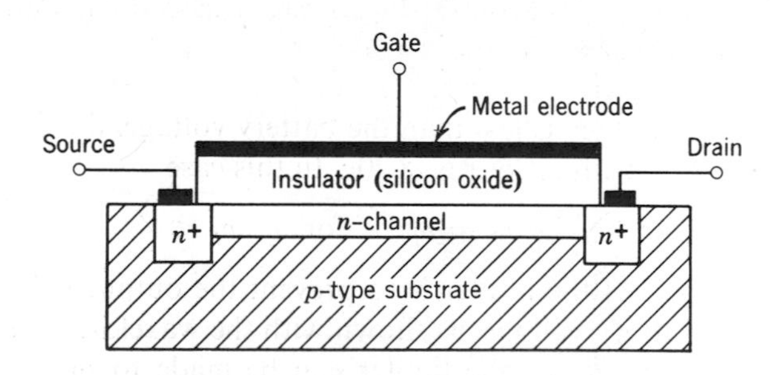

Figure 9.17
n-channel MOS field-effect transistor structure.

The gate-channel structure in the MOSFET is basically a parallel-plate capacitor. When a negative voltage is applied from gate to source, as shown in Fig. 9.18, a net negative charge appears on the gate electrode with a corresponding positive charge produced by forming a depletion region at the insulator-channel interface. The magnitude of the net charge densities and the width of the depletion region depend on the magnitude of v_{GS} and, hence, control of the conductance between the drain and source terminals is achieved. When v_{GS} is sufficiently negative, the entire n-channel will become depleted producing the pinch-off condition. Thus, the MOS structure will operate in the depletion mode with a set of drain characteristics essentially the same as those shown in Fig. 9.16.

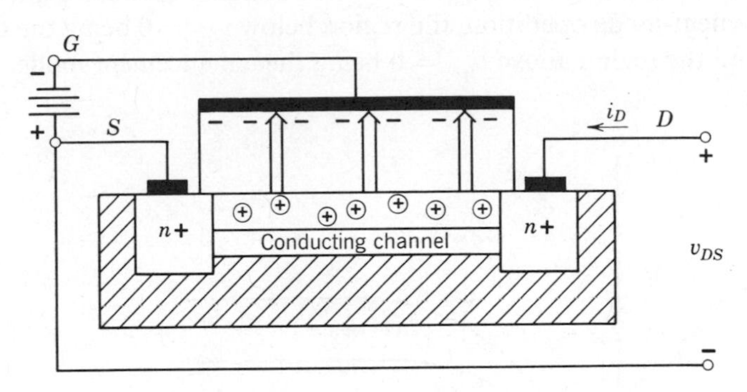

Figure 9.18
MOSFET operating in the depletion mode.

The MOS structure, however, does not have the restriction against positive values of v_{GS} that the JFET does, since there is no problem of a forward-biased p-n junction. The insulating layer prevents gate current for either polarity of v_{GS}. If a positive value of v_{GS} is applied, as shown in Fig. 9.19, a net positive charge will appear at the gate electrode and a net negative charge

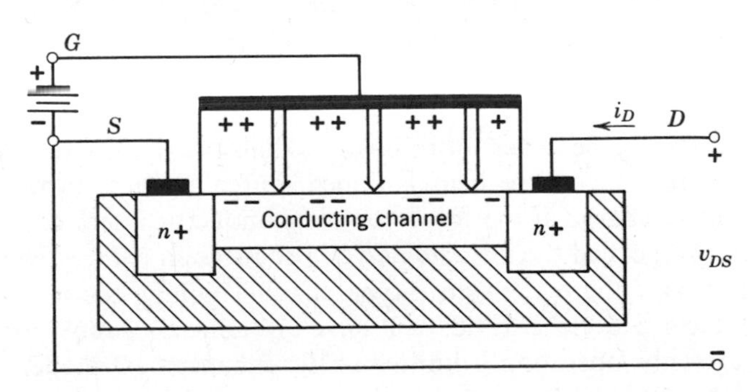

Figure 9.19
MOSFET operating in the enhancement mode.

at the channel insulator interface. This net negative charge is mobile and is *in addition to the normal free-electron concentration* that is neutralized by the fixed ionized impurities. These additional carriers increase the drain-to-source conductance. The amount of increase is controlled by the gate-to-source voltage. This mode of operation, where the carrier concentration is increased above the equilibrium concentration, is called the *enhancement mode.*

The drain characteristics in the enhancement mode resemble those in the depletion mode in shape, and produce increased drain current with increasing v_{GS} (see Fig. 9.20). The curve $v_{GS} = 0$ is the boundary between depletion- and enhancement-mode operation, the region below $v_{GS} = 0$ being the depletion mode and the region above $v_{GS} = 0$ being the enhancement mode.

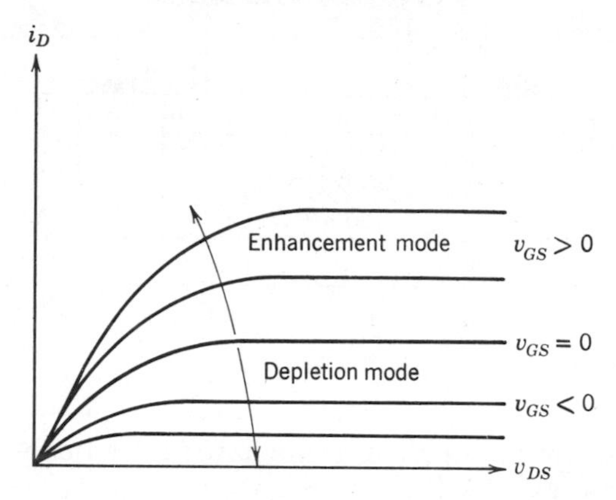

Figure 9.20
Output characteristic of an *n*-channel MOSFET exhibiting both depletion and enhancement mode operation.

A second type of MOSFET, Fig. 9.21*a*, is constructed by omitting the *n*-type channel from the structure in Fig. 9.17. In this device, with $v_{GS} = 0$ there is a negligible current flow between drain and source for applied v_{DS}, since one or the other of the p-n^+ junctions will be reverse biased. However, when v_{GS} becomes more positive than the threshold voltage, V_T, electrons are drawn from the heavily doped source region to form a layer of mobile negative charge at the insulator-semiconductor interface. This layer of free electrons, called the *inversion layer*, produces an *induced channel* that behaves just as if there were an *n*-type channel joining the n^+ drain and source contacts. Since the device will have an induced channel for positive values of v_{GS} only, operation is limited to the enhancement mode.

The drain characteristics for this structure are again similar to those in Figs. 9.16 and 9.20, with the exception that the curve for zero drain current occurs for $v_{GS} \leq V_T$, where V_T is a positive voltage. Figure 9.22 contains a typical set of characteristic curves for an induced channel MOSFET.

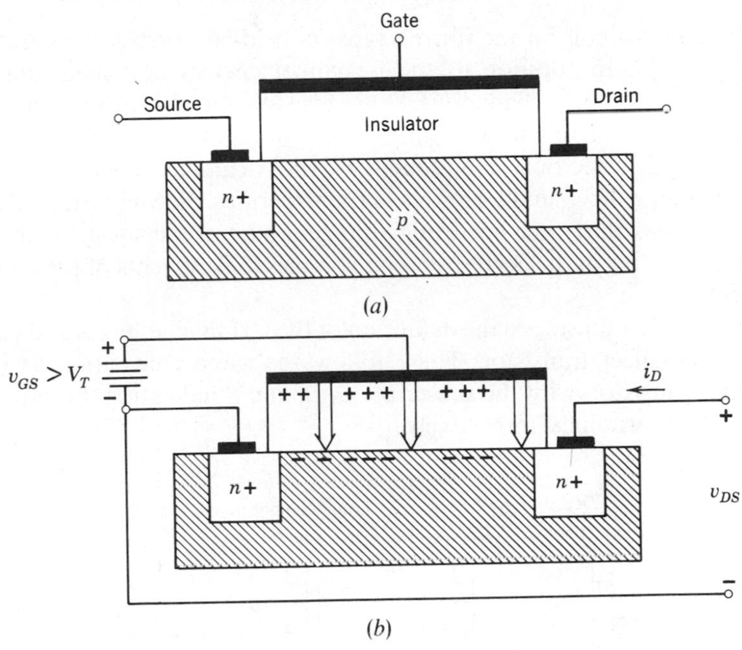

Figure 9.21
An n-channel MOSFET that operates only in the enhancement mode. (a) Structure. (b) Applied v_{GS} produces an induced channel.

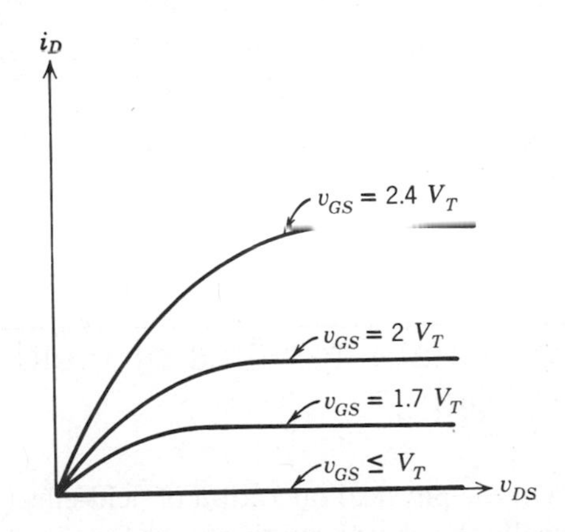

Figure 9.22
Output characteristics of an n-channel MOSFET that operates only in the enhancement mode.

9.2.3 Circuit Symbols and Terminal Variables of Field-Effect Transistors

The circuit symbols for the various types of field-effect transistors are shown in Fig. 9.23*a*. In addition to the *n*-channel devices described above, the circuit symbols for *p*-channel FETs are also shown; the distinction is in the direction of the arrow in the symbol. As with bipolar transistors, *p*-channel FETs are the electrical compliment to the *n*-channel devices, formed by interchanging the *n*- and *p*-regions in the structures shown in Figs. 9.13, 9.17, and 9.21*a*. Similarly, the operating voltages and currents for *p*-channel units are the negative of the corresponding voltages and currents of the *n*-channel device.

Figure 9.23*b* illustrates the definition of the terminal voltages and currents for a field-effect transistor. These follow the same convention as for the bipolar transistor, with the subscripts *G*, *D*, and *S* indicating the gate, drain, and source terminals, respectively.

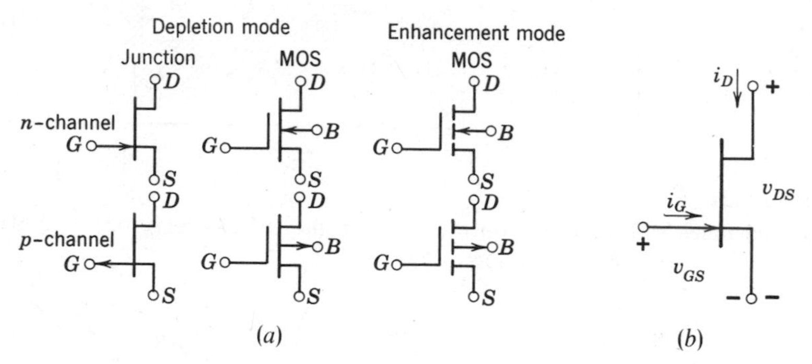

Figure 9.23

(*a*) Field-effect transistor circuit symbols. D: Drain terminal, S: Source terminal, G: Gate terminal, B: Bulk or substrate connection, often connected to source terminal internally. (*b*) Terminal variables.

9.2.4 The *v-i* Characteristics of Field-Effect Transistors

In our discussion of the physical operation of field-effect transistors we have presented graphically the characteristic curves for the device. In this section we present the volt-ampere equations that lead to those characteristic curves. We will not attempt to derive the equations from the physical electronics, but instead, concentrate on interpreting the results of the derivations in preparation for their application to practical circuit problems.

Examination of the common-source output characteristics (see Figs. 9.16, 9.20, or 9.22) shows that there are two distinct regions of operation: Region I, where i_D depends on both v_{GS} and v_{DS}, and Region II, where i_D depends on v_{GS} but is independent of v_{DS}. In a depletion-mode device, the dividing line between these two regions is determined by whether or not pinch-off has occurred at some point in the channel. In units operating in the enhancement mode only, the dividing line is determined by the value of v_{DS} which causes the induced channel to disappear at some point between source and drain. Thus, we expect to find two separate equations for drain current corresponding to these two regions of operation.

In Region I, the analysis of an MOS device that can operate in both the depletion and enhancement modes yields for the drain current

$$i_D = K\left[(v_{GS} - V_P)v_{DS} - \frac{v_{DS}^2}{2}\right] \quad \text{for} \quad \begin{cases} v_{DS} \leq v_{GS} - V_P \\ v_{GS} - V_P \geq 0 \end{cases} \tag{9.15}$$

The first inequality, $v_{DS} \leq v_{GS} - V_P$, insures that no point in the channel is pinched off because of excessive drain-to-source voltage, while the second inequality insures that no point is pinched-off because of the applied gate-to-source voltage. The constant K involves material constants and geometry parameters, and is best determined by experimental measurement. Figure 9.24*a* shows a plot of Eq. 9.15 subject to the limits placed on v_{DS} and v_{GS} by the inequalities, and represents the *v-i* characteristics for the device in Region I.

In Region II, the drain current remains constant at that value of i_D which corresponds to the beginning of pinch-off in Eq. 9.15. This current is found by setting v_{DS} equal to $v_{GS} - V_P$ and substituting into Eq. 9.15 to eliminate v_{DS}:

$$\begin{aligned} i_D &= K\left[(v_{GS} - V_P)(v_{GS} - V_P) - \frac{(v_{GS} - V_P)^2}{2}\right] \\ &= \frac{K}{2}(v_{GS} - V_P)^2 \quad \text{for} \begin{cases} v_{DS} > v_{GS} - V_P \\ v_{GS} - V_P \geq 0 \end{cases} \end{aligned} \tag{9.16}$$

This portion of the *v-i* characteristic is shown in Fig. 9.24*b*.

The complete set of characteristic curves are shown in Fig. 9.24*c* with the dashed line given by

$$v_{DS} = v_{GS} - V_P \tag{9.17}$$

indicating the boundary between Regions I and II. Also indicated are methods of evaluating the device constants V_P and K. By applying v_{DS} of sufficient

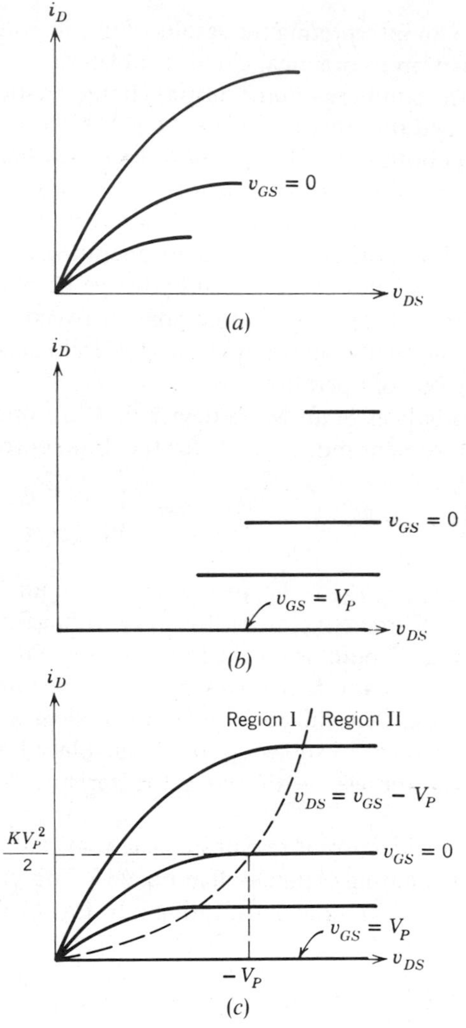

Figure 9.24
v-i characteristic curves for an ideal MOSFET that operates in both the depletion and enhancement modes. (*a*) Region I. (*b*) Region II. (*c*) Complete characteristics.

magnitude to insure operation in Region II, v_{GS} is adjusted to make i_D just go to zero. Equation 9.16 then shows that this value of v_{GS} is equal to V_P. Then by setting $v_{GS} = 0$, the resulting drain current is

$$i_D = \frac{KV_P^2}{2} \quad \text{for} \quad v_{GS} = 0 \tag{9.18}$$

from which K may be determined. The quantity $KV_P^2/2$ is often denoted on manufacturers' data sheets as I_{DSS}.

Notice that the FET control parameter is the gate-to-source voltage, v_{GS}. Since the gate is insulated from the remaining structure, the gate current is essentially zero. (The extremely small leakage current in the gate lead, on the order 10^{-10} to 10^{-15} amperes, is negligible for our purposes.) Thus, gate current is not a suitable control parameter. Compare this with the bipolar transistor, where either base current or base-to-emitter voltage is used as a control parameter to develop the characteristic curves.

The physical electronics of the junction field-effect transistor produces v-i characteristic equations that differ in algebraic form from Eqs. 9.15 and 9.16. However, numerically they differ from Eqs. 9.15 and 9.16 by only a few percent. Thus, we may employ these results with JFETs also provided that we observe the restriction $v_{GS} < 0$ in n-channel devices to insure the gate junction is reverse biased. The determination of V_P and K is identical for both types of device.

The v-i characteristic equations for the MOSFET that operates in the enhancement mode only are algebraically identical to Eqs. 9.15 and 9.16 with the replacement of the pinch-off voltage V_P by the threshold voltage V_T.

$$i_D = K\left[(v_{GS} - V_T)v_{DS} - \frac{v_{DS}^2}{2}\right] \quad \text{for} \begin{cases} v_{DS} \le v_{GS} - V_T \\ v_{GS} > V_T \end{cases} \tag{9.19}$$

$$i_D = \frac{K}{2}(v_{GS} - V_T)^2 \quad \text{for} \begin{cases} v_{DS} > v_{GS} - V_T \\ v_{GS} > V_T \end{cases} \tag{9.20}$$

The main difference between Eqs. 9.15 and 9.16 and Eqs. 9.19 and 9.20 is the fact that for an n-channel device V_P is a negative voltage while V_T is positive. Thus, the ranges of allowable values of v_{GS} and v_{DS}, as dictated by the inequalities, differ in the two types of MOS field-effect transistors.

The determination of V_T and K for an enhancement-mode device can be made in a similar fashion to the depletion-mode units. First, a sufficiently large v_{DS} is applied to insure the validity of Eq. 9.20. When v_{GS} is increased to the point where the drain current just begins to flow, then that value of v_{GS} is equal to V_T. Finally, increasing v_{GS} to $2V_T$ and noting the resulting value of i_D determines the value of K from Eq. 9.20.

$$i_D = \frac{KV_T^2}{2} \quad \text{for} \quad v_{GS} = 2V_T \tag{9.21}$$

The characteristic curves resulting from Eqs. 9.19 and 9.20, along with the method of evaluating V_T and K, are shown in Fig. 9.25.

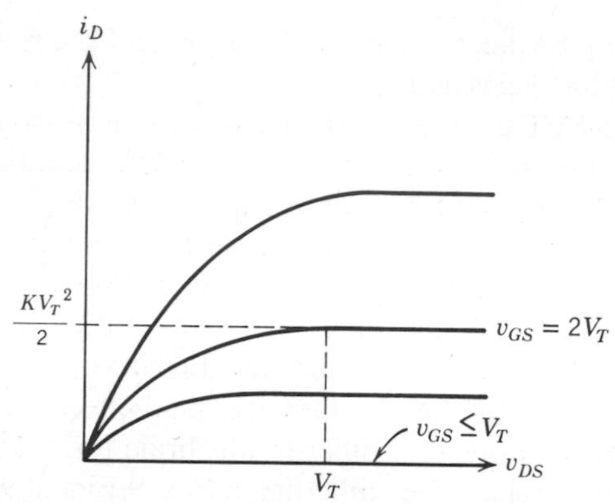

Figure 9.25
v-i characteristic for an ideal MOSFET that operates only in the enhancement mode.

9.2.5 Maximum Limits of Voltage and Current

As with diodes and bipolar transistors, the voltage and current limitations for an FET depend on its power dissipation capability and breakdown effects. The power dissipated in an FET is given by the product of v_{DS} and i_D. Thus for a given maximum dissipation we have

$$v_{DS} i_D \leq P_{D,max} \tag{9.22}$$

This equation is analogous to Eq. 9.13, and plots as a pair of first- and third-quadrant hyperbolas on the output characteristics.

The maximum allowable voltages are usually governed by the reverse breakdown voltage of the *p-n* gate junction in the JFET, or the breakdown voltage of the thin insulating layer in the MOSFET. Breakdown effects are evidenced on the characteristic curves by an abrupt increase in drain current with increasing v_{DS}. In a MOSFET, breakdown usually results in permanent destruction of the integrity of the insulating layer. These voltage limits are normally specified on the manufacturer's data sheet.

Field-effect transistors are basically current limiting devices, and the maximum current permitted is normally governed by the power dissipation capability of the unit. However, in enhancement mode operation, large gate voltages may permit currents of sufficient magnitude to melt the internal connecting wires. Thus the manufacturer will also specify a maximum permissible value of drain current. Figure 9.26 contains a typical boundary of the first-quadrant safe operating region of an FET.

Because of the extremely high input resistance at the gate of an MOSFET, it can be damaged by the accumulation of excess static charge. To avoid

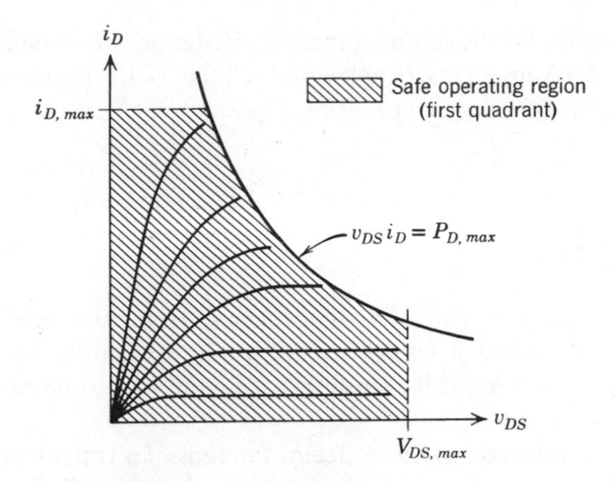

Figure 9.26
Safe operation region (first quadrant) for a field-effect transistor.

damage, the leads should remain shorted together when the device is not being used, the device should be handled by the case and not the leads, and units should not be inserted into circuits while the power is applied.

REFERENCES

R9.1 Paul E. Gray and Campbell L. Searle, *Electronic Principles: Physics, Models, and Circuits* (New York: John Wiley, 1969), Chapters 7, 9, and 10.

R9.2 Paul D. Ankrum, *Semiconductor Electronics* (Englewood Cliffs, N.J.: Prentice-Hall, 1971), Chapters 5 and 10.

R9.3 James M. Feldman, *The Physics and Circuit Properties of Transistors* (New York: John Wiley, 1972).

R9.4 Ben G. Streetman, *Solid State Electronic Devices* (Englewood Cliffs, N.J.: Prentice-Hall, 1972).

R9.5 Carl David Todd, *Junction Field-Effect Transistors* (New York: John Wiley, 1968).

R9.6 Richard S. C. Cobbold, *Theory and Applications of Field-Effect Transistors* (New York: Wiley-Interscience, 1970).

R9.7 Bruce D. Wedlock and James K. Roberge, *Electronic Components and Measurements* (Englewood Cliffs, N.J.: Prentice-Hall, 1969), Chapter 20 on thermal ratings.

QUESTIONS

Q9.1 Why can the relatively small base current in a bipolar transistor be considered a *control variable*? Equivalently, why in Fig. 9.4 does $i_B = 0$ logically require that i_E and i_C also be zero?

Q9.2 The magnitude of the collector current of a transistor operating in the active region is limited by the rate of diffusion of minority carriers across the base. Can you think of other natural processes that are diffusion-limited?

Q9.3 The reverse bias on the collector-base junction maintains the base minority-carrier concentration at the junction near zero. What happens to the minority-carrier concentration when this junction is no longer reverse biased (i.e., in saturation)?

Q9.4 Do minority carriers play any significant role in the operation of JFET? Of a MOSFET?

Q9.5 FET gate currents are small. For what kind of signal source might this be important?

CHAPTER TEN
LARGE-SIGNAL TRANSISTOR CIRCUITS

10.0 INTRODUCTION

The circuit applications of the different kinds of transistors are so diverse that they require treatises devoted solely to their analysis. Therefore, the examples selected in the remainder of this text represent only the most fundamental types of transistor uses, and are chosen as much for their value in illustrating methods of analysis as for their specific applicability to signal-processing problems.

One can classify transistor circuits roughly into two classes: the large-signal circuits, in which the variations of voltages and currents are sufficiently great to require that the full, nonlinear device characteristics be considered; and small-signal circuits, in which the signal components of voltages and currents are sufficiently small to permit the use of local linear models of device behavior. The boundary is occasionally fuzzy, and depends on the accuracy with which one needs to know or predict the operating characteristics of a particular circuit. In this chapter, we explore circuits of the large-signal variety, admitting at the outset that we are willing to sacrifice some accuracy in the representation of device characteristics in order to carry out a simple analysis of circuit performance.

10.1 THE COMMON-EMITTER AMPLIFIER

10.1.1 A Graphical Approach

We turn our attention first to the common-emitter amplifier of Fig. 10.1. The emitter is "common" to both the input and output circuits, which accounts for the name of the connection. Although the circuit of Fig. 10.1 is quite simple, the Thévenin input circuit, consisting of v_S and R_S, and the Thévenin output circuit, consisting of V_{CC} and R_L, may actually represent linear networks of great complexity. We shall use the graphical methods introduced in Section 8.2.1 to analyze the behavior of this circuit.

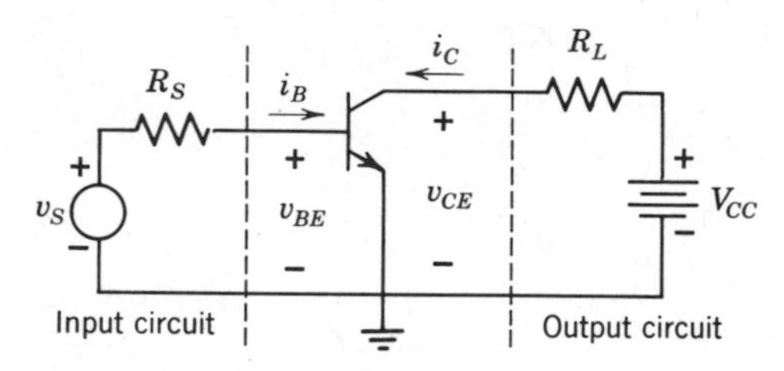

Figure 10.1
Common-emitter amplifier configuration.

From our discussion in Chapter 9, we know that between the base and emitter terminals, the transistor has the *v-i* characteristic of a semiconductor diode. Figure 10.2*a* shows a typical *n-p-n* silicon transistor input characteristic illustrating this diode behavior. Four *input load lines* are shown on the input characteristic, each drawn with slope $-1/R_S$, and each one intercepting the v_{BE} axis at a different value of the source voltage. In the example shown, we have used a value of $R_S = 20\ \text{k}\Omega$. The four load lines will be used below to trace the operation of the circuit as if the input voltage were changing by 0.4 V steps from v_{S1} to v_{S2} to v_{S3} and finally to v_{S4}.

The load lines in Fig. 10.2*a* were obtained from KVL for the input circuit, written in terms of v_{BE} and i_B.

$$v_{BE} = v_S - i_B R_S \tag{10.1}$$

This linear equation plots as a straight line in the v_{BE}-i_B plane, has slope $-1/R_S$, and intercepts the v_{BE} axis at the open-circuit Thévenin voltage v_S. The intercept on the i_B axis is

$$i_B = \frac{v_S}{R_S} \qquad \text{for} \qquad v_{BE} = 0 \tag{10.2}$$

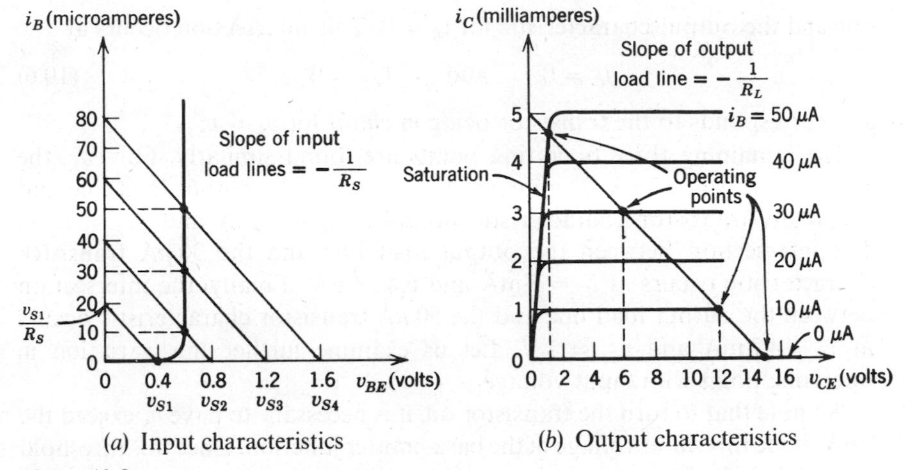

(*a*) Input characteristics (*b*) Output characteristics

Figure 10.2
Graphical analysis of common-emitter amplifier.

The intersection of each load line with the input characteristic gives the operating point for that value of input voltage. Thus in our example, the base current is zero when v_S is equal to v_{S1} (cutoff), and increases to 10 μA, 30 μA, and finally to 50 μA as v_S increases to v_{S2}, v_{S3}, and v_{S4}.

We now turn to the output circuit, where the graphical analysis proceeds in a similar way. First, we express KVL for the linear output circuit in terms of v_{CE} and i_C.

$$v_{CE} = V_{CC} - i_C R_L \tag{10.3}$$

This is a linear equation relating v_{CE} and i_C, and will plot as a straight line with slope $-1/R_L$ in the v_{CE}-i_C plane. The two intercepts for this *output load line* are

$$v_{CE} = V_{CC} \qquad \text{for} \qquad i_C = 0 \tag{10.4}$$

and

$$i_C = \frac{V_{CC}}{R_L} \qquad \text{for} \qquad v_{CE} = 0 \tag{10.5}$$

The load line in Fig. 10.2*b* has been plotted for a specific choice $V_{CC} = 15$ V and $R_L = 3\ \text{k}\Omega$.

To find the output operating point corresponding to each value of v_S, we must locate the intersection of the output load line with the transistor characteristic corresponding to the correct value of base current. Thus, for v_{S1}, which yielded $i_B = 0$, we find the intersection between the output load line and the output characteristic for $i_B = 0$. This intersection occurs at

$$i_C = 0 \qquad \text{and} \qquad v_{CE} = V_{CC} \tag{10.6}$$

and corresponds to the transistor being in *cutoff* for $v_S = v_{S1}$.

The remaining three operating points are found similarly. For v_{S2}, the base current was 10 μA. The intersection between the output load line and the 10 μA transistor characteristic occurs at $i_C = 1$ mA and $v_{CE} = 12$ V. The intersection between the output load line and the 30 μA transistor characteristic occurs at $i_C = 3$ mA and $v_{CE} = 6$ V. Finally, the intersection between the output load line and the 50 μA transistor characteristic occurs at $i_C = 4.7$ mA and $v_{CE} = 1$ V. Let us examine further this variation in operating point with input voltage.

We note that to turn the transistor on, it is necessary to have v_S exceed the 0.6 V diode threshold voltage of the base-emitter junction. Once this threshold is exceeded, the base current increases steadily in response to an increase in v_S. As the base current increases, the output operating point moves up the load line from the cutoff characteristic ($i_B = 0$) into the active-gain region. Eventually, v_{CE} decreases until the collector-base junction becomes forward-biased, resulting in saturation with a small, nearly constant v_{CE}. In the example discussed, v_{S2} and v_{S3} produce operation in the linear region, while v_{S4} produces operation in the saturation region. Any increase in v_S beyond v_{S4} *still* produces operation at the same saturation point ($i_C =$ 4.7 mA and $v_{CE} = 1$ V), because all characteristics for $i_B > 50$ μA continue to pass through that same point of intersection with the load line.

In the active-gain region, the transistor functions as an amplifier. In our example, the base current increased from 10 to 30 μA in response to a change

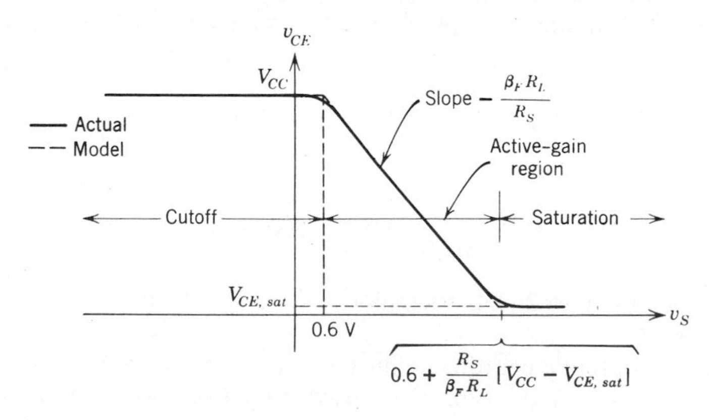

Figure 10.3
Common-emitter amplifier transfer characteristic.

of v_S from v_{S2} to v_{S3}. The corresponding change in collector current was from 1 mA to 3 mA, representing a current gain of 100. The relationship between voltage changes is more complicated. As v_S increases above 0.6 V, v_{CE} begins to drop. In our example, an *increase* of about 0.4 V from v_{S2} to v_{S3} produces a *decrease* in v_{CE} from 12 V to 6 V. The magnitude of the *change* in voltage is amplified by a factor of $6/0.4 = 15$, but the directions of the input and output voltage changes are opposite. Thus, this amplifier is operating with an *inverting voltage gain of* -15.

A sketch of the overall *transfer characteristic* relating the input voltage v_S to the output voltage v_{CE} is shown in Fig. 10.3. For $v_S < 0.6$ V, the transistor is cutoff. As v_S increases, v_{CE} drops from its cutoff value of V_{CC} toward its saturation value of about 1 V. The slope of the transfer characteristic in the active-gain region is about -15, and corresponds to the voltage gain of the amplifier. That slope is labeled with the notation $-\beta_F R_L/R_S$ in Fig. 10.3. To understand why that particular algebraic expression approximates the slope of the actual curve, let us examine in the next section a piecewise linear model of the bipolar junction transistor.

10.1.2 Large-Signal Model for the Bipolar Junction Transistor

We can construct a model for the forward characteristics of a bipolar junction transistor ($v_{CE} > 0$) by combining an exponential diode (to represent the base-emitter junction) with a dependent current source (to represent the collector action of the reverse-biased collector-base junction). We recall from Chapter 9 that in the active-gain region, the collector current is independent of v_{CE}, and is related to the base current by

$$i_C = \beta_F i_B \tag{10.7}$$

Thus a dependent current source in the collector lead with magnitude $\beta_F i_B$ will correctly represent the active-gain transistor characteristics. A model which combines a dependent current source with an exponential diode is shown in Fig. 10.4*a*. The label I_{ES} indicates the reverse saturation current of the base-emitter diode.

Fortunately the circuit model of Fig. 10.4*a* happens also to represent the cutoff region of operation. If v_{BE} is less than 0.6 V, the base current will be zero, so that the collector current will be also zero. Thus the model of Fig. 10.4*a* works both for the cutoff and active-gain regions. We will see momentarily, however, that the model does not correctly represent saturation.

Figure 10.4*b* shows a piecewise linear version of this model. The exponential diode has been replaced by an ideal diode in series with a 0.6 V battery.

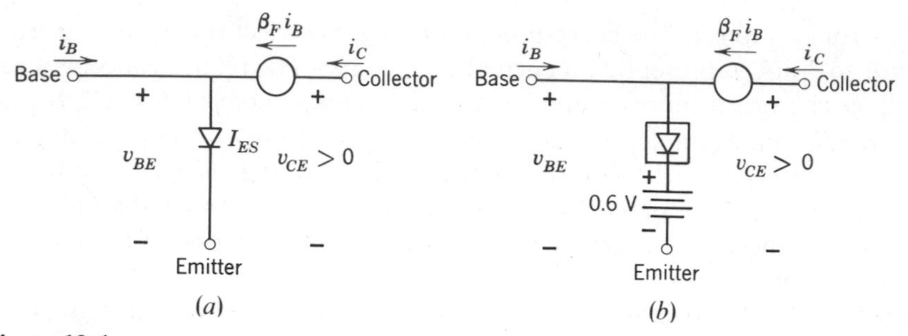

Figure 10.4
Large signal *n-p-n* transistor models.

A comparison between a set of actual transistor characteristics and the characteristics calculated from the model of Fig. 10.4*b* is shown in Fig. 10.5. Although the model's characteristics (which are simply a set of parallel horizontal lines) closely resemble the first-quadrant behavior of the actual transistor, they do not exhibit the saturation drop toward $i_C = 0$ as v_{CE} approaches zero. It would be perfectly possible to add more elements to the model of Fig. 10.4*b* to represent saturation region of the characteristics.

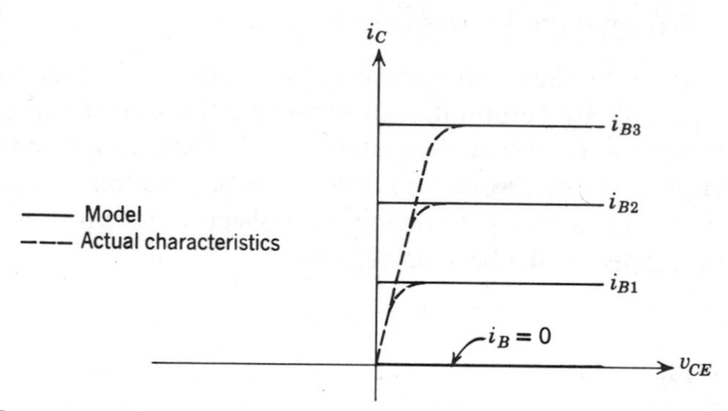

Figure 10.5
Output characteristics of model in Fig. 10.4*b*.

However, it is often simpler to work with the less cumbersome model for cutoff and the active-gain region, and use a model in which $v_{CE} = V_{CE,sat}$ or even $v_{CE} \approx 0$ to represent saturation. Let us illustrate the use of these models to calculate an approximate transfer characteristic for the common-emitter amplifier (Fig. 10.3).

Figure 10.6 shows a common-emitter amplifier in which the piecewise linear model of Fig. 10.4*b* has replaced the transistor. If $v_{BE} < 0.6$ V, the

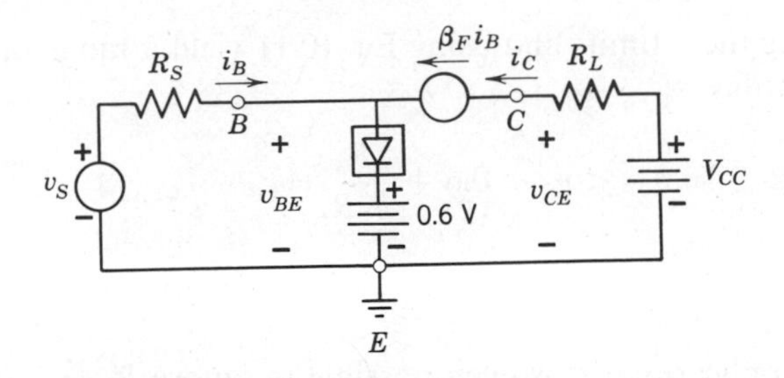

Figure 10.6
Common-emitter amplifier with large-signal model.

ideal diode is reverse biased and the transistor is cutoff. If $i_B > 0$, the ideal diode is forward biased, becoming a short circuit, and v_{BE} becomes equal to 0.6 V. Thus for $i_B > 0$,

$$i_B = \frac{v_S - 0.6}{R_S} \tag{10.8}$$

From the dependent current source

$$i_C = \beta_F i_B = \frac{\beta_F(v_S - 0.6)}{R_S} \tag{10.9}$$

Using KVL, we find that

$$v_{CE} = V_{CC} - i_C R_L \tag{10.10}$$

and substituting for i_C in Eq. 10.10 we obtain for the active-gain region

$$v_{CE} = V_{CC} - \left(\frac{\beta_F R_L}{R_S}\right)(v_S - 0.6) \tag{10.11}$$

This modeled active-gain transfer characteristic is shown in Fig. 10.3, along with an actual transfer characteristic. Note that the slope is $-(\beta_F R_L/R_S)$. For our example, $R_L = 3\ \text{k}\Omega$, $R_S = 20\ \text{k}\Omega$, and $\beta_F = 100$ (from the characteristics). This slope has the value of -15, as estimated in Section 10.1.1 from the graphical analysis.

The validity of Eq. 10.11 is limited to the active-gain region, which requires

$$v_S > 0.6\ \text{V} \tag{10.12}$$

and

$$v_{CE} > V_{CE,sat} \tag{10.13}$$

Combining these limits and using Eq. 10.11 yield a range of v_S for active-region operation

$$0.6 < v_S < 0.6 + \frac{R_S}{\beta_F R_L}(V_{CC} - V_{CE,sat}) \tag{10.14}$$

Any larger value of v_S produces saturation, while any smaller value produces cutoff.

In some applications, it is even possible to ignore $V_{CE,sat}$, and model the saturation region with $V_{CE,sat} = 0$. For rough estimates of circuit performance, it is reasonable to extend the model active-gain region all the way to $v_{CE} = 0$, then to insert an actual value for $V_{CE,sat}$ to see whether any important feature of circuit performance would be modified.

10.1.3 An Example

If the levity can be pardoned, we can shed further "light" on the use of the common-emitter configuration and the large-signal transistor models by examining the lamp-driver circuit of Fig. 10.7. In this circuit, an *n-p-n* transistor in the common-emitter configuration is used in conjunction with a 10 V supply to light a 500 Ω lamp whenever the input signal v_A has a 1 V pulse. The transistor is represented by three parameters, the forward drop on the base-emitter junction ($V_{BE} = 0.6$ V), the current gain β_F, and the saturation voltage $V_{CE,sat}$. These last two quantities are listed on transistor data sheets, although the symbol h_{FE} is often used for β_F.

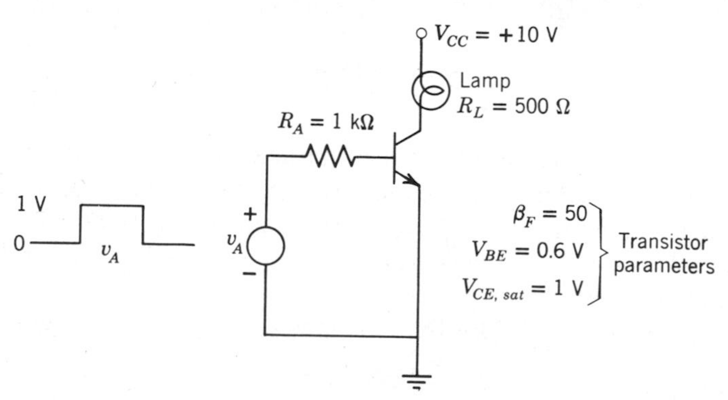

Figure 10.7
A transistor switch is used to light the 500 Ω lamp when $v_A = 1$ V. Observe the use of the node-voltage notation to represent the V_{CC} supply.

We assume that the v_A-R_A combination is the Thévenin equivalent of a source network that produces a 1 V signal under open-circuit conditions. We assume that the lamp has a nominal resistance of 500 Ω and requires a current of 15 mA or more for proper illumination. There is no way for the source network by itself to provide enough current to light the lamp. The transistor, however, acting as a control element, allows current to flow from the V_{CC} supply. Under these circumstances, the source network need only supply the *base* current to the transistor, which is roughly $1/\beta_F$ or 1/50 of the current in the lamp.

Analysis of this circuit proceeds as follows. First, for $v_A = 0$, the transistor is cut off, so that the lamp current is zero. When $v_A = 1$ V, we can use the equations developed in the last section to calculate the lamp current. The base current from Eq. 10.8 is

$$i_B = \frac{(1 - 0.6)\text{ V}}{1\text{ k}\Omega} = 0.4\text{ mA} \tag{10.15}$$

The collector current, *assuming no saturation*, then is from Eq. 10.9

$$i_C = 50 \times 0.4 = 20\text{ mA} \tag{10.16}$$

This current is certainly larger than the 15 mA needed to light the lamp. However, we must be careful to check whether our assumption about saturation is correct.

The way to check on saturation is to calculate v_{CE} from Eq. 10.11 and see whether it exceeds $V_{CE,sat}$. In this case we obtain from Eq. 10.11

$$v_{CE} = 10 - \frac{(50)(500)}{1000}(1 - 0.6) = 0\text{ V} \tag{10.17}$$

This calculated value of 0 V being slightly less than $V_{CE,sat}$ for the transistor indicates that the transistor has just barely entered the saturation region. Therefore, we should revise our assumption, and analyze the circuit assuming saturation.

If the transistor is saturated, then the collector current is given by

$$I_{C,sat} = \frac{V_{CC} - V_{CE,sat}}{R_L} \tag{10.18}$$

In this case, the value of the current is

$$I_{C,sat} = \frac{10 - 1}{500} = 18\text{ mA} \tag{10.19}$$

which is still greater than the 15 mA required. Thus the circuit of Fig. 10.7, which switches between cutoff and saturation in response to a 1 V signal, will light the lamp.

10.2 THE COMMON-COLLECTOR OR EMITTER-FOLLOWER CONNECTION

10.2.1 Emitter-Follower Amplifier

A second important transistor amplifier configuration is the *common-collector* or *emitter-follower* circuit, shown in Fig. 10.8*a*. The origin of "common collector" will not be clear until we discuss the small-signal characteristics of this circuit. The name emitter-follower will be explained forthwith.

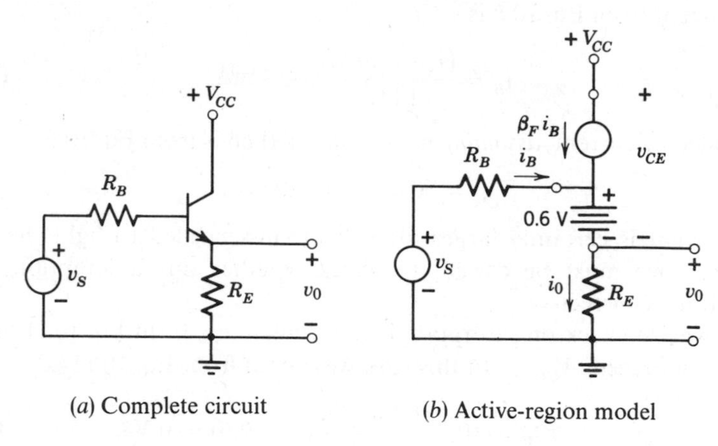

Figure 10.8
Emitter-follower amplifier.

Let us determine the large-signal transfer characteristic by replacing the transistor with the large-signal active-gain region model, yielding the circuit shown in Fig. 10.8*b*. Remember that this model is only valid where $v_{CE} > V_{CE,sat}$ and $i_B \geq 0$. When either of these inequalities is violated, the calculations based on the circuit model will not correspond to the actual circuit behavior of the transistor. We assume that the inequalities are satisfied and then determine the limits on the active-gain region.

From the circuit of Fig. 10.8*b* we can use KCL to obtain

$$i_O = i_B + \beta_F i_B = (1 + \beta_F) i_B \tag{10.20}$$

from which we find

$$v_O = i_O R_E = (1 + \beta_F) R_E i_B \tag{10.21}$$

Next we write KVL around the input circuit loop

$$v_S - i_B R_B - 0.6 - v_O = 0 \tag{10.22}$$

Elimination of i_B between Eqs. 10.21 and 10.22 results in the required voltage-transfer characteristic,

$$v_O = \frac{(1 + \beta_F)R_E}{R_B + (1 + \beta_F)R_E}(v_S - 0.6) \tag{10.23}$$

This equation is plotted in Fig. 10.9 as the active-gain portion of the transfer characteristic.

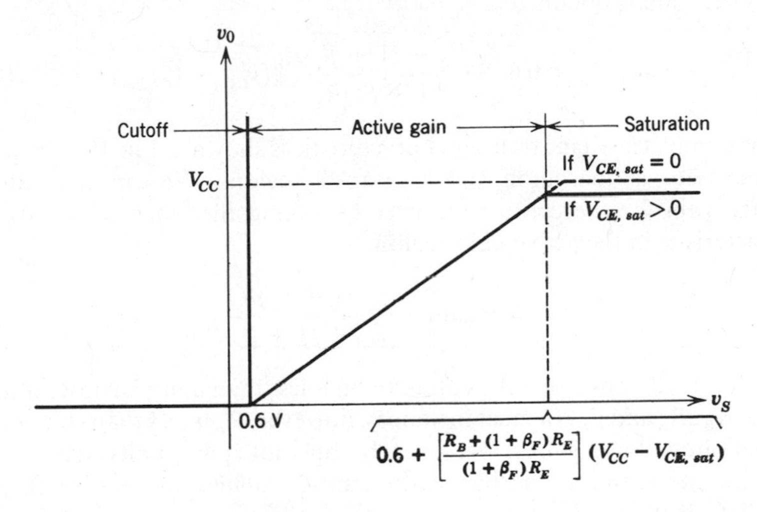

Figure 10.9
Large signal emitter-follower transfer characteristic. The output saturates at $(V_{CC} - V_{CE,sat})$, which is nearly equal to V_{CC}.

To determine the limits on the range of validity of Eq. 10.23 we first note that

$$v_{CE} = V_{CC} - v_O \tag{10.24}$$

Thus, the condition $v_{CE} > V_{CE,sat}$ requires that $v_O \leq (V_{CC} - V_{CE,sat})$. If we substitute $v_O = (V_{CC} - V_{CE,sat})$ into Eq. 10.23 we find that the maximum input voltage for active-gain operation is

$$v_{S,max} = 0.6 + \left[\frac{R_B + (1 + \beta_F)R_E}{(1 + \beta_F)R_E}\right](V_{CC} - V_{CE,sat}) \tag{10.25}$$

If v_S exceeds $v_{S,max}$, the transistor will be saturated and v_0 will remain constant at the voltage $(V_{CC} - V_{CE,sat})$, which is nearly equal to V_{CC}.

By eliminating v_0 between Eqs. 10.21 and 10.22, we obtain the dependence of i_B on v_S.

$$i_B = \frac{v_S - 0.6}{R_B + (1 + \beta_F)R_E} \tag{10.26}$$

To satisfy the condition $i_B > 0$, Eq. 10.26 shows that v_S must be greater than 0.6 V. If v_S is less than 0.6 V, the transistor will be cut off, i_B will be zero, and Eq. 10.21 shows that v_0 will also be zero. Thus the range of v_S for which active-gain operation is obtained is given by

$$0.6 < v_S < 0.6 + \frac{R_B + (1 + \beta_F)R_E}{(1 + \beta_F)R_E}(V_{CC} - V_{CE,sat}) \tag{10.27}$$

The complete voltage transfer characteristic is shown in Fig. 10.9, including the behavior in the cutoff and saturation regions. We can calculate the voltage gain for the emitter follower by taking the slope of the transfer characteristic in the active-gain region.

$$\text{Voltage gain} = \frac{(1 + \beta_F)R_E}{R_B + (1 + \beta_F)R_E} \tag{10.28}$$

Equation 10.28 shows that the voltage gain is less than unity for any combination of R_B, R_E and β_F. In most instances, however, R_B is less than $(1 + \beta_F)R_E$, so that the voltage gain is very nearly, but not quite, unity. Thus, in an approximate sense, the output at the emitter *follows* the input with unity gain. This is the origin of the name *emitter follower*.

Although the emitter follower does not produce voltage gain, it does produce current gain of magnitude β_F, and therein lies its usefulness. Two examples of this application are presented in the following sections.

10.2.2 A Transistorized Current Source

Figure 10.10*a* shows a circuit in which a transistor is connected in the emitter-follower configuration with a load resistor R_L added in the collector lead. The topology of the circuit looks a little different, but this is only because the V_{CC} supply is serving two functions, that of supply for the collector circuit, and that of input signal through the R_1-R_2 divider. The circuit is redrawn in Fig. 10.10*b* with these two functions separated, and with the input circuit replaced by its Thévenin equivalent.

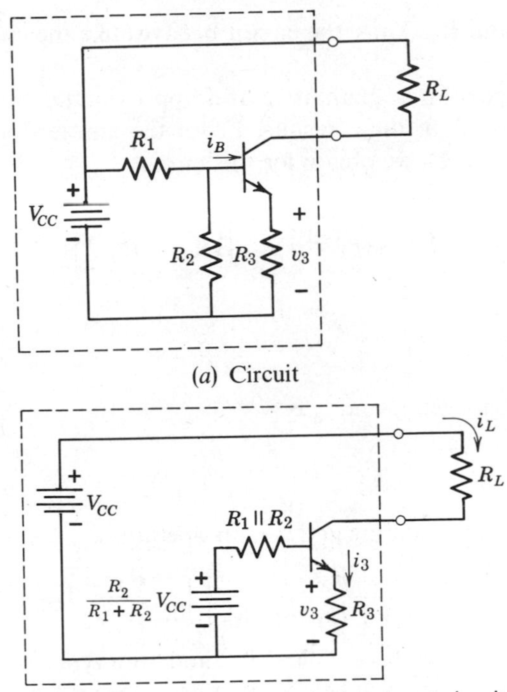

(*a*) Circuit

(*b*) Redrawn with Thévenin equivalent input circuit

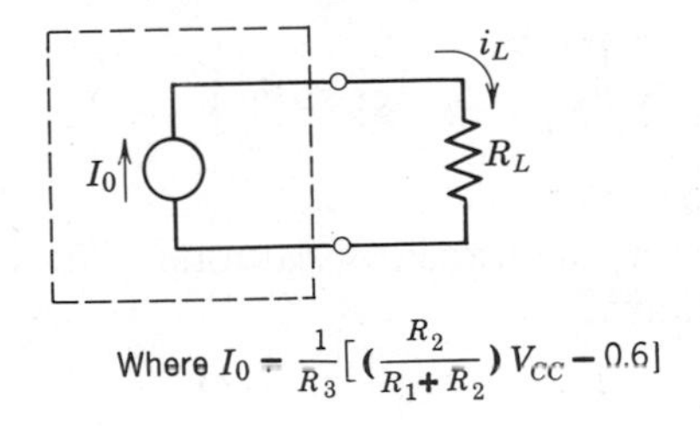

Where $I_0 = \frac{1}{R_3}[(\frac{R_2}{R_1 + R_2}) V_{CC} - 0.6]$

(*c*) Model for $0 < v_L < \left(\frac{R_1}{R_1 + R_2} V_{CC} + 0.6 - V_{CE,sat}\right)$

Figure 10.10
A transistorized current source.

In qualitative terms, here is how the circuit functions. The voltage across R_3 is nearly equal to the constant input voltage $[R_2/(R_1 + R_2)]V_{CC}$ through the emitter-follower action just discussed. By Ohm's law, the current in R_3 is also constant. If we can assume active-gain-region operation, the collector current i_L must be equal to i_3 to within a few percent, because α_F for a transistor is nearly unity (see Eq. 9.5). Therefore, i_L is constant, and is determined only by V_{CC}, R_1, R_2, and R_3. Thus, the circuit behaves like the current source in Fig. 10.10*c*.

The above discussion is qualitative and approximate. Now let us work through a derivation of these results. From the emitter-follower transfer characteristic, Eq. 10.23, we obtain for this circuit

$$v_3 = \left[\frac{(\beta_F + 1)R_3}{(R_1 \| R_2) + (\beta_F + 1)R_3}\right]\left(\frac{R_2}{R_1 + R_2}V_{CC} - 0.6\right) \tag{10.29}$$

The current i_3 is

$$i_3 = \frac{v_3}{R_3} \tag{10.30}$$

and the relation between i_L and i_3 is

$$i_L = \alpha_F i_3 = \frac{\beta_F}{\beta_F + 1} i_3 \tag{10.31}$$

Therefore we obtain, assuming active-gain operation,

$$i_L = \frac{1}{R_3}\left[\frac{\beta_F R_3}{(R_1 \| R_2) + (\beta_F + 1)R_3}\right]\left(\frac{R_2}{R_1 + R_2}V_{CC} - 0.6\right) \tag{10.32}$$

For R_3 of a size comparable to R_1 and R_2, and for a typically large β_F (50 or more), the term in square brackets is unity to within a few percent. In this case, we obtain

$$i_L = \frac{1}{R_3}\left(\frac{R_2}{R_1 + R_2}V_{CC} - 0.6\right) \tag{10.33}$$

which is constant and independent of R_L.

If R_L is too large, however, the transistor saturates, and Eq. 10.33 is no longer valid. By KVL, we have

$$V_{CC} - i_L R_L - v_{CE} - v_3 = 0 \tag{10.34}$$

To avoid saturation, therefore, we must have

$$V_{CE,sat} < v_{CE} = V_{CC} - i_L R_L - v_3 \tag{10.35}$$

Therefore, the boundary for saturation occurs when

$$i_L R_{L,max} = V_{CC} - v_3 - V_{CE,sat} \tag{10.36}$$

Substituting for i_L and v_3, Eq. 10.36 reduces to

$$R_L < R_3\left[\frac{\left(\dfrac{R_1}{R_1 + R_2}\right)V_{CC} + 0.6 - V_{CE,sat}}{\left(\dfrac{R_2}{R_1 + R_2}\right)V_{CC} - 0.6}\right] \tag{10.37}$$

Let us humanize these equations with actual numbers. Suppose $V_{CC} = 12$ V, $R_1 = 500\ \Omega$, $R_2 = R_3 = 100\ \Omega$, and $V_{CE,sat} \approx 0$. In this case we get a current source of magnitude

$$i_L = \frac{1}{0.1}\left(\frac{1}{6} \times 12 - 0.6\right) = 14\ \text{mA} \tag{10.38}$$

for

$$R_L < 0.1\frac{(5/6) \times 12 + 0.6}{(1/6) \times 12 - 0.6} = 0.76\ \text{k}\Omega = 760\ \Omega \tag{10.39}$$

10.2.3 A Regulated dc Power Supply

Figure 10.11*a* shows a variation of the emitter-follower theme, this time combined with a Zener diode operating in reverse breakdown to produce a regulated dc output voltage across a load resistor R_L. Once again, the V_{CC} supply is playing a dual role, both as collector supply and as bias supply for both the base input to the transistor and the Zener diode. If we assume that the Zener is operating in reverse breakdown, and if we assume that the Zener impedance R_Z is much less than R_1 (see Section 8.3.4 and Fig. 8.23), we can

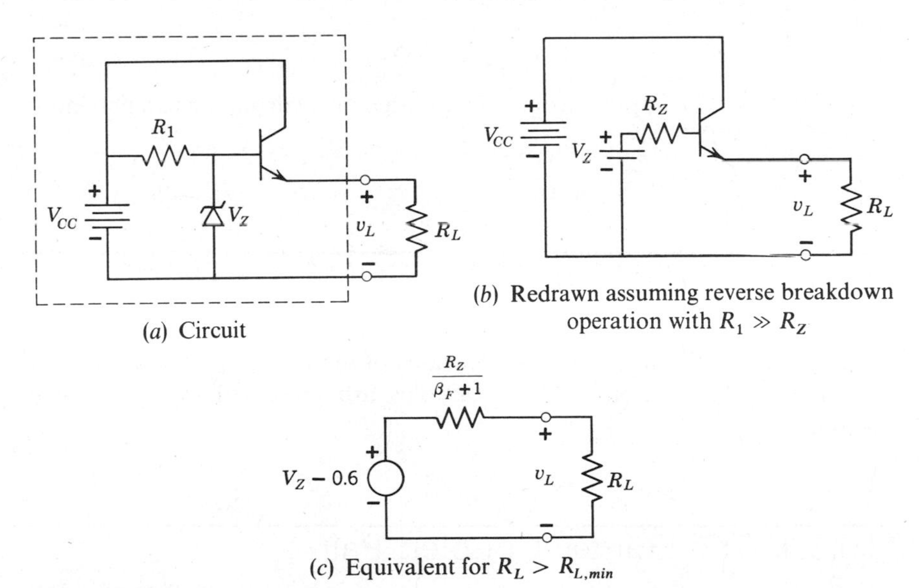

Figure 10.11
A regulated dc power supply.

redraw this circuit as in Fig. 10.11*b*. Drawing on our analysis of the emitter follower, we have immediately from Eq. 10.23 that

$$v_L = \left[\frac{(\beta_F + 1)R_L}{R_Z + (\beta_F + 1)R_L}\right](V_Z - 0.6) \tag{10.40}$$

which for $(\beta_F + 1)R_L \gg R_Z$ is effectively independent of R_L. Thus the emitter follower is acting like a dc voltage source of magnitude $V_Z - 0.6$. The output voltage in our approximate analysis is also independent of V_{CC}. Thus any ripple that might be present on the V_{CC} supply is strongly attenuated by this regulator. Emitter followers can be used in this way to obtain well-filtered dc supply voltages at relatively modest cost.

The equivalent circuit of Fig. 10.11*c* shows the Thévenin resistance of the regulated supply is $R_Z/(\beta_F + 1)$. Derivation of this resistance from Eq. 10.40 is left as an exercise. Also left as an exercise is the derivation that the regulator works only for $R_L > R_{L,min}$ where

$$R_{L,min} = \frac{R_1(V_Z - 0.6)}{(\beta_F + 1)(V_{CC} - V_Z)} \tag{10.41}$$

As a practical case, assume $V_{CC} = 12$ V, $V_Z = 6.2$ V, $R_Z = 10\ \Omega$, $R_1 = 1$ kΩ, and $\beta_F = 50$. In that case the regulated output voltage is

$$v_L = 5.6 \text{ V}$$

for

$$R_{L,min} = 20\ \Omega$$

Equivalently, this supply can deliver regulated load current up to a maximum of $v_L/R_{L,min} = 280$ mA.

10.3 TWO-TRANSISTOR COMBINATIONS

There are many two-transistor combinations that are so useful and that occur so often that they have acquired identities of their own. This section introduces two combinations: other examples are presented in the problem section.

10.3.1 The Emitter-Coupled Pair

The emitter-coupled pair, shown in its simplest form in Fig. 10.12, contains two transistors connected at the emitters. Although the example treated here

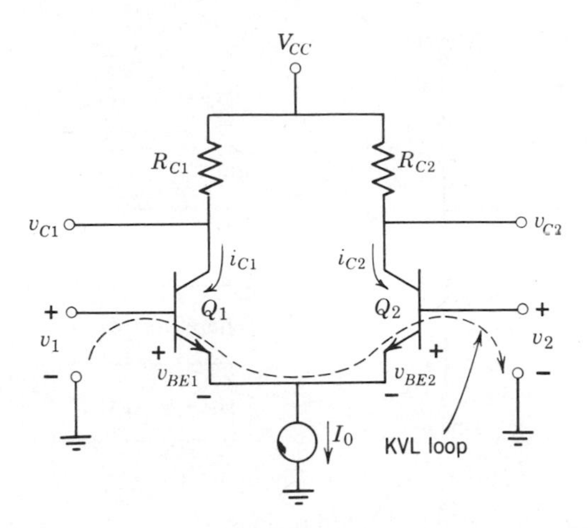

Figure 10.12
Emitter-coupled pair.

contains a pair of bipolar junction transistors, an equivalent-circuit configuration involving field-effect transistors (the source-coupled pair) is also widely used. The basic characteristics of the circuit are similar in either case.

In analyzing the operation of this circuit, we assume that both transistors are identical. This would be a risky assumption if one were building this circuit from discrete devices. In an integrated circuit, however, the two transistors can be fabricated simultaneously on the same substrate, so that their parameters are nearly identical.

The emitter-coupled pair is one of the few circuits we shall encounter in which a piecewise-linear large-signal transistor model is *not* adequate to describe the operation of the circuit. The reason, as we shall see below, is that it is the difference between the v_{BE}'s of the two transistors that controls their operating point. Therefore, the piecewise-linear transistor model, which presumes v_{BE} to be constant, is not a sufficiently accurate model. Instead, we shall use an exponential diode model shown in Fig. 10.13. The $\beta_F i_B$ dependent current source used in previous examples is represented in this model by its exact equivalent, the $\alpha_F i_E$ dependent current source. The assumption that both transistors are identical simply means that the parameters I_{ES} and α_F are the same in the two transistors. This model successfully represents both the active-gain and cutoff regions. It will be necessary, however, to examine saturation separately.

Let us return to Fig. 10.12 and examine the circuit. The two transistors are biased with a collector supply V_{CC} and a current source I_0. Either base can serve as an input, and outputs can be obtained from either or both collectors. The operating points of the transistors can be found from KCL at the emitter

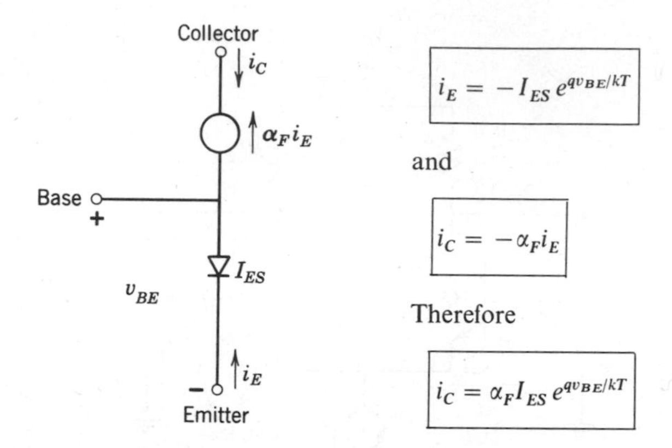

Figure 10.13
Transistor model for active-gain and cutoff regions.

node and from KVL around the loop indicated in Fig. 10.12. Taking KCL at the emitter node first, we note that the sum of the emitter currents is constant.

$$I_0 = -(i_{E1} + i_{E2}) \tag{10.42}$$

In the absence of saturation, since α_F is presumed very close to unity, Eq. 10.42 can be rewritten in terms of the collector currents as

$$I_0 = i_{C1} + i_{C2} \tag{10.43}$$

Next, we write KVL around a loop that goes from ground through the v_1 node, through both emitter-base junctions and back to ground through the v_2 node. The result of KVL around this loop is that any *difference* between the two input voltages shows up as a *difference* between the v_{BE}'s of the two transistors:

$$v_{BE1} - v_{BE2} = v_1 - v_2 \tag{10.44}$$

Let us consider the qualitative implications of these two relations (Eqs. 10.43 and 10.44). The sum of the collector currents is fixed by I_0, while the difference in the v_{BE}'s is determined by the difference in the input voltages. If the input voltages happen to be equal, then the v_{BE}'s are equal. Furthermore, if the v_{BE}'s are equal, the fact that the transistors are identical requires that the two collector currents be equal. Thus, if v_1 and v_2 are equal, the current from the current source must divide equally between the two transistors. That is,

$$\text{If} \quad v_1 = v_2 \quad \text{then} \quad i_{C1} = i_{C2} = \frac{I_0}{2} \tag{10.45}$$

Consider now what happens if v_1 and v_2 are not equal. If, for example, v_1 is greater than v_2, Eq. 10.44 shows that v_{BE1} must be greater than v_{BE2}. Since the transistors are identical, whichever transistor has the greater v_{BE} must have the greater collector current. Therefore, i_{C1} must be greater than i_{C2}. But since the sum of i_{C1} and i_{C2} must remain constant, as i_{C1} becomes greater than $I_0/2$, i_{C2} must become smaller than $I_0/2$. Thus the current I_0 divides between the two transistors according to the unbalance in the v_{BE}'s imposed at the inputs. We shall now develop a quantitative relationship that exhibits this property.

The transistor model of Fig. 10.13 has the characteristic that

$$i_{C1} = \alpha_F I_{ES} \, e^{qv_{BE1}/kT} \tag{10.46}$$

and

$$i_{C2} = \alpha_F I_{ES} \, e^{qv_{BE2}/kT} \tag{10.47}$$

If we substitute from Eq. 10.44 for v_{BE1}, Eq. 10.46 becomes

$$i_{C1} = (\alpha_F I_{ES} \, e^{qv_{BE2}/kT}) \, e^{q(v_1 - v_2)/kT} \tag{10.48}$$

The term in parentheses equals i_{C2}. Therefore

$$i_{C1} = i_{C2} \, e^{q(v_1 - v_2)/kT} \tag{10.49}$$

Substituting Eq. 10.49 into Eq. 10.43, we find

$$I_0 = i_{C2}(1 + e^{q(v_1 - v_2)/kT}) \tag{10.50}$$

This can be rearranged to yield

$$i_{C2} = \frac{I_0}{1 + e^{q(v_1 - v_2)/kT}} \tag{10.51}$$

and

$$i_{C1} = \frac{I_0}{1 + e^{-q(v_1 - v_2)/kT}} \tag{10.52}$$

Finally, we obtain the output voltages

$$v_{C2} = V_{CC} - \frac{I_0 R_{C2}}{1 + e^{q(v_1 - v_2)/kT}} \tag{10.53}$$

$$v_{C1} = V_{CC} - \frac{I_0 R_{C1}}{1 + e^{-q(v_1 - v_2)/kT}} \tag{10.54}$$

These equations are valid provided there is no saturation; that is, v_{C1} must not drop below v_1 by more than 0.1 or 0.2 V, and v_{C2} must not drop below v_2 by more than 0.1 or 0.2 V.

10.3.2 Difference-Mode Amplification

The division of current between the two transistors (Eqs. 10.51 and 10.52) is determined by the difference between v_1 and v_2. Therefore, the output voltages (Eqs. 10.53 and 10.54) also depend only on the difference between v_1 and v_2. For this reason, the emitter-coupled pair is often called a *differential amplifier*. The difference between v_1 and v_2 constitutes a *difference-mode* input signal.

The transfer characteristic between the difference-mode input $(v_1 - v_2)$ and one of the outputs (v_{C2}) is shown in Fig. 10.14. Two cases are illustrated.

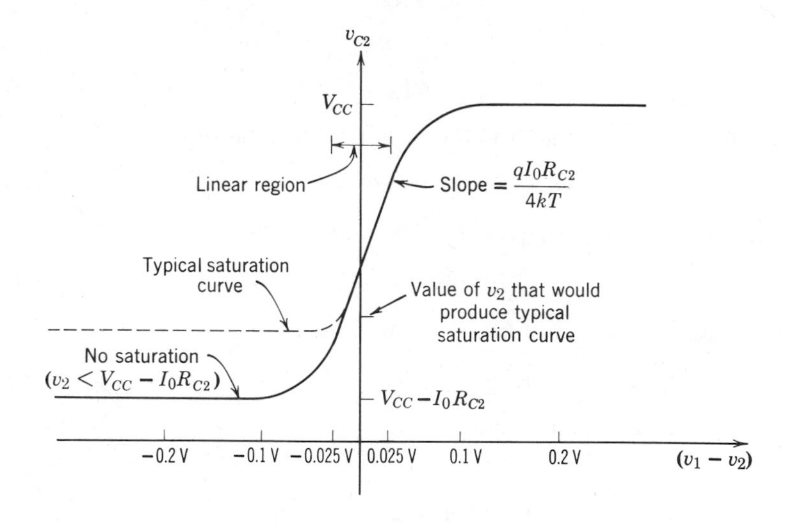

Figure 10.14

Emitter-coupled pair transfer functions. Solid curve: $v_2 < V_{CC} - I_0R_{C2}$. No saturation. Dashed curve: $v_2 > V_{CC} - I_0R_{C2}$. In this case, Q_2 saturates when $v_1 - v_2$ gets negative enough to bring v_{C2} below v_2 by more than a few tenths of a volt.

The solid curve assumes no saturation, and is thus a plot of Eq. 10.53. The upper and lower limits of v_{C2} are at V_{CC} (with Q_2 cutoff) and at $V_{CC} - I_0R_{C2}$ (with Q_2 carrying the entire current from the I_0 source). Within about ± 25 mV of the origin, the relation between v_{C2} and $v_1 - v_2$ is linear. The slope of the transfer characteristic in the linear region is $qI_0R_{C2}/4kT$. This slope represents the linear-region difference-mode gain of the amplifier.

If the no-saturation curve is to apply, then v_{C2} must never drop below v_2 by more than 0.1 or 0.2 V. The dashed curve in Fig. 10.14 illustrates how the transfer characteristic might be modified if v_2 were actually larger than the limiting value of $V_{CC} - I_0 R_{C2}$. As $v_1 - v_2$ becomes negative, transistor Q_2 saturates, with the result that v_{C2} drops only to a value that is a few tenths of a volt less than v_2.

Since the linear-region gain is proportional to the bias current I_0, the gain of the emitter-coupled pair can be adjusted over a wide range simply by changing the bias current. A circuit for accomplishing the current-source biasing is shown in Fig. 10.15. The role of the current source is played by a third transistor Q_3 whose collector supplies the current I_0 (see Section 10.2.2).

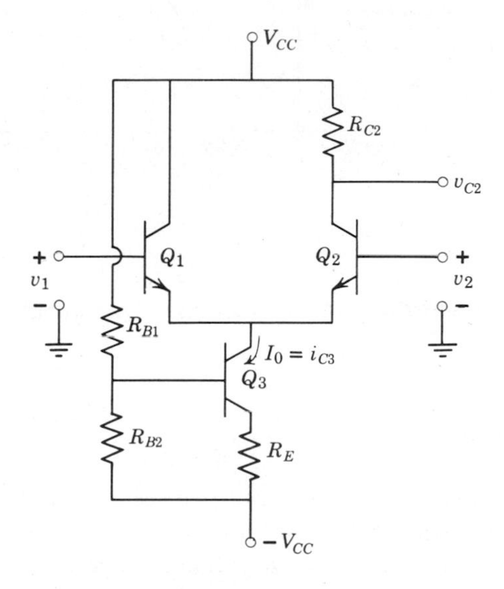

Figure 10.15
An emitter-coupled pair with a transistor current source Q_3. If only the v_{C2} output is used, R_{C1} can be set to zero.

As the operating point of Q_3 is changed, the bias current changes. The effect of such a change is shown in Fig. 10.16, in which output curves for v_{C2} have been plotted for three different choices of bias current. Notice that for $v_1 - v_2$ very positive, all three curves saturate at the positive supply V_{CC}. For $v_1 - v_2$ very negative, however, the limit of v_{C2} (which equals $V_{CC} - I_0 R_{C2}$) depends on I_0. Note further that the slopes of the three curves at $v_1 - v_2 = 0$ are different, being in fact proportional to I_0. Because of this dependence of linear-region gain on bias current, the emitter-coupled pair finds wide usage in automatic-gain-control (AGC) applications and in modulation circuits, a topic covered in Chapter 15.

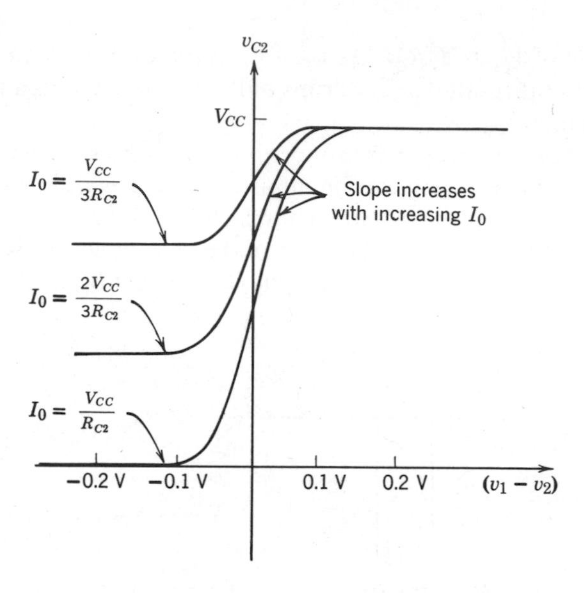

Figure 10.16
Emitter-coupled pair transfer characteristics for three different bias currents.

10.3.3 Common-Mode Rejection

We have already commented on the fact that when v_1 and v_2 are equal, the bias current divides equally between the identical transistors Q_1 and Q_2 with the result that the output voltages do not depend on v_1 or v_2. Voltages that appear equally at both inputs are called *common-mode voltages* (as distinguished from the difference-mode voltages of the preceding section). The emitter-coupled pair, and the related circuit involving FETs, has the property that a common-mode voltage applied to the inputs produces no change in output. This remarkable cancellation of common-mode signals is called *common-mode rejection.* Op-amps almost always have an emitter-coupled pair (or source-coupled pair) as an input circuit to obtain a large common-mode rejection.

In actual devices there is always some degree of imperfection in the match between two transistors. Therefore, actual devices do not show the perfect common-mode rejection of our somewhat idealized model. A measure of the quality of the rejection of common-mode signals is called the *common-mode rejection ratio*, abbreviated CMRR (or often, simply CMR). The definition of CMR as introduced in Section 5.4.6 is general, and applies to these circuits.

Although the common-mode rejection just discussed refers to variations of the input signals about zero, these circuits show good common-mode rejection even if v_1 and v_2 differ appreciably from zero. The circuit, however, does impose limitations on how large v_1 and v_2 can be before the common-mode rejection breaks down. Consider the circuit of Fig. 10.15 in which the bias current is derived from a transistor current source. If v_1 and v_2 become too negative, Q_3 will saturate. Once Q_3 saturates, further decreases in v_1 and v_2 will cause Q_1 and Q_2 to cut off, producing a marked effect on the output. In the other direction, if v_1 and v_2 get more than a few tenths of a volt above v_{C1} or v_{C2}, the emitter-coupled pair will saturate, once again with resulting changes in output. Thus the common-mode rejection characteristic and the useful amplification region is limited to a range of input voltages.

10.3.4 Complementary-Symmetry Pairs

We have just seen how two transistors can be combined in the emitter-coupled pair to provide symmetrical voltage amplification of positive or negative difference-mode signals. This section describes in qualitative fashion how a pair of complementary transistors, one *n-p-n* the other *p-n-p*, can be used to provide symmetrical current amplification. The complementary symmetry pair is shown in simplest form in Fig. 10.17. Q_1 is an *n-p-n* transistor connected in the emitter-follower configuration, Q_2 is a *p-n-p* transistor, also connected in the emitter-follower configuration. The two inputs are tied together, and the load R_L is shared.

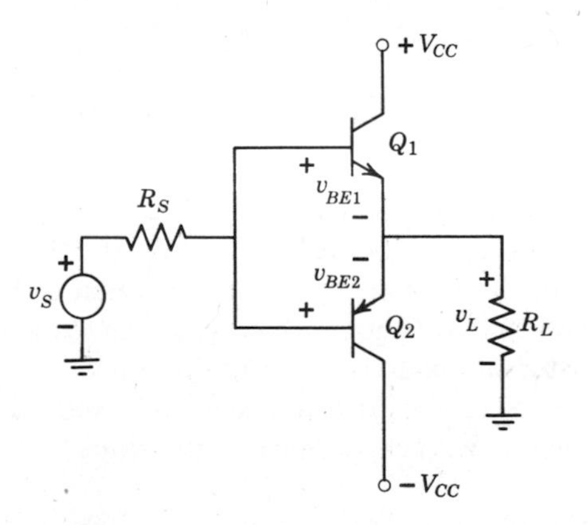

Figure 10.17
Complementary-symmetry pair.

The transfer characteristic for this circuit is sketched in Fig. 10.18 using the piecewise-linear transistor models appropriate to *n-p-n* and *p-n-p* transistors. The easiest way to keep them straight is to remember that the variables of interest (v_{BE}, i_B, v_{CE}, and i_C) are all positive for the *n-p-n* transistor in the active-gain region, and are all negative for the *p-n-p* transistor in the active-gain region. Thus, for $v_S > +0.6$ V, Q_1 turns on and we get the transfer characteristic in the right half-plane of Fig. 10.18. In exactly analogous fashion, for $v_S < -0.6$ V, Q_2 turns on, and we get the transfer characteristic in the left half-plane of Fig. 10.18. Notice, however, that there is a region between -0.6 and $+0.6$ V where neither Q_1 nor Q_2 is conducting. The resulting transfer characteristic is zero in this region, and serious distortion, called *crossover distortion*, would result from the use of this circuit as it is.

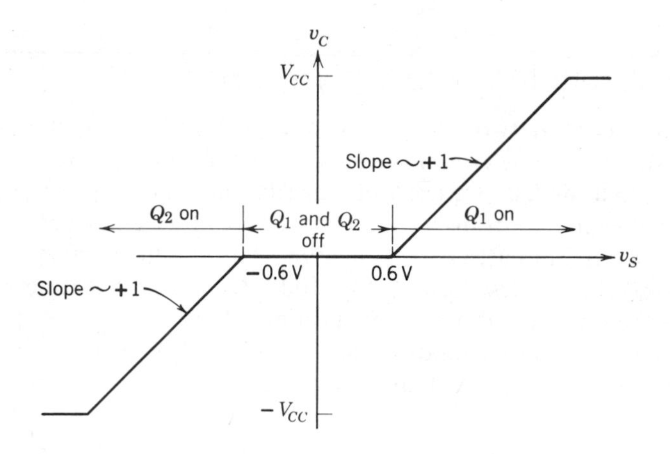

Figure 10.18
Crossover distortion in the complementary-symmetry pair.

Figure 10.19 shows schematically two ways of eliminating this crossover distortion. In Fig. 10.19*a*, diodes are used to bias the two transistors so that Q_1 and Q_2 are both in the active-gain region when v_S is zero. As v_S goes positive, Q_2 turns off, and as v_S goes negative, Q_1 turns off, but there is always at least one transistor in the active-gain region for v_S near zero. The diodes significantly reduce the crossover distortion, and produce a transfer characteristic that is much more nearly linear with slope $+1$ in the vicinity of $v_S = 0$.

A second method illustrates the use of an op-amp in a feedback loop to compare the output voltage v_L against the input signal v_S. Remember that

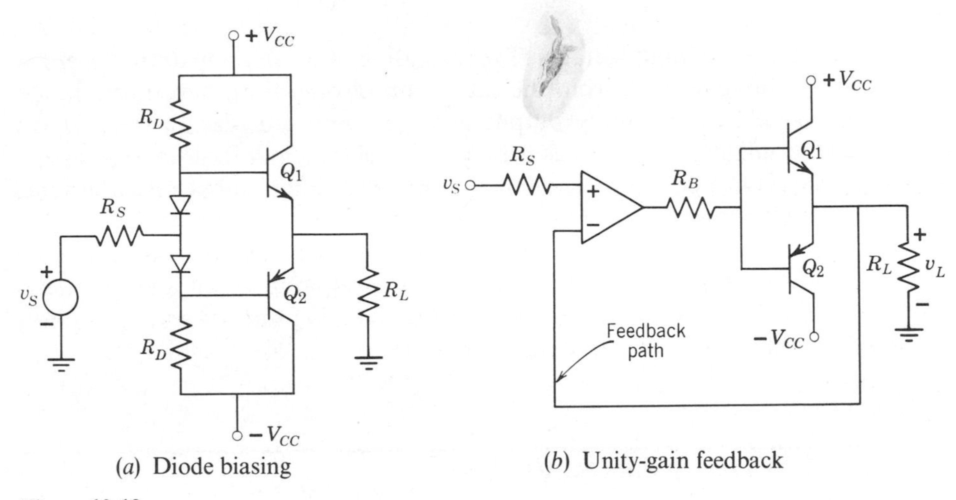

(*a*) Diode biasing

(*b*) Unity-gain feedback

Figure 10.19
Two ways to reduce crossover distortion.

the open-loop gain of the op-amp is enormous, 10,000 or more. This means that the voltage applied to the bases of the two transistors from the op-amp's output terminal will have magnitude sufficient to turn on either Q_1 or Q_2 as soon as v_S and v_L differ by as much as 0.6 V/10,000, or 60 μV. If $v_S = 0.1$ V, for example, which is in the crossover distortion region, the output of the op-amp goes up toward 0.6 V just enough to turn Q_1 and drive enough current through R_L to bring v_L to within 60 μV of 0.1 V. Thus, feedback is used to compare the output to the input, and apply a correction (here to the voltage on the transistor bases) until input and output agree.

10.3.5 A Comment: Understanding Op-Amps

A typical commercial op-amp consists of an emitter-coupled pair (or source-coupled pair) for initial amplification followed by one or more amplification stages that might be of a variety of types (common emitters and emitter-coupled pairs being most common). The output stage very often consists of a complementary-symmetry pair in which a combination of diode biasing and feedback is used to eliminate crossover distortion. Thus with the circuits we have already discussed in hand, it becomes possible to understand at least qualitatively many of the characteristics of actual op-amps.

The differential amplification and excellent common-mode rejection come from the input emitter-coupled pair. The limitations on allowed common-mode input voltage arise because various transistors in either the input stage or in subsequent stages can be driven into saturation or cutoff by excessive variation of the input voltages. The saturation of the op-amp output near the supply voltages arises from the saturation of one of the transistors in the complementary-symmetry output pair. The nonzero bias currents at the input terminals are needed as base currents of the input transistors to assure active-region operation. The input offset voltage and input offset currents arise from imperfect matching of the input pair of transistors and from biasing imperfections. Although we have yet to discuss issues related to input and output resistance and issues related to op-amp speed and frequency response, with a few basic circuit concepts involving transistors, many quirks of the "black box" begin to make sense.

10.4 THE FET SWITCH

The uses of the various kinds of field-effect transistors (FET) are in many ways similar to the uses of the bipolar junction transistors (BJT). For example, the large-signal transfer characteristic of a common-source amplifier (take an *n-p-n* common-emitter amplifier and replace the transistor with an enhancement mode *n*-channel FET) will be similar in many ways to the transfer characteristic of an *n-p-n* common-emitter amplifier. There is a cutoff region, an active-gain region, and a saturation region, and many of the same circuit functions can be performed. In several important ways, however, FETs are not equivalent to BJTs, and it is these features that we wish to emphasize here. First, the FET behaves like an open circuit at its gate terminal. The MOSFET has an insulator between the gate and channel, while the JFET gate is back-biased with respect to the channel (in normal operation). As a result, the FET in normal operation draws no current (or *almost* no current) from a source network and is therefore extremely useful in applications where it is necessary not to load down a source network that has a high Thévenin resistance. Second, the FET output characteristics are rather symmetrical about the origin. That is, for small signals ($|v_{DS}| \ll |V_P|$) the FET channel looks like a pure resistance, with no nonlinearities or saturation voltage present. As a result, the FET is particularly useful as a low-power voltage-controlled switch for low-level signals. We shall discuss this application further below, in full recognition that we are of necessity being very incomplete.

10.4.1 The Need For A Graphical Approach

Unlike the BJT, where we could make an extremely simple model of the device, in many uses the FET defies the use of simple models made of ordinary circuit elements (such as diodes and dependent current sources). Therefore, one must often use an algebraic representation of the device characteristics as derived in Section 9.2, or one must be willing to work with graphical methods from graphical representations of the device characteristics.

Figure 10.20*a* shows the output characteristics of an *n*-channel JFET. Notice that the characteristics have been extended into the third quadrant. Figure 10.20*b* shows an expanded scale view of the region around the origin, where $|v_{DS}| \ll |V_P|$. Notice that in this region the characteristics are relatively

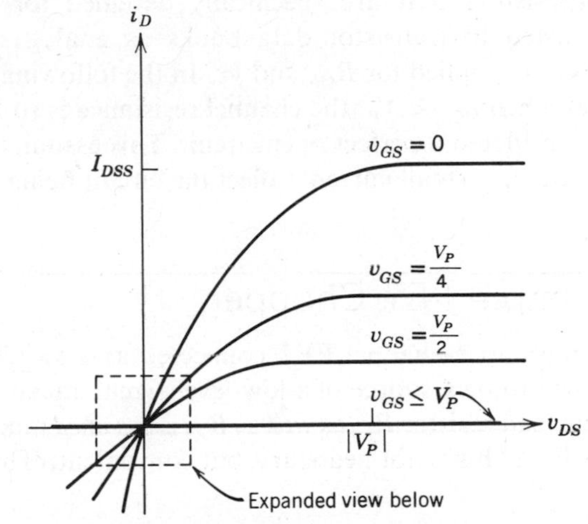

(*a*) Complete output characteristic curves for an *n*-channel JFET

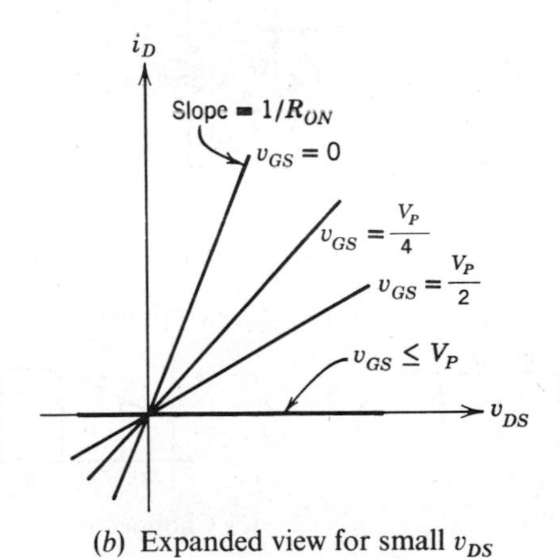

(*b*) Expanded view for small v_{DS}

Figure 10.20
Characteristics of an *n*-channel JFET.

linear and all pass through the origin. That is, for small values of v_{DS}, the output characteristics of an FET are just those of a voltage-controlled resistor. If, for example, we change the value of v_{GS} from a value near zero, at which the slope of the v_{DS}-i_D characteristic is largest, to a value more negative than V_P, at which the v_{DS}-i_D characteristic lies along the $i_D = 0$ axis, the effective resistance of the channel can be made to change from some minimum value R_{ON} to some very large value. Thus, by alternating between zero gate voltage ($v_{GS} = 0$) and pinchoff ($v_{GS} < V_P$) we can use the channel resistance of the FET as an "ON–OFF" switch.

Field-effect transistors that are specifically designed for this kind of application are listed in transistor data books as analog switches, and numerical values are supplied for R_{ON} and V_P. In the following examples we shall assume that when $v_{GS} < V_P$, the channel resistance is so large that the channel can be considered a perfect open circuit. This assumption generally is valid, unless extremely small currents affect the circuit being analyzed.

10.4.2 A Simple FET Chopper

Figure 10.21*a* shows an *n*-channel FET connected as a switch. The v_A-R_A network is assumed to be a source of a low-level signal, much less than $|V_P|$ in magnitude. The load resistor R_L, as well as R_A, is *assumed* much larger than the R_{ON} for the FET. (This is not necessary, but convenient.) The drive signal

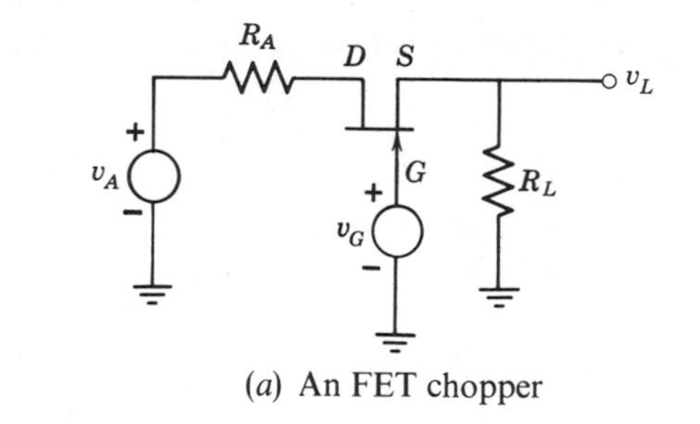

(*a*) An FET chopper

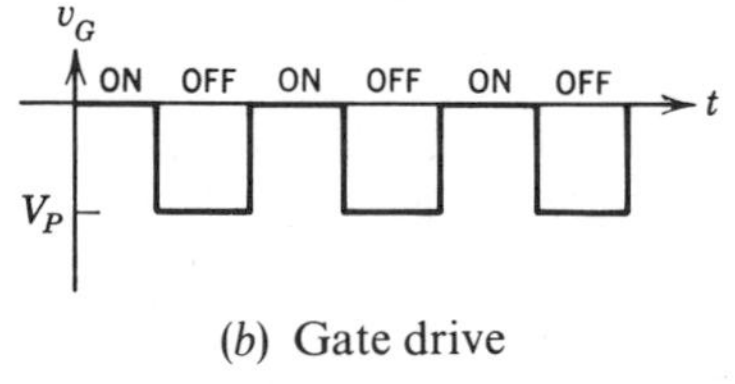

(*b*) Gate drive

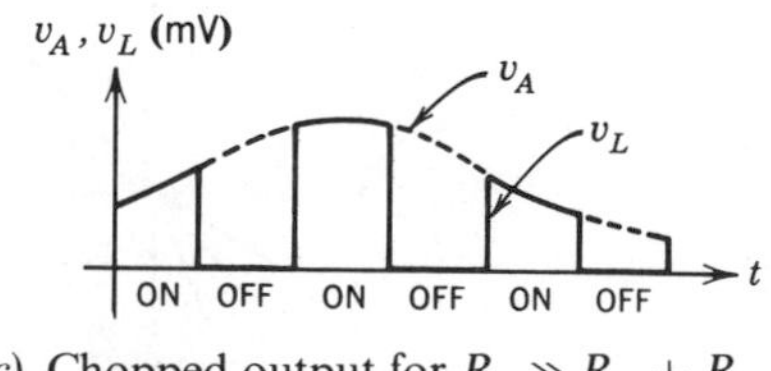

(*c*) Chopped output for $R_L \gg R_A + R_{ON}$

Figure 10.21
A single-channel FET chopper.

for the gate, v_G in the figure, differs from v_{GS} only by the amount of signal from v_A that appears across R_L. Since v_A is by assumption very small compared to $|V_P|$, v_G is here equal to v_{GS} for all practical purposes.

If the negative-going square wave of Fig. 10.22*b* is applied to the gate while a signal is present from v_A, the output waveform will be a *chopped* or *sampled* version of v_A. Specifically, when $v_{GS} = 0$, we have by voltage divider action

$$v_L = \frac{R_L}{R_L + R_A + R_{ON}} v_A$$

or (10.55)

$$v_L \approx v_A \qquad \text{for} \qquad R_L \gg R_A + R_{ON}$$

and when $v_{GS} < V_P$, we must have $v_L = 0$ because the drain-to-source connection is an open circuit.

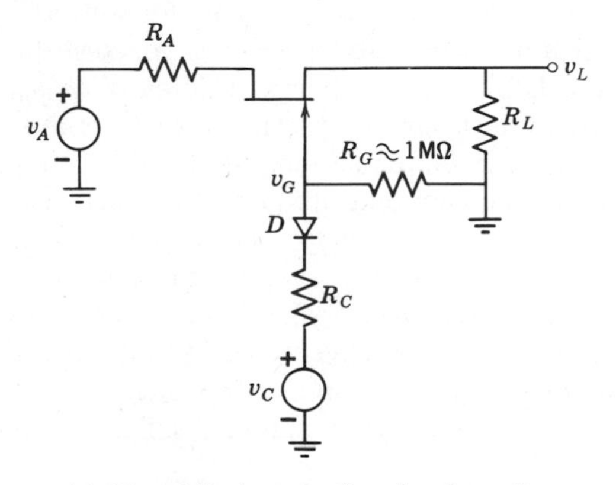

(*a*) Use of diode and referral resistor R_G

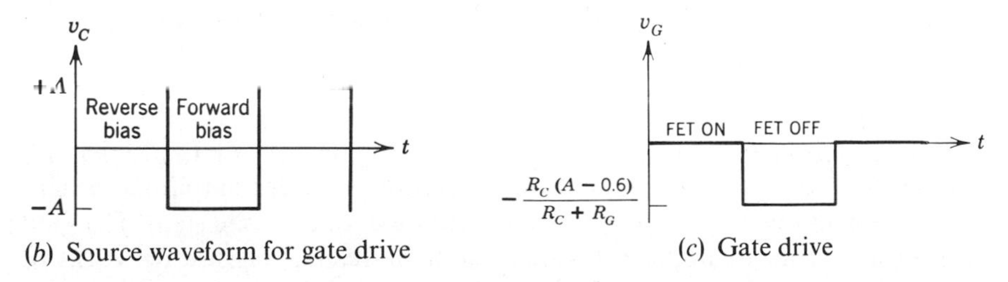

(*b*) Source waveform for gate drive

(*c*) Gate drive

Figure 10.22
Use of a diode and referral resistor to assure $v_{GS} = 0$ when FET is ON.

It is usually necessary to construct a voltage waveform for the gate drive v_G from other more conventional waveforms, such as square waves (see

(Fig. 10.22). The source v_C supplies a square wave of amplitude $\pm A$, which is coupled to the FET gate through a diode. At the same time, the FET gate is tied to ground through a large resistor R_G, the *referral resistor*. When v_C becomes positive (we assume A is greater than the magnitude of V_P, and is therefore much greater than v_A), the diode becomes reverse biased, and conducts no current. The FET gate, being tied to ground through R_G, comes to ground potential, and v_{GS} is held near zero. When v_C goes negative, the diode becomes forward biased. Current flows from ground, through R_G and the diode, to the v_C source. The voltage on the gate is obtained from a voltage divider relation:

$$v_G = -\frac{R_G}{R_C + R_G}(A - 0.6) \tag{10.56}$$

To use this drive circuit, the negative excursion of v_G (Eq. 10.56) must exceed the pinchoff voltage of the FET.

Choppers have two principal uses. The first is for converting slowly varying waveforms to high-frequency ac waveforms for drift-free amplification. These applications fall under the generic heading of *chopper amplifiers* and are discussed further in Chapter 18. The second application is of fundamental importance in communications applications. If the signal varies slowly compared to the chopping frequency applied to the gate, it is not necessary to sample v_A very often to recover the original waveform once again with suitable filters. It becomes possible, therefore, to sample in sequence many such slowly varying waveforms and to send samples of each down a single transmission line. This process is called *multiplexing* of signals, and since it involves time-sequential sampling, it is called *time-division multiplexing*. We discuss now a two-channel FET analog multiplexer.

10.4.3 A Two-Channel Analog Multiplexer

Figure 10.23 shows a two-channel analog multiplexer, in which alternate samples of two low-level source waveforms, v_A and v_B appear at the output node v_M. The equal but opposite gate drives for the two FETs are derived from a sine wave, using a comparator (to produce a large-amplitude square wave) and an inverter (to produce a square wave of opposite sign). The gate drive waveforms are applied through the diode-referral resistor networks of Fig. 10.22 to insure that v_{GS} is zero during the ON cycle, and not some positive voltage that would inject carriers into the channel and produce a signal error. Figure 10.24 shows a demultiplexer that accepts the waveform v_M, and applies the waveform to two alternately switched *sample-and-hold* networks, one for v_A, the other for v_B. The drive waveforms for the demultiplexer must

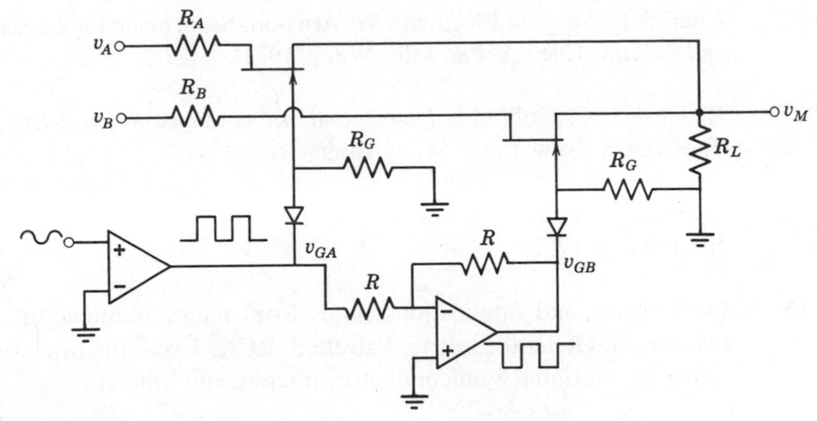

Figure 10.23
A two-channel multiplexer.

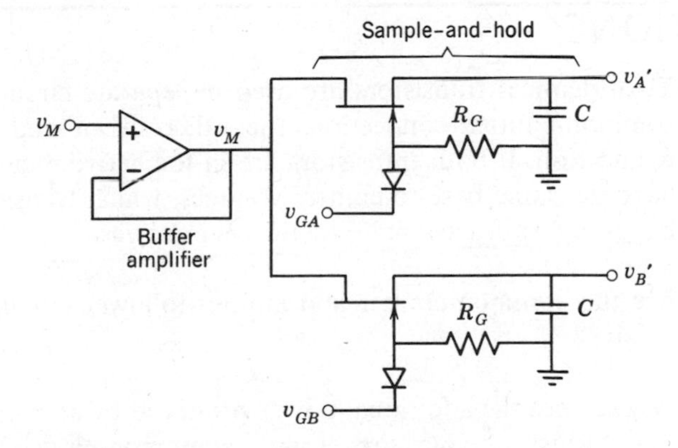

Figure 10.24
A synchronous demultiplexer for the multiplexer of Fig. 10.23.

be accurately synchronized with the multiplexer drives v_{GA} and v_{GB}, or else the alternate multiplexed samples of v_A and v_B will not get switched to the correct sample-and-hold network at the correct time. Further analysis of this network is reserved for the problem section.

References

R10.1 Paul E. Gray and Campbell L. Searle, *Electronic Principles: Physics, Models, and Circuits* (New York: John Wiley, 1969), Chapter 2 and Chapters 7 to 10.

R10.2 Charles L. Alley and Kenneth W. Atwood, *Semiconductor Devices and Circuits* (New York: John Wiley, 1971).

R10.3 Richard S. C. Cobbold, *Theory and Applications of Field-Effect Transistors* (New York: Wiley-Interscience, 1970).

R10.4 Carl David Todd, *Junction Field Effect Transistors* (New York: John Wiley, 1968).

R10.5 Data sheets and applications notes from major manufacturers (Motorola, General Electric, Fairchild, RCA, Texas Instruments, Siliconix, National Semiconductor, Intersil, and others).

QUESTIONS

Q10.1 Two identical transistors are used in separate circuits, one in a common-emitter connection, the other in an emitter-follower connection. If both transistors are in the active region and both have the same base-to-emitter voltages, which transistor has the larger collector current? Explain your answer.

Q10.2 Are the common-emitter and emitter-follower circuits active or passive?

Q10.3 Where does the additional power produced by an amplifier come from? Illustrate your answer with a common-emitter amplifier.

Q10.4 Where along a particular load line is the power dissipation maximum? Minimum?

Q10.5 Which is likely to dissipate the least power, a transistor switch or a transistor amplifier operating in the linear region? Explain.

EXERCISES

E10.1 (a) Using the collector characteristics in Fig. E10.1*c* and assuming $V_{BE} = 0.6$ V, find the operating point of the transistor (V_{CE}, I_C) in the circuit in Fig. E10.1*a*.

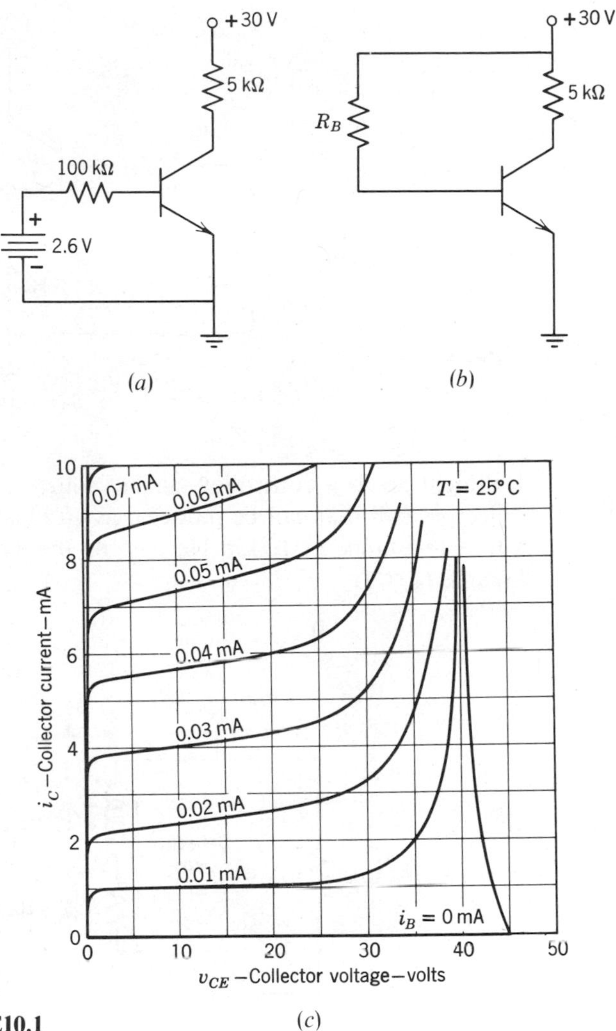

Figure E10.1

(b) It is desired to obtain the base current from the +30 V supply, eliminating the 2.6 V battery. Find R_B in Fig. E10.1*b* such that the operating point of part *a* is maintained.

E10.2 Plot the large-signal transfer characteristic (v_O vs. v_I) for the emitter-follower amplifier shown in Fig. E10.2, indicating the cutoff, active-gain, and saturation regions. Estimate the voltage gain v_O/v_I in the active-gain region.

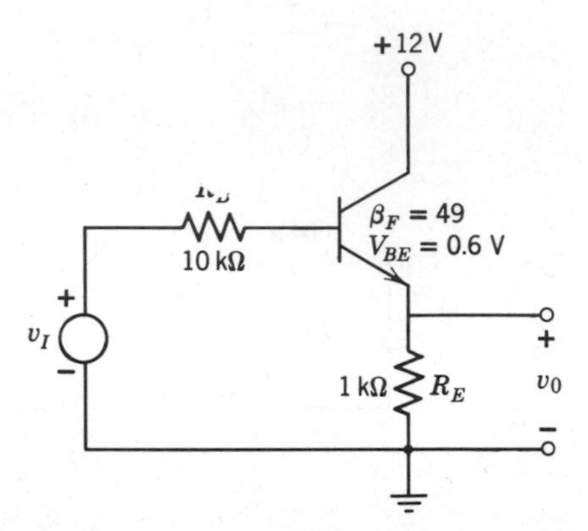

Figure E10.2

E10.3 A student needs a controlled current source to drive a magnetic deflection coil that can be modeled as an ideal inductor in series with a resistance of 100 Ω. He designs the circuit in Fig. E10.3. Assume $I_B \ll I_1$.

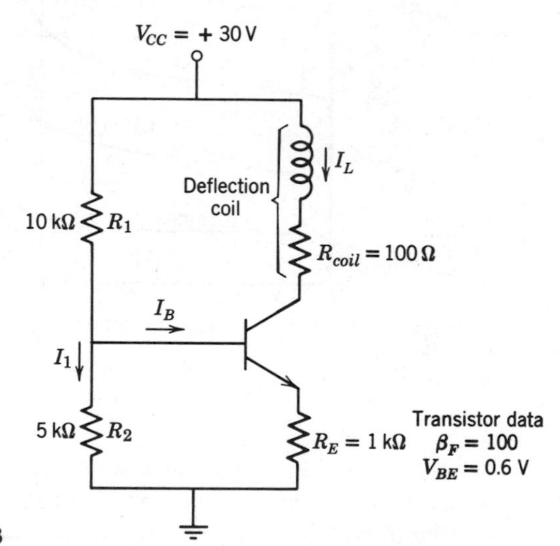

Figure E10.3

(a) For the above circuit, find the current in the coil, I_L. Call it I_0.

(b) Leaving the rest of the circuit unchanged, pick a *new value of* R_1 so that the coil current becomes $I_L = 2I_0$.

E10.4 How should you select the value of R_3 so that the network in Fig. E10.4b will have the same v-i characteristic as the network in Fig. E10.4a? (Assume $R_2 < \beta_F R_1$.) Explain your answer.

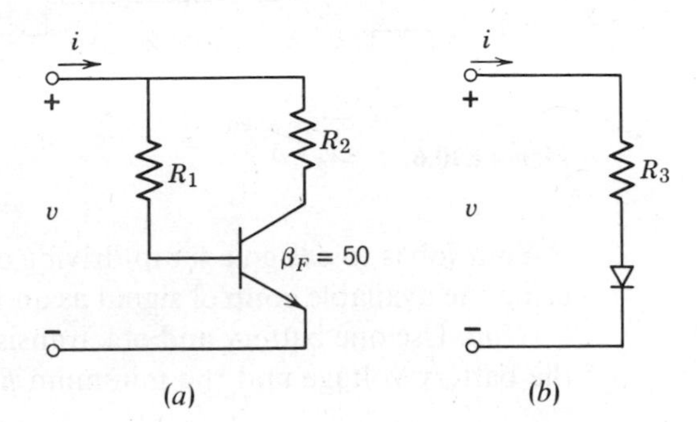

Figure E10.4

E10.5 Plot v_2 versus v_S for the circuit of Fig. E10.5. Label carefully all features of your plot, and indicate the values of v_S at which the transistor will enter the cutoff and saturation regions of operation.

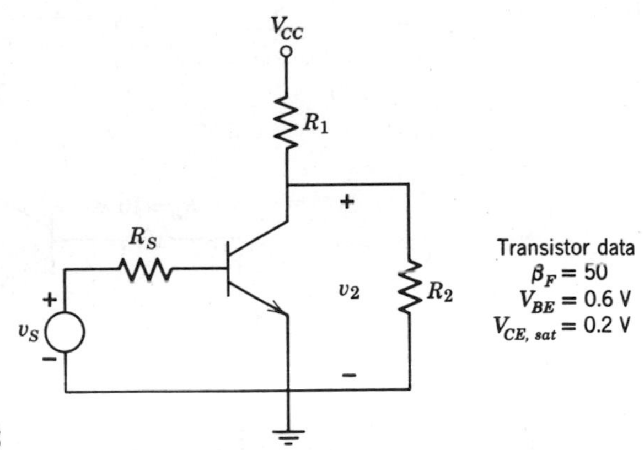

Figure E10.5

E10.6 You are given a light bulb that has a resistance of 600 Ω and a requirement for at least 10 mA of current for proper lighting (Fig. E10.6).

At the same time, you have a control circuit that has a Thévenin resistance of 1 kΩ and an open-circuit voltage that changes from 0 to 1 V when the lamp is to be turned on.

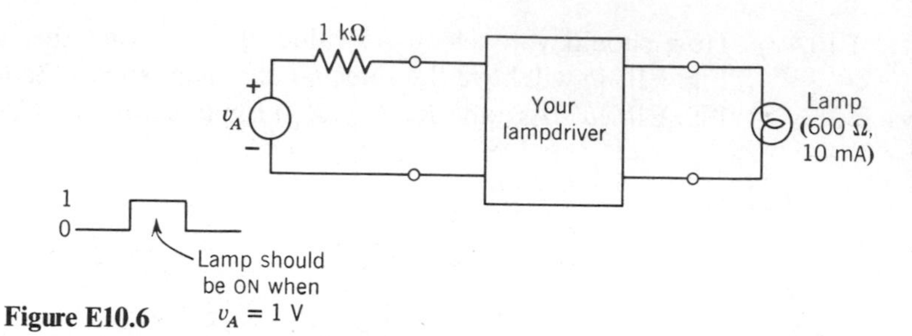

Figure E10.6

Your job is to design a lamp-driving circuit that lights the lamp using the available control signal as an input.

(*Hint*. Use one battery and one transistor, but be sure to specify the battery voltage and the minimum β_F for the transistor.)

E10.7 (a) What are the limits on v_B for transistor operation in the linear region in Fig. E10.7?
(b) Sketch and dimension v_E versus v_B and v_C versus v_B in the linear region.
(c) For what value of v_B is $i_C = 1$ mA?

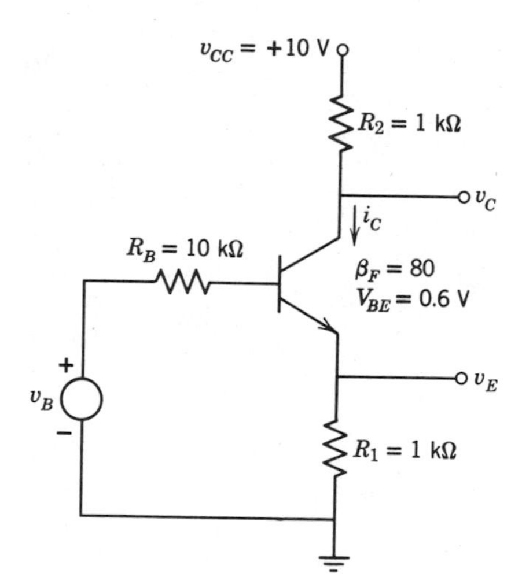

Figure E10.7

E10.8 (a) In Fig. E10.8 with $V_i = 0$, use the large-signal model of Fig. 10.4*b* to find the operating point (V_{CE}, I_C) of the transistor.
(b) Sketch and dimension the output waveform $v_0(t)$ for V_i = 100 mV.
(c) Repeat part *b* for V_i = 2 V.

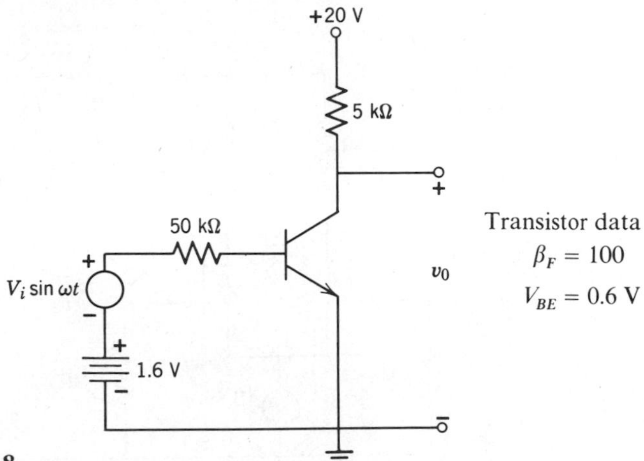

Figure E10.8

E10.9 It is claimed that the bias circuit in Fig. E10.9 *guarantees* that the transistor is in the active region. *Show* whether this claim is right or wrong.

Assume that the transistor is characterized in the active region by β_F and V_{BE} with $V_{BE} < V_{CC}$.

Assume further that in saturation, the collector-to-emitter voltage is less than V_{BE}.

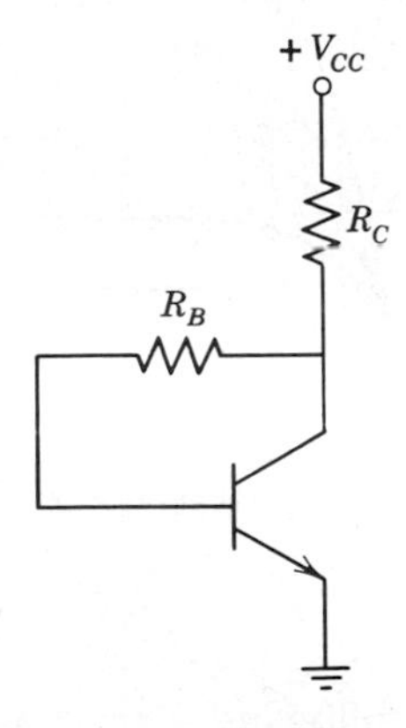

Figure E10.9

E10.10 The differential amplifier shown in Fig. E10.10*a* uses an FET front end and a *p-n-p* transistor. Q_1 and Q_2 are identical transistors with the drain characteristics in Fig. E10.10*b*. Q_3 has a β_F of 100.

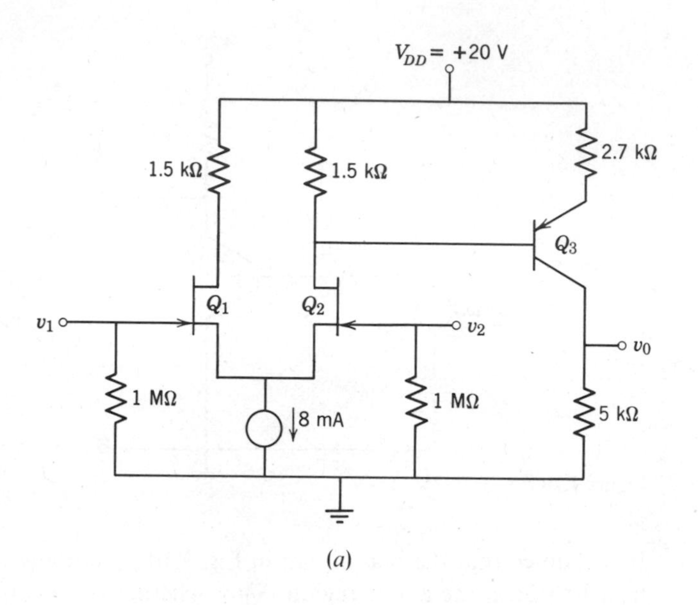

(a)

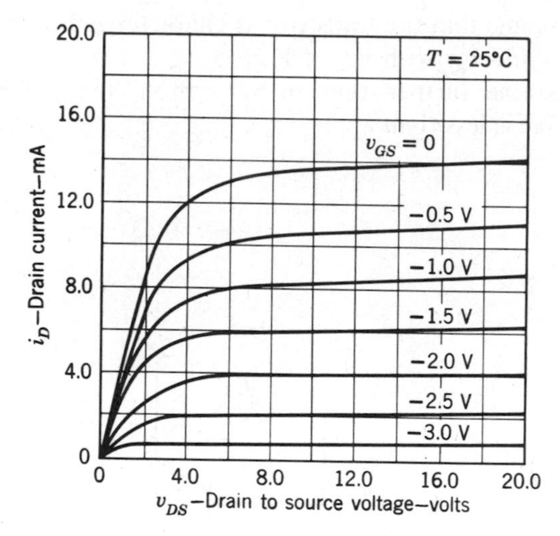

Figure E10.10 (b)

Find the following operating point information, assuming $v_1 = v_2 = 0$. State clearly all approximations used..

(a) Find V_{DS} for Q_2.

(b) Find V_{GS} for Q_2.

(c) Find V_0.

PROBLEMS

P10.1 The current in an inductor cannot change instantaneously. Show in Fig. P10.1 using the characteristics of Fig. E10.1*c* that when v_A drops from 1 to 0 V, the transistor is driven into breakdown along the $i_B = 0$ characteristic, with v_{CE} rising to about V_{CEO}. Show how a free-wheeling diode in parallel with the inductor can prevent such breakdown (see Section 8.3.4).

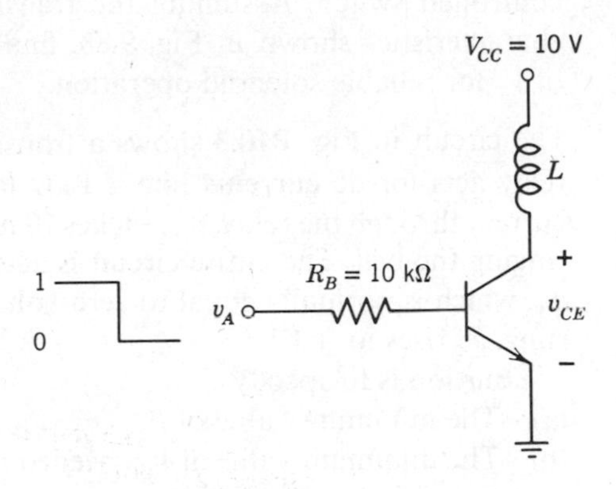

Figure P10.1

P10.2 It is desired to operate a solenoid valve on a pipeline located 30 miles from the control station. The solenoid requires 25 mA of current to operate reliably and the coil may be modeled for dc currents by a resistor of 200 Ω.

(a) One method would be to feed power directly to the solenoid through two 30 mile lengths of wire, as shown schematically in Fig. P10.2*a*. If each 30 mile length of wire has a resistance of 1500 Ω, find the voltage V_0 required for reliable solenoid operation.

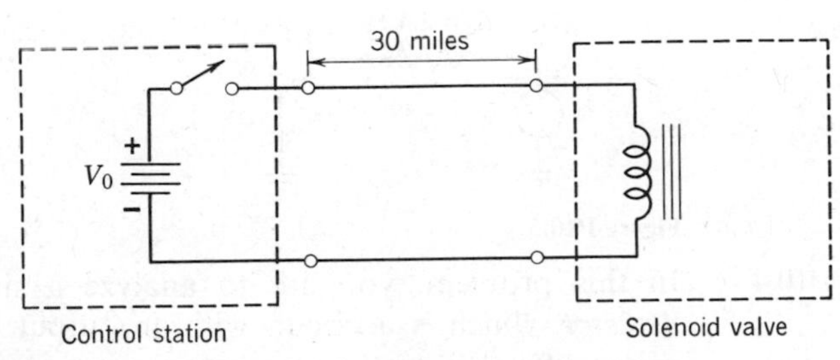

Figure P10.2*a*

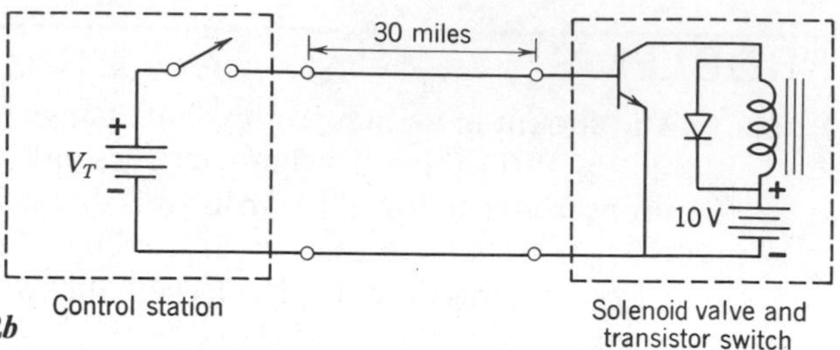

Figure P10.2b

(b) An alternate method is to use a transistor as a remote-controlled switch. Assuming the transistor has the output characteristics shown in Fig. 9.8*b*, find the minimum value of V_T for reliable solenoid operation.

P10.3 The circuit in Fig. P10.3 shows a transistor driving a relay. The relay acts for dc currents like a 1 kΩ load resistance. When the current through the relay, i_C, reaches 10 mA, the relay switch closes, ringing the bell. The entire circuit is activated by an alarm signal v_A, which is normally equal to zero volts. When an alarm is to be rung, v_A rises to +1 V.

Your job is to specify

(a) The minimum value of β_F.

(b) The minimum value of V_{CC} needed to guarantee alarm operation.

You may assume for design purposes that the transistor has a V_{BE} of 0.6 V and a $V_{CE,sat}$ of 0.3 V.

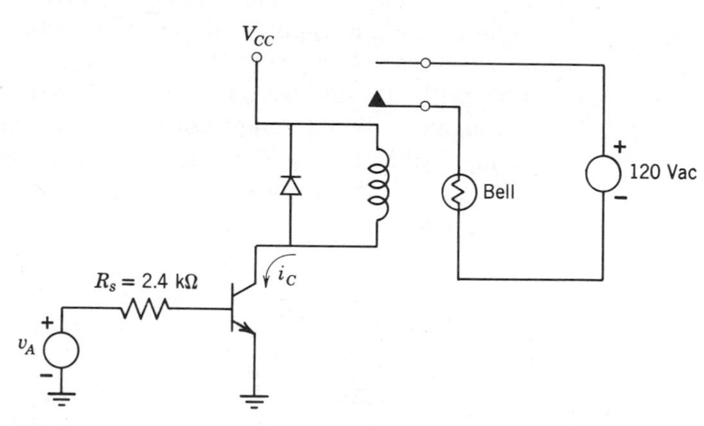

Figure P10.3

P10.4 In this problem, you are to analyze a simple *phase-sensitive detector*, which is a circuit with an output that depends on the phase difference or time lag between two periodic signals.

Assume that v_A is the 5 V square wave shown in Fig. P10.4 and that v_B is an identical square wave but delayed by a time τ. You can assume $0 \leq \tau \leq T/2$.

(a) Using the large-signal transistor parameters shown in the diagram, sketch and dimension v_C as a function of time.

(b) An ideal linear phase sensitive detector would have the dc component of v_C exactly proportional to τ. Compare this circuit with the ideal case by plotting the magnitude of the dc component of v_C versus τ.

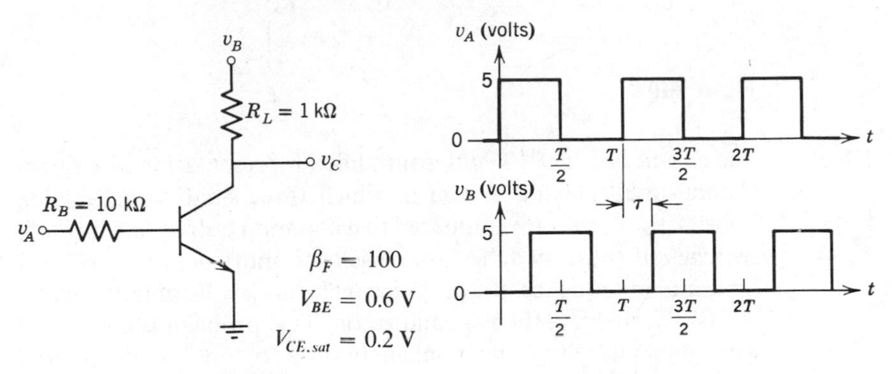

Figure P10.4

P10.5 Prove that the output resistance of the regulated power supply of Section 10.2.3 is $R_Z/(\beta_F + 1)$.

P10.6 Prove Eq. 10.41.

P10.7 The two-transistor connection of Fig. P10.7 behaves just like a very high β *n-p-n* transistor. Plot i_O versus v_O for various values of i_1. Consider only $v_O \geq 0, i_1 \geq 0$.

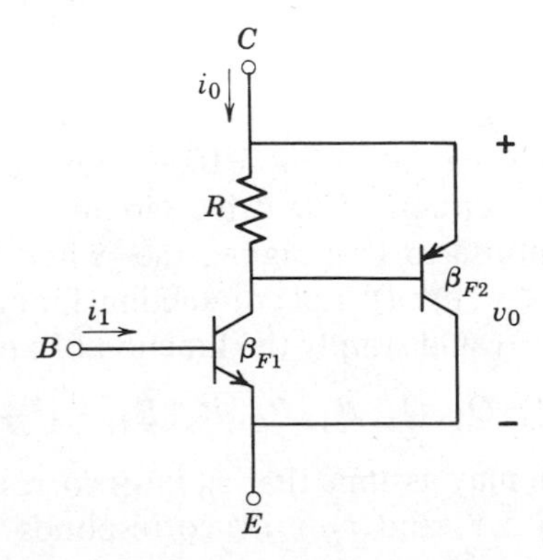

Figure P10.7

P10.8 The two-transistor connection in Fig. P10.8 is called the Darlington connection. It behaves like a very high β *n-p-n* transistor. Plot i_O versus v_O for various values of i_1. Consider only $v_O \geq 0$, $i_1 \geq 0$.

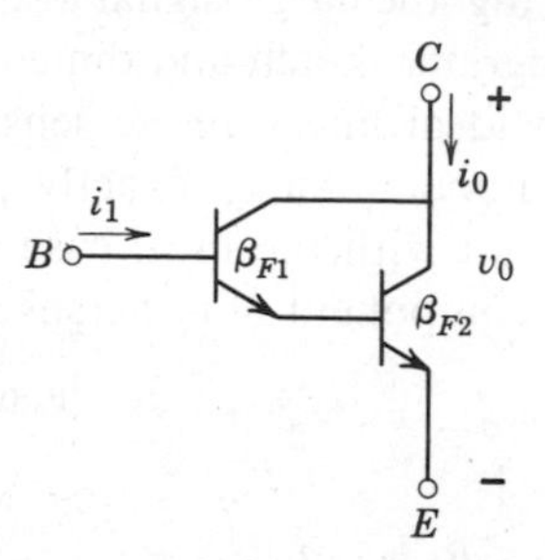

Figure P10.8

P10.9 The circuit in Fig. P10.9*a* is a simplified representation of a three-channel multiplexing system in which three small-signal analog sources, v_A, v_B, v_C, are connected to a common output line through switches. If these switches are closed for short periods, one at a time, and in sequence, the output waveform v_O will contain *samples* of first v_A, then v_B, then v_C, and so on. This problem is concerned with understanding the implementation of one of the switch channels using solid-state devices.

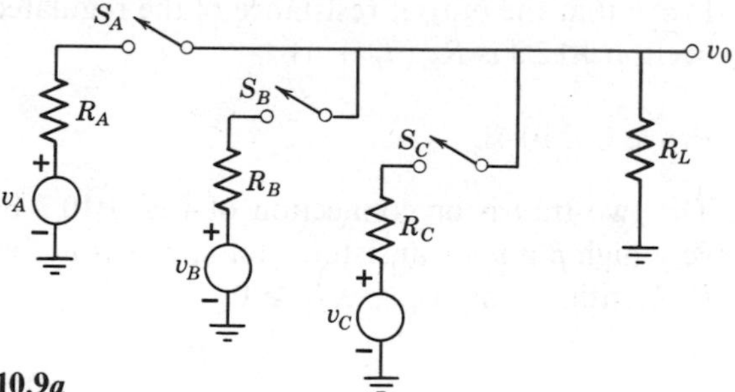

Figure P10.9*a*

The circuit in Fig. P10.9*b* shows a FET switch together with its drive circuit. The drive circuit accepts a digital signal v_D. In response to that signal, the switch is either ON (conducting) for v_D LOW, or OFF (not conducting) for v_D HIGH. Your task is to explain *clearly* and simply the function of each of these 10 components:

$$Q_A, Q_1, Q_2, R_1, R_2, R_3, R_4, R_5, +5\text{ V supply}, -15\text{ V supply}$$

You may assume that v_D HIGH corresponds to a range between 2.4 and 5 V, and v_D LOW corresponds to a range between 0.0 and

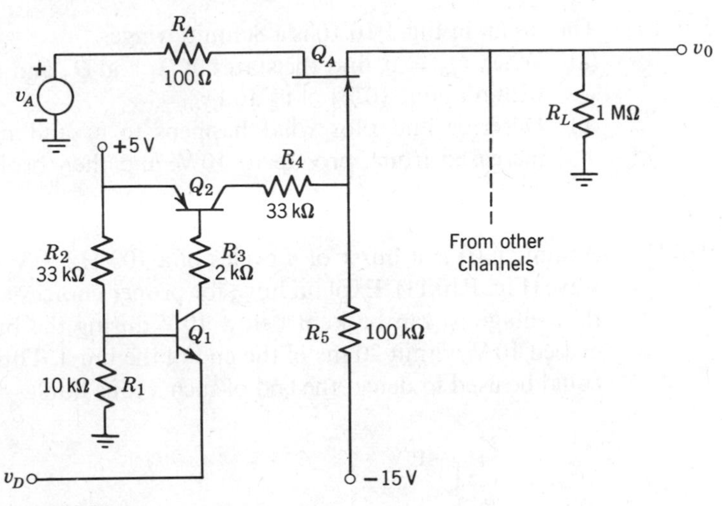

Figure P10.9*b*

0.4 V. The transistors Q_1 and Q_2 have β_F of 50 and have forward drops on the base-emitter junction of 0.6 V when turned on. Q_A has a pinch-off voltage of -6 V, and a small-signal drain-to-source resistance of 100 Ω when $v_{GS} = 0$. For this problem v_A can be considered a small signal, that is $|v_A| < 10$ mV, and the other channels can be ignored.

As a method of approach, consider how the circuit might fail in its task if any of the 10 components listed above were either (a) removed, (b) replaced by short circuits, or (c) assigned very different numerical values.

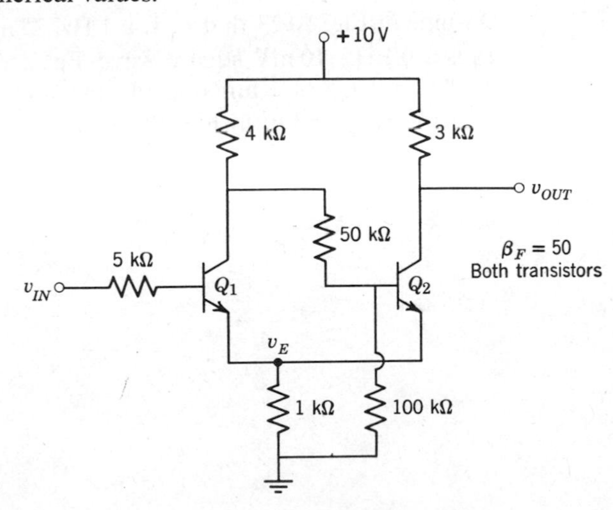

Figure P10.10

P10.10 The circuit in Fig. P10.10 is a Schmitt trigger.

(a) When $v_{IN} = 0$, find the states of Q_1 and Q_2 and the values (to within about 10%) of v_E and v_{OUT}.

(b) Describe and plot what happens to v_E and v_{OUT} as v_{IN} is increased from zero up to 10 V, and then brought back to zero.

P10.11 Assume $v_S(t)$ is a burst of n cycles of a 100 Hz, 1 V peak, square wave (Fig. P10.11). Explain how, for proper choices of R_1 and R_2, the voltage v_C can be kept below 10 V during the burst, but will exceed 10 V within 20 ms of the end of the burst. Thus this circuit could be used to detect the end of such a tone burst.

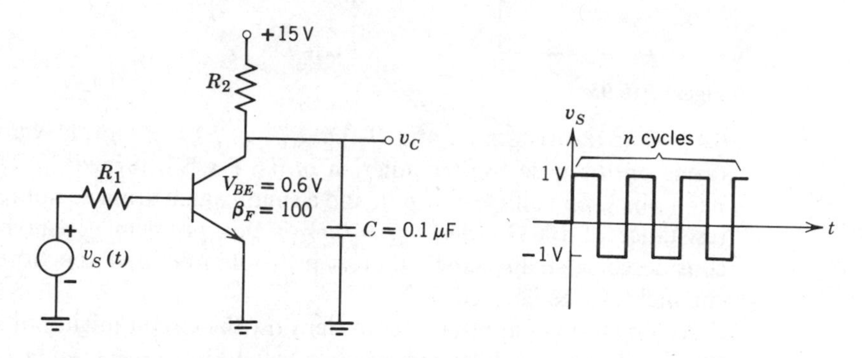

Figure P10.11

P10.12 Assume in Fig. 10.23 that v_A is a 1 Hz, 25 mV sine wave and that v_B is a 0.1 Hz, 10 mV square wave. For a chopping frequency of 10 Hz, plot v_M as a function of time, and plot the demodulated waveforms v'_A and v'_B in Fig. 10.24.

CHAPTER ELEVEN
SMALL-SIGNAL MODELS AND CIRCUITS

11.0 INTRODUCTION

In the previous chapter, we discussed a variety of circuits in which approximate representations of overall device characteristics—graphical, algebraic, and a piecewise linear circuit model—were used to obtain acceptable estimates of the large-signal circuit behavior. There are many other circuit situations, most notably those involving the amplification of small analog signals, in which the global device characteristics are not of interest; instead, one is interested in circuit behavior within narrow ranges of voltage and current. If these ranges are sufficiently restricted we can make linear models of device performance that are restricted in range of applicability but that are much more precise than the overall piecewise linear models of Chapter 10. Furthermore, since circuits that contain only linear elements are much simpler to analyze than circuits containing nonlinearities, it becomes possible to make these small-signal models somewhat more complex than the corresponding large-signal models without an intolerable increase in the difficulty of the analysis problem.

This chapter deals with the formulation and use of small-signal models for circuits containing nonlinear elements, particularly transistors. There are two closely related issues. The first is how to formulate a model for a particular device when the range of operating points is known. The second, called the biasing problem, is how to design a circuit that will produce and maintain device operating points in the correct range, regardless of variations in device characteristics due to both temperature changes and manufacturing tolerances.

11.1 INCREMENTAL SIGNALS AND VARIABLES

11.1.1 A Small-Signal Amplifier

Figure 11.1*a* shows a common-emitter amplifier circuit. The large-signal transfer characteristic (see Section 10.1) is given in Fig. 11.1*b*. Within the heavily shaded region, the transfer characteristic is fairly linear. Any small variation of v_B within this region produces a corresponding variation of v_C. The ratio between these variations is called the *incremental gain* of the amplifier:

$$\text{Incremental gain} = \frac{\text{variation in output}}{\text{variation in input}} \tag{11.1}$$

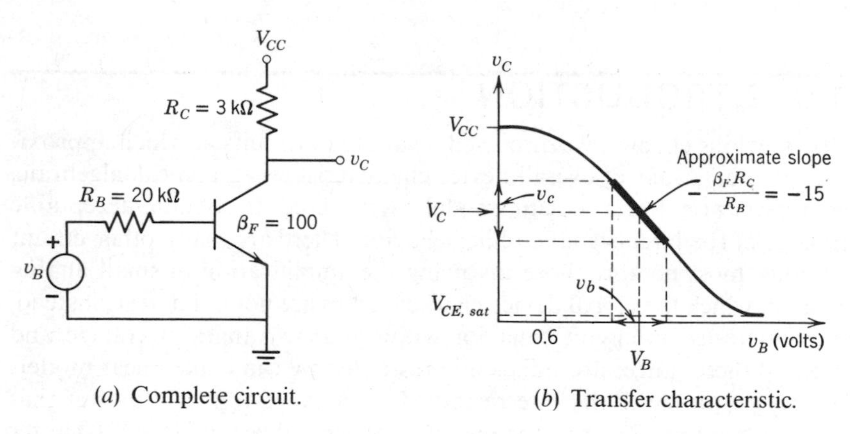

(*a*) Complete circuit. (*b*) Transfer characteristic.

Figure 11.1
A common-emitter amplifier.

In the example of Fig. 11.1, the incremental gain will be given by the *slope* of the transfer characteristic in the shaded region. Our large-signal analysis of this circuit in Section 10.1 showed that this slope was approximately equal to $-\beta_F R_C/R_B$. For the values shown this slope is equal to -15. Therefore, we can write the approximate relation

$$\text{Variation in } v_C = (-15)\ (\text{variation in } v_B) \tag{11.2}$$

Thus the common-emitter circuit of Fig. 11.1*a*, operating in the shaded portion of the transfer characteristic, functions as an *inverting amplifier* with an incremental gain of about -15.

11.1.2 Some Important Notational Shorthand

In our discussion, we shall not wish to use the cumbersome name "variation in_____" to indicate a small time-varying component of a waveform. Therefore, in accordance with widely used standards, we shall adopt an important shorthand notation. *This notation requires meticulous attention to the use of capital and lower-case symbols.* Although this is occasionally frustrating for the beginner, the resulting simplification of all formulas and discussions is well worth the effort.

When discussing small-signal circuits, we shall represent all variables as consisting of a dc or average part *plus* a small time-varying part. We use different symbols for each component, as illustrated for v_B below.

$$\begin{aligned} v_B \quad &= \quad V_B \quad + \quad v_b \\ \text{Total} &= \text{operating point} + \text{incremental signal} \end{aligned} \tag{11.3}$$

Total variables require lower-case variables with capital subscripts. The dc component of a waveform, which is called the *operating point*, is denoted with a capital variable and a capital subscript. The small time-varying component of a waveform, which is called the *incremental signal*, uses a lower-case variable and a lower-case subscript. These subtle distinctions between symbols must be noted with care.

Figure 11.2 shows typical waveforms for the amplifier example of Fig. 11.1 (the voltage axes, however, are *not* to scale). The operating points, V_B and V_C, are indicated, as well as the incremental signals v_b and v_c. Using this notation, we can write a succinct version of the voltage gain relationship of Eq. 11.2.

$$v_c = -15v_b \tag{11.4}$$

The notation shows clearly that the incremental gain relates incremental input signals to incremental output signals.

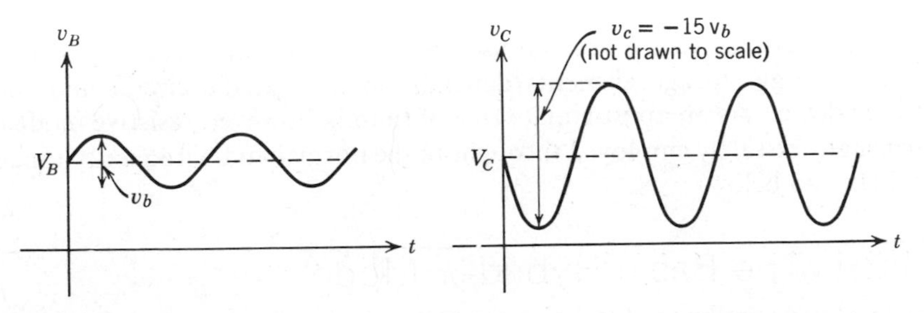

Figure 11.2
Waveforms for Fig. 11.1 (voltage not to scale).

11.1.3 The Concept of an Incremental Model

An *incremental model* is a circuit model for a device that correctly represents the relations among incremental input and output variables. The form of the incremental model and some of the numerical parameters in the model may depend strongly on the operating point of the device. For example, we have just seen that with an operating point in the active-gain region, the common-emitter amplifier of Fig. 11.1 has an incremental voltage gain of about -15. If the operating point of the transistor were in the cutoff region, however, an incremental variation of v_B would not produce any change in v_C. In this extreme case, therefore, v_c would be zero and the incremental voltage gain would be zero. Similarly, in saturation, there would be an insignificant incremental variation in v_C in response to a small change in v_B. The incremental model for a device, therefore, must be tailored to the correct operating point for the specific circuit at hand. One must first use large-signal analysis to find the operating point and then develop an incremental model appropriate to that operating point.

In the next section we shall be concerned exclusively with models that apply to the active-gain regions of the devices of interest. Therefore, in discussing the models, we shall assume that appropriate bias networks have been designed to produce operating points in the active-gain region. In Section 11.3, we shall discuss some of the problems that arise in designing these bias networks.

11.2 LOW-FREQUENCY SMALL-SIGNAL TRANSISTOR MODELS

We shall now develop small-signal models for bipolar and field-effect transistors constructed entirely out of resistive elements. These models apply as long as the time variation of any signal of interest is not too rapid. In Section 11.4 we will include some energy storage elements in the models, which extend their range of validity to much higher frequencies. The boundary between where one can use resistive models and where one must include energy storage elements depends on the specific circuit and the specific device. As an approximate rule of thumb, however, resistive models can usually be safely employed throughout the normal audio frequency range (10 kHz and below).

11.2.1 The Basic Hybrid-π Model

To develop a small-signal resistive model for the bipolar transistor, we begin with the nearly linear large-signal behavior of the device in the active-gain region. Our large-signal model for the active-gain region included a current source in the collector lead with an approximate proportionality between base current and collector current

$$i_C = \beta_F i_B \tag{11.5}$$

If we know that i_B will always be close to a particular operating point I_B, then we can be much more precise in how we represent this proportionality. That is, if

$$i_B = I_B + i_b \tag{11.6}$$

and

$$i_C = I_C + i_c \tag{11.7}$$

we can continue to use the approximate large-signal model for the operating points

$$I_C = \beta_F I_B \tag{11.8}$$

but can allow for some local variation in the proportionality constant with operating point, and define a new parameter, β_0, exactly analogous to β_F, but specifically relating the incremental signals

$$i_c = \beta_0 i_b \tag{11.9}$$

The only difference between β_0 and β_F is that β_0 represents the incremental current gain at a particular operating point, while β_F represents a single global average of current gains over all operating points. The numerical values of β_0 are generally close to β_F. Indeed, as a first estimate for β_0 at *any* operating point, it is safe to use the given value of β_F. When β_0 is listed separately on transistor data sheets, it is usually denoted h_{fe}, as opposed to h_{FE} for β_F. Be sure to look at the subscripts of these symbols.

The second characteristic of the bipolar junction transistor is the exponential relation between the base-to-emitter voltage and the various currents (e.g., base current, emitter current, collector current). For example, in the active-gain region with $\alpha_F \approx 1$, the relation between total collector current and total v_{BE} for an *n-p-n* transistor is (from Eq. 9.11)

$$i_C = I_{ES}\, e^{qv_{BE}/kT} \tag{11.10}$$

We can make an incremental version of this exponential relation by substituting

$$v_{BE} = V_{BE} + v_{be} \tag{11.11}$$

where V_{BE} is the operating point (about 0.6 V) and v_{be} is an incremental signal. Substituting into Eq. 11.10, we obtain

$$i_C = I_{ES}\, e^{qV_{BE}/kT}\, e^{qv_{be}/kT} \tag{11.12}$$

If we define a small signal by the condition that

$$|v_{be}| \ll \frac{kT}{q} \tag{11.13}$$

then the exponential factor can be approximated by using the first two terms in a Taylor's expansion.

$$e^{qv_{be}/kT} \cong 1 + \frac{qv_{be}}{kT} \tag{11.14}$$

Substituting into Eq. 11.12 yields

$$i_C = I_{ES}\, e^{qV_{BE}/kT} + (I_{ES}\, e^{qV_{BE}/kT})\left(\frac{q}{kT}\right) v_{be} \tag{11.15}$$

The first term in this sum is the dc operating point

$$I_C = I_{ES}\, e^{qV_{BE}/kT} \tag{11.16}$$

The second term of Eq. 11.15 is the incremental component of the collector current. The first factor in this term is simply I_C. Therefore

$$i_c = \frac{qI_C}{kT} v_{be} \tag{11.17}$$

Finally, we can relate this incremental component of collector current to a corresponding incremental component of base current by using Eq. 11.9.

$$i_b = \left(\frac{1}{\beta_0}\right)\left(\frac{qI_C}{kT}\right) v_{be} \tag{11.18}$$

The equivalent derivation for a *p-n-p* transistor leads to identical results, but with $|I_C|$ instead of I_C appearing in Eqs. 11.17 and 11.18. Since I_C is positive for *n-p-n* transistors operating in the linear region, we can use $|I_C|$ for both *n-p-n* and *p-n-p* transistors.

By using the relations of Eq. 11.18 and either Eq. 11.9 or 11.17 we can now construct models that correctly represent the behavior of the incremental signals. One such model is shown in Fig. 11.3*a*. The proportional relation of Eq. 11.9 is represented by a dependent current source in the collector lead, and the proportionality between i_b and v_{be} of Eq. 11.18 is represented by a

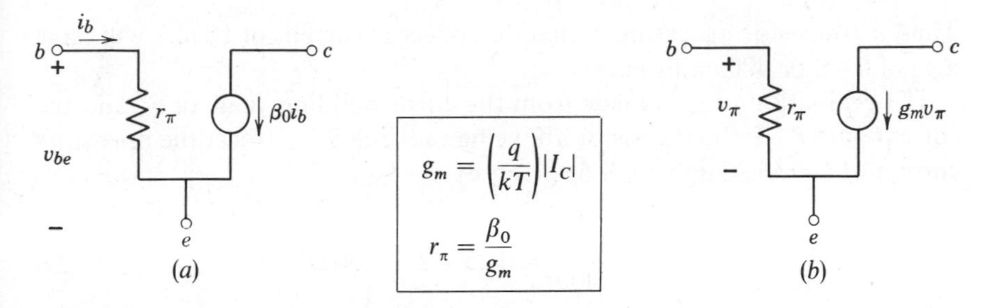

Figure 11.3
Two equivalent versions of the basic hybrid-π model.

resistor between the base and emitter. The resistor is denoted by r_π, and has the value

$$\frac{1}{r_\pi} = \frac{i_b}{v_{be}} = \frac{1}{\beta_0}\left(\frac{q|I_C|}{kT}\right) \tag{11.19}$$

An equivalent version of this model, called the *hybrid-π model* for historical reasons, is shown in Fig. 11.3*b*. Here, the dependent current generator is made proportional to v_{be} as in Eq. 11.17 or to its equivalent, the voltage across r_π. The proportionality constant is called the *transconductance*, and is defined as

$$g_m = \frac{q|I_C|}{kT} \tag{11.20}$$

Physically, the transconductance represents the incremental change in the collector current in response to an incremental change in base-to-emitter voltage. Using the transconductance, r_π can be defined as

$$r_\pi = \frac{\beta_0}{g_m} \tag{11.21}$$

Notice that the value of g_m for *any transistor* depends only upon the dc collector current I_C, the junction temperature T, and the constants k and q. The determination of the operating point I_C remains a large-signal, nonlinear problem, and must be attacked either graphically or through an appropriate large-signal model, but once I_C is known, g_m may be calculated directly. Using the room-temperature value of kT/q, g_m may be evaluated numerically as follows:

$$g_m(\text{S}) = 0.04\, I_C(\text{mA}) \tag{11.22}$$

or

$$g_m(\text{mS}) = 40\, I_C(\text{mA}) \tag{11.23}$$

Thus, a transistor operating with a dc collector current of 10 mA will have a g_m of 0.4 S or 400 millisiemens.

The evaluation of r_π is made from the corresponding value of g_m and the current gain β_0. If the transistor above has a value $\beta_0 = 100$ at the operating current of 10 mA, then r_π will be given by

$$r_\pi = \frac{100}{400} = 0.25\ \text{k}\Omega = 250\ \Omega \qquad (11.24)$$

11.2.2 The Common-Emitter Amplifier

Let us now use the hybrid-π model to examine the small-signal properties of the common-emitter amplifier shown in Fig. 11.4. The circuit contains an incremental signal source v_i with a Thévenin source resistance R_S. This signal is coupled to the transistor through a *coupling capacitor* C_1. We assume that this capacitor is sufficiently large to appear as a short circuit for the time-varying signal v_i. (A more precise discussion of this issue is reserved for Chapter 13.) This same coupling capacitor, however, is an open circuit for dc current. Therefore, with this capacitor in the circuit, all of the dc current from the V_{CC} supply flows through the *bias resistor* R_B and into the base lead in the transistor, and thus consitutes the operating point current I_B for the amplifier. The load resistance R_L is connected between the

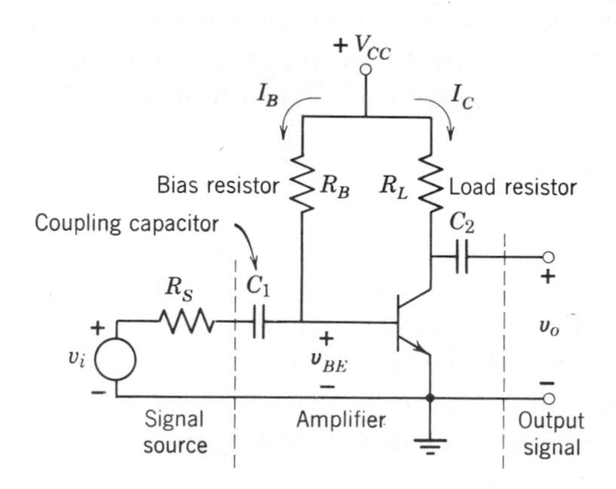

Figure 11.4

Common-emitter amplifier. The coupling capacitors C_1 and C_2 are open circuits for the large-signal operating-point currents, I_B and I_C.

collector terminal and the V_{CC} supply, and the incremental output signal v_o is coupled out through a second coupling capacitor C_2. This capacitor prevents the dc component of the current through R_L from flowing in the output leads. Thus the operating point currents, I_B and I_C, can be calculated from the large-signal model very simply.

$$I_B = \frac{V_{CC} - V_{BE}}{R_B} = \frac{V_{CC} - 0.6}{R_B} \tag{11.25}$$

and

$$I_C = \beta_F I_B = \beta_F\left(\frac{V_{CC} - 0.6}{R_B}\right) \tag{11.26}$$

With this value of I_C, we can calculate g_m from Eq. 11.20, and by using the value of β_0 for the specific transistor (or using the value of β_F as an approximation), the value of r_π can be determined from Eq. 11.21.

The next step in the analysis is to convert the total circuit of Fig. 11.4 into an equivalent incremental circuit that correctly represents the relations among the incremental variables. This incremental circuit will include the incremental signal source network, the incremental hybrid-π model for the transistor, *and* the incremental equivalents of the rest of the circuit components. All resistors are linear and appear in the incremental circuit exactly as they appear in the total circuit. The treatment of independent dc sources, however, such as the V_{CC} supply, requires some explanation.

There is no incremental variation of the V_{CC} supply voltage, since it is a dc quantity. Therefore, if we are to replace the V_{CC} supply by an incremental equivalent, we must replace it by a zero-volt incremental voltage source. But a zero-volt voltage source is a short circuit. Therefore, the incremental equivalent of a dc voltage source is simply a short circuit. Similarly, in networks containing independent dc current sources, the incremental equivalent of a dc current source would be an open circuit.

Figure 11.5*a* shows an incremental circuit for the amplifier of Fig. 11.4. The transistor has been replaced, terminal-for-terminal, by its incremental model, and the dc supply has been replaced by its incremental equivalent, a short circuit to ground. Since we assume the coupling capacitors C_1 and C_2 are short circuits for the time-varying signal v_i, we can redraw the circuit as in Fig. 11.5*b*, with these capacitors replaced by short circuits, and with a somewhat more compact circuit topology.

To determine the voltage gain of the amplifier we proceed as follows. The output voltage v_o is given by

$$v_o = -g_m R_L v_\pi \tag{11.27}$$

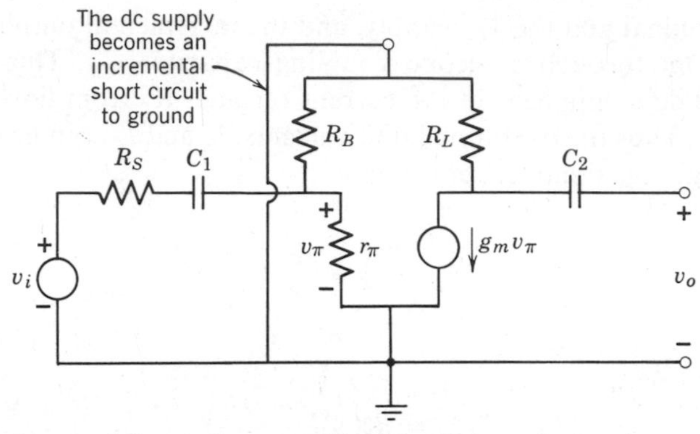

(*a*) The transistor and dc supply are replaced by their equivalent incremental models

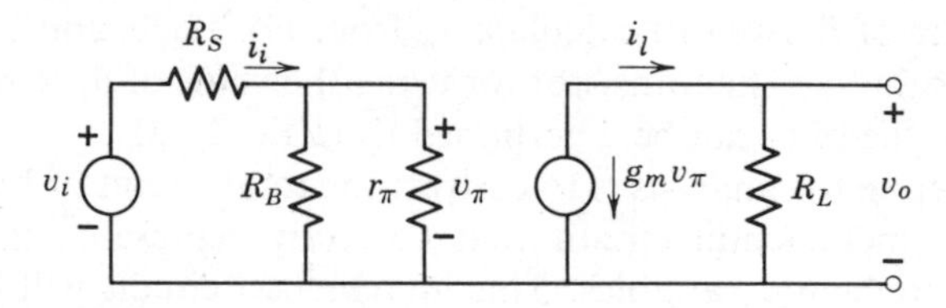

(*b*) Redrawn, with the coupling capacitors replaced by short circuits

Figure 11.5
Incremental circuits for the common-emitter amplifier of Fig. 11.4.

We now need to determine v_π in terms of v_i. From the voltage divider between R_S and the parallel combination of r_π and R_B

$$v_\pi = \frac{(r_\pi \| R_B)}{R_S + (r_\pi \| R_B)} v_i \tag{11.28}$$

Substitution of Eq. 11.28 into Eq. 11.27 then yields the voltage gain A_v.

$$A_v = \frac{v_o}{v_i} = -\frac{g_m R_L (r_\pi \| R_B)}{R_S + (r_\pi \| R_B)} \tag{11.29}$$

The minus sign in Eq. 11.29 shows that there will be a *phase inversion* between the output and input signals. That is, whenever the input goes positive, the output goes negative, and vice versa.

In designing bias networks for amplifiers, bias resistors such as R_B are usually chosen to be very large compared to r_π. If this is done, then Eq. 11.29 reduces to

$$A_v = -\frac{g_m r_\pi R_L}{R_S + r_\pi} = -\frac{\beta_0 R_L}{R_S + r_\pi} \tag{11.30}$$

where we have used the relationship $\beta_0 = g_m r_\pi$ to obtain the second form. Notice the similarity between Eq. 11.30 and the approximate gain $-\beta_F R_L/R_S$ that would be obtained from a large-signal analysis. The major difference is a more accurate treatment of the input characteristic through the inclusion of r_π in the incremental model.

Given the element values and β_0 it is easy to calculate the voltage gain from Eq. 11.29 or Eq. 11.30. Typically, the voltage gain for a common-emitter circuit is between -10 and -100. If a voltage gain greater than -100 is calculated, it should be treated with some skepticism, since a more complex hybrid-π model than the one shown in Fig. 11.3 is needed to accurately represent the transistor when the voltage gain becomes so large.

The incremental current gain can also be calculated by using the circuit shown in Fig. 11.5*b*. The incremental current through R_L is

$$i_l = -g_m v_\pi \tag{11.31}$$

and v_π is related to i_i through the parallel combination of R_B and r_π:

$$v_\pi = i_i(R_B \| r_\pi) \tag{11.32}$$

Thus, the current gain becomes

$$A_i = \frac{i_l}{i_i} = -g_m(R_B \| r_\pi) \simeq -\beta_0 \tag{11.33}$$

where the approximate form assumes $R_B \gg r_\pi$. Equation 11.33 shows that the current gain for a common-emitter amplifier depends mainly on the value of β_0 and will, therefore, usually be in the range of -50 to -200.

Finally, we can also calculate the incremental input and output resistances. The incremental input resistance R_i is the incremental resistance seen by the Thévenin signal source looking to the right of the dashed line (Fig. 11.4). From Fig. 11.5*b*, this is seen to be

$$R_i = R_B \| r_\pi \simeq r_\pi \tag{11.34}$$

The incremental output resistance R_o is the Thévenin resistance seen looking into the output terminals. To compute this resistance, we refer to the circuit of Fig. 11.6. As in all networks involving dependent sources, the Thévenin resistance can be determined by applying a voltage v_t and calculating the current i_t with the independent sources (here v_i) set equal to zero. With $v_i = 0$, v_π is also zero and, thus, no current flows in the dependent current source. Therefore

$$i_t = \frac{v_t}{R_L} \tag{11.35}$$

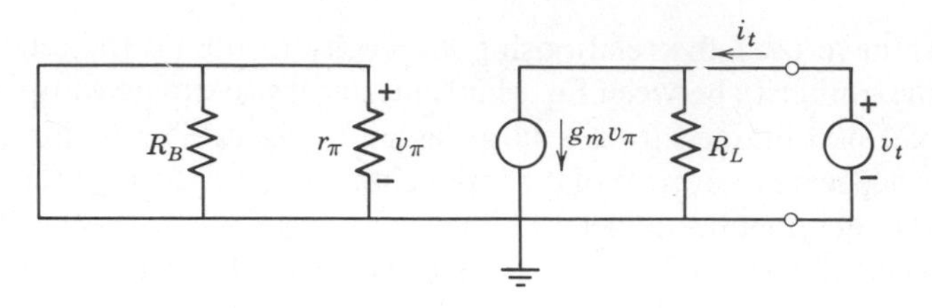

Figure 11.6
Circuit for calculation of common-emitter incremental output resistance.

and the incremental output resistance is simply

$$R_o = \frac{v_t}{i_t} = R_L \tag{11.36}$$

In summary, we see that the common-emitter amplifier can have both substantial voltage and current gain, has an incremental input resistance on the order of r_π, and has an incremental output resistance equal to the collector load resistor R_L.

11.2.3 A Numerical Example

To gain some familiarity with the actual numerical results pertinent to the common-emitter amplifier, let us consider the case where the parameters of the amplifier in Fig. 11.4 have these values:

$$V_{CC} = +12\text{ V}$$
$$R_B = 470\text{ k}\Omega$$
$$R_L = 5\text{ k}\Omega$$
$$R_S = 1\text{ k}\Omega$$
$$\beta_F = 50$$

From Eq. 11.26, the dc collector current is

$$I_C = 50\frac{(12 - 0.6)\text{ V}}{470\text{ k}\Omega} = 1.2\text{ mA} \tag{11.37}$$

We can now evaluate the hybrid-π parameters. By using Eq. 11.23 we obtain

$$g_m = 40 \times 1.2 = 48\text{ ms} \tag{11.38}$$

Since no separate information is available on β_0 we will assume $\beta_0 = \beta_F$ Thus from Eq. 11.21 we have

$$r_\pi = \frac{50}{48 \text{ mS}} \cong 1 \text{ k}\Omega \tag{11.39}$$

We now have numerical values for all the parameters contained in the gain and resistance expressions derived in Section 11.2.2. Note that $r_\pi = 1 \text{ k}\Omega$ is much less than $R_B = 470 \text{ k}\Omega$, so that the approximate results that neglect R_B when in parallel with r_π are valid. On that basis, the voltage gain is

$$A_v = -\frac{\beta_0 R_L}{R_S + r_\pi} = -\frac{50 \times 5}{1 + 1} = -125 \tag{11.40}$$

Since this is somewhat greater than -100, we know that we are pushing the model to the limit of its validity. The actual gain may be a trifle less than that given by Eq. 11.40.

The current gain is

$$A_i = -\beta_0 = -50 \tag{11.41}$$

The input resistance is given by

$$R_i = r_\pi = 1 \text{ k}\Omega \tag{11.42}$$

and the output resistance is

$$R_o = R_L = 5 \text{ k}\Omega \tag{11.43}$$

These numerical values are typical of what one can expect from the single-stage common-emitter amplifier.

Before leaving this example, let us calculate the maximum value of the input voltage v_i for which the small-signal analysis remains valid. From Eq. 11.28, assuming $(R_B \| r_\pi) \cong r_\pi$,

$$v_i = \frac{R_S + r_\pi}{r_\pi} v_\pi \tag{11.44}$$

To keep the small-signal model valid, we must restrict v_π to $|v_\pi| \ll kT/q$. We recall that $kT/q = 25 \text{ mV}$. Thus, if we allow a maximum $|v_\pi|$ of 5 mV, v_i is restricted to

$$|v_i| \leq \left(\frac{1 + 1}{1}\right) \times 5 \approx 10 \text{ mV} \tag{11.45}$$

This equation defines "how small is small" for this particular amplifier circuit.

11.2.4 The Emitter-Follower Amplifier

We next apply the low-frequency hybrid-π model to the analysis of the emitter-follower or common-collector amplifier, whose large-signal characteristics were presented in Section 10.2. The typical emitter-follower connection on which we shall base our analysis is shown in Fig. 11.7. As with the common emitter, a Thévenin equivalent form of the signal source is employed, and we shall assume that both the input and output coupling capacitors are short circuits at the signal frequencies.

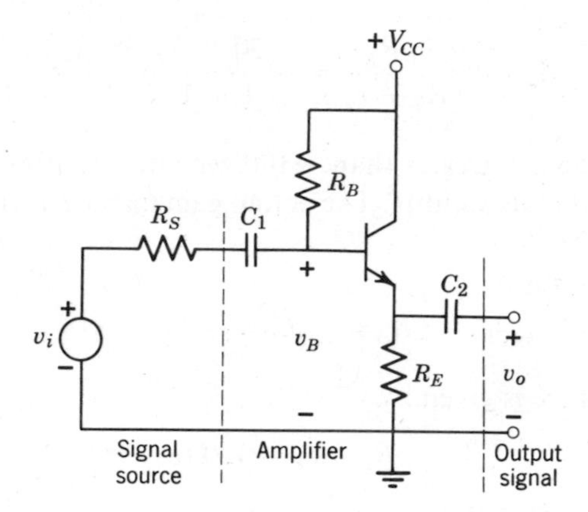

Figure 11.7
Emitter-follower amplifier.

Construction of the incremental circuit model for the emitter follower is accomplished by substituting the hybrid-π model for the transistor and replacing all dc voltage sources by short circuits. The resulting circuit is shown in Fig. 11.8. Notice that in the *incremental* circuit, the collector is

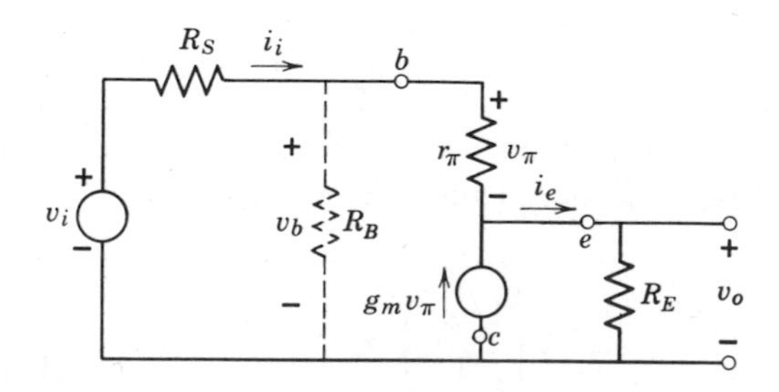

Figure 11.8
Emitter-follower incremental circuit. The collector is *common* to the incremental input and output circuits, which explains the name *common collector* for this connection.

common to both the input and output circuits. This explains the name *common collector* that is sometimes used for this connection. The base bias resistor R_B is shown dotted in Fig. 11.8 to indicate its correct position in the model. In the following analysis, however, we shall ignore R_B to focus on the essential properties of the emitter follower without unnecessary algebraic complications. Generally, R_B has a negligible effect on the voltage gain and output resistance but can substantially modify the current gain and input resistance of the emitter follower. These effects are studied in the problems at the end of the chapter.

From the circuit of Fig. 11.8, neglecting R_B, we can write the following equations:

$$v_o = i_e R_E \tag{11.46}$$

$$v_\pi = i_i r_\pi \tag{11.47}$$

$$i_e = i_i + g_m v_\pi \tag{11.48}$$

Substituting Eq. 11.47 in Eq. 11.48

$$i_e = i_i(1 + g_m r_\pi) = i_i(1 + \beta_0) \tag{11.49}$$

and combining Eq. 11.49 with Eq. 11.46 we obtain

$$v_o = i_i(1 + \beta_0)R_E. \tag{11.50}$$

The value of i_i is determined by noting that

$$v_i = i_i R_S + i_i r_\pi + v_o \tag{11.51}$$

and using Eq. 11.50 to eliminate v_o we obtain

$$i_i = \frac{v_i}{R_S + r_\pi + (1 + \beta_0)R_E} \tag{11.52}$$

Finally, the elimination of i_i between Eq. 11.50 and Eq. 11.52 yields for the voltage gain

$$A_v = \frac{v_o}{v_i} = \frac{(1 + \beta_0)R_E}{R_S + r_\pi + (1 + \beta_0)R_E} \tag{11.53}$$

Examination of Eq. 11.53 shows one of the important properties of the emitter follower: the voltage gain is less than unity. Unity voltage gain can be approached by making

$$(1 + \beta_0)R_E \gg R_S + r_\pi \tag{11.54}$$

Since the inequality (Eq. 11.54) is not difficult to achieve, the emitter follower is often said to have a unity voltage gain. Note that the voltage gain in

Eq. 11.53 differs from the gain estimated from the large-signal model (Eq. 10.28) primarily by the inclusion of r_π.

The current gain of the emitter follower can be found directly from Eq. 11.49.

$$A_i = \frac{i_e}{i_i} = 1 + \beta_0 \tag{11.55}$$

Thus, while the amplifier does not provide any voltage gain, it does have a substantial current gain capability, and this is its primary utilization. However, the current gain given in Eq. 11.55 neglects the effect of R_B, which in most cases is not negligible. Typically, Eq. 11.55 will be optimistic by about a factor of two. For the details of this question, see Problem P11.4.

We next calculate the input and output resistances of the emitter follower. The input resistance as seen by the signal source is

$$R_i = \frac{v_i}{i_i} = \frac{v_\pi + v_o}{i_i} \tag{11.56}$$

Substitution of Eq. 11.47 and Eq. 11.50 into Eq. 11.56 yields the required result

$$R_i = r_\pi + (1 + \beta_0)R_E \tag{11.57}$$

The important point contained in Eq. 11.57 is that the dependent generator has the effect of multiplying the emitter resistance R_E by a factor $(1 + \beta_0)$, which may be as large as 100 or more, Equation 11.57 thus shows that for *input circuit calculations* we may construct an equivalent circuit by replacing the parallel combination of the dependent generator and R_E by a resistor of value $(1 + \beta_0)R_E$ as shown in Fig. 11.9. If we apply the inequality, Eq. 11.54, to Eq. 11.57 we see that the input resistance is approximately

$$R_i \approx (1 + \beta_0)R_E \tag{11.58}$$

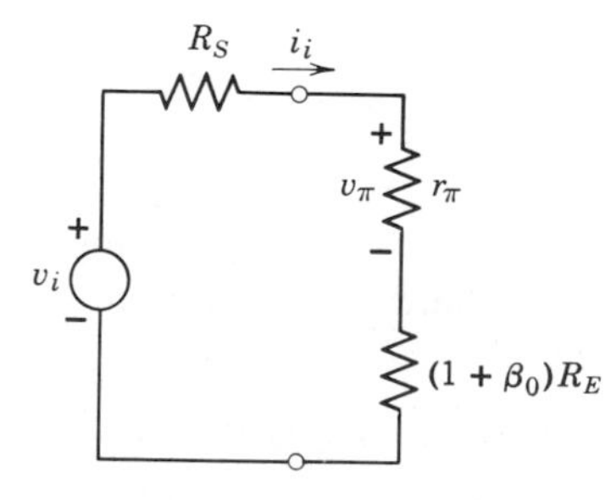

Figure 11.9
Equivalent input circuit for emitter-follower input calculations.

This illustrates the second important property of the emitter follower: the ability to transform a small resistance in the emitter circuit to a much larger value as seen by the input circuit. This property is a direct result of the current gain capability of the amplifier, and thus the emitter follower is frequently used to reduce the loading effect of a low-resistance load on the voltage gain of a preceding stage of amplification. Comparison of Fig. 11.9 with Fig. 11.8 shows that the true emitter-follower input resistance is the value given in Eq. 11.57 or Eq. 11.58 paralleled by R_B. As explored in Problem P11.5, R_B is typically on the order of $(1 + \beta_0)R_E$ so that it cannot be neglected in an accurate calculation of R_i.

To find the output impedance of the emitter follower we must compute the Thévenin resistance seen looking into the output terminals. To this end, we replace v_i in Fig. 11.8 with a short circuit and apply a test voltage v_t to the output terminals. The resulting incremental circuit is shown in Fig. 11.10.

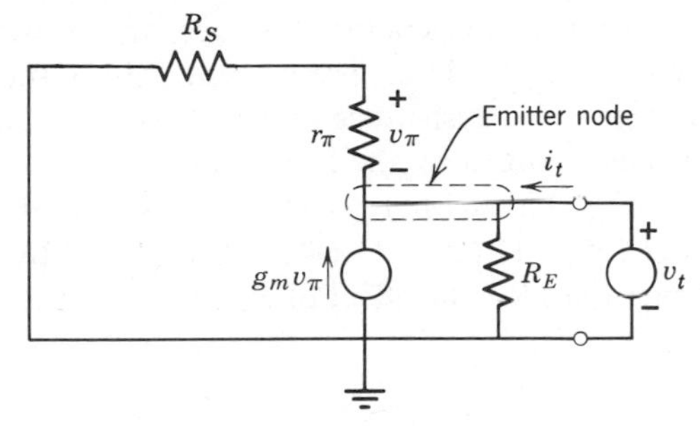

Figure 11.10
Circuit for calculation of emitter-follower incremental output resistance.

The required output resistance is then given by

$$R_o = \frac{v_t}{i_t} \qquad (11.59)$$

and we must now find i_t. Applying KVL to the emitter node we have

$$i_t = \frac{v_t}{R_E} - g_m v_\pi + \frac{v_t}{R_S + r_\pi} \qquad (11.60)$$

The voltage v_π is simply related to v_t through the r_π-R_S voltage divider:

$$v_\pi = -\frac{r_\pi}{R_S + r_\pi} v_t \qquad (11.61)$$

The minus sign in Eq. 11.61 arises from the assumed definition of positive v_π. The substitution of Eq. 11.61 into Eq. 11.60 and the collection of terms yield the desired result:

$$R_o = R_E \left|\right| \left(\frac{R_S + r_\pi}{1 + \beta_0} \right) \tag{11.62}$$

In most cases,

$$R_E \gg \frac{R_S + r_\pi}{1 + \beta_0} \tag{11.63}$$

so that the output resistance is given by

$$R_o = \frac{R_S + r_\pi}{1 + \beta_0} \tag{11.64}$$

Note that here the dependent generator reduces the resistance in the input circuit by the factor $(1 + \beta_0)$. Thus, the current gain of the emitter follower produces both a high input resistance and a low output resistance. In each case, the resistance is changed by the factor $(1 + \beta_0)$.

Comparison of Fig. 11.9 with Fig. 11.8 shows that the presence of R_B would require R_S in Eqs. 11.62 to 11.64 to be replaced by $R_S \| R_B$. Since R_B is often much larger than R_S, the effect of R_B on R_o is often very small.

11.2.5 An FET Source Follower

At low frequencies, the input of an FET looks like an open circuit. For this reason, FET small-signal amplifiers are particularly useful with source networks that have high Thévenin resistances. One example of an FET amplifier is shown in Fig. 11.11*a*. It is a *source follower*, the FET analog of the emitter follower.

Unlike the bipolar transistor, there is no simple piecewise linear model that can be used to determine the FET operating point. It must be found either from analytical or graphical methods (see Section 11.3.3). Once the operating point is known, however, it is possible to construct a low-frequency incremental model for the FET. For example, one can determine an *incremental transconductance* from the graphical characteristics in the vicinity of the operating point by finding how much change in i_D would be produced by a 1 V change in v_{GS}. Beyond pinchoff, where the output characteristics are horizontal, the drain-to-source connection can be modeled as a dependent current source. Combining the open-circuit input with the current-source

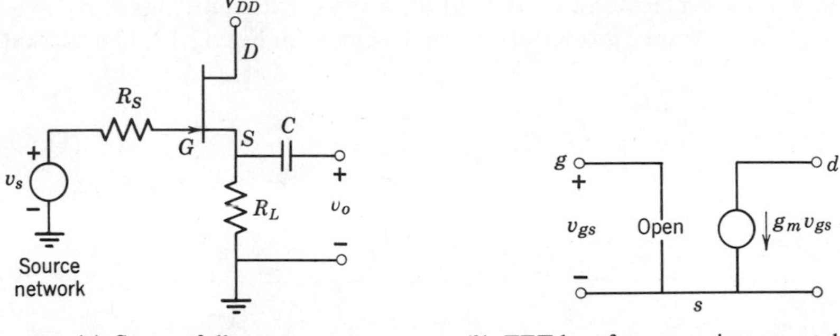

(a) Source-follower

(b) FET low-frequency incremental model

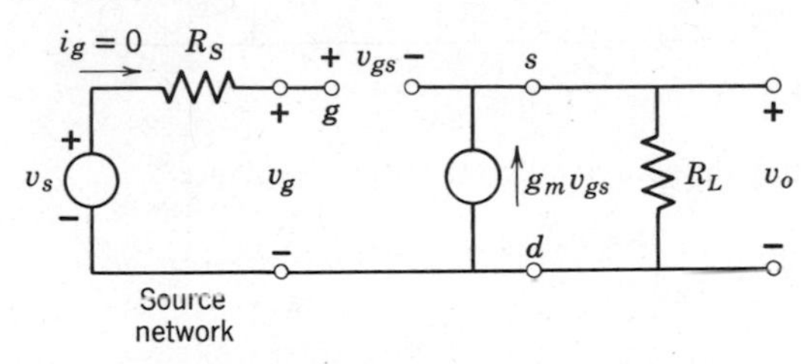

(c) Complete incremental circuit

Figure 11.11
A source-follower amplifier.

output gives the model of Fig. 11.11*b*, where the incremental transconductance g_m is defined above. Many FET data sheets list numerical values for g_m, under the symbol g_{fs}. Typical values are in the range 0.5 to 5 mS.

By using this incremental model, one can construct the incremental circuit of Fig. 11.11*c*. The Thévenin input resistance of the amplifier is clearly infinite in this model. The voltage gain and output resistance are obtained as follows. From the circuit we see that

$$v_{gs} = v_g - v_o \tag{11.65}$$

But since $i_g = 0$, $v_g = v_s$. Furthermore,

$$v_o = g_m R_L v_{gs} \tag{11.66}$$

Combining Eqs. 11.65 and 11.66 yields

$$v_o = g_m R_L (v_s - v_o) \tag{11.67}$$

which can be solved for v_o

$$v_o = \left(\frac{g_m R_L}{1 + g_m R_L} \right) v_s \tag{11.68}$$

This voltage gain is less than unity, and only approaches unity for $g_m R_L \gg 1$.

The output resistance is obtained from the circuit of Fig. 11.12. The current i_t is

$$i_t = \frac{v_t}{R_L} - g_m v_{gs} \tag{11.69}$$

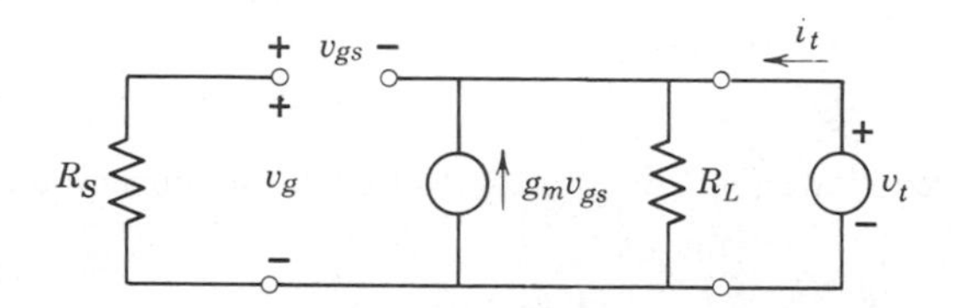

Figure 11.12
Circuit for calculating the source-follower output resistance.

But from Eq. 11.65

$$v_{gs} = -v_t \tag{11.70}$$

Combining these results yields

$$R_o = \frac{v_t}{i_t} = \left(\frac{R_L}{1 + g_m R_L}\right) \tag{11.71}$$

If $g_m R_L$ is large, the output resistance approaches $1/g_m$, which has values in the range from 200 to 2000 Ω for typical FETs. This is not as low as the output resistances that can be obtained with emitter followers. As a result, applications calling for very low output resistances use emitter followers instead of source followers.

11.3 THE BIASING PROBLEM

We have made several passing references to the issue of *biasing*, that is, the establishment of correct operating points for active devices. A full discussion of how the design of bias circuits is carried out is beyond the scope of this text. We intend here to identify the issues, and to illustrate some typical examples of bias circuits.

11.3.1 Factors Influencing Bias Design

There are four major factors that influence amplifier bias design. First, one must be certain not to exceed device ratings on power, voltage, and current. Second, one must take account of variations in device parameters due to unit-to-unit differences and due to the effect of temperature. Third, allowance must be made for the output signal amplitude. For linear operation, the dc operating point must be placed far enough from cutoff and saturation to insure that the anticipated signal excursions to either side of the operating point do not cause the device to cut off or saturate. Finally, since small-signal parameters depend on operating point, specific requirements for the incremental device parameters may restrict the choice of the dc operating point. Discussion of this last issue can be found in the more advanced texts on transistor circuits.

The current, voltage, and power limitations for the various devices were discussed in Chapter 9. Normally, for signal-level applications, the currents and voltages are small enough so that one *can* find a transistor with sufficient power-handling capabilities. In applications involving large currents, such as power amplifiers, the power ratings of the transistors may be of paramount importance. We shall, however, assume that transistors with adequate power ratings are available for the application of interest.

The question of avoiding saturation and cutoff is illustrated in Fig. 11.13*a*, which shows a typical voltage transfer characteristic for a common-emitter

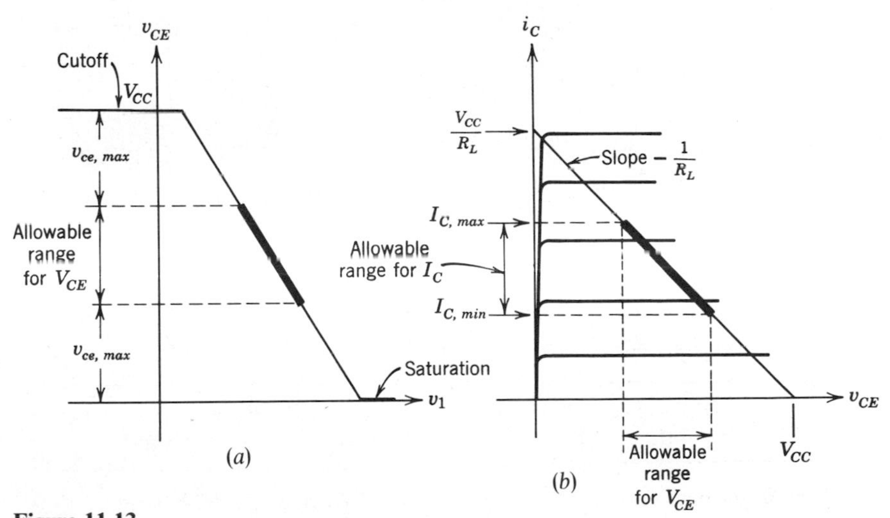

Figure 11.13
Graphical constructions showing allowable ranges for the operating point of the common-emitter amplifier.

amplifier. If the incremental component of the output signal, v_{ce}, has a maximum magnitude $v_{ce,max}$, then the dc operating point V_{CE} must not lie within $v_{ce,max}$ of either saturation or cutoff. This limitation on V_{CE} produces a corresponding limitation on the dc collector current (Fig. 11.13*b*). The biasing problem then reduces to guaranteeing that I_C will fall in the proper range.

What makes the biasing problem a real problem is the fact that β_F varies widely from unit to unit, factors of four or five being typical. In addition, β_F is a strong function of temperature, often doubling over the range of allowed junction temperatures for the device. Since in many applications the range of allowed variation in I_C might be no wider than a factor of 2, it is clear that variations in β_F by factors 4 or 8 might produce unacceptably large changes in operating point. Adding to the β_F problem is a similar, but smaller variation in V_{BE}, the range of variation with temperature and from unit-to-unit being on the order of 0.1 volt at the same nominal collector current.

11.3.2 Biasing Bipolar Transistors

We have already encountered in Fig. 11.4 a simple biasing scheme for the common-emitter amplifier. Since the common-emitter example includes all of the basic issues, we shall restrict our attention to that case. The biasing

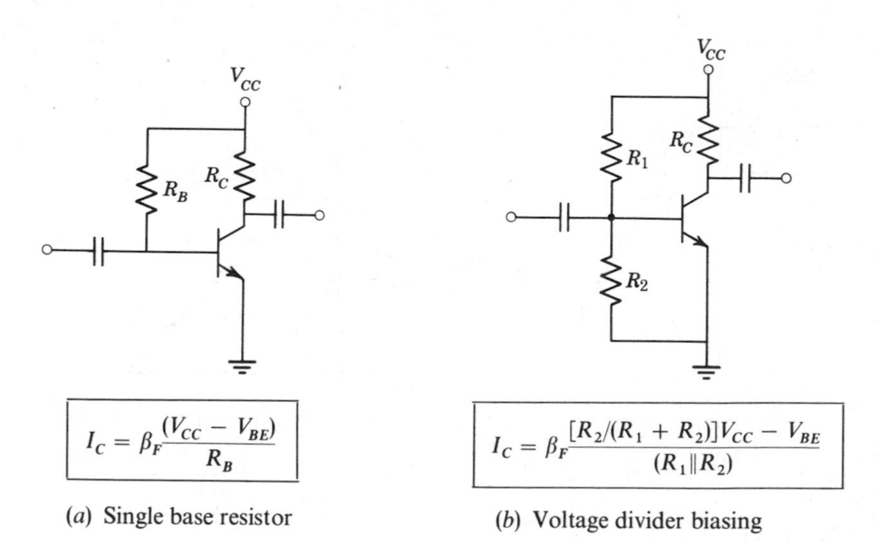

$$I_C = \beta_F \frac{(V_{CC} - V_{BE})}{R_B}$$

$$I_C = \beta_F \frac{[R_2/(R_1 + R_2)]V_{CC} - V_{BE}}{(R_1 \| R_2)}$$

(*a*) Single base resistor

(*b*) Voltage divider biasing

Figure 11.14
Two bias schemes that are sensitive to variations in β_F.

scheme of Fig. 11.4 is reproduced in Fig. 11.14*a*. The operating point for that design is given in Eq. 11.26, and is repeated below:

$$I_C = \beta_F \frac{(V_{CC} - V_{BE})}{R_B} \tag{11.72}$$

Notice that the operating point collector current I_C is directly proportional to β_F. Therefore, if β_F varies by a factor of 5, the collector current I_C will also vary by a factor of five. In many applications, this is acceptable. In many others, it is not. Therefore, more elaborate and more stable schemes are often needed.

An interim approach to improved operating point stability is illustrated in Fig. 11.14*b*. Here, through the action of the R_1-R_2 voltage divider, the Thévenin equivalent dc source becomes $[R_2/(R_1 + R_2)]\, V_{CC}$, and the Thévenin equivalent dc bias resistance becomes $R_1 \| R_2$. The resulting collector current is

$$I_C = \beta_F \frac{\left[\left(\dfrac{R_2}{R_1 + R_2}\right) V_{CC} - V_{BE}\right]}{R_1 \| R_2} \tag{11.73}$$

a result that still suffers from the problem of Fig. 11.14*a*. The operating point is still proportional to β_F, and might be too unstable.

Figure 11.15*a* shows a solution that works for most single transistor bias design problems. A resistor has been added to the emitter circuit. We recall from Section 10.2 that with an emitter resistor present, the voltage drop across the emitter resistor is relatively insensitive to β_F. In fact, the insertion of the emitter resistor converts the bias network into a large-signal emitter-follower with the following operating point

$$I_C = \left[\frac{\beta_F}{(R_1 \| R_2) + (\beta_F + 1)R_E}\right]\left[\left(\frac{R_2}{R_1 + R_2}\right) V_{CC} - V_{BE}\right] \tag{11.74}$$

Now the operating point expression has β_F in both the numerator and denominator. If R_E and the divider resistors R_1 and R_2 are chosen such that $(\beta_F + 1)R_E \gg (R_1 \| R_2)$, then the dependence of I_C on β_F almost disappears. As an approximate rule of thumb, one picks R_E such that the voltage drop across R_E for the desired collector current is about 1 to 2 volts, or so that $R_E \approx 0.2\, R_C$, and then selects the R_1-R_2 resistor pair so that

$$(\beta_{F,min} + 1)R_E > (R_1 \| R_2) \gtrsim 5 r_{\pi,max} \cong 5 \frac{\beta_{F,max}}{(qI_{C,max}/kT)} \tag{11.75}$$

where $\beta_{F,min}$ and $\beta_{F,max}$ are the smallest and largest expected values of β_F, and $I_{C,max}$ is the largest allowable value of I_C. The resulting operating point is usually acceptably stable to variations both in β_F and V_{BE}.

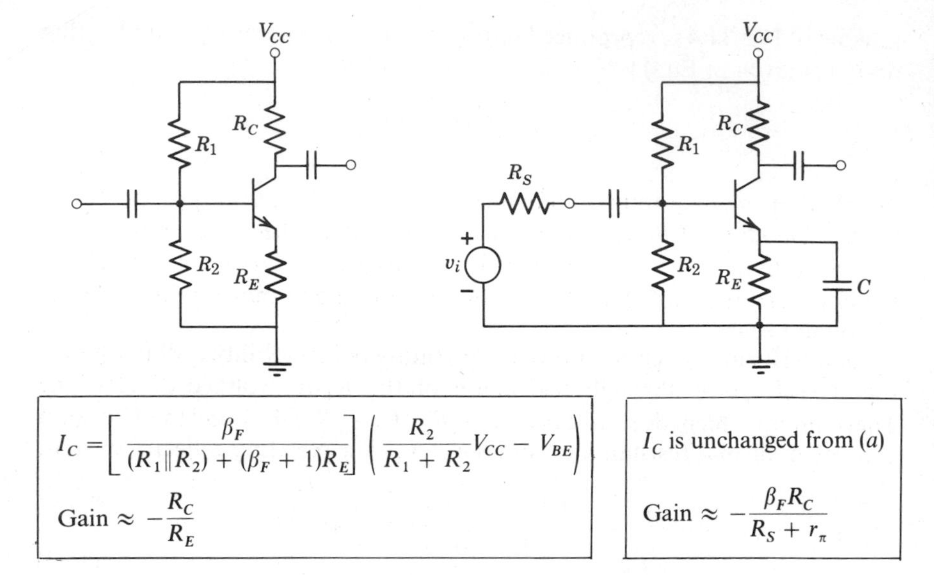

(*a*) Unbypassed emitter resistor

(*b*) Bypass capacitor added

Figure 11.15
Use of an emitter resistor to improve stability of the operating point.

This improved stability is not without its cost. The incremental gain of this circuit is

$$A_v \approx -\frac{R_C}{R_E} \tag{11.76}$$

because R_E and R_C carry about the same incremental current, and the presence of R_E makes the incremental collector current the same as for the emitter follower. The problem is that the emitter resistor represents a form of *negative feedback*. As we recall from our discussion in the context of op-amp circuits, negative feedback reduces gain and increases stability.

If the signal to be amplified is a time-varying signal, we can use a capacitor to short out the feedback effect of R_E for the incremental signals without destroying the dc stability produced by R_E. Figure 11.15*b* shows the use of this so-called *bypass capacitor*. For dc currents, the capacitor is an open circuit, and the operating point is that calculated for Fig. 11.15*a*. For ac currents, however, the capacitor is a short circuit. Incrementally, therefore, the emitter is still connected to ground, so the characteristic voltage gain on the order of $-\beta_F R_C/(R_S + r_\pi)$ can still be realized. Thus, using a capacitor, we have been able to provide separate current paths for ac and dc currents.

This technique is often useful in treating bias currents and signal currents independently.

Although the bias network of Fig. 11.15*b* is useful for single-stage amplifiers, it is rarely used for multiple-transistor low-frequency amplifiers. The requirement for coupling capacitors between one stage of amplification and the next and for bypass capacitors across each emitter resistor have two undesirable effects: first, increased cost, and, second, a degrading of the response of the amplifier to very-low-frequency signals. As a result, biasing arrangements for amplifiers involving many transistors often couple transistors directly to one another, and then use overall feedback to maintain operating-point stability. In many cases, the feedback path incorporates capacitors in a fashion analogous to the bypass capacitor of Fig. 11.15*b* in order that the ac gain of the amplifier not be reduced by the dc feedback. An example is shown in Fig. 11.16. This circuit will not be analyzed in detail. The principle, however, is that if either of the β_F's increase, the dc feedback through R_F produces a compensating change in the base current of Q_1. At the same time, the signal bypass capacitor permits a small-signal gain in excess of 1000 for β_0's near the nominal value of 100.

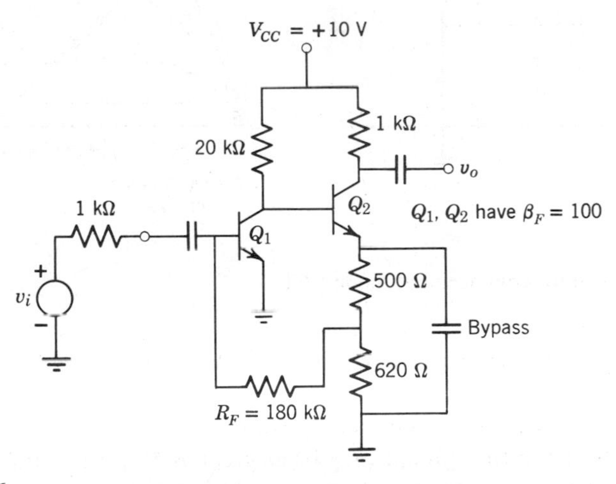

Figure 11.16
A two-stage direct-coupled amplifier with feedback biasing.

11.3.3 Biasing Field-Effect Transistors

Bias networks for field-effect transistors are generally simpler than for bipolar devices. FETs are less temperature dependent and draw virtually zero gate current when operating in the active-gain region. Although it is possible to construct a large-signal FET model, it is somewhat more complex, and hence less convenient, than the active-gain region bipolar transistor model. For this reason we shall concentrate on the use of the graphical output characteristics in designing the bias networks.

From the discussion of the output characteristics of an enhancement mode MOSFET contained in Section 9.2.4, it is clear that both the drain and gate voltages must be of the same polarity for linear operation in the active-gain region. Thus, both voltages may be derived from a single power supply. One such arrangement is shown in Fig. 11.17*a*. The drain characteristics for the MOSFET are shown in Fig. 11.17*b*.

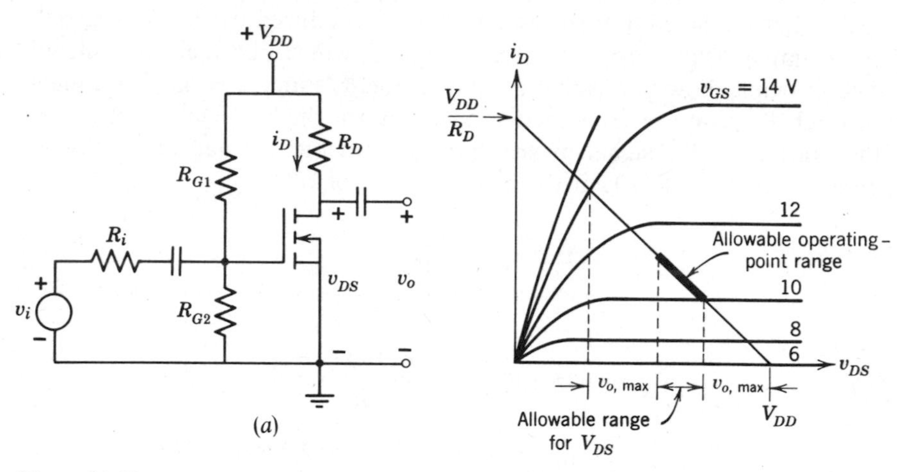

Figure 11.17
Bias arrangement for enhancement mode FETs.

The first step in a simple bias design is to construct the load line from the given values of the supply voltage V_{DD} and drain resistor R_D. Next, the limits imposed by the required peak signal swings, $v_{o,max}$ are entered along the v_{DS} axis. Note that although v_{DS} may be as great as V_{DD}, the minimum value is usually substantially greater than zero. At the lower values of v_{DS}, the characteristics crowd along the load line, resulting in nonlinear amplification. Judgment has to be excerised, therefore, in picking the lower acceptable limit for v_{DS}. Finally, between the limits imposed by $v_{o,\, max}$ there will be a range of allowable operating points lying on the load line.

Once the allowable range for the operating point has been found, the corresponding value for V_{GS} can be read directly from the characteristics. In Fig. 11.17*b*, a value of V_{GS} between 10 and 11 volts is acceptable. Since the gate is effectively an open circuit, and since the input coupling capacitor prevents the dc current from flowing back to the input source v_i, the value of V_{GS} derived from V_{DD} will be given by the simple voltage divider relationship:

$$V_{GS} = \left(\frac{R_{G2}}{R_{G1} + R_{G2}} \right) V_{DD} \tag{11.77}$$

The parallel combination of R_{G1} and R_{G2} represents a loading effect on v_i and, hence, is kept large compared to the Thévenin resistance R_I looking back toward v_i. In practice $R_{G1} \| R_{G2}$ can be kept in the range of 10^5 to 10^7 Ω without difficulties. In this case, if we select $R_{G1} = R_{G2} = 1$ MΩ, the 20 V supply will produce $V_{GS} = 10$ V, which intersects the load line within the allowable range.

In contrast to the enhancement mode case, the depletion mode FET requires voltages of opposite polarity at the drain and gate terminals to be biased in the linear amplification region. This can be accomplished by placing a resistor in series with the source terminal and returning the gate to ground through a resistor as shown in Fig. 11.18, an arrangement that is called *self-bias*. As with the bipolar transistor, a bypass capacitor can be used in parallel with R_S to prevent reduction in the amplifier gain.

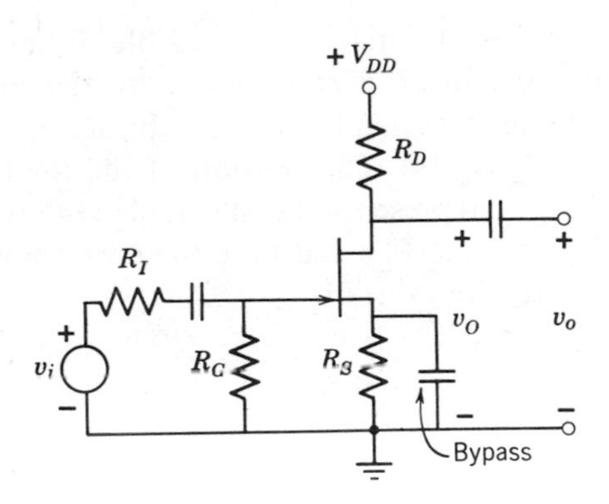

Figure 11.18
Bias arrangement for depletion mode FETs.

The design of the circuit is accomplished as follows. First, the load line is constructed as shown in Fig. 11.19, assuming $R_S \ll R_D$ so that the slope is simply $-1/R_D$. Then the minimum and maximum values of I_D are entered on the graph. The minimum value is

$$I_{D,min} = \frac{v_{o,max}}{R_D} \tag{11.78}$$

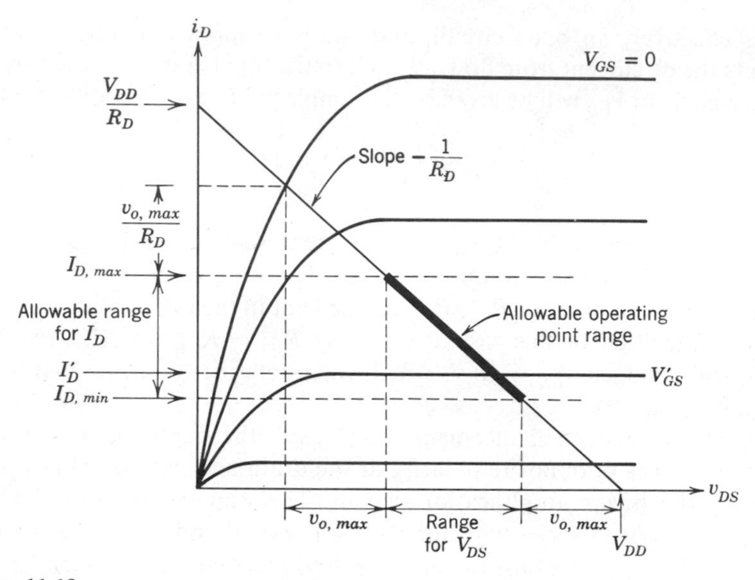

Figure 11.19
Initial design of depletion-mode bias circuit.

In a depletion-mode unit, the maximum possible drain current occurs at $v_{GS} = 0$. Thus, from the drain current given by the intersection of the $v_{GS} = 0$ characteristic and the load line, we subtract $v_{o,max}/R_D$ to arrive at the value of $I_{D,max}$ (Fig. 11.19). The portion of the load line between the limits of $I_{D,min}$ and $I_{D,max}$ represents the allowable region for the operating point. We now select a value V'_{GS} and note the corresponding value of I'_D. From these we calculate the value of R_S

$$R_S = \frac{-V'_{GS}}{I'_D} \tag{11.79}$$

The value of V'_{GS} is established by the drop produced by the drain current through R_S, since there is no current through and hence no voltage across the resistor R_G.

If the value of R_S is much smaller than R_D, the design is then complete. If, however, the value of R_S is comparable to or greater than R_D, it is necessary to check the values of V'_{GS} and I'_D as follows. We construct a new load line with slope $-1/(R_D + R_S)$, and from the intersection of this load line and the characteristic curve for $V_{GS} = 0$ we construct a new value of $I_{D,max}$ (see Fig. 11.20). This produces a new region of allowable operation on the new load line. If V'_{GS} intersects this region, the design is satisfactory. If not, we pick new values of V'_{GS} and I'_D that do intersect the new allowed region of operation

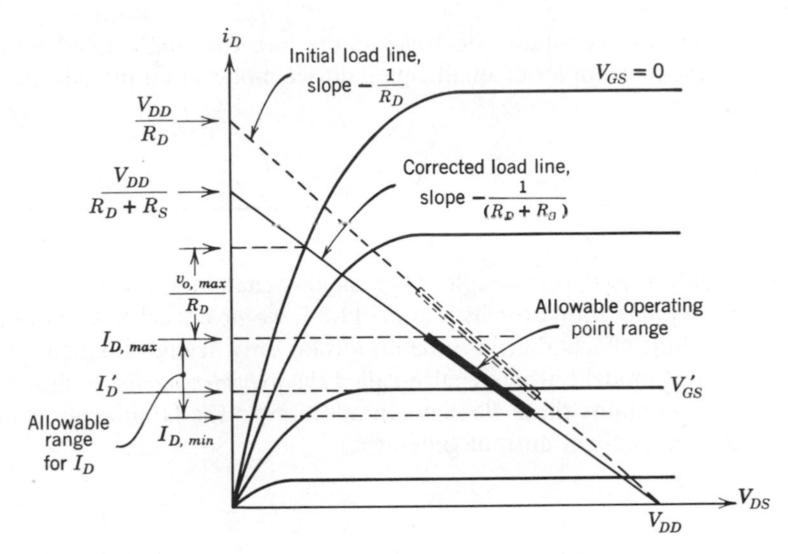

Figure 11.20
Correction for R_S in design of depletion-mode FET bias circuit.

and calculate a new value of R_S with Eq. 11.79. Then, the checking proceedure is repeated. In most cases a single iteration is adequate to meet the design requirements.

11.4 MODELS WITH ENERGY STORAGE

The resistive small-signal device models discussed thus far provide both simple and accurate representations of device behavior as long as the rate of variation of the incremental signals is not too fast. At frequencies in the high audio range and above, the *speed* of the device can become critical in determining the actual response to an excitation. For example, the gate of a MOSFET is actually connected to the channel through a capacitor. This capacitor must be charged and discharged as the gate-to-source voltage is changed. The charging and discharging introduce time delays and finite response times.

If we are to represent accurately the time delays and their effects on signals, we must allow for the inherent energy-storage and charge-storage features of device operation. Large-signal models that include these capacitive effects are called *charge-control models*, and are discussed in advanced texts. For small-signal applications, the capacitive effects can be represented by the addition of constant, ideal capacitors into the small-signal model. We now present examples of small-signal device models that include energy storage.

11.4.1 The Hybrid-π Model Revisited

When we introduced our simple r_π-g_m small-signal model for the linear region of a bipolar transistor in Section 11.2.1, we promised to expand the model to include effects that become important only at high frequencies. A more complete model,[1] which is also called the *hybrid-π model*, is shown in Fig. 11.21. It contains three elements in addition to the familiar resistance r_π and the g_m dependent current generator.

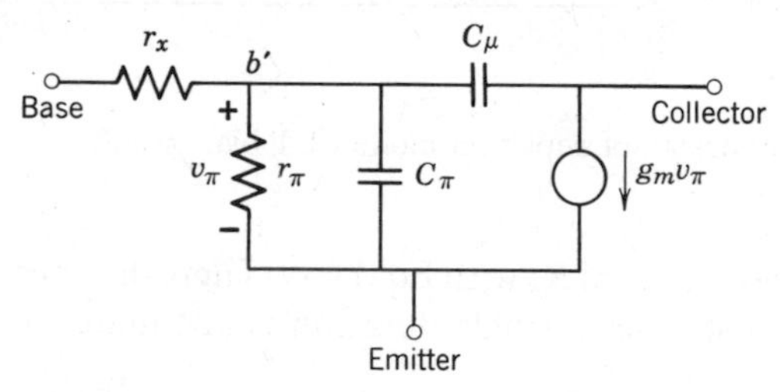

Figure 11.21
The hybrid-π model including energy-storage elements.

The first new element, r_x, represents the resistance of that part of the base region between the base terminal and the active base region, the latter being confined to that portion of the base between the collector and emitter through which minority carriers diffuse. Its value may vary between a few ohms and about 100 ohms depending on the specific transistor and on the operating point. Since it is not an energy storage element, we could have added it to the transistor model earlier. Its primary importance, however, is at high frequencies, as it provides one of the paths for charging up the two internal capacitances, C_π and C_μ.

The capacitors C_π and C_μ have their origin in the physical characteristics of the base region and the surrounding junctions. C_μ, for example, represents the capacitance of the back-biased collector-base junction. As the collector-base voltage changes, charge must be added or subtracted from the space-

[1] It is possible to construct even more elaborate (and more accurate) hybrid-π models than the one discussed in this text.

charge layer at the junction. Therefore, the space-charge layer itself acts like a capacitance: a change in voltage must be accompanied by flow of charge. The value of C_μ ranges from about 2 pF in the very best high frequency transistors up to 5 or 10 pF in typical signal transistors; it depends somewhat on the value of V_{CE}.

The C_π capacitance, which appears in parallel with r_π between the emitter and the node b' in Fig. 11.21, arises from a combination of sources. One contribution to C_π is the emitter-base junction capacitance, which behaves similar to the collector-base junction capacitance. The second, and more important contribution to C_π at moderate and large values of I_C, is from the storage of excess minority carriers in the neutral-base region between the two junctions. This contribution to C_π is illustrated in Fig. 11.22, which shows a schematic representation of the distribution of minority carriers through the base region for two slightly different values of the emitter-base voltage, v_1 and v_2. The forward bias on the emitter-base junction injects many minority carriers near the emitter junction, while the reverse bias on the collector-base junction keeps the concentration of minority carriers near that junction small. The total number of minority carriers "stored" in the base region for a given emitter-base voltage is represented by the total area under the appropriate curve. Therefore, the crosshatched area *between* the curve for v_1 and the curve for v_2 represents the *change* in total minority carriers that must be made when the emitter-base voltage changes from v_1 to v_2. To keep electrical neutrality in the base region when v_1 changes to v_2, an equivalent amount of majority charge must flow into or out of the base terminal. The network equivalent of this phenomenon is a capacitance, because a change in the emitter-base voltage requires a net flow of charge. The total capacitance, C_π, is the sum of this stored-charge capacitance and the space-charge capacitance of the emitter-base junction. The value of C_π depends approximately linearly on the transistor collector current with typical values lying in the range of several hundred picofarads.

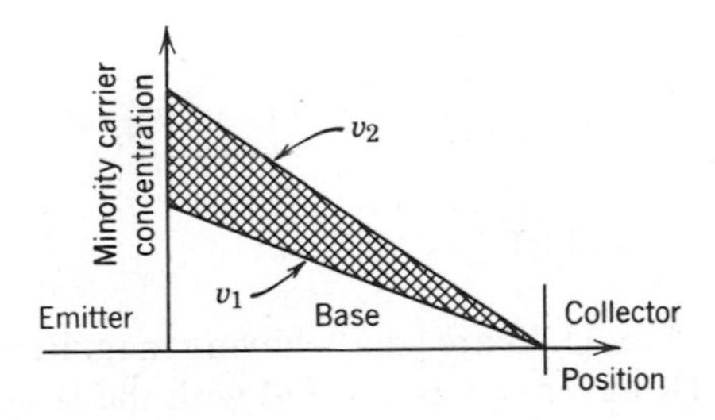

Figure 11.22
The minority carrier contribution to C_π.

Although the elements of the hybrid-π model described above depend on the operating point of the transistor, they *do not* depend on the frequency of the waveform being applied to the transistor. This fact makes the hybrid-π model very useful. There are several other incremental models discussed in the literature (such as the h-parameter and y-parameter; see Section 11.4.3), but these models always turn out to depend on frequency and must therefore be represented in graphical form. The hybrid-π parameters can be obtained from the transistor data sheet, and once obtained, the RC network of Fig. 11.21 represents the transistor's operation at any frequency of interest.

When one constructs the incremental circuit for an amplifier with the hybrid-π model including the capacitors, one is faced with a network problem that is beyond the scope of the analysis techniques we have developed thus far. Therefore, we defer the study of actual circuits containing this more elaborate model until Chapter 13, where we will be able to discuss both the method of determining numerical values of the hybrid-π parameters from a manufacturer's data sheet, and the effect of the capacitive elements on the response of amplifiers at high frequencies.

11.4.2 An FET Incremental Model

As mentioned earlier, the gate of an FET is capacitively coupled to the channel either through a dielectric-filled capacitor in the case of a MOSFET, or the space-charge-layer capacitance of the back-biased gate-to-channel junction in a JFET. In either case, Fig. 11.23 shows an incremental model that includes gate-to-channel capacitors.

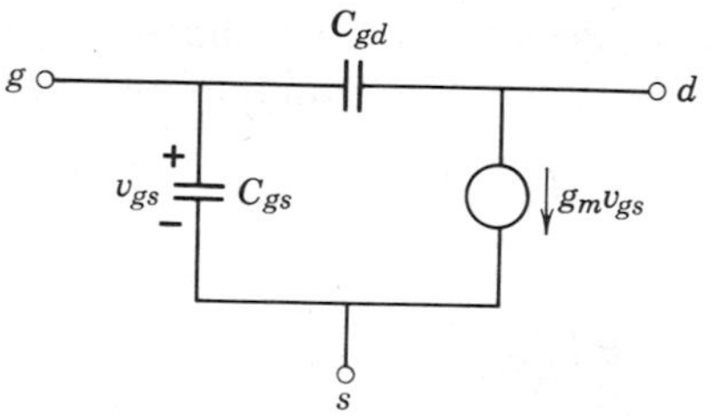

Figure 11.23
An incremental model for an FET.

The capacitors in Fig. 11.23 are small, having magnitudes on the order of several picofarads. However, when coupled with the large Thévenin source resistances often encountered in FET applications, the resulting time-constants can severely limit the response to rapidly varying signals.

11.4.3 Empirical Models: *h*-parameters and *y*-parameters

As linear network theory grew and matured, several elegant and highly formalized methods were developed for handling so-called *two-port networks*: networks with an input terminal pair and an output terminal pair. It was a natural development for these formal methods of network description to be adopted for the representation of the small-signal behavior of active devices. Six different yet equivalent sets of generalized two-port parameters can be defined. Two of these are outlined below.

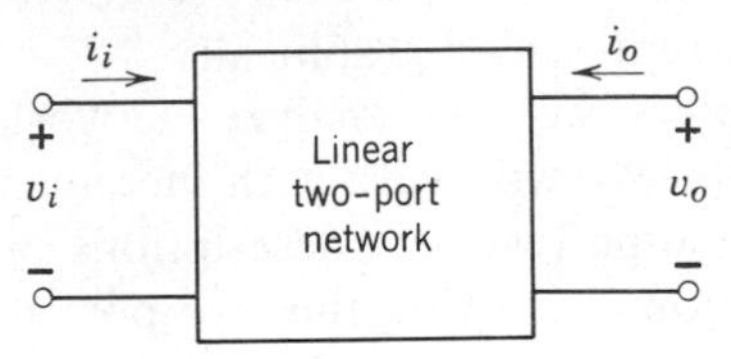

Figure 11.24
A general two-port network.

A general two-port network is shown in Fig. 11.24, with input and output terminals labeled with incremental variables. Since the network is linear, there exist sets of linear equations relating any pair of variables to the other pair. For example, if v_i and v_o are considered as independent variables, one defines the *y-parameters* as follows:

$$\begin{aligned} i_i &= y_i v_i + y_r v_o \\ i_o &= y_f v_i + y_o v_o \end{aligned} \tag{11.80}$$

The y parameters, y_i, y_r, y_f, and y_o have the dimensions of a conductance. They are called *admittance* parameters for reasons that will become clear in Chapter 13. If the two-port network happens to contain a small-signal amplifier circuit containing the low-frequency, resistive hybrid-π model, one can solve the algebraic equations to relate the y-parameters to the elements of the networks. But if the two-port network contains an amplifier that is intended to operate at frequencies at which the energy-storage components of device models must be included, then interpretation of the y-parameters gets very cumbersome. The y-parameters become functions of frequency, and must be represented in graphical form. Many manufacturers publish such graphs as part of the transistor data sheets. The y-parameters are useful primarily in design problems involving relatively narrow ranges of frequency, such as in tuned amplifiers used in radio, TV and other communications applications.

A similar set of two-port parameters, the *h-parameters*, use i_i and v_o as the independent variables.

$$v_i = h_i i_i + h_r v_o$$
$$i_o = h_f i_i + h_o v_o \tag{11.81}$$

As with the *y*-parameters, there are algebraic relations between the *h*-parameters and the elements of the low-frequency, resistive, hybrid-π model. For example, h_{fe} for a small-signal transistor model in the common-emitter configuration[2] happens to equal β_0, while h_{ie} equals r_π. When capacitive effects must be included, however, the *h*-parameters become functions of frequency and must be represented graphically.

In succeeding chapters we will analyze networks containing several energy storage elements. We will work with incremental models instead of with the more cumbersome two-port descriptions of circuits, but we will point out simple relations involving the two-port parameters when they exist.

REFERENCES

R11.1 Paul E. Gray and Campbell L. Searle, *Electronic Principles: Physics, Models, and Circuits* (New York: John Wiley, 1969), Chapters 7 to 10.

R11.2 Richard S. C. Cobbold, *Theory and Applications of Field-Effect Transistors* (New York: Wiley-Interscience, 1970).

QUESTIONS

Q11.1 If the active region of a transistor is restricted to $v_{CE} > V_{CE,sat}$ how can one explain a negative value of an incremental output waveform?

Q11.2 What would be the incremental model for an exponential diode if the diode is biased with a forward current I_D? (Assume $I_D \gg I_S$.)

---

[2] The second subscript denotes that terminal which is common to input and output.

Q11.3 Where does the power gain come from in an incremental circuit?

Q11.4 Estimate how many successive stages of amplification are needed in an op-amp?

Q11.5 Suppose a common-emitter transistor amplifier is biased at the operating point along the load line that corresponds to maximum power dissipation in the transistor. Does the average power dissipated in the transistor increase or decrease when a sinusoidal signal is applied to the amplifier? What about the average power dissipated in the load resistance?

EXERCISES

E11.1
(a) Evaluate the parameters r_π and g_m of the low-frequency incremental model for the transistor in the amplifier of Fig. E11.1.
(b) Draw the incremental circuit for the amplifier and calculate the voltage gain v_o/v_s. Assume the capacitors are short circuits for the incremental signals.
(c) What is the peak magnitude of v_s for which the incremental model remains valid?

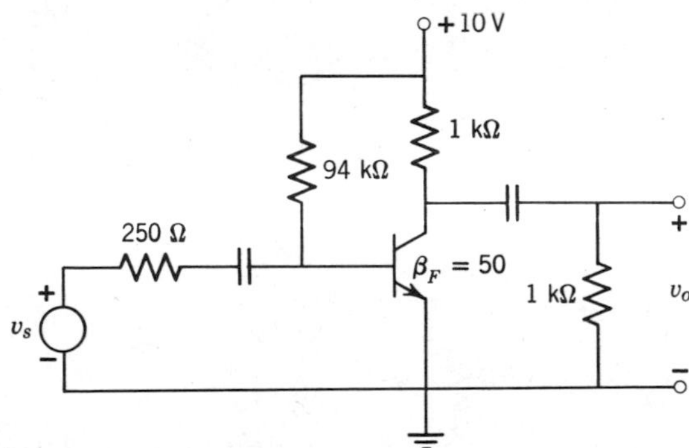

Figure E11.1

E11.2 (a) Find R_B in Fig. E11.2 to set the operating point of the transistor at $I_C = 10$ mA.

(b) Assume that C is a short circuit for the ac signal v_s and is an open circuit for the dc bias currents.
  (i) Sketch the incremental circuit of the amplifier.
  (ii) Estimate to within 20% the incremental voltage gain

$$A = \frac{v_o}{v_s}$$

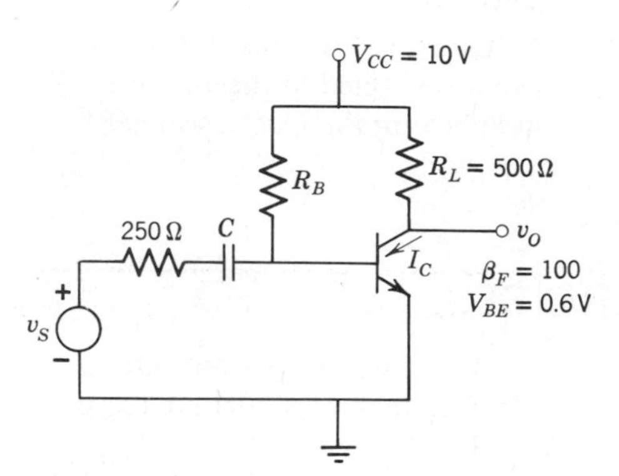

Figure E11.2

E11.3 A circuit for a triode vacuum tube amplifier is shown in Fig. E11.3 along with its incremental circuit model. Assuming that the amplifier is operating in the active-gain region with $\mu = 40$ and $r_p = 24\ \text{k}\Omega$, find the incremental voltage gain v_o/v_s. Assume that the capacitors are short circuits for the incremental signals.

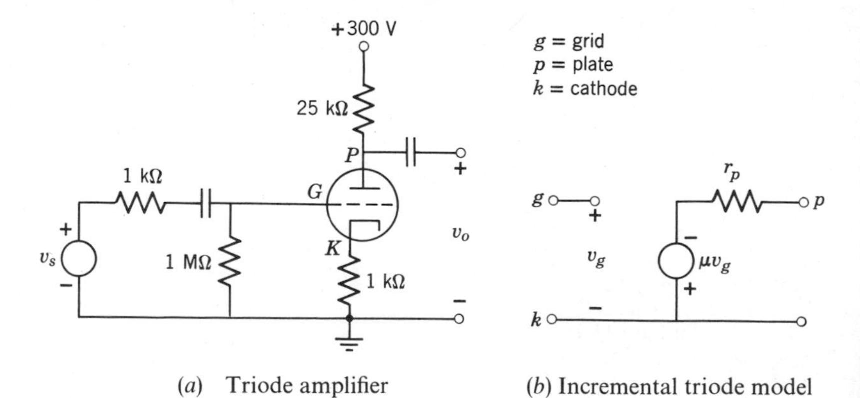

(*a*) Triode amplifier

(*b*) Incremental triode model

Figure E11.3

E11.4 A common-emitter amplifier is required with these specifications:

$$V_{CC} = 15 \text{ V}$$

$$v_{o,\,max} = 5 \text{ V}$$

$$R_L = 2 \text{ k}\Omega$$

Assuming that transistors are selected for β_F in the range of 60 to 80, design a bias network using a single base resistor as shown in Fig. 11.4.

E11.5 Find the incremental voltage gain, v_o/v_s in Fig. E11.5. Assume that all capacitors are short circuits for the incremental signals.

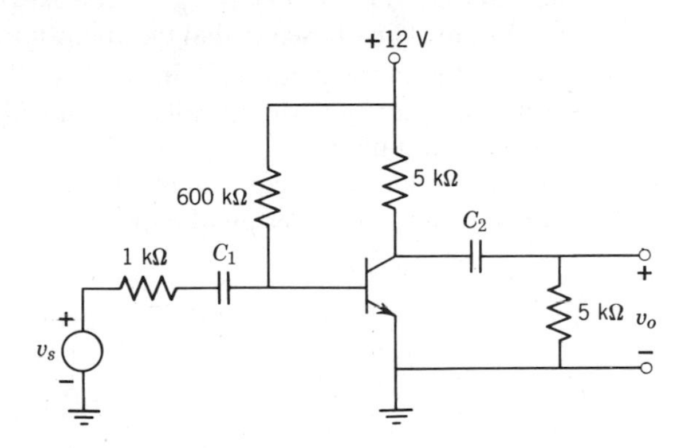

Figure E11.5
$\beta_F = \beta_0 = 100$. C_1, C_2 : coupling capacitors.

E11.6 The v_p source together with R_p comprise a small-signal Thévenin equivalent of a photomultiplier tube (v_p is the open-circuit response to a small variation in incident light intensity.) Because of the high Thévenin resistance ($R_p = 1 \text{ M}\Omega$), an FET follower is used as the first stage of a detector circuit. You may assume that the operating point has already been determined and that the incremental model for the FET shown in Fig. 11.11*b* may be used.

(a) Draw the complete incremental circuit assuming C is a short circuit for the incremental signals.

(b) Find the incremental voltage transfer ratio v_s/v_p. Use numerical values in your final answer.

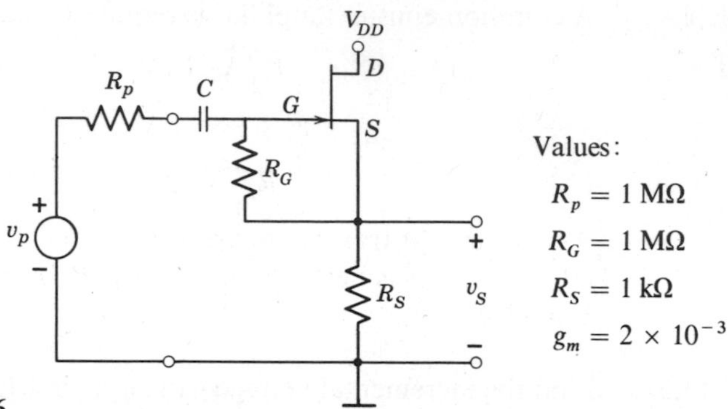

Figure E11.6

E11.7 Assume in Fig. E11.7 that $v_B = V_B + v_b$, where v_b is an incremental signal. Assume V_B is chosen so that the operating point is $I_C = 1$ mA.

(a) Draw the incremental circuit, labeling the two incremental outputs v_e and v_c. Numerically evaluate all parameters of the incremental model.

(b) Find the Thévenin equivalent of the incremental circuit as seen from the v_c-v_e terminal pair.

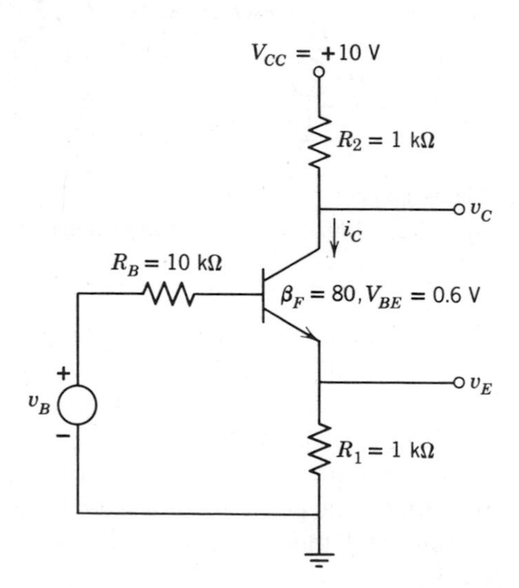

Figure E11.7

E11.8 In the circuit in Fig. E11.8 all transistors have $V_{BE} = 0.6$ V and $\beta_F = 100$. Assume that all capacitors are short circuits for the incremental signals.

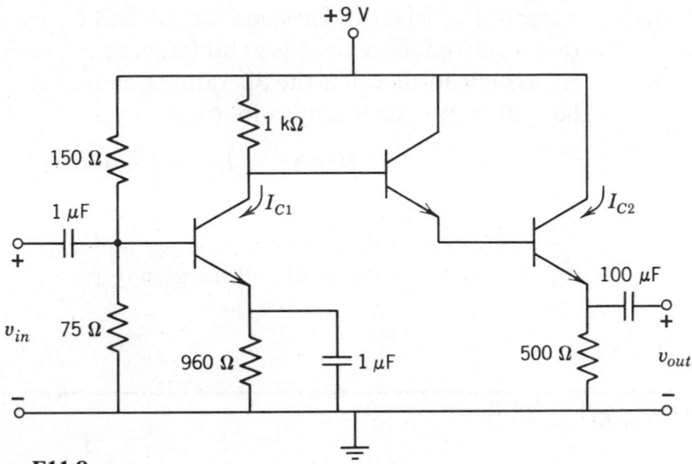

Figure E11.8

(a) What are the bias currents I_{C1} and I_{C2}?
(b) Draw the incremental circuit.
(c) What is the incremental input resistance?
(d) What is the incremental voltage gain v_{out}/v_{in} ?

E11.9 Transistor Q_1 and Q_2 are connected in what is called a "Darlington" connection, a connection that behaves in a large-signal sense like a "supertransistor" with $\beta_{super} \approx \beta_F^2$. Find the operating point collector currents I_{C1} and I_{C2}.

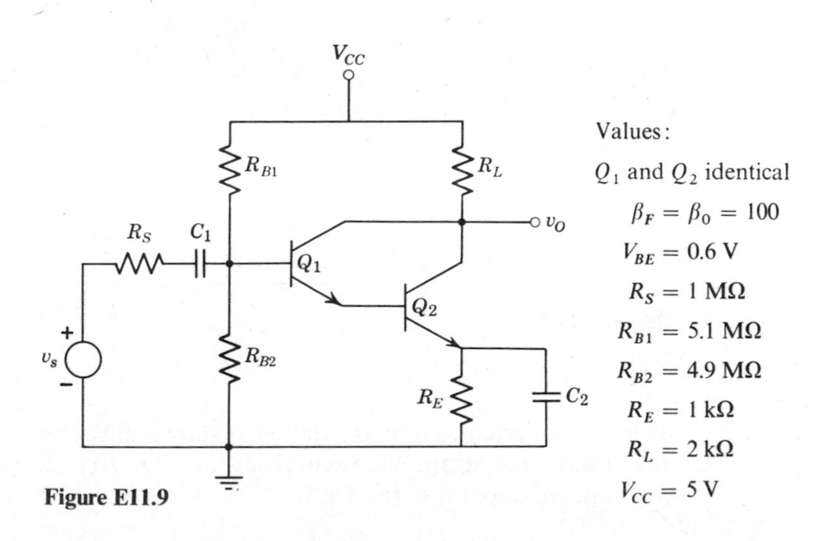

Figure E11.9

E11.10 Refer to Fig. E11.9 and assume that C_1 and C_2 are large enough to be considered short circuits at all frequencies present in the source v_s. Assume further that the operating points of the two transistors have already been found as follows:

$$I_{C1} = 10\ \mu\text{A} \qquad I_{C2} = 1\ \text{mA}$$

(a) Draw the complete incremental circuit labeling all elements. Determine values for $r_{\pi 1}$, $r_{\pi 2}$, g_{m1} and g_{m2}.
(b) Find the incremental voltage gain v_o/v_s.

PROBLEMS

P11.1 (a) Find the Thévenin equivalent for the incremental behavior at the output terminals of the emitter-follower amplifier shown in Fig. P11.1. Assume the capacitor is a short circuit for the incremental signals.
(b) Evaluate the parameters in your Thévenin equivalent for $V_{CC} = +15$ V. $R_{B1} = R_{B2} = 22$ kΩ, $R_E = R_S = 1$ kΩ, and $\beta_F = \beta_0 = 100$.

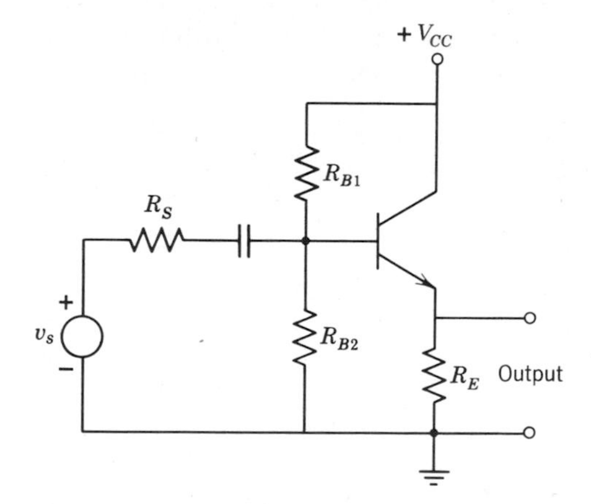

Figure P11.1

P11.2 A FET connected as a phase splitter is shown in Fig. P11.2*a*.
(a) Using the drain characteristics in Fig. P11.2*b*, find the operating point of the FET.

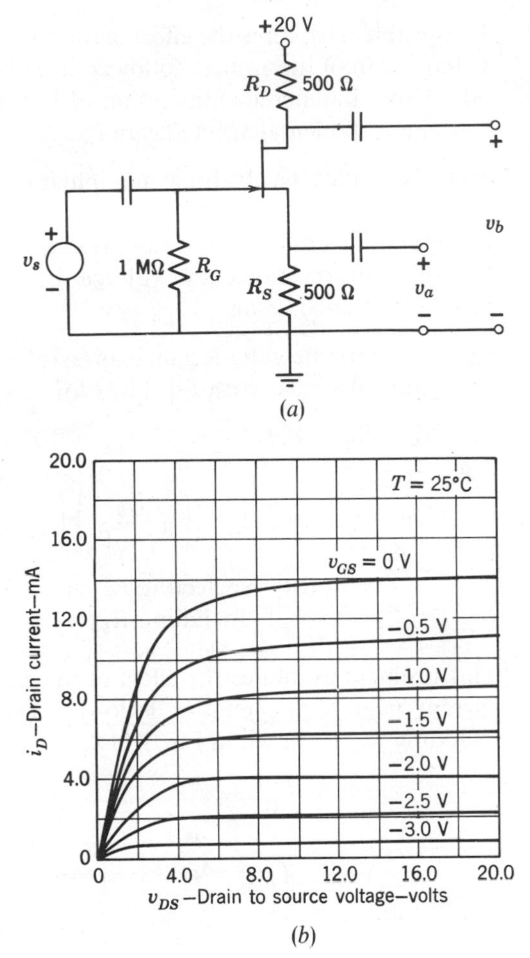

Figure P11.2 (b)

(b) Noting that for small changes in operating point, Δi_D and Δv_{GS},

$$g_m = \frac{\Delta i_D}{\Delta v_{GS}}$$

graphically evaluate g_m for the FET at the operating point of part *a*.

(c) Use the incremental model to calculate the voltage gains v_a/v_s and v_b/v_s. Assume the capacitors are short circuits for the incremental signals.

(d) Find the incremental Thévenin equivalent circuits that represent the behavior at the output terminals (v_a and v_b).

P11.3 This problem explores the effect of the base-bias resistor R_B on the voltage gain of the emitter-follower amplifier.

(a) Using the incremental circuit of Fig. 11.8 including R_B, find the incremental voltage gain v_o/v_i.

Hint. To reduce the algebraic manipulations, first form a Thévenin equivalent for v_i, R_S, and R_B. The circuit is then reduced to the form for which the solution Eq. 11.53 is valid, and substitution of the Thévenin resistance and voltage source values into Eq. 11.53 yields the desired result.

(b) Compare the voltage gain expression found in part *a* with the gain calculated with Eq. 11.53 for this case:

$R_S = R_E = 1\ \text{k}\Omega$, $\beta_0 = 99$, $V_{CC} = 10\ \text{V}$, and $V_B = 5\ \text{V}$.

Answer. (a) $A_v = \left(\dfrac{R_B}{R_S + R_B}\right)\left(\dfrac{(1+\beta_0)R_E}{R_S\|R_B + r_\pi + (1+\beta_0)R_E}\right)$

(b) Neglecting R_B, $A_v = 0.985$.
Including R_B, $A_v = 0.953$.

P11.4 This problem examines the effect of the base-bias resistor on the current gain of the emitter follower. The incremental circuit, including R_B, is shown in Fig. P11.4.

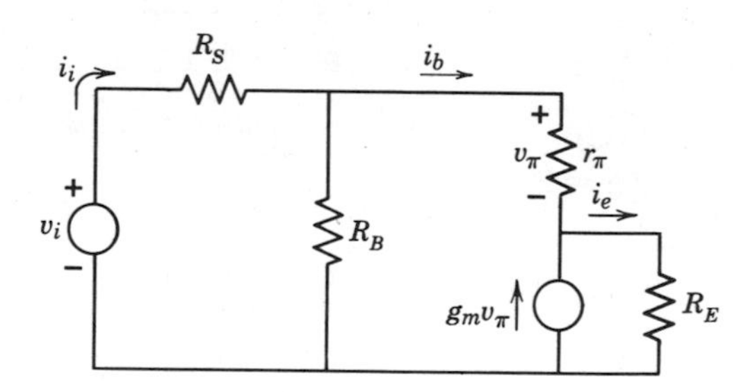

Figure P11.4

(a) Find the current ratio i_e/i_b.

(b) By using the equivalent input circuit of Fig. 11.9, find the current ratio i_b/i_i.

(c) Use the results of parts *a* and *b* to find the current gain i_e/i_i.

(d) With $R_E = 1\ \text{k}\Omega$, $R_B = 100\ \text{k}\Omega$, $r_\pi = 0.5\ \text{k}\Omega$, and $\beta_0 = 99$, compare the current gain expression derived in part *c* with the current gain derived in Eq. 11.55.

Answer. (a) $\dfrac{i_e}{i_b} = 1 + \beta_0$

(b) $\dfrac{i_b}{i_i} = \dfrac{R_B}{R_B + r_\pi + (1 + \beta_0)R_E}$

(c) $\dfrac{i_e}{i_i} = \dfrac{(1 + \beta_0)R_B}{R_B + r_\pi + (1 + \beta_0)R_E}$

(d) Including R_B: $\dfrac{i_e}{i_i} = 50$

Eq. 11.55: $\dfrac{i_e}{i_i} = 100$

P11.5 This problem is concerned with the effect of R_B on the input resistance of the emitter-follower amplifier shown in Fig. 11.7. For the calculations you may assume

$$\beta_F = \beta_0, \quad (1 + \beta_0)R_E \gg r_\pi, \quad \text{and} \quad V_{BE} = 0.$$

(a) Express the base voltage V_B in terms of R_B, R_E, β_F, and V_{CC}.
(b) Express R_i in terms of R_B, R_E, and β_0.
(c) Eliminate R_B between the expressions in parts *a* and *b* to yield R_i in terms of β_0, R_E, V_B, and V_{CC}.
(d) If the output voltage v_o is required to have peak incremental voltages $v_{o,max}$, what is the maximum possible input resistance R_i?
(e) Find maximum possible value of R_i for this case:

$R_E = 1\ \text{k}\Omega$, $\beta_0 = 99$, $V_{CC} = 15$ V, and $v_{o,max} = 5$ V.

Compare this result with the result obtained from Eq. 11.57.

Answer. (a) $V_D = \dfrac{(1 + \beta_0)R_E V_{CC}}{(1 + \beta_0)R_E + R_B}$

(b) $R_i = R_B \| (1 + \beta_0)R_E$

(c) $R_i = (1 + \beta_0)R_E \left(\dfrac{V_{CC} - V_B}{V_{CC}}\right)$

(d) $R_{i,max} = (1 + \beta_0)R_E \left(\dfrac{V_{CC} - v_{o,max}}{V_{CC}}\right)$

(e) Maximum possible: $R_i = 67\ \text{k}\Omega$

From Eq. 11.57: $R_i = 100\ \text{k}\Omega$

P11.6 Using the drain characteristics in Fig. P11.2*b*, use a graphical construction technique to find the operating point (I_D, V_{DS}) for the self-biased FET amplifier shown in Fig. P11.6.

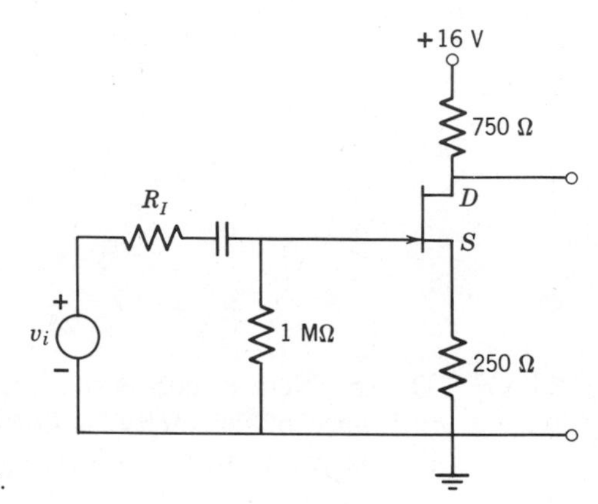

Figure P11.6
Self-biased FET amplifier.

P11.7 Using the drain characteristics in Fig. P11.2*b* find a value for R_S that will permit $v_{o,\,max} = 4$ V in the FET amplifier in Fig. P11.7.

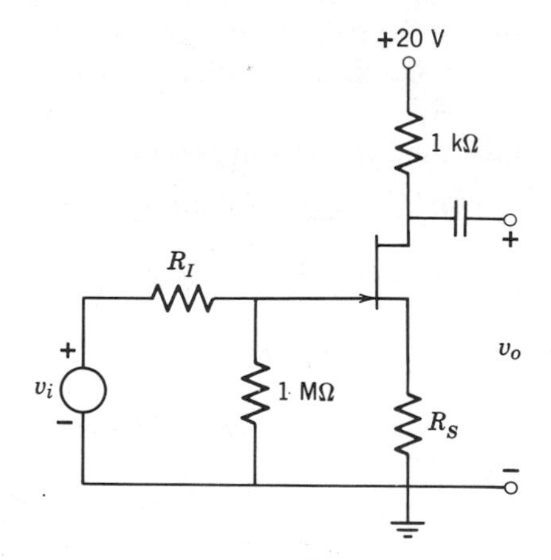

Figure P11.7

P11.8 The circuit in Fig. P11.8*a* is frequently used in the design of operational amplifiers.

(a) Assuming that $v_a = v_b = 0$, find the operating point of each transistor.

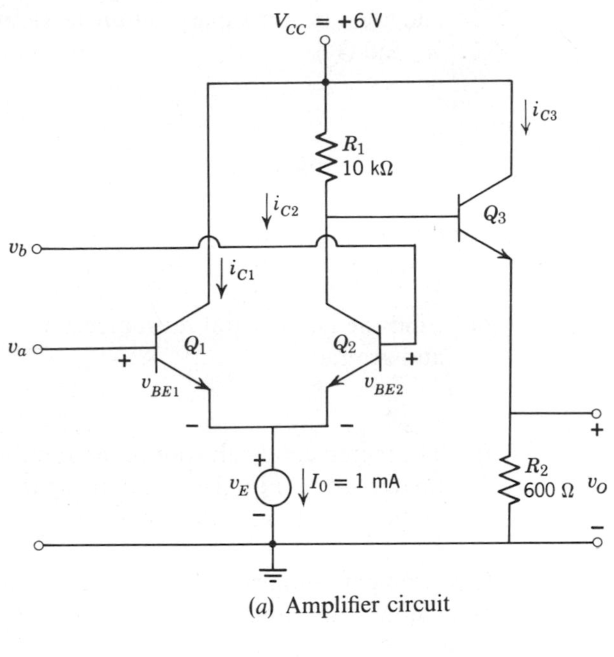

(*a*) Amplifier circuit

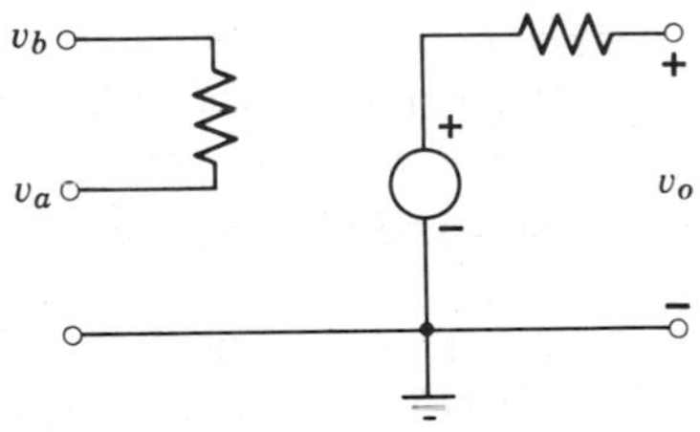

Figure P11.8

(*b*) Amplifier model

Answer.

$$I_{C1} = 0.5 \text{ mA}, V_{CE1} = 6.6 \text{ V}.$$
$$I_{C2} = 0.5 \text{ mA}, V_{CE2} = 1.6 \text{ V}.$$
$$I_{C3} = 0.67 \text{ mA}, V_{CE3} = 5.6 \text{ V}.$$

(b) Draw an incremental circuit for the amplifier and evaluate r_π and g_m for each transistor.

(c) Using the incremental circuit from part *b*, find the voltage gains, v_o/v_a and v_o/v_b. (*Hint.* Since the circuit is linear, the output must be of the form

$$v_o = A_1 v_a + A_2 v_b$$

and we can use superposition to simplify the evaluation of A_1 and A_2).

Answer. $v_o/v_a = 81.$

$v_o/v_b = -81.$

(d) Find the incremental resistance between the input terminals *a-b*.

Answer. 10 kΩ.

(e) Find the incremental resistance between the output terminal and ground.

Answer. 112 Ω.

(f) The incremental behavior of the amplifier is to be modeled by the circuit in Fig. P11.8*b*. Label the circuit elements based on your analysis of the amplifier.

P11.9 This problem concerns a common-base amplifier (Fig. P11.9). Assume the capacitors are open circuits for dc and short circuits for ac.

Assume $\beta_F = \beta_0 = 100$ and $V_{BE} = 0.6$ V in the active-gain region.

(a) Determine the dc collector current, I_C.

(b) Draw the ac incremental circuit using the g_m-r_π model for the transistor. Determine numerical values for both g_m and r_π.

(c) Calculate the incremental voltage gain, current gain, input resistance, and output resistance.

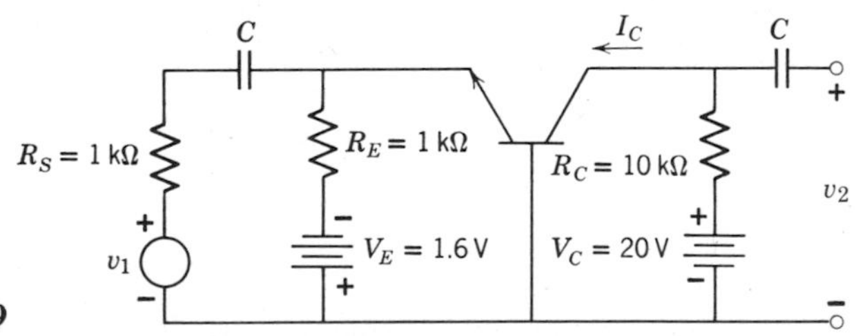

Figure P11.9

P11.10 In integrated circuit designs, diodes are often used in the transistor bias network. Consider the example in Fig. P11.10.

Assuming a silicon transistor and an exponential silicon semiconductor diode:

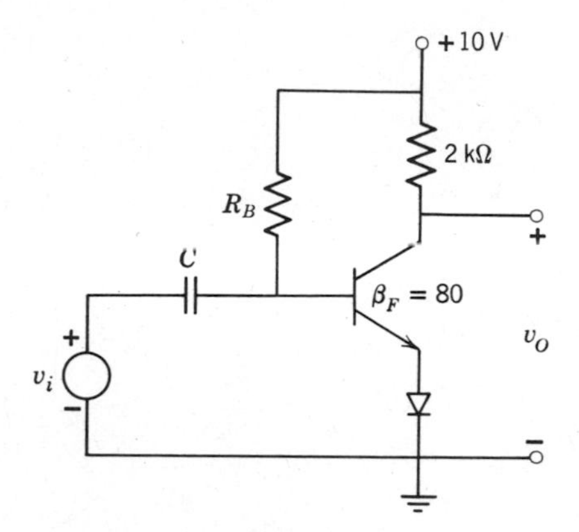

Figure P11.10

(a) Find the value of R_B that establishes the dc collector current at 2 mA.

(b) At the operating point given in part *a* and assuming the capacitor is an incremental short, find the incremental voltage gain v_o/v_i for the amplifier. (See also Question Q11.2.)

P11.11 Find the operating points in Fig. 11.16. Draw the incremental circuit and find the small signal ac voltage gain v_o/v_i.

PART C

LINEAR CIRCUITS AND SYSTEMS

CHAPTER TWELVE
LINEAR-SYSTEM RESPONSE

12.0 A RETURN TO THEORY

Our repertory of circuits and devices has been greatly expanded in Part B. It is now time to expand our ability to describe how signals are processed by these circuits. In Part C we shall be concerned specifically with *linear networks*; these are networks that can be represented as interconnections of ideal sources, resistances, capacitances, inductances, and transformers. As we have just learned in Chapter 11, any nonlinear device can, for sufficiently restricted ranges of voltage and current, be represented by an incremental model. Thus, networks containing linear incremental models of transistors, op-amps, or other elements are included in the class of linear networks.

This chapter presents a general, formal method for finding the response of linear networks to a wide variety of excitations. Although our analysis techniques will be discussed entirely in the context of electrical circuits, the techniques can be applied to any linear system. There are many situations in mechanics, optics, acoustics, even in economics and business, that can be modeled by networks of idealized linear elements, analogous to the electrical circuits discussed in this chapter. A working knowledge of the methods for analyzing linear systems will enhance the appreciation of almost any technical subject.

As a prelude to the formal consideration of linear networks, some useful properties of complex numbers and complex exponential functions are presented. Then the concepts of complex impedance and system function are introduced and used to find the natural responses of networks and the total responses of networks to a general class of excitations.

12.1 COMPLEX NUMBERS AND COMPLEX FREQUENCIES

12.1.1 Complex Numbers

The square root of -1 is an *imaginary number*, usually denoted by i. To prevent confusion between imaginary numbers and current, engineering usage has adopted the letter j to represent $\sqrt{-1}$. That is, by definition of j,

$$(j)^2 = -1 \tag{12.1}$$

Additional imaginary numbers can be formed by multiplying j by any real number. Thus, if b is a real number, jb is an imaginary number. The square of any imaginary number is negative.

$$(jb)^2 = (j)^2 b^2 = -b^2 \tag{12.2}$$

A *complex number* is formed by adding a real number to an imaginary number. Thus, if a and b are real numbers, a complex number c can be formed:

$$c = a + jb \tag{12.3}$$

The pair of real numbers, a and b, are called the *real part* of c and the *imaginary part* of c, respectively. The real part is written:

$$\text{Re}\{c\} = a \tag{12.4}$$

while the imaginary part is written

$$\text{Im}\{c\} = b \tag{12.5}$$

Note that by convention the imaginary part is equal to b, *not* jb.

Two complex numbers, c_1 and c_2, can be added and multiplied as in normal algebra as long as one remembers that $j^2 = -1$. That is,

$$\begin{aligned} c_1 + c_2 &= (a_1 + jb_1) + (a_1 + jb_2) \\ &= (a_1 + a_2) + j(b_1 + b_2) \end{aligned} \tag{12.6}$$

and

$$\begin{aligned} c_1 c_2 &= (a_1 + jb_1)(a_2 + jb_2) \\ &= (a_1 a_2 - b_1 b_2) + j(a_1 b_2 + a_2 b_1) \end{aligned} \tag{12.7}$$

12.1.2 The Complex Plane

Complex numbers can be represented as points in a *complex plane*. Figure 12.1 shows three different complex numbers plotted in such a plane. The *real axis* is labeled by σ, while the *imaginary axis* is labeled by $j\omega$. When σ and ω are used to signify the real and imaginary parts of a complex number, s is commonly used to represent the complex number itself.

$$s = \sigma + j\omega \tag{12.8}$$

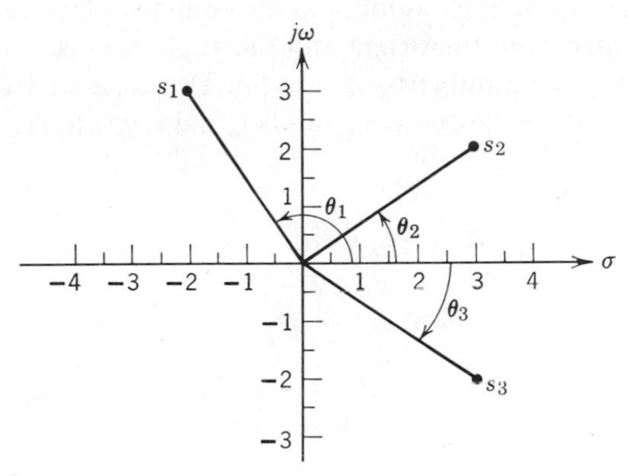

Figure 12.1
Numbers in the s-plane.

The plane with these coordinate labels is called the s-plane.

Consider the number s_1 in Fig. 12.1. In complex notation, we write

$$s_1 = -2 + j3 \tag{12.9}$$

The real part of s_1 is negative (-2), which puts s_1 in what is called the *left half-plane*. The other two numbers, s_2 and s_3, are written

$$\begin{aligned} s_2 &= 3 + j2 \\ s_3 &= 3 - j2 \end{aligned} \tag{12.10}$$

The real parts of s_2 and s_3 are positive ($+3$), which put them in the *right half-plane*. (A complex number with a real part equal to zero lies on the imaginary axis. It is not considered to be in either half-plane.)

The *complex conjugate* of a number is obtained by replacing j by $-j$. If s is the complex number in Eq. 12.8, the complex conjugate of s is denoted by s^* and is written

$$s^* = \sigma - j\omega \tag{12.11}$$

Examining the pair of numbers s_2 and s_3, we see that

$$s_2 = s_3^* \tag{12.12}$$

and

$$s_3^* = s_2 \tag{12.13}$$

A pair of numbers that are complex conjugates of each other, such as s_2 and s_3 above, is called a *complex conjugate pair*, or more simply, a *complex pair*.

The position of a complex number in the complex plane can also be specified by its distance from the origin and the angle between the real axis and the arrow joining the number to the origin. The distance from the origin is called the *magnitude* of the complex number, and is written $|s|$. By the Pythagorean theorem

$$|s| = \sqrt{(\text{Re}\{s\})^2 + (\text{Im}\{s\})^2} = \sqrt{\sigma^2 + \omega^2} \tag{12.14}$$

The *angle* of the complex number, written $\angle s$, is given by

$$\angle s = \tan^{-1}\left(\frac{\text{Im}\{s\}}{\text{Re}\{s\}}\right) = \tan^{-1}\frac{\omega}{\sigma} \tag{12.15}$$

Returning to the three numbers plotted in Figure 12.1, the magnitudes are all equal.

$$|s_1| = |s_2| = |s_3| = \sqrt{(2)^2 + (3)^2} = \sqrt{13} \tag{12.16}$$

The angles of the three numbers are all different. For example,

$$\angle s_1 = \theta_1 = \tan^{-1}(-\tfrac{3}{2}) = 123.7° \tag{12.17}$$

while

$$\angle s_2 = \theta_2 = \tan^{-1}(\tfrac{2}{3}) = 33.7° \tag{12.18}$$

By convention, positive angles are measured from the real axis in a counterclockwise direction. Therefore θ_3 in Fig. 12.1 must be considered to be a negative angle. That is,

$$\angle s_3 = \theta_3 = -\theta_2 = -33.7° \tag{12.19}$$

Members of a complex pair always have equal magnitudes and angles that differ only in sign.

12.1.3 Exponential Notation

If we denote $\angle s$ by θ, simple trigonometry yields

$$\begin{aligned} \text{Re}\{s\} &= |s| \cos \theta \\ \text{Im}\{s\} &= |s| \sin \theta \end{aligned} \tag{12.20}$$

Therefore, we can write

$$s = |s| (\cos \theta + j \sin \theta) \tag{12.21}$$

This particular way of writing complex numbers can be reduced to a shorthand notation using an important property of the exponential function, called Euler's theorem. If α is a real number, then

$$e^{j\alpha} = \cos \alpha + j \sin \alpha \tag{12.22}$$

This relation, which must be memorized thoroughly, allows us to write Eq. 12.21 in this very concise form:

$$s = |s| \, e^{j\theta} \tag{12.23}$$

This exponential notation for complex numbers is particularly convenient when dealing with products, quotients, or complex conjugates. For example, if s_1 and s_2 are two complex numbers with angles θ_1 and θ_2,

$$s_1 s_2 = (|s_1| \, e^{j\theta_1})(|s_2| \, e^{j\theta_2}) = |s_1| |s_2| \, e^{j(\theta_1 + \theta_2)} \tag{12.24}$$

while

$$\frac{s_1}{s_2} = \frac{|s_1|}{|s_2|} \cdot \frac{e^{j\theta_1}}{e^{j\theta_2}} = \frac{|s_1|}{|s_2|} \, e^{j(\theta_1 - \theta_2)} \tag{12.25}$$

Finally, if

$$s = |s| \, e^{j\theta} \tag{12.26}$$

then

$$s^* = |s| \, e^{-j\theta} \tag{12.27}$$

and

$$s^* s = |s|^2 \, e^{j(\theta - \theta)} = |s|^2 \tag{12.28}$$

In Chapter 13 we will have frequent need for the magnitude of complex numbers. Equation 12.28 provides one convenient method for finding this quantity.

12.1.4 Complex Frequencies

When a complex number s appears in the exponential time function e^{st}, s is called the *complex frequency*. Since from Eqs. 12.8 and 12.22

$$e^{st} = e^{\sigma t}\, e^{j\omega t} = e^{\sigma t}(\cos \omega t + j \sin \omega t) \tag{12.29}$$

it follows that

$$\begin{aligned} \text{Re}\{e^{st}\} &= e^{\sigma t} \cos \omega t \\ \text{Im}\{e^{st}\} &= e^{\sigma t} \sin \omega t \end{aligned} \tag{12.30}$$

Depending on the values of σ and ω, the real or imaginary parts of e^{st} can represent a wide variety of waveforms. Let us consider some of them.

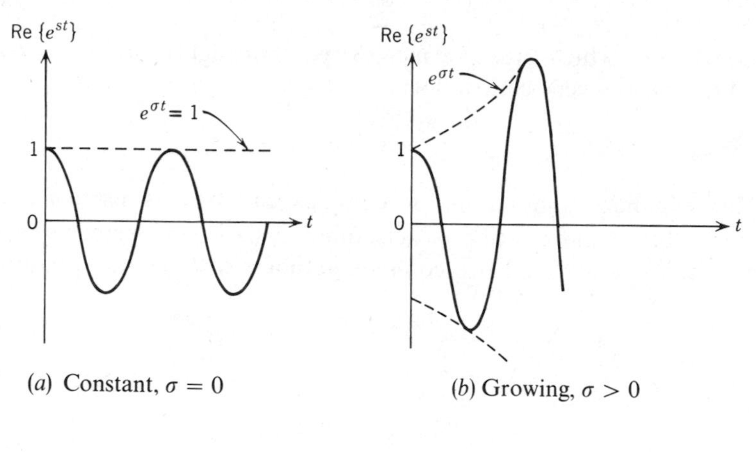

(*a*) Constant, $\sigma = 0$

(*b*) Growing, $\sigma > 0$

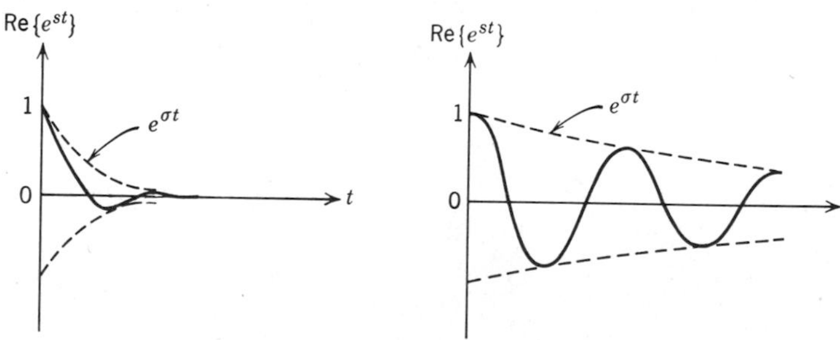

(*c*) Strongly damped, $\sigma < 0$ with $|\sigma| > |\omega|$

(*d*) Weakly damped, $\sigma < 0$ with $|\sigma| < |\omega|$

Figure 12.2
Exponential waveforms for complex *s*.

When $s = 0$ (both σ and ω being zero), e^{st} is identically equal to unity (a real number). Thus, the trivial case $s = 0$ can be used to represent a dc or constant waveform.

Next, if s is real ($\omega = 0$ but $\sigma \neq 0$), e^{st} is the exponential time function encountered in Chapter 6. For $\sigma > 0$ (s on the positive real axis) the exponential grows with time, while for $\sigma < 0$ (s on the negative real axis), the exponential decays with time.

For $\omega \neq 0$, both the real and imaginary parts of e^{st} have sinusoidal time dependencies. Figure 12.2 displays four typical waveforms of $\text{Re}\{e^{st}\}$ for $t > 0$. The $\sigma = 0$ case (s on the imaginary axis) corresponds to a constant amplitude sinusoid of frequency ω. When $\sigma > 0$ (s in the right half-plane), the amplitude of the sinusoid grows exponentially, while for $\sigma < 0$ (s in the left half-plane) the amplitude decays exponentially. It is important to recognize the qualitative difference between the *strongly damped* case of Fig. 12.2*c* and the *weakly damped* case of Fig. 12.2*d*. Strong damping occurs when $|\sigma| > |\omega|$; the exponential varies so quickly compared to the period of the sinusoid that the waveform dies out before one recognizable full cycle. Weak damping occurs when $|\sigma| < |\omega|$; the sinusoid takes many periods to die out. This latter waveform is often called *ringing*, since the sound from a struck bell resembles a weakly damped sinusoid.

The properties of the various complex frequencies and their associated waveforms are summarized in Fig. 12.3. In the remaining sections of this chapter, we shall use complex numbers and complex frequencies to assist our analysis of linear networks.

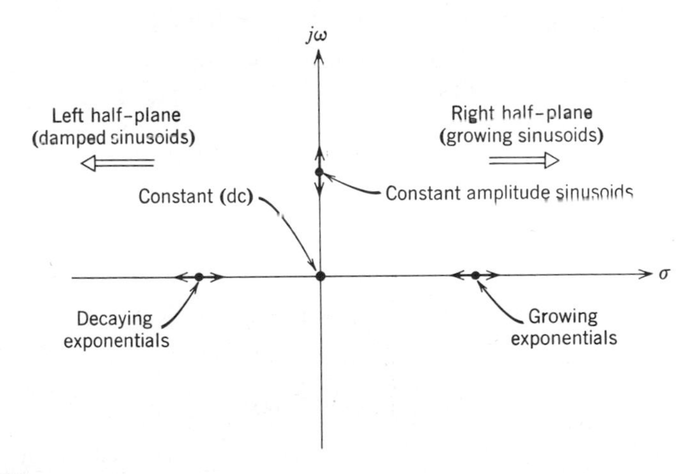

Figure 12.3
The complex frequency plane.

12.2 RESPONSE TO AN e^{st} SOURCE

12.2.1 Representing Real Sources in e^{st} Form

Because of the wide variety of waveforms that can be represented by the real and imaginary parts of e^{st}, the response of a network to a general e^{st} source contains an enormous amount of information about the behavior of that network.

The first problem we encounter is that physical voltages and currents must be represented by real variables, whereas e^{st} is usually a complex number. Therefore, as a standard practice, the waveform of the physical voltage or current must correspond either to the real part *or* to the imaginary part of an e^{st} complex waveform.

To represent various amplitudes of real waveforms, we introduce the concept of a *complex amplitude*.[1] If A_o is a complex number, with magnitude $|A_o|$ and angle ϕ, then the quantity

$$A_o e^{st} = |A_o| e^{\sigma t} e^{j(\omega t + \phi)} \tag{12.31}$$

is a complex waveform with real and imaginary parts, as shown below:

$$\begin{aligned} \mathrm{Re}\{A_o e^{st}\} &= |A_o| e^{\sigma t} \cos(\omega t + \phi) \\ \mathrm{Im}\{A_o e^{st}\} &= |A_o| e^{\sigma t} \sin(\omega t + \phi) \end{aligned} \tag{12.32}$$

The angle ϕ is called the *phase angle*. It specifies the starting angle (at $t = 0$) for the cosine and sine functions. Thus, for example, if

$$v_O = 5\, e^{-t/2} \sin(1000t + \pi/6) \tag{12.33}$$

we could represent it as

$$v_O = \mathrm{Im}\{(5\, e^{j\pi/6})(e^{(-1/2 + j1000)t})\} \tag{12.34}$$

Comparing with Eqs. 12.31 and 12.32, we have

$$\begin{aligned} A_o &: \begin{cases} |A_o| = 5 \\ \phi = \pi/6 \end{cases} \\ s &: \begin{cases} \sigma = -\frac{1}{2} \\ \omega = 1000 \end{cases} \end{aligned} \tag{12.35}$$

[1] In our shorthand notation, complex amplitudes are denoted by capital letters with lower-case subscripts.

The reason for wishing to represent real waveforms in e^{st} form is shown in Fig. 12.4. In most cases, the direct calculation of the response to a source waveform like Eq. 12.33 is a cumbersome job. By contrast, we shall see shortly that finding the response to the complex source $A_o e^{st}$ is very easy. Therefore, we can simplify our task if we first express v_O as $\text{Re}\{A_o e^{st}\}$ (or $\text{Im}\{A_o e^{st}\}$ as in Eq. 12.34 above). We then find the complex response to the complex source $A_o e^{st}$. Finally, we obtain the real response $v_R(t)$ by taking the corresponding real or imaginary part of the complex response. If the source $v_O(t)$ was originally expressed as $\text{Re}\{A_o e^{st}\}$, the real response $v_R(t)$ is the real part of the complex response. If the source $v_O(t)$ was originally expressed as $\text{Im}\{A_o e^{st}\}$, the imaginary part of the response is used.

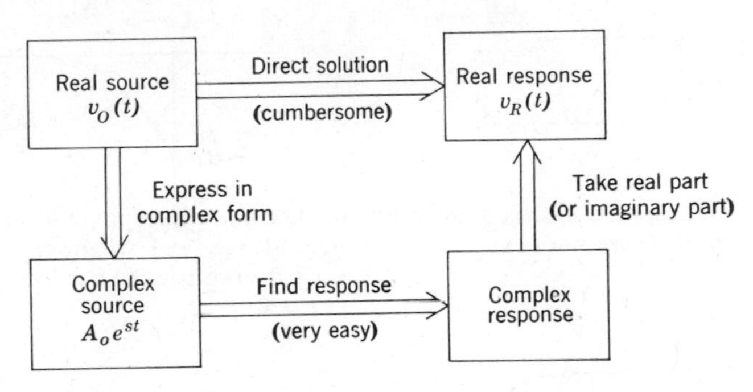

Figure 12.4
Use of complex sources and responses.

Finally, we shall assume that the $A_o e^{st}$ source waveforms are turned on at some convenient time, usually $t = 0$. To remind us of this fact, we shall multiply $A_o e^{st}$ by the unit step, $u_{-1}(t)$. Since $u_{-1}(t)$ is zero before $t = 0$ and unity thereafter, its presence does not alter the real or imaginary parts of $A_o e^{st}$ for $t > 0$. Thus a general complex waveform initiated at $t = 0$ would be written $u_{-1}(t)A_o e^{st}$. Notice that a step waveform can also be represented in this way by choosing $s = 0$.

12.2.2 Forced and Natural Responses

When the e^{st} method is used for representing sources and responses, it is extremely convenient to break up the total response into two parts. The *forced response* is the specific response to the excitation, and has the same e^{st} time dependence as the source waveform. The *natural response*, which was introduced in Chapter 6 in the context of step responses in *RC* and *LR*

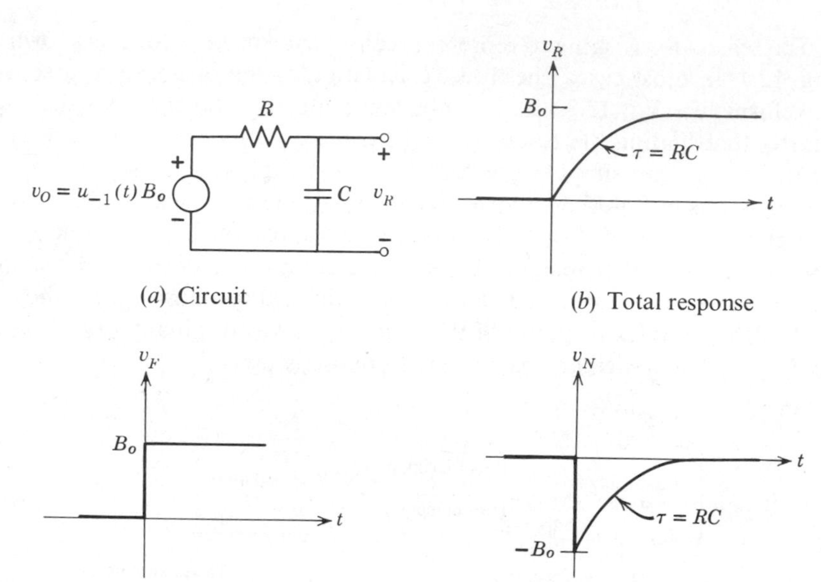

(*c*) The forced response has the same s as the source. In this case $s = 0$

(*d*) The natural response has its own time dependence, and whatever amplitude is needed to satisfy the continuity condition at $t = 0$

Figure 12.5
Forced and natural responses.

circuits, has its own characteristic time dependence, and has whatever amplitude is needed to satisfy continuity conditions on the energy storage elements. Here is an example.

A step source driving an RC network is shown in Fig. 12.5*a*. The total response, which represents exponential charging of the capacitor to the step voltage B_o, is shown in Fig. 12.5*b*. The time constant $\tau = RC$ determines the time dependence of the natural response. Notice that the form of the natural-response time dependence, $e^{-t/\tau}$, can be represented in exponential form $e^{s_a t}$ with the specific choice $s_a = -(1/RC)$. As we shall see, all natural responses of linear networks can be represented with e^{st} waveforms. The algebraic decomposition of the total response into its components is written

$$v_R = [u_{-1}(t)B_o] + [-u_{-1}(t)B_o e^{-t/\tau}]$$

$$\text{Total response} = \text{forced response} + \text{natural response} \qquad (12.36)$$

$$(s = 0) \qquad\qquad (s_a = -1/RC)$$

Figures 12.5*c* and *d* show separate graphs of the forced response and the natural response. Notice that the time dependence of the forced response is the same as that of the source (a dc step in this example) while the natural response has its own characteristic time dependence and an amplitude determined by the requirement that the voltage across the capacitor not change instantaneously at $t = 0$.

12.2.3 Complex Response to e^{st}: The System Function

We shall now find the complex response to a complex source for the circuit of Fig. 12.6. Assuming that v_O is of the form $u_{-1}(t)A_o e^{s_o t}$, we know that v_O is zero for all times before $t = 0$. Therefore, we know that before $t = 0$, the capacitor voltage v_C must also be zero. Because the capacitor voltage cannot change instantaneously, the presence of $u_{-1}(t)$ in the source waveform establishes an *initial condition* for the capacitor voltage,

$$v_C = 0 \quad \text{at} \quad t = 0^+ \tag{12.37}$$

As in Chapter 6, we shall solve the differential equation in detail for the example of Fig. 12.6, and in doing so, we shall identify those characteristics of the solution that will enable us to skip differential equations in the future. The differential equation from KVL is (see Section 6.3.3).

$$RC\frac{dv_C}{dt} + v_C = v_O \tag{12.38}$$

where for $t > 0$,

$$v_O = A_o e^{s_o t} \tag{12.39}$$

The total solution to Eq. 12.38 consists of a forced response to the $A_o e^{st}$ waveform plus enough natural response to make the total solution satisfy the initial condition Eq. 12.37. Let us begin by finding the forced response, v_F.

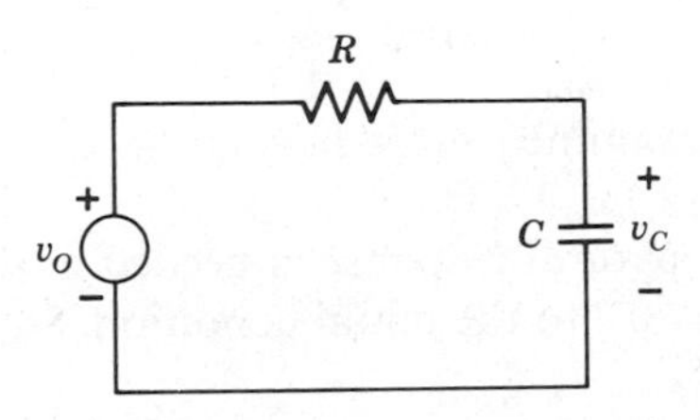

Figure 12.6

If we assume that the forced complex response can be expressed in e^{st} form, we try a solution for Eq. 12.38 such as

$$v_F = A_f\, e^{s_f t} \tag{12.40}$$

Notice that for the exponential functions, the rate of change of v_F is simply s_f times v_F. That is,

$$\frac{dv_F}{dt} = A_f s_f\, e^{s_f t} = s_f v_F \tag{12.41}$$

Substitution in Eq. 12.38 yields

$$(RCs_f + 1)A_f\, e^{s_f t} = A_o\, e^{s_o t} \tag{12.42}$$

For this equation to be satisfied for all times, we must choose

$$s_f = s_o \tag{12.43}$$

In this case, the exponentials cancel out of the equation, and the differential equation is reduced to an algebraic equation in which s_o replaces s_f

$$(RCs_o + 1)A_f = A_o \tag{12.44}$$

Solving for A_f and substituting in Eq. 12.40,

$$v_F = \left(\frac{1}{1 + RCs_o}\right) A_o\, e^{s_o t} = \left(\frac{1}{1 + RCs_o}\right) v_O \tag{12.45}$$

Notice that the forced response has the same time dependence as the source, and has a complex amplitude given by

$$A_f = \left(\frac{1}{1 + RCs_o}\right) A_o \tag{12.46}$$

The function in parentheses, which relates the complex amplitude of the source to the complex amplitude of the forced response, is called the *system function.* Since it depends on the value of s, the system function is written with s as a dependent variable, the symbol $H(s)$ being commonly used. For the example in Eq. 12.46, the system function is written

$$H(s) = \frac{1}{1 + RCs} \tag{12.47}$$

System functions are examined more fully in Section 12.3. But first, let us complete the solution of Eq. 12.38.

To see whether any natural response is needed in the total response, we must compare v_F at $t = 0^+$ to the initial condition. Since

$$v_F = \frac{A_o}{1 + RCs_o} \qquad \text{at} \qquad t = 0^+ \tag{12.48}$$

the natural response v_N is needed, and must satisfy the condition

$$v_N = -\frac{A_o}{1 + RCs_o} \quad \text{at} \quad t = 0^+ \tag{12.49}$$

so that the total response ($v_C = v_N + v_F$) can satisfy the initial condition of zero (Eq. 12.37).

To find the natural response, we solve Eq. 12.38 with the source term set to zero.

$$RC\frac{dv_N}{dt} + v_N = 0 \tag{12.50}$$

In the spirit of seeking exponential solutions, we assume

$$v_N = A_n e^{s_n t} \tag{12.51}$$

and substitute in Eq. 12.50 obtaining

$$(RCs_n + 1)A_n e^{s_n t} = 0 \tag{12.52}$$

The only way to satisfy Eq. 12.52 (aside from the trivial case $A_n = 0$) is to require

$$RCs_n + 1 = 0 \tag{12.53}$$

or

$$s_n = -\frac{1}{RC} \tag{12.54}$$

Therefore, invoking the initial condition on v_N (Eq. 12.49) and substituting for s_n in Eq. 12.51, we have

$$v_N = -\left(\frac{A_o}{1 + RCs_o}\right) e^{-t/RC} \tag{12.55}$$

The total response is the sum of v_N and v_F.

$$v_C = u_{-1}(t)\left(\frac{A_o}{1 + RCs_o}\right)(e^{s_o t} - e^{-t/RC}) \tag{12.56}$$

The complex frequency of the natural response (s_n in our example) is called the *natural frequency* of the network. Let us examine the algebraic equation Eq. 12.53 from which this natural frequency was determined. Comparing Eq. 12.53 with Eq. 12.47, we see that the *natural frequency* s_n *makes the denominator of the system function vanish.* This is a general result. The natural frequencies of any linear system are those values of s that cause the denominator of the system function to vanish. In Section 12.3, we shall learn how to find the system function for any network. Once this system function is obtained, both the forced response to e^{st} and the natural response can be found with the algebraic relations summarized as below.

1. The complex amplitude of the forced response is equal to the complex amplitude of the source times the system function evaluated at the same value of s as the source.
2. The natural frequencies are the values of s that make the denominator of the system function vanish.
3. The total response is the forced response plus enough natural response to satisfy any initial conditions.

12.2.4 An Example

Let us use the results in the last section to find the response of the network of Fig. 12.6 to

$$\begin{aligned} v_O &= u_{-1}(t)A_o \cos(\omega_o t) \\ &= \mathrm{Re}\,\{u_{-1}(t)A_o\, e^{j\omega_o t}\} \end{aligned} \tag{12.57}$$

For this waveform, shown in Fig. 12.7a, A_o is real, and $s_o = j\omega_o$.

The forced response for $t > 0$ is found by taking the real part of Eq. 12.45

$$\begin{aligned} v_F &= \mathrm{Re}\left\{\frac{A_o}{1 + j\omega_o RC}\, e^{j\omega_o t}\right\} \\ &= \mathrm{Re}\left\{\frac{A_o}{\sqrt{1 + (\omega_o RC)^2}}\, e^{-j\phi}\, e^{j\omega_o t}\right\} \end{aligned} \tag{12.58}$$

where

$$\phi = \tan^{-1}(\omega_o RC) \tag{12.59}$$

To simplify this particular problem, let us assume that

$$\omega_o RC \ll 1 \tag{12.60}$$

(In Chapter 13, we shall take up this problem without any restricting assumptions.) In this case, ϕ is very small and we can approximate v_F (see Fig. 12.7b) as

$$v_F = A_o \cos(\omega_o t) \tag{12.61}$$

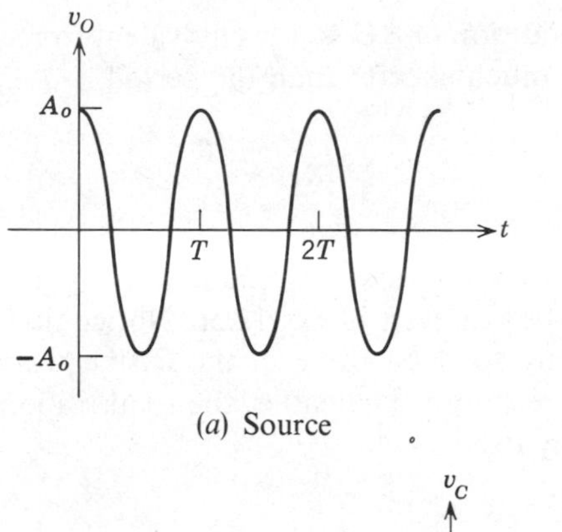

(a) Source

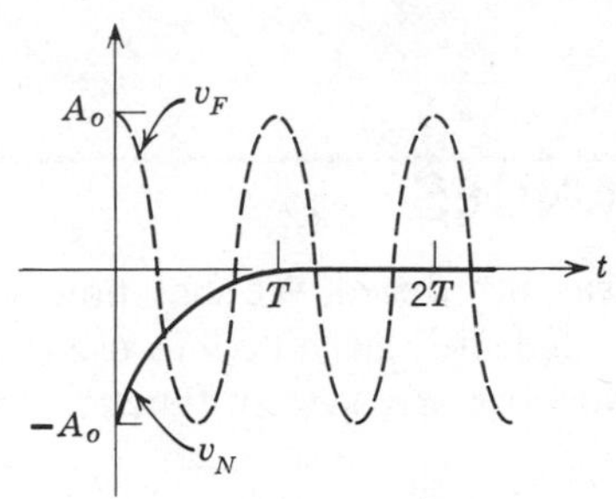

(b) Forced and natural responses for $RC \ll T/2\pi$

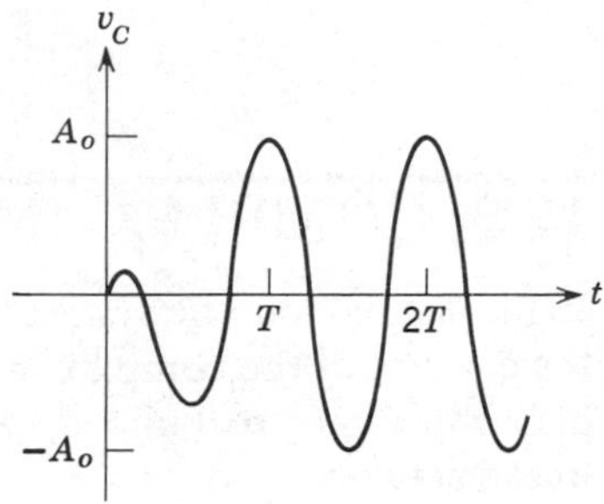

(c) Total response

Figure 12.7
Response to a sudden applied sinusoid.

The natural response is found by taking the real part of Eq. 12.55

$$v_N = -\text{Re}\left\{\frac{A_o}{\sqrt{1 + (\omega_o RC)^2}} e^{-j\phi} e^{-t/RC}\right\} \tag{12.62}$$

With the same simplifying assumption (Eq. 12.60), this becomes

$$v_N = -A_o e^{-t/RC} \tag{12.63}$$

The natural response is plotted in Fig. 12.7*b*. Notice that the exponential decay is very rapid compared to the period T. Recalling that

$$T = \frac{2\pi}{\omega_o} \tag{12.64}$$

and using Eq. 12.60, we must have

$$\omega_o RC = \frac{2\pi RC}{T} \ll 1 \tag{12.65}$$

Therefore, the condition $\omega_o RC \ll 1$ is equivalent to requiring that the decay time constant be much shorter than the period

$$RC \ll \frac{T}{2\pi} \tag{12.66}$$

The total response is the sum of v_F and v_N (see Fig. 12.7*c*). Equivalently, the total response is the real part of Eq. 12.56. Notice that because of the short time constant, only the first cycle of the cosine wave is distorted by the transient natural response. Thereafter, the total response is the same as the forced response by itself.

12.3 COMPLEX IMPEDANCES

In the last section, the *system function* was introduced. We shall now develop the concept of the complex impedance, and shall learn how to use the complex impedance to build up and characterize any system function for any linear network.

12.3.1 Impedance

When the source function is a current and the response is a voltage, the system function is called an *impedance*. Figure 12.8 illustrates this case. We know that

$$v_L = L\frac{di_O}{dt} \tag{12.67}$$

When an e^{st} current is used,

$$i_O = \text{Re}\,\{I_o\, e^{s_o t}\} \tag{12.68}$$

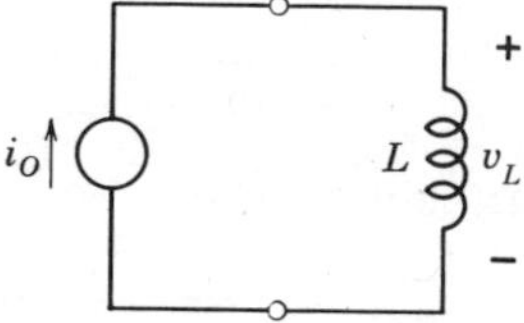

Figure 12.8

the forced response v_F is

$$v_F = \text{Re}\left\{L\frac{di_O}{dt}\right\} = \text{Re}\,\{s_o L I_o e^{s_o t}\} \tag{12.69}$$

The *impedance system function* is the ratio of the complex voltage amplitude to the complex current amplitude in the forced response to an e^{st} source. Impedance is given the symbol Z and has the dimension of ohms. We see from Eq. 12.69 that the impedance Z_L for an inductance, written as a function of s, is

$$Z_L = \frac{\text{complex amplitude of response}}{\text{complex amplitude of source}} = sL \tag{12.70}$$

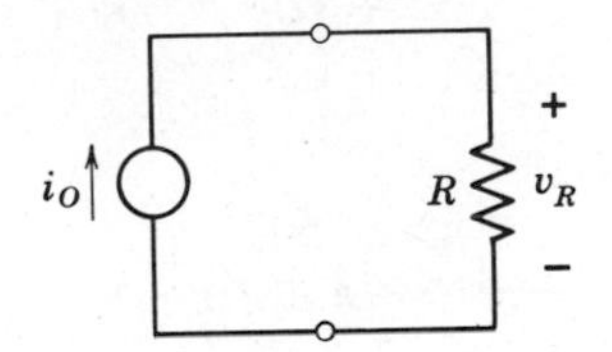

Figure 12.9

In Fig. 12.9, a current source drives a resistance R. By Ohm's law,

$$v_R = Ri \tag{12.71}$$

for any time dependence of i, including e^{st}. Therefore, the impedance of a resistance is independent of s.

$$Z_R = R \tag{12.72}$$

Figure 12.10 shows a current source driving the series combination of R and L. Using KVL, Eqs. 12.69, and 12.71, we obtain

$$v = v_R + v_L = Ri_O + L\frac{di_O}{dt} \tag{12.73}$$

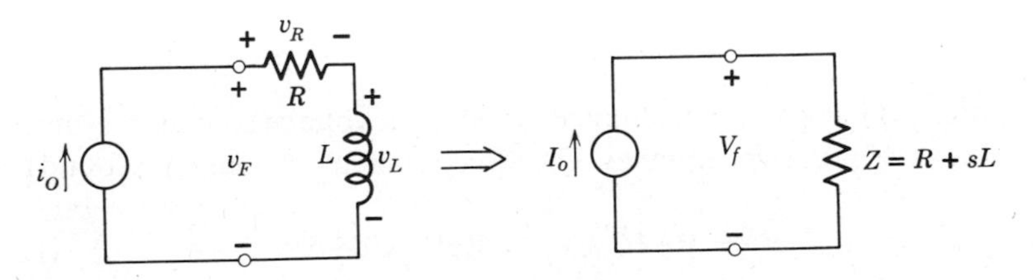

Figure 12.10
Impedances in series.

When i has e^{st} time dependence the complex amplitudes for the forced response V_f and the current I_o are related by

$$V_f = (R + sL)I_o$$

Therefore the total impedance Z (generally written with the resistance symbol) is the sum of the two impedances in series. *Impedances in series add*, just as resistances do.

A corollary to this result is that the *voltage divider relations are valid for impedances* just as they were for resistances. Thus, using Eq. 12.73 the complex amplitude of V_r is

$$V_r = I_o R = \left(\frac{V_f}{R + sL}\right) R = \frac{R}{R + sL} V_f \tag{12.74}$$

or

$$V_r = \frac{Z_R}{Z_R + Z_L} V_f \tag{12.75}$$

and

$$V_l = \frac{Z_L}{Z_R + Z_L} V_f = \frac{sL}{R + sL} V_f \tag{12.76}$$

12.3.2 Admittance

When the source is a voltage and the response is a current, the system function is called an *admittance*. It is given the symbol Y, and has the dimension of siemens. For a resistance, the admittance is the same as the conductance.

$$Y_R = \frac{i}{v} = \frac{1}{R} = G \tag{12.77}$$

Notice that by definition the admittance of an element is the reciprocal of its impedance.

Figure 12.11 shows a capacitance driven by a voltage source, presumed to have $\text{Re}\{V_o e^{s_o t}\}$ time dependence. The forced current response i_F is found from

$$i_F = \text{Re}\left\{C\frac{dv_o}{dt}\right\} = \text{Re}\{(s_o C)V_o e^{s_o t}\} \tag{12.78}$$

The admittance system function for a capacitance, therefore, is

$$Y_C = sC \tag{12.79}$$

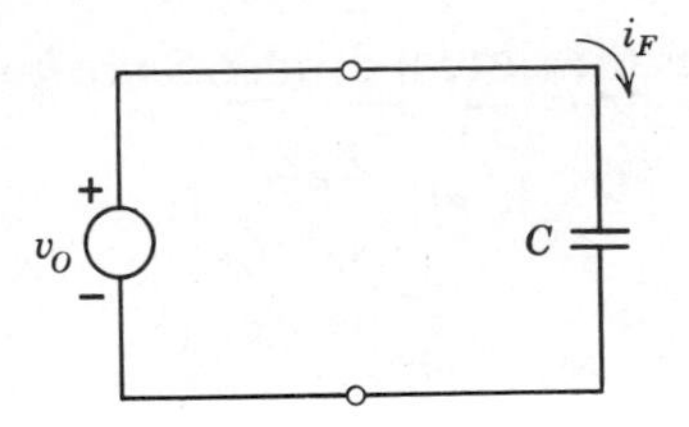

Figure 12.11

Since the impedance is the reciprocal of the admittance, we must have

$$Z_C = \frac{1}{Y_C} = \frac{1}{sC} \tag{12.80}$$

Figure 12.12 shows a parallel connection of elements. The complex amplitude of the total current is

$$\begin{aligned} I_o = I_r + I_c &= GV_o + sCV_o \\ &= (G + sC)V_o \end{aligned} \tag{12.81}$$

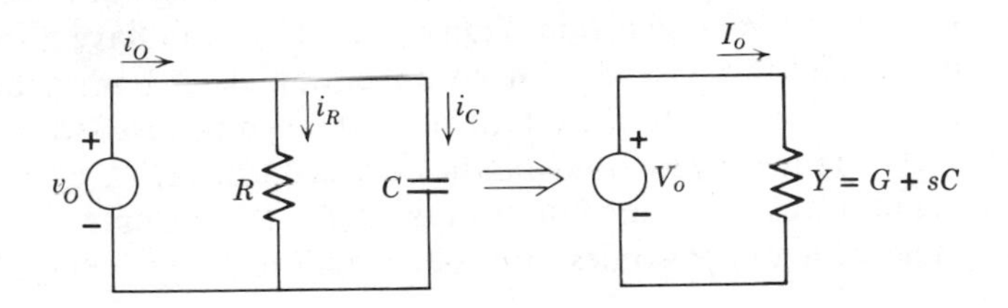

Figure 12.12
Admittances in parallel.

The total admittance (generally written with the conductance circuit symbol) is the sum of admittances for each element. Thus *admittances in parallel add.*

Equation 12.81 can also be expressed as an equivalent impedance:

$$\frac{V_o}{I_o} = Z = \frac{1}{G + sC} = \frac{1}{(1/R) + sC} \tag{12.82}$$

Manipulating terms, we obtain the familiar parallel combination, but this time for impedances:

$$Z = \frac{R(1/sC)}{R + (1/sC)} = \frac{Z_R Z_C}{Z_R + Z_C} \tag{12.83}$$

A final result involves the current divider. Since from Eqs. 12.82 and 12.83

$$V_o = \frac{Z_R Z_C}{Z_R + Z_C} I_o \tag{12.84}$$

and since

$$I_r = \frac{V_o}{Z_R} \tag{12.85}$$

we obtain the usual form of the current divider, but with impedances instead of resistances:

$$I_r = \frac{Z_C}{Z_R + Z_C} I_o \tag{12.86}$$

12.3.3 Transfer Functions

When the source and response are in different parts of a network, the system function is called a *transfer function.* Transfer functions can have dimensions of impedance or admittance, or can be dimensionless if both source and response are the same kind of quantity. The generic symbol used for a transfer function is $H(s)$. Transfer functions can be obtained from KVL and KCL by using the admittances and impedances of the network elements.

An example of a dimensionless transfer function is the voltage divider Eq. 12.74.

$$V_r = \left(\frac{R}{R + sL}\right) V_f \tag{12.87}$$

The transfer function is in this case

$$H(s) = \frac{V_r}{V_f} = \frac{R}{R + sL} \tag{12.88}$$

Similarly, the current divider of Eq. 12.86 has a transfer function

$$H(s) = \frac{I_r}{I_o} = \frac{Z_C}{Z_R + Z_C} = \frac{1}{1 + sRC} \tag{12.89}$$

The importance of the complex impedance and admittance is that for any linear network, it is possible to find any system function of interest by using KVL, KCL, and the generalized version of Ohm's law derived from the complex impedances and admittances of the individual elements. For an impedance

$$V = Z(s)I \tag{12.90}$$

and for any element

$$Y(s) = \frac{1}{Z(s)} \tag{12.91}$$

No differential equations are required to find the system functions. Use of the complex impedance reduces KVL and KCL to algebraic equations.

12.3.4 Poles and Zeros

Let us find the voltage transfer function for the network of Fig. 12.13. Using the voltage-divider relation and the parallel-impedance relation,

$$\frac{V_2}{V_1} = H(s) = \frac{R_2}{R_2 + [R_1/(1 + sCR_1)]} = \frac{R_2(1 + sCR_1)}{R_1 + R_2 + sCR_1R_2} \tag{12.92}$$

This transfer function has s in both numerator and denominator. The natural response for this network can be obtained by two methods. Using the results of Chapter 6, the natural response for a network with one energy storage element is an exponential decay with time constant R_TC where R_T is the Thévenin equivalent resistance as seen from the terminals of the capacitance. That resistance, obtained with v_1 set to zero is

$$R_T = R_1 \| R_2 \tag{12.93}$$

Thus

$$\tau = C\frac{R_1R_2}{R_1 + R_2} \tag{12.94}$$

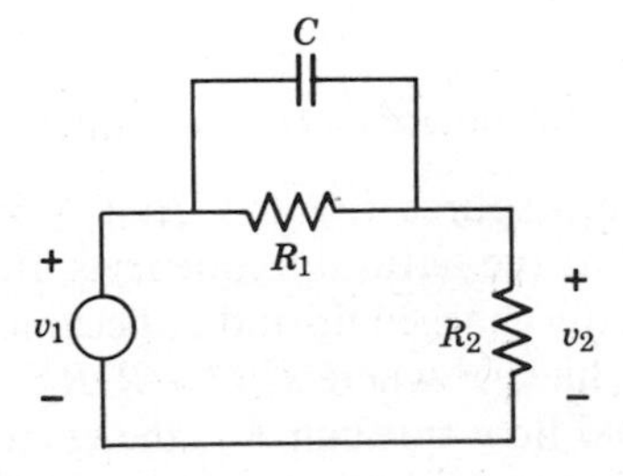

Figure 12.13

An alternative method is to examine the denominator of $H(s)$. The value of s that causes the denominator to vanish corresponds to the natural response. This value is

$$s_a = -\frac{R_1 + R_2}{R_1 R_2 C} \tag{12.95}$$

yielding a natural response

$$e^{-s_a t} = e^{-t/\tau} \tag{12.96}$$

where

$$\tau = C\frac{R_1 R_2}{R_1 + R_2} \tag{12.97}$$

the same value as found in Eq. 12.94. This value of s, for which the denominator vanishes, is called a *pole* of $H(s)$. When s nears s_a, the denominator approaches zero and the magnitude of $H(s)$ becomes infinite. Thus the roots of the denominator, which are the natural responses of the network, correspond to the poles, or infinities of the system function.

Another value of s that is important is the value

$$s_1 = -\frac{1}{R_1 C} \tag{12.98}$$

For this particular value the *numerator* of $H(s)$ is zero, and s_1 is called a *zero* of the system function. In general, the roots of the denominator polynomial locate the system function poles, while the roots of the numerator polynomial locate the system function zeros.

The physical meaning of poles and zeros is as follows. The system function will have a pole at every value of s for which the response can exist in the absence of any source excitation. Crudely speaking, since

$$\text{Response} = H(s) \times \text{source} \tag{12.99}$$

a finite response for zero source requires $H(s) \to \infty$. These infinities have already been identified as the natural frequencies. Referring to Fig. 12.13, if the capacitor were initially charged up and v_1 became zero, the charge would decay naturally to zero like $e^{s_a t}$ where $s_a = -R_1 R_2 C/(R_1 + R_2)$. During the discharge, current would flow through R_2; the response voltage, v_2, would be nonzero, following the natural-response time variation, $e^{s_a t}$.

For the natural response of a network to be finite, it must have a decaying (instead of a growing) time dependence. To insure a finite response, the natural frequencies of a network must lie in the left half of the s-plane.

The physical meaning of the zero is opposite to that of a pole. A zero corresponds to a particular excitation that can exist without any response. Consider, for example, in Fig. 12.14, what kinds of voltages v_1 are consistent

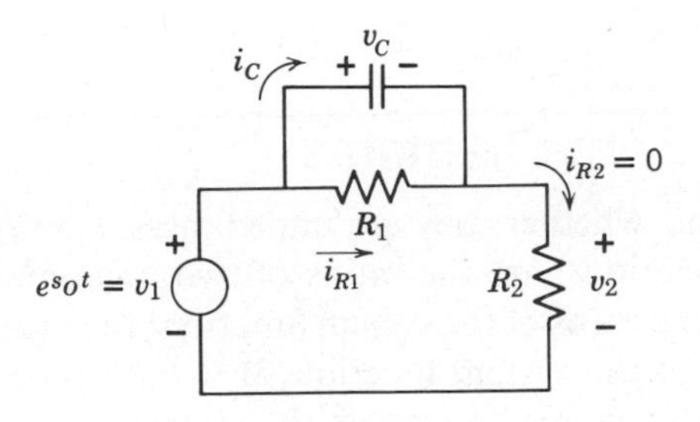

Figure 12.14
Circuit demonstrating the meaning of a zero.

with no response ($v_2 = 0$). For there to be no response, i_{R_2} has to be zero. In that case both KVL

$$v_C = v_1 \tag{12.100}$$

and KCL

$$i_C + i_{R_1} = 0 \tag{12.101}$$

must be satisfied. Let us determine, for $v_1 = e^{s_1 t}$, what value of s_1 is consistent with these equations

$$i_{R_1} = \frac{v_C}{R_1} = \frac{v_1}{R_1} = \frac{e^{s_1 t}}{R_1} \tag{12.102}$$

Furthermore

$$i_C = C\frac{dv_C}{dt} = s_1 C\frac{e^{s_1 t}}{R_1} \tag{12.103}$$

Therefore, to satisfy

$$i_{R_1} + i_C = \left(s_1 C + \frac{1}{R_1}\right) e^{s_1 t} = 0 \tag{12.104}$$

we must choose

$$s_1 = -\frac{1}{R_1 C} \tag{12.105}$$

Physically, this means that the capacitor can discharge through R_1 at the rate e^{-t/R_1C} without producing any output response ($v_2 = 0$). If v_1 drives the network with that particular time dependence, no current flows through R_2 and no voltage is developed at the output. The transfer function has a *zero* for that value of s.

Zeros, unlike poles, are not confined to the left half of the s-plane.

12.3.5 Pole-Zero Diagrams

All system functions, whether they are impedances, admittances, or dimensionless transfer functions, are the ratios of two polynomials. The roots of the numerator are the zeros of the system function; the roots of the denominator are the poles of the system function. It should be possible, therefore, by factoring the polynomials, to express the system function in very concise form.

$$Z(s),\ Y(s),\ \text{or}\ H(s) = K\frac{(s - s_1)(s - s_2)\ldots(s - s_m)}{(s - s_a)(s - s_b)\ldots(s - s_n)} \tag{12.106}$$

where

$s_1, s_2, \ldots, s_m$ are the zeros

$s_a, s_b, \ldots, s_n$ are the poles

K is a constant

If one can specify where all the zeros are and where all the poles are, the system function is completely determined except for the constant K. The information on the pole and zero locations is usually summarized graphically in what is called a *pole-zero diagram.* An "×" is plotted in the s-plane at each pole location; "0" is plotted in the s-plane for each zero. Thus the system function

$$H(s) = \frac{R_2(1 + sCR_1)}{R_1 + R_2 + sCR_1R_2} \tag{12.107}$$

has the pole-zero diagram shown in Fig. 12.15. Notice that

$$|s_a| > |s_1| \tag{12.108}$$

because

$$\frac{1}{R_1 \| R_2} > \frac{1}{R_1} \tag{12.109}$$

Therefore s_a occurs further out on the negative real axis.

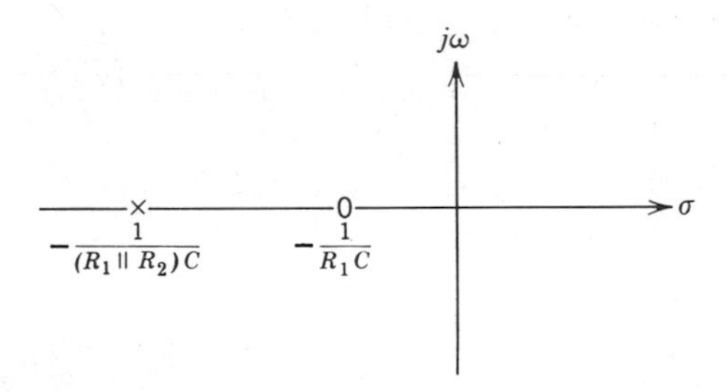

Figure 12.15
Pole-zero diagram for the circuit of Fig. 12.13.

Since poles and zeros are the roots of polynomials in s with real coefficients, poles and zeros must either be real numbers or occur in complex conjugate pairs. Figure 12.16 shows a pole-zero diagram with one complex pair of poles and one complex pair of zeros.

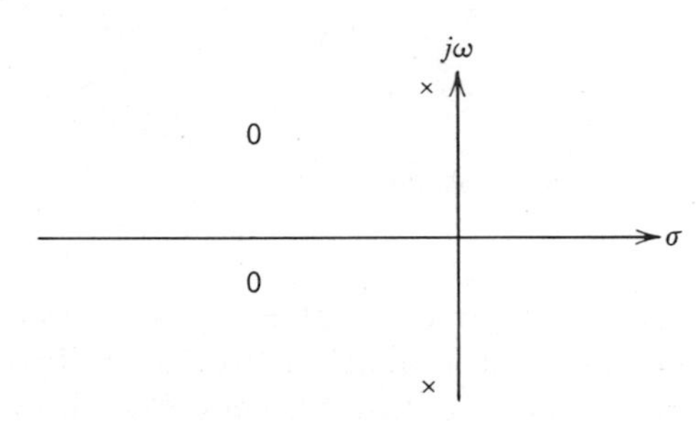

Figure 12.16
Complex conjugate pairs of poles and zeros.

The pole-zero diagram, which neatly catalogs most of the information on the system function, contains *no* information on the value of K. This constant must always be determined from consideration of the particular network. But many different networks might give rise to the same pole-zero plot. Consider, for example, the two networks in Figure 12.17. If one were to choose

$$\frac{1}{R_1 C} = \frac{R_2}{L} \tag{12.110}$$

the two networks would have identical transfer characteristics, even though they might look very different when constructed from physical components. This example stresses the great usefulness of the system function and pole-zero diagram in extracting the essential features of network behavior, and

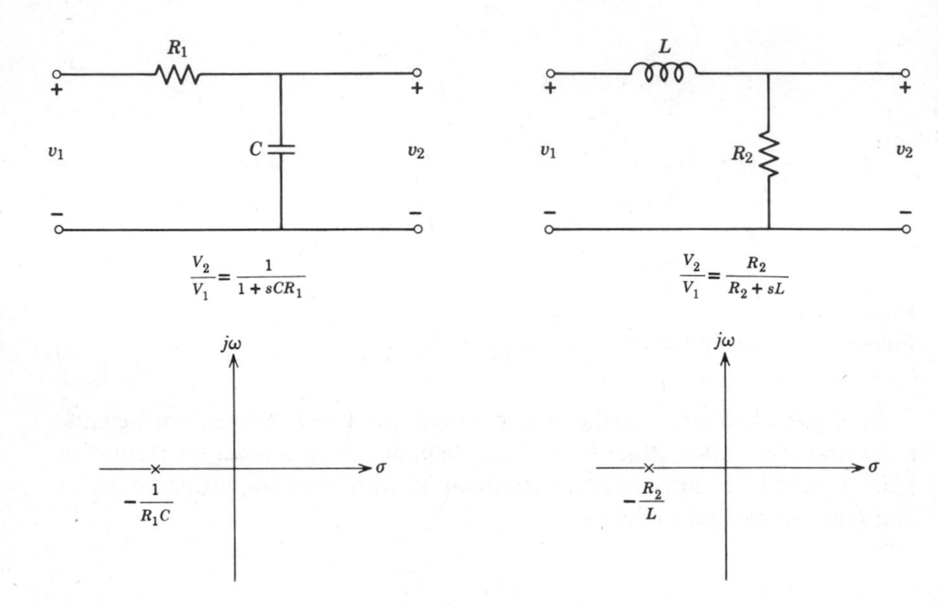

Figure 12.17
Two networks with similar pole-zero diagrams.

making comparisons between different networks possible. Furthermore, with the use of op-amps and feedback networks (see Problems P12.1 and P12.2), it becomes possible to construct networks with a wide variety of system functions.

As we remarked before, the system function, in addition to providing the natural frequencies for the network, enables us to determine the forced response to any of the e^{st} excitations. Two particular forced responses are of principal interest: the step response ($s = 0$) and the sinusoidal response ($s = j\omega$). Simple step responses have already been examined briefly. Sinusoidal responses are the subject of the next chapter.

REFERENCES

R12.1 Amar G. Bose and Kenneth N. Stevens, *Introductory Network Theory* (New York: Harper & Row, 1965).

R12.2 Hugh H. Skilling, *Electrical Engineering Circuits* (New York: John Wiley, 1957).

QUESTIONS

Q12.1 Can you prove that the two roots of a quadratic equation with real coefficients either are both real or form a complex pair?

Q12.2 Explain the relation between an actual voltage or current waveform and the complex voltage or current used to represent it. Is the physical voltage or current ever imaginary?

Q12.3 Is the imaginary part of an imaginary number imaginary?

Q12.4 Is the real part of a complex number real?

Q12.5 If θ is the angle of complex number s, what is the angle of s^2? Of $\sqrt{s}$?

Q12.6 Can there be a forced response without a source? A natural response?

Q12.7 What determines how large the natural response is in any particular case? What about the forced response?

Q12.8 What determines the time dependence of the forced response? Of the natural response?

Q12.9 In general, is the impedance of a network element a real number or a complex number?

Q12.10 Will a network spontaneously exhibit waveforms corresponding to its natural response? (Consider the stored energy when formulating your answer).

EXERCISES

E12.1 For two complex numbers A and B show that

(a) $|A| \times |B| = |A \times B|$

(b) $\frac{|A|}{|B|} = \frac{|A|}{|B|}$

(c) $\text{Re}\{A \pm B\} = \text{Re}\{A\} \pm \text{Re}\{B\}$

E12.2 For two complex numbers A and B in Cartesian form, $A = a_1 + ja_2$ and $B = b_1 + jb_2$, find expressions for

(a) $\text{Re}\{AB\}$

(b) $\text{Im}\left\{\dfrac{A}{B}\right\}$

(c) $\text{Re}\left\{\dfrac{1}{A + B}\right\}$

E12.3 Repeat E12.2 for the polar form $A = A_o\, e^{j\alpha}$ and for $B = B_o\, e^{j\beta}$.

E12.4 Does $\text{Re}\left\{\dfrac{1}{A}\right\} = \dfrac{1}{\text{Re}\{A\}}$?

E12.5 Express in the form $\text{Re}\{A\, e^{s_a t}\}$ and also in the form $\text{Im}\{B\, e^{s_b t}\}$:

(a) $5 \cos 3t$

(b) $2 \cos\left(t - \dfrac{\pi}{4}\right)$

(c) $4 \sin\left(t + \dfrac{\pi}{4}\right)$

(d) $M \cos(\omega t + \theta)$

E12.6 Express using either the real part or the imaginary part of an e^{st} function

(a) $e^{-t/\tau}$

(b) $e^{\alpha t} \sin \omega t$

(c) $e^{-t/2} \cos\left(3t + \dfrac{\pi}{4}\right)$

(d) $e^{-\omega t} \cos \omega t$

E12.7 Plot the following waveforms.

(a) $\text{Re}\{e^{j3t}\}$
(b) $\text{Im}\{e^{-4t}\}$
(c) $\text{Im}\{e^{j3t}\}$
(d) $\text{Re}\{e^{-4t}\}$
(e) $\text{Re}\{e^{-t/5}\, e^{j\pi t}\}$
(f) $\text{Im}\{e^{-t/5}\, e^{j\pi t}\}$
(g) $\text{Re}\{4\, e^{j\pi/4}\, e^{j3t}\}$
(h) $\text{Im}\{10\, e^{-j\pi/3}\, e^{-2t + j\pi t}\}$

E12.8 Find the system function V_o/V_i for the circuits in Fig. E12.8, and plot the pole-zero diagrams.

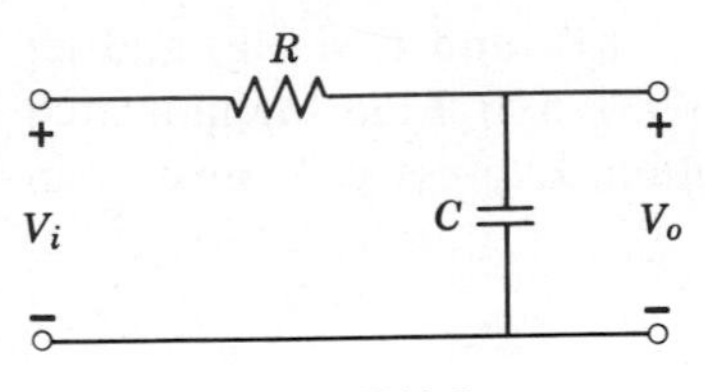

Figure E12.8*a*

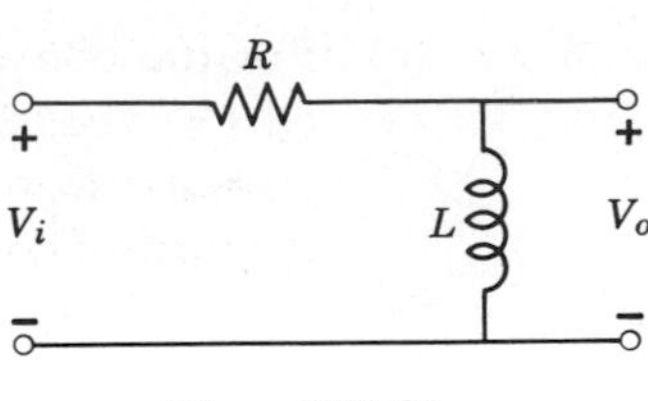

Figure E12.8*b*

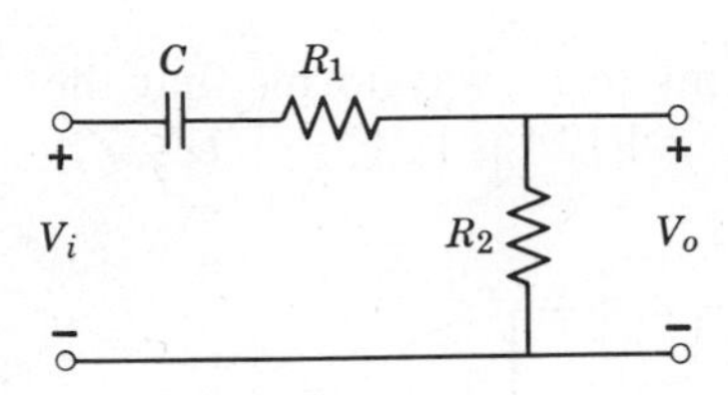

Figure E12.8*c*

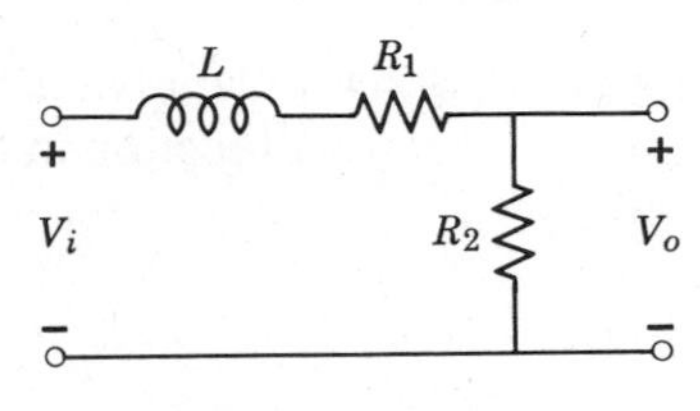

Figure E12.8*d*

E12.9 Plot the pole-zero diagrams for the following system functions assuming $\alpha, \beta, \gamma > 0$.

(a) $H(s) = \dfrac{2s^2 + 6s}{s^2 + 6s + 5}$

(b) $H(s) = \dfrac{k(s + \gamma)}{(s + \alpha)^2 + \beta^2}$

(c) $H(s) = \dfrac{k}{s^2 + \alpha^2}$

(d) $H(s) = \dfrac{1}{s^2 + s + 1}$

E12.10 (a) Find the system function relating v_o to the source for the networks in Fig. E12.10.

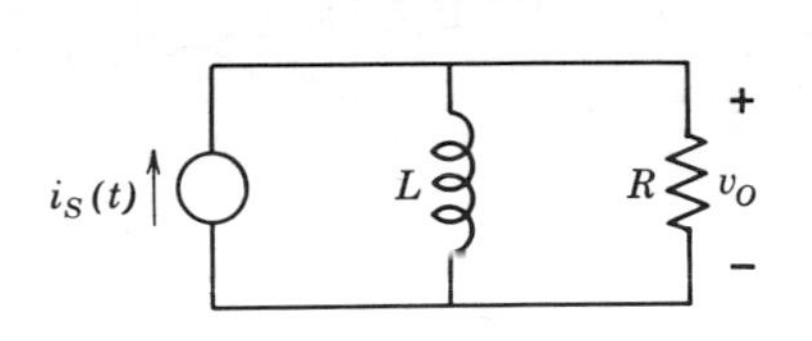

Figure E12.10*a*

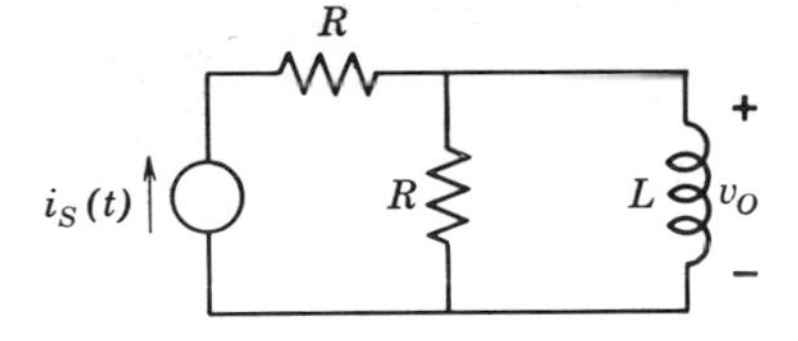

Figure E12.10*b*

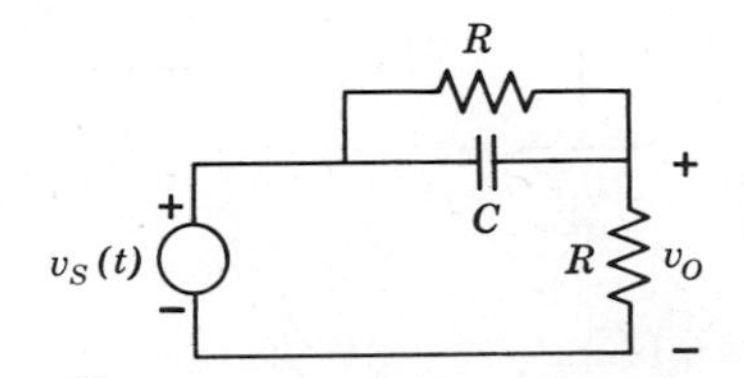

Figure E12.10*c*

(b) For the above, let $R = 1\,\Omega$, $L = 1$ H, and $C = 1$ F, and let $v_S(t) = 3\cos 2t$ and $i_S(t) = \cos(2t + \pi/4)$. Find the indicated response from the system function. Express your answer in the form $A\cos(\omega t + \phi)$.

PROBLEMS

P12.1 Show by applying op-amp arguments to e^{st} waveforms that the transfer function in the network in Fig. P12.1 is $V_o/V_i = -(Z_f/Z_s)$.

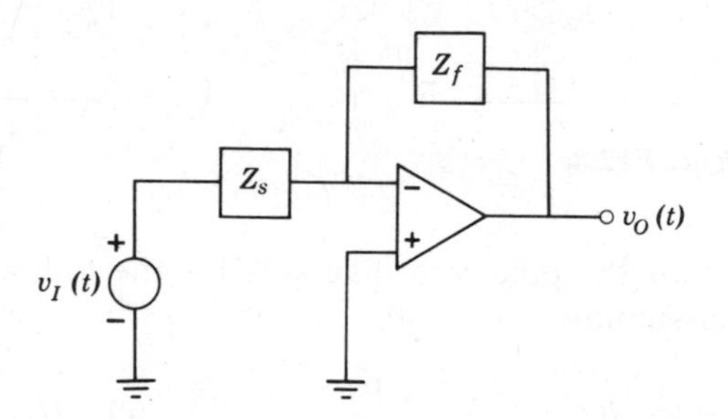

Figure P12.1

P12.2 Find the transfer functions and plot pole-zero diagrams for the networks in Figs. P12.2. (See P12.1.)

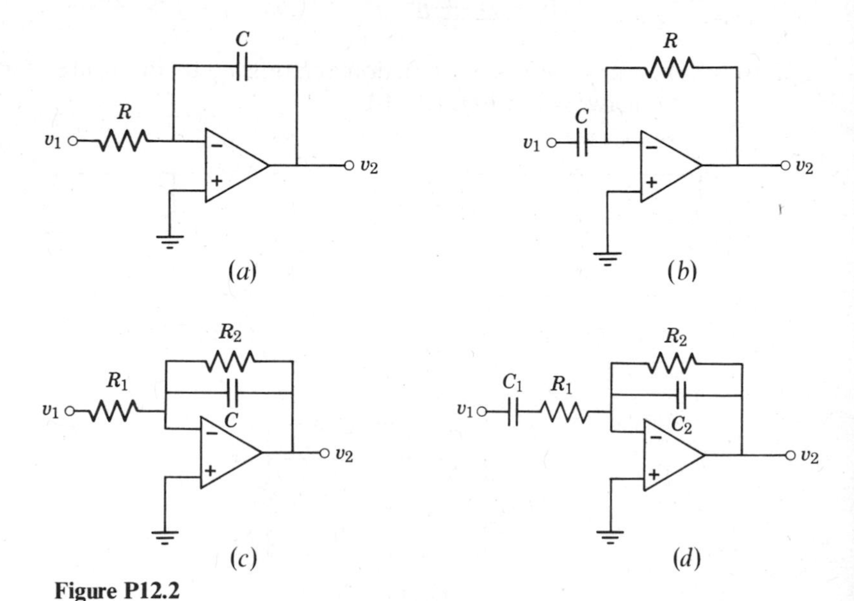

Figure P12.2

P12.3 Construct at least two different networks having system functions (impedance, admittance, or transfer function) with the pole-zero diagrams in Fig. P12.3.

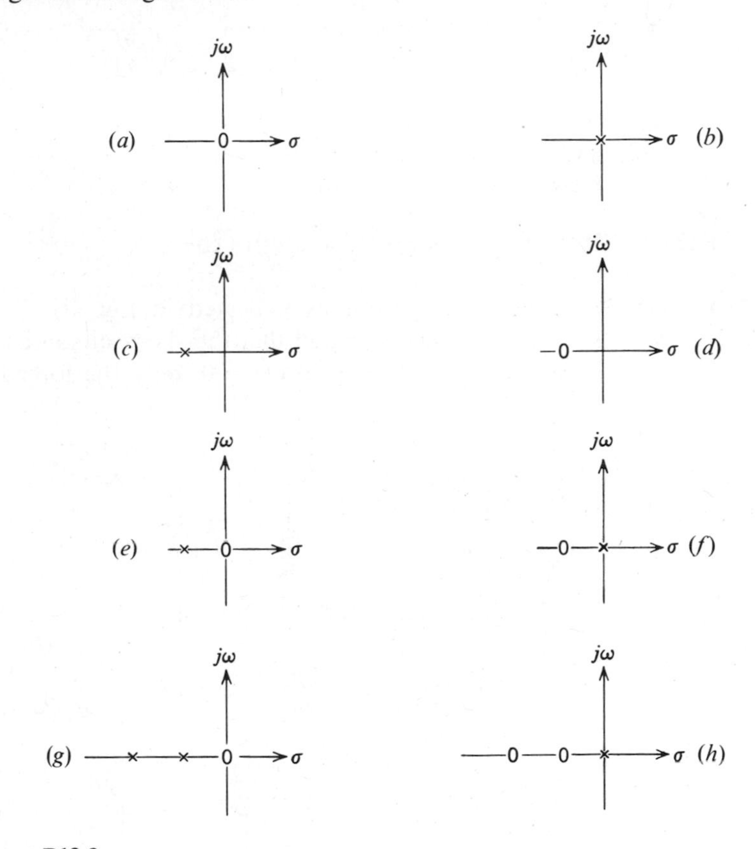

Figure P12.3

P12.4 For the network of Fig. P12.4, plot the pole-zero diagram of the system function V_o/V_i for $4R^2C/L \ll 1$ and for $4R^2C/L \gg 1$.

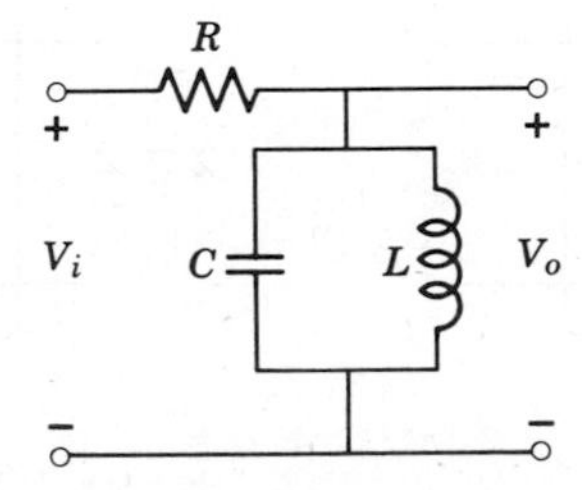

Figure P12.4

P12.5 Find the total responses indicated in the networks in Fig. P12.5 for $v_S(t) = Au_{-1}(t)$. Identify forced and natural responses.

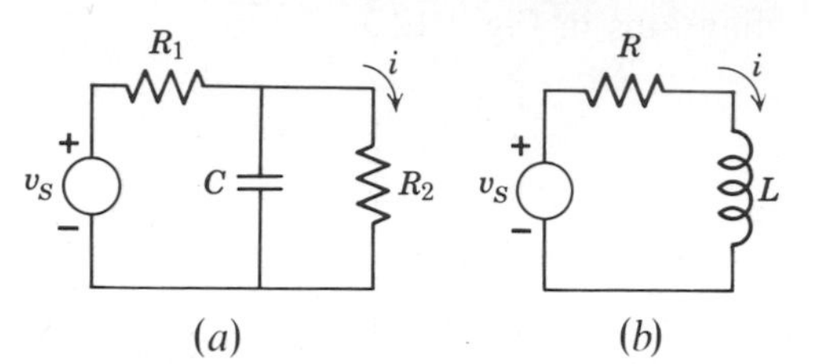

Figure P12.5

P12.6 Repeat P12.5 for $v_S(t) = u_{-1}(t)A\cos 5\pi t$.

P12.7 Find the total responses indicated in Fig. P12.7. Assume the switch closes at $t = 0$, and that the $A\cos\omega t$ source has been on long enough so that before $t = 0$, only the forced response is present.

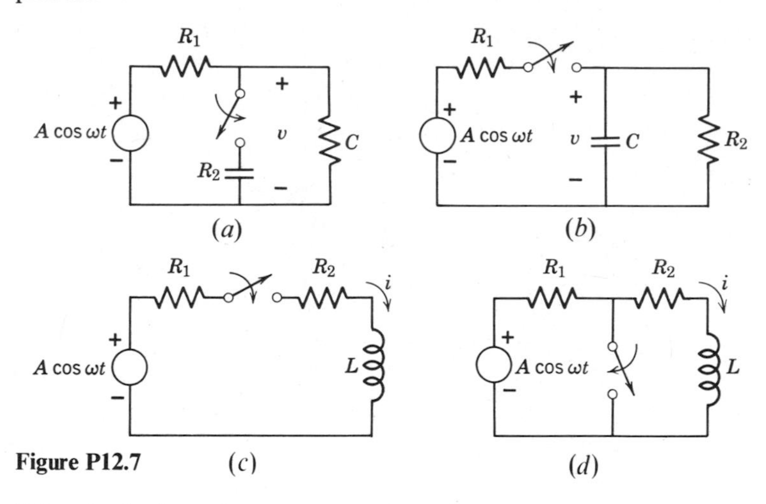

Figure P12.7

P12.8 What is the impedance seen at the primary, assuming the transformer is ideal (Fig. P12.8)?

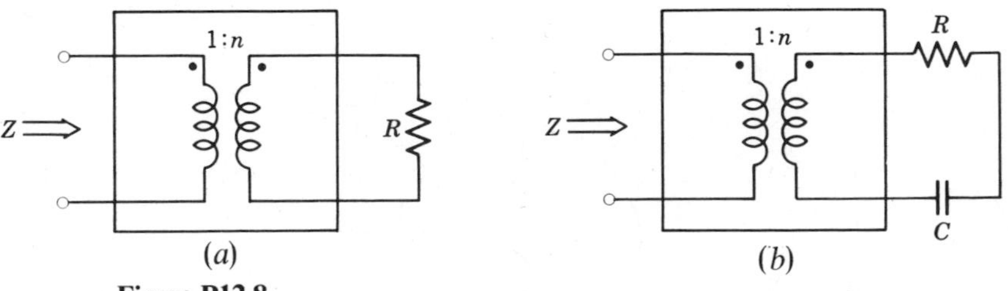

Figure P12.8

P12.9 A transducer for measuring wind velocity w is intended to deliver a voltage proportional to w. Because of an error in the design of the transducer, it actually produces a voltage v_A of the form

$$v_A = Aw + B\frac{dw}{dt}$$

where A and B are known constants.

You are to design a circuit that accepts v_A as an input and that delivers an output voltage that is *proportional* to w.

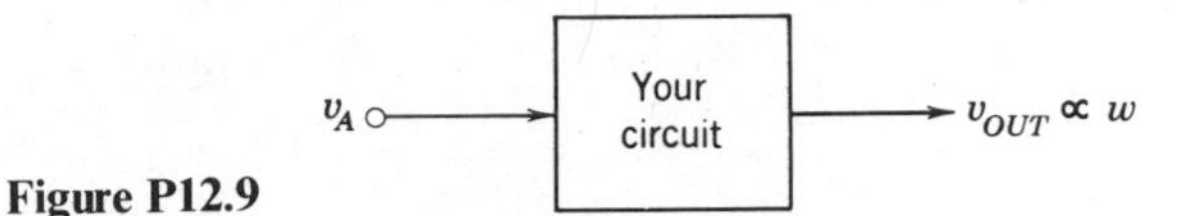

Figure P12.9

P12.10 The equivalent circuit in Fig. P12.10 represents a superconducting magnet, where L is the inductance of the magnet, and R_P is parasitic resistance in the leads to the magnet. R_S is a 500 $\mu\Omega$ resistor (yes, micro-ohms) used to monitor the current i_0.

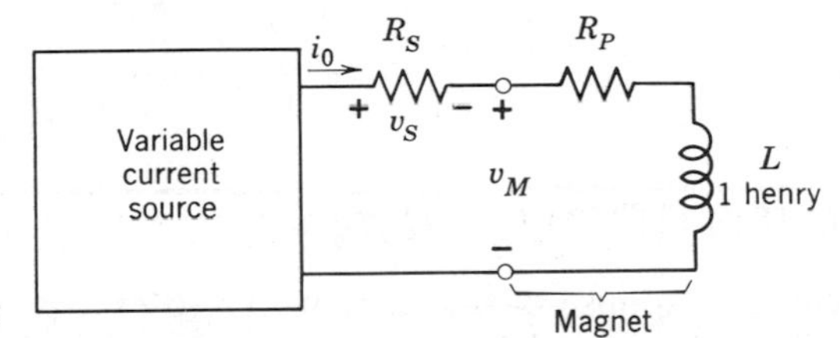

Figure P12.10

The trouble with a superconductor is that it must be kept very cold. If the cooling system fails, and the superconductor warms up through a transition temperature, the magnet wire abruptly changes from superconducting to its normal-resistance state.

You are to design a circuit to detect the onset of this "normal transition" and indicate an alarm condition that might be used to prevent excessive Joule heating. The normal transition can be recognized as a rapid rise in the value of R_P above its usual value.

The problem is that when i_0 is changed, the voltage appearing across the inductor causes v_M to change. At the same time, it is very desirable to detect a normal transition as soon as any significant rise in R_P is detected. Your circuit, therefore, which should use v_S and v_M as inputs, should output +5 volts into a 100 Ω load if R_P rises from 0.002 Ω to 0.1 Ω but should *not* respond in the superconducting state ($R_P = 0.002\ \Omega$) when i_0 is changed at a rate of up to ±100 A/s. The usual range of operating points is $100 \leq i_0 \leq 1000$ A.

CHAPTER THIRTEEN
FREQUENCY RESPONSE IN LINEAR SYSTEMS

13.0 INTRODUCTION

The general response of linear networks to excitations of the form e^{st} has now been developed. In this chapter, we proceed to a more detailed examination of the linear-system response for the particular case of sinusoidal excitations ($s = j\omega$).

There are several reasons for separating the sinusoidal excitations from the other exponential excitations. The primary reason is that the forced response to a sinusoidal excitation is not damped ($\sigma = 0$). Thus, while turning on the sinusoidal excitation usually produces some transient response in a system, these transients eventually die away. Thereafter, the forced sinusoidal response persists indefinitely. Therefore, one can speak of the *sinusoidal steady state*, in which the forced sinusoidal response to a sinusoidal excitation can be examined independent of the precise conditions under which the excitation was established.

A second reason for giving sinusoidal excitations special attention is that almost any waveform can be represented as a superposition of sinusoids. Therefore, if one has a thorough understanding of the response of a linear system to sinusoidal excitations, one can use the appropriate superposition of these responses to obtain the total forced response to almost any excitation.

13.1 BASIC CONCEPTS

13.1.1 Definition of Frequency Response

The sinusoidal steady-state response of a linear system as a function of the frequency of the excitation is a fundamental characteristic of that linear system. It is called the *frequency response.*

The frequency response of any linear system can be obtained directly from the system function. Suppose that a source waveform is given by

$$x(t) = A \cos(\omega t + \phi) \tag{13.1}$$

where A is the amplitude, ω is the angular frequency and, ϕ is the phase angle. We can write Eq. 13.1 in the form

$$x(t) = \text{Re}\{A\, e^{j\phi}\, e^{j\omega t}\} \tag{13.2}$$

where $(A\, e^{j\phi})$ is the complex amplitude of the $e^{j\omega t}$ excitation. Note that since we are concerned here only with the forced response, we do not have to be particularly careful about specifying when or how the sinusoidal excitation was established.

If we denote the steady-state forced response by $y(t)$, we can write

$$y(t) = \text{Re}\{H(j\omega)A\, e^{j\phi}\, e^{j\omega t}\} \tag{13.3}$$

where $H(j\omega)$ is written to emphasize the fact that when discussing frequency responses, one always evaluates the system function $H(s)$ at the particular value $s = j\omega$. Once the value of s is specified in this fashion, the system function $H(j\omega)$ becomes an ordinary complex number, with a magnitude and an angle. Denoting the magnitude of $H(j\omega)$ by $|H(j\omega)|$ and the angle of $H(j\omega)$ by θ, we obtain an equivalent expression for $y(t)$.

$$y(t) = \text{Re}\{(|H(j\omega)|A)\, e^{j(\phi+\theta)}\, e^{j\omega t}\} \tag{13.4}$$

Carrying through to find the real part, the steady-state response $y(t)$ is

$$y(t) = |H(j\omega)|A \cos[\omega t + (\theta + \phi)] \tag{13.5}$$

where, for comparison, the source waveform was

$$x(t) = A \cos(\omega t + \phi) \tag{13.6}$$

Comparing the source $x(t)$ and the response $y(t)$, we can observe three significant features characteristic of the sinusoidal response of linear systems:

1. The *frequency* of the response is the same as the frequency of the source.
2. The *amplitude* of the response is equal to the amplitude of the source *multiplied by* the magnitude of the system function $|H(j\omega)|$.

3. The *phase angle* of the response is equal to the phase angle of the source *plus* the angle of the system function.

Thus, a knowledge of how $|H(j\omega)|$ and θ vary with frequency is sufficient to determine the steady-state response to any sinusoidal excitation. These quantities, the magnitude and phase angle of $H(j\omega)$, comprise the frequency response of the linear system.

13.1.2 An Example: The Low-Pass Filter

Consider the network of Fig. 13.1, with system function[1]

$$\frac{V_2}{V_1} = H(s) = \frac{1}{1 + sRC} \tag{13.7}$$

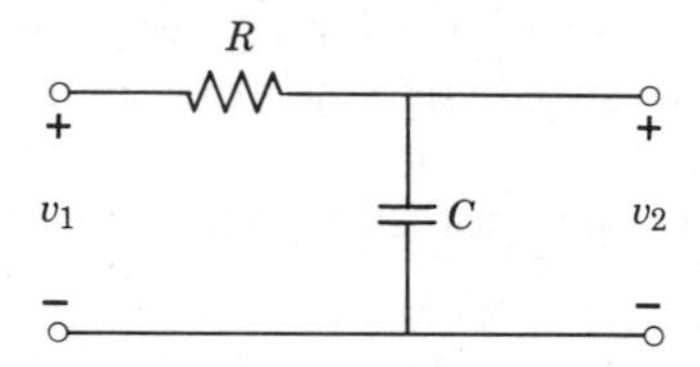

Figure 13.1
A low-pass filter.

Suppose ω_o is a particular frequency, and suppose

$$v_1(t) = A \sin\left(\omega_o t + \frac{\pi}{4}\right) \tag{13.8}$$

The steady-state response $v_2(t)$ can be determined from the system function. Substituting $s = j\omega$ into Eq. 13.7, we obtain

$$H(j\omega) = \frac{1}{1 + j\omega RC} \tag{13.9}$$

Multiplying numerator and denominator by $(1 - j\omega RC)$ yields

$$H(j\omega) = \frac{1 - j\omega RC}{1 + (\omega RC)^2} \tag{13.10}$$

[1] Remember that with our notational shorthand, a lower-case variable v will have an upper-case complex amplitude V. The system function is the ratio of the complex amplitude of the forced response to the complex amplitude of the e^{st} source.

Thus

$$\text{Re}\{H(j\omega)\} = \frac{1}{1 + (\omega RC)^2} \tag{13.11}$$

and

$$\text{Im}\{H(j\omega)\} = \frac{-\omega RC}{1 + (\omega RC)^2} \tag{13.12}$$

The magnitude and angle of $H(j\omega)$ are thus given by

$$|H(j\omega)| = [H(j\omega) \times H^*(j\omega)]^{1/2} = \frac{1}{\sqrt{1 + (\omega RC)^2}} \tag{13.13}$$

and

$$\theta = \angle H(j\omega) = \tan^{-1}\left(\frac{\text{Im}\{\ \}}{\text{Re}\{\ \}}\right) = -\tan^{-1}\omega RC \tag{13.14}$$

The steady-state response $v_2(t)$ is then obtained from Eq. 13.5. The expressions for $|H(j\omega)|$ and θ in Eqs. 13.13 and 13.14 are evaluated at the specific frequency $\omega = \omega_o$, and then are substituted into Eq. 13.5. The result for $v_2(t)$ is

$$v_2(t) = \frac{A}{\sqrt{1 + (\omega_o RC)^2}} \sin\left(\omega_o t + \frac{\pi}{4} - \tan^{-1}\omega_o RC\right) \tag{13.15}$$

Notice that for low frequencies ($\omega_o RC \ll 1$), v_2 and v_1 are equal. As the frequency increases, the magnitude of v_2 decreases and its phase is shifted relative to v_1. A network of this type, which passes low frequencies but attenuates high frequencies, is called a *low-pass filter*.

13.1.3 Resistive and Reactive Elements

The polar form of $H(j\omega)$ (i.e., magnitude and phase angle) was used in the previous section to describe the frequency response. In the particular cases of impedance and admittance system functions, however, the Cartesian form of $H(j\omega)$ (real and imaginary parts) are often used, and are given special names. These special names, defined below, apply to impedances and admittances only when dealing with sinusoidal excitations.

A steady-state impedance $Z(j\omega)$ can be written in the form

$$Z(j\omega) = R(\omega) + jX(\omega) \tag{13.16}$$

where $R(\omega)$, which is the real part of $Z(j\omega)$, is called the *resistance*, and $X(\omega)$, which is the imaginary part of $Z(j\omega)$ is called the *reactance*. Both $R(\omega)$ and $X(\omega)$ are real quantities with the dimension of ohms.

Let us determine the steady-state impedance for the network shown in Fig. 13.2. The system function impedance is

$$Z(s) = R_o + sL \tag{13.17}$$

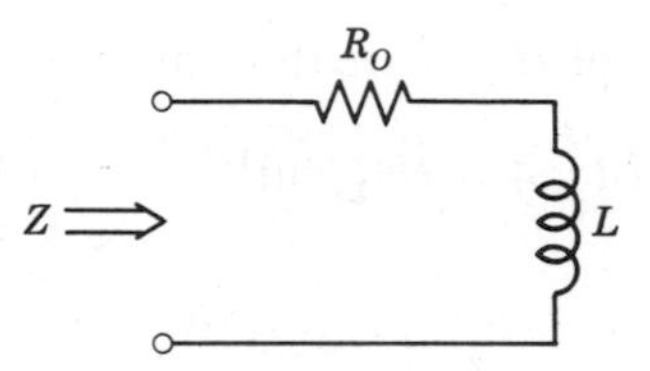

Figure 13.2
Impedances in series.

In the sinusoidal steady state,

$$Z(j\omega) = R_o + j\omega L \tag{13.18}$$

Taking the real and imaginary parts, we obtain the resistance R and the reactance X:

$$\begin{aligned} R(\omega) &= R_o \\ X(\omega) &= \omega L \end{aligned} \tag{13.19}$$

Notice that for this particular circuit, the steady-state resistance is a constant, but the steady-state reactance depends on the value of the frequency. In other circuits, however, both R and X may depend on frequency.

A similar set of quantities is defined for an admittance. A steady-state admittance, $Y(j\omega)$ can be written in the form

$$Y(j\omega) = G(\omega) + jB(\omega) \tag{13.20}$$

where $G(\omega)$, which is the real part of $Y(j\omega)$, is called the *conductance* and $B(\omega)$, which is the imaginary part of $Y(j\omega)$, is called the *susceptance*. As with resistance and reactance, both $G(\omega)$ and $B(\omega)$ are real quantities. The dimensions of conductance and susceptance are siemens (or mhos).

An example of a complex admittance is shown in Fig. 13.3. The admittance system function is

$$Y(s) = G_o + sC \tag{13.21}$$

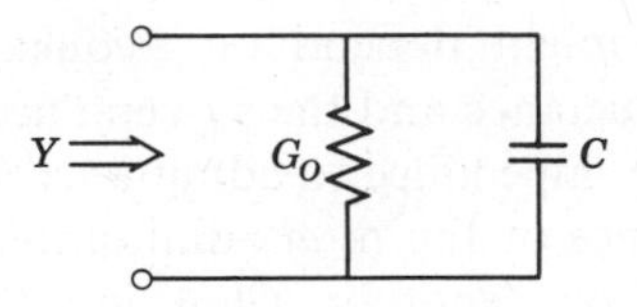

Figure 13.3
Admittances in parallel.

where G_O is the conductance of the resistive element. In the sinusoidal steady state,

$$Y(j\omega) = G_O + j\omega C \tag{13.22}$$

Taking real and imaginary parts, the conductance and susceptance are obtained:

$$\begin{aligned} G(\omega) &= G_O \\ B(\omega) &= \omega C \end{aligned} \tag{13.23}$$

A more interesting example can be constructed if we analyze the *admittance* of the network in Fig. 13.2. Since the impedance system function for this network was

$$Z(s) = R_O + sL \tag{13.24}$$

we may obtain immediately the admittance system function

$$Y(s) = \frac{1}{Z(s)} = \frac{1}{R_O + sL} \tag{13.25}$$

In the sinusoidal steady state, the admittance becomes

$$Y(j\omega) = \frac{1}{R_O + j\omega L} = \frac{R_O - j\omega L}{R_O^2 + (\omega L)^2} \tag{13.26}$$

Taking real and imaginary parts, the conductance is

$$G(\omega) = \frac{R_O}{R_O^2 + (\omega L)^2} \tag{13.27}$$

while the susceptance is given by

$$B(\omega) = \frac{-\omega L}{R_O^2 + (\omega L)^2} \tag{13.28}$$

The above examples illustrate some important characteristics of steady-state impedances and admittances. First, in a general network, any of the steady-state quantities might depend on frequency. Thus, in the above example, both the conductance and the susceptance change with frequency.

Second, in converting impedance to admittance (or the reverse) one must invert the *total* impedance or the *total* admittance. That is, while $Y(s)$ and $Z(s)$ are reciprocals of one another, $G(\omega)$ and $R(\omega)$ are *not* reciprocals. Neither do $X(\omega)$ and $B(\omega)$ have a reciprocal relation.

Finally, in each of the preceding examples, the imaginary parts of both the impedance and the admittance arise because of the presence of either an inductance or a capacitance. If, for example, the inductance in Fig. 13.2 were to become zero, the reactance of that circuit (Eq. 13.19) would be zero. Equivalently, the susceptance of that same circuit (Eq. 13.28) would also be zero. In fact, all of the frequency-dependent terms in any of the impedances or admittances above involve either a capacitance or an inductance. If the capacitances and inductances were zero, there would be no frequency dependence in either R or G, and both X and B would be zero. For this reason, inductances and capacitances are called *reactive elements*. They give rise to nonzero reactance (or nonzero susceptance). Only networks containing these reactive elements have interesting frequency responses. If there are no reactive elements, the system function is always real (zero phase angle) and independent of frequency (a constant amplitude factor). Networks with reactive elements, on the other hand, can have system functions with amplitude and phase angles that depend on frequency. Many detailed examples are taken up in the following sections.

13.2 PHASE-MAGNITUDE PLOTS (BODE PLOTS)

One concise way of summarizing the frequency dependence of the magnitude and phase of the system function when $s = j\omega$ is to display both quantities graphically. Such plots are often called *phase-magnitude plots*, or Bode plots.

13.2.1 Log-Log and Semilog Coordinates

The most convenient coordinates for making phase-magnitude plots are determined by the polynomial nature of the system function itself. In general, a system function in the sinusoidal steady state has the form

$$H(j\omega) = K\frac{(j\omega - s_1)(j\omega - s_2)(\dots)}{(j\omega - s_a)(j\omega - s_b)(\dots)} \qquad (13.29)$$

where

$$s_1, s_2, \ldots, \text{are the zeros,}$$

$$s_a, s_b, \ldots, \text{are the poles}$$

and where K is the scale constant. From this expression, the magnitude of $H(j\omega)$ is simply

$$|H(j\omega)| = |K| \times |j\omega - s_1| \times |j\omega - s_2| \times |\ldots| \times \frac{1}{|j\omega - s_a|} \times \frac{1}{|\ldots|} \tag{13.30}$$

That is, the magnitude of $H(j\omega)$ is a *product* of the magnitudes of each factor, some from zeros and some from poles. Therefore, one can calculate the frequency-dependent function $|H(j\omega)|$ from a knowledge of the behavior of each factor.

The simplest way to build up a sequence of factors is to work with the logarithm of $|H(j\omega)|$:

$$\log|H(j\omega)| = \log|K| + \log|j\omega - s_1| + \ldots + \log\frac{1}{|j\omega - s_a|} + \ldots \tag{13.31}$$

Notice that the logarithm of $|H(j\omega)|$ is a *sum* of terms. Therefore, if one plots the logarithm of the magnitude of each factor in $|H(j\omega)|$, the logarithm of the total magnitude can be found by simple graphical addition.

When plotting the logarithm of quantities, we do not wish continually to consult tables of logarithms. Graph paper is available with coordinate axes ruled off logarithmically; when one plots a number on the scale, the position of the point on the scale corresponds to the logarithm of the number. Figure 13.4 shows an example of graph paper with both axes ruled off with logarithmic coordinates. Anticipating our use of this kind of paper for plotting $|H(j\omega)|$ versus ω, the abscissa has been labeled with frequency units while the ordinate has been labeled with possible values of $|H(j\omega)|$.

The use of the logarithmic scale for the frequency is not absolutely necessary. But since the frequency range of interest often covers many decades (factors of 10) the log scale becomes very convenient. Furthermore, as we shall see, the use of a log scale for ω makes the plots of $|H(j\omega)|$ versus ω take on particularly simple forms.

A similarly useful graph paper is available for plotting the phase angle of $H(j\omega)$. Recalling that when complex numbers are multiplied together their angles add, one sees immediately that

$$\angle H(j\omega) = \angle(j\omega - s_1) + \ldots + \angle\left(\frac{1}{j\omega - s_a}\right) + \ldots \tag{13.32}$$

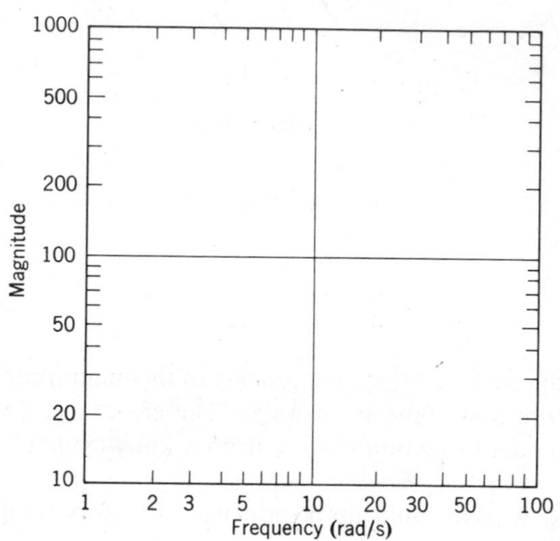

Figure 13.4
Log-log coordinates.

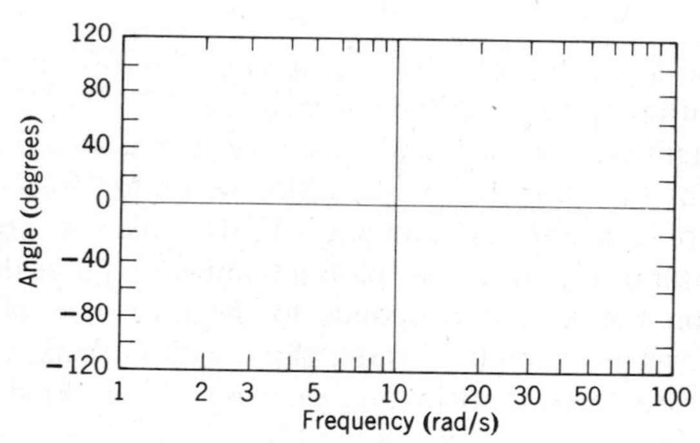

Figure 13.5
Semilog coordinates.

Thus, by plotting the frequency dependence of the angle for each simple term, some from poles, others from zeros, one can use graphical addition to build up the frequency dependence of $\angle H(j\omega)$.

Figure 13.5 shows an example of semilogarithmic paper that is convenient for plotting $\angle H(j\omega)$ versus ω. A linear scale is used for the angle of $H(j\omega)$, since the building up of the total angle involves addition, not multiplication. A logarithmic scale has been used for the frequency axis so that the plots of angle versus frequency can be easily correlated with the plots of magnitude versus frequency.

Because $\angle H(j\omega)$ is an odd function of frequency and $|H(j\omega)|$ is an even function of frequency, they are totally determined by their values for $\omega \geq 0$. Therefore $\angle H(j\omega)$ and $|H(j\omega)|$ are commonly plotted only for $\omega \geq 0$.

13.2.2 Decibels

Before taking up the evaluation of each of the simple terms in either Eq. 13.31 or Eq. 13.32, one last set of definitions is needed. Many of the quantities we deal with involve ratios. In particular, voltage transfer functions and current transfer functions are dimensionless ratios with magnitudes either greater than unity (indicating "gain") or magnitudes less than unity (indicating "attenuation" or "loss"). Engineers have developed a special unit called the *decibel* (abbreviated dB) for dealing with dimensionless ratios. Since the decibel unit is in common use, it is essential that its meaning be well understood.

The decibel is properly defined in terms of ratios of power. Figure 13.6 illustrates the definition. A system, shown in the box, has input resistance R_I and is connected to a load resistance R_L. The power into the system, P_I, is given by v_I^2/R_I, while the power delivered by the system into the load, P_L, is given by v_L^2/R_L. The power gain of the system in decibels is defined as

$$\text{Power gain (dB)} = +10 \log \frac{P_L}{P_I} \tag{13.33}$$

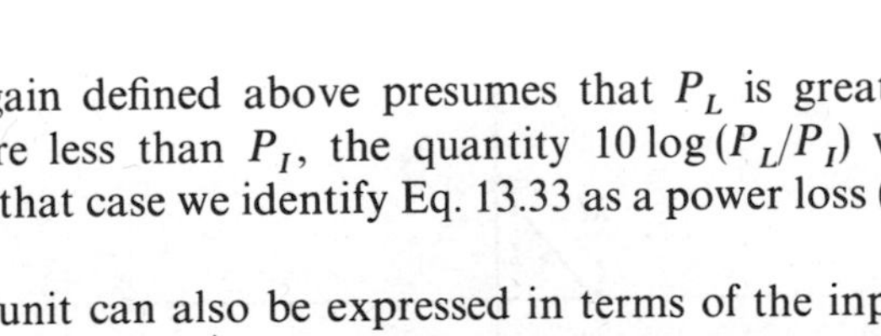

Figure 13.6

Actually, the gain defined above presumes that P_L is greater than P_I. If instead P_L were less than P_I, the quantity $10 \log (P_L/P_I)$ would actually be negative. In that case we identify Eq. 13.33 as a power loss (also measured in dB).

The decibel unit can also be expressed in terms of the input and output voltages. Using the explicit forms of P_I and P_L, Eq. 13.33 can be written

$$\text{Power gain (dB)} = 10 \log \left[\frac{(v_L^2/R_L)}{(v_I^2/R_I)} \right] = 20 \log \left| \frac{v_L}{v_I} \right| + 10 \log \frac{R_I}{R_L} \tag{13.34}$$

If the input and load resistance happen to be equal, then $\log(R_I/R_L) = 0$ and Eq. 13.34 reduces to an expression for the voltage gain in decibels

$$\text{Voltage gain (dB)} = 20 \log \left| \frac{v_L}{v_I} \right| \tag{13.35}$$

As an example, a voltage gain of magnitude 10 is represented by a gain of +20 dB. It has become common practice to ignore the fact that Eq. 13.35 only represents power gain in decibels when R_I and R_L are equal. Instead, engineers have adopted Eq. 13.35 to express voltage gain in decibels without regard to the values of R_I and R_L. Under these conditions, the decibel of voltage gain carries no information about power gain. Therefore, when one encounters a voltage gain in dB, one should assume that it refers to Eq. 13.35, and inquire separately about the question of power gain.

13.2.3 A Helpful Graphical Construction

Each term in Eqs. 13.31 and 13.32 for $|H(j\omega)|$ and $\angle H(j\omega)$ involves a quantity of the form $(j\omega - s_o)$, where ω is the frequency, presumed variable, and s_o is a constant, locating either a pole or a zero. Since $(j\omega - s_o)$ is itself a complex number, it must have a well defined magnitude and angle, which we shall denote by

$$M = |j\omega - s_o| \tag{13.36}$$

and

$$\psi = \angle(j\omega - s_o) \tag{13.37}$$

Figure 13.7 shows a graphical construction that is very helpful in visualizing the quantities M and ψ. An arrow is drawn from s_o (taken as a negative real number in this example) to a point on the imaginary axis, labeled by $j\omega'$ in

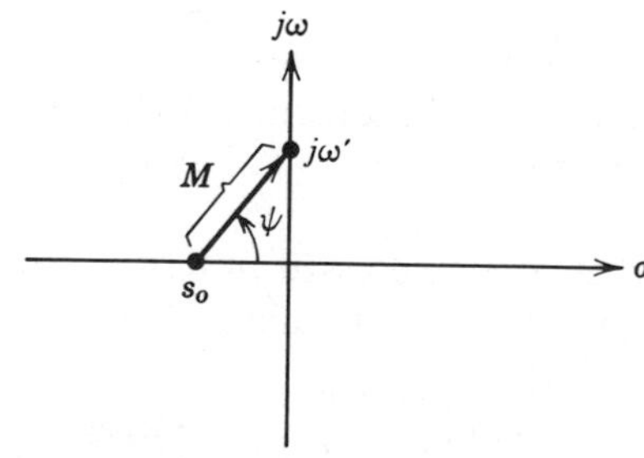

Figure 13.7
A graphical construction.

the figure. The length of this arrow is equal to M, while the angle between the $+\sigma$ direction and the arrow is equal to ψ.

If s_o corresponds to a zero (such as s_1 in Eqs. 13.31 and 13.32), the factor M enters directly into the magnitude of $H(j\omega)$, while the angle ψ simply adds to the total angle. If, however, s_o corresponds to a pole (such as s_a in Eqs. 13.31 and 13.32), the quantity $(j\omega - s_o)$ appears in the denominator of $H(j\omega)$. Therefore, the contribution to the total magnitude of $H(j\omega)$ is a factor $1/M$, and the contribution to the total angle is $-\psi$. The entire problem of determining the frequency response in a linear system can now be reduced to that of finding how the magnitude M and the angle ψ vary with frequency for the various possible locations of poles and zeros.

Since the phase-magnitude plot for an arbitrary system can be built up graphically from the phase-magnitude plots for individual poles and zeros, one should develop a catalog of basic solutions and learn them thoroughly. There are only three kinds of simple factors one might encounter. The pole or zero in question is either (1) at the origin, (2) on the real axis, or (3) part of a complex pair. These are discussed in the following sections.

13.2.4 Pole or Zero at the Origin

Figure 13.8 illustrates the magnitude and phase for a system function with a zero at the origin:

$$H(s) = s \tag{13.38}$$

The arrow shown lies along the $j\omega$ axis for all frequencies. Therefore, for positive ω, the magnitude M is exactly equal to ω and the angle ψ is a

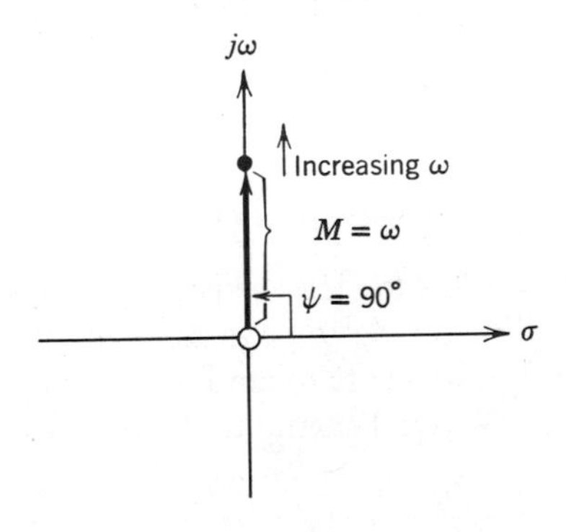

Figure 13.8
A zero at the origin.

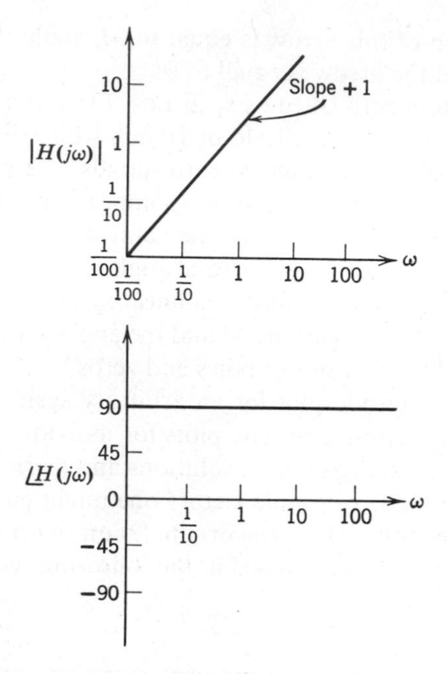

Figure 13.9
Phase-magnitude plot for a zero at $s = 0$.

constant, $+90°$. The phase-magnitude plot for this case is shown in Fig. 13.9. The phase angle is, of course, a constant, but the plot of log magnitude versus log frequency is shown as a straight line with "slope" + 1. Let us examine the analytical expression for $|H(j\omega)|$ to show why the magnitude plot has this particular form.[2]

Substituting $j\omega$ for s in Eq. 13.38 and finding the magnitude, we obtain

$$|H(j\omega)| = \omega \tag{13.39}$$

from which we see that

$$\log |H(j\omega)| = (1) \cdot (\log \omega) \tag{13.40}$$

It is clear from Eq. 13.40 that $\log |H(j\omega)|$ is proportional to $\log \omega$. Therefore, one expects a straight line on log-log coordinates. In addition, Eq. 13.40 has been written with a (1) in brackets to emphasize that this particular straight line has a slope of $+1$. (We shall encounter below straight lines with different slopes).

---

[2] When two variables are related by a power law, such as $y = x^n$, the "slope" of y versus x plotted on log-log coordinates is equal to n.

The magnitude plot for the zero at $s = 0$ illustrates another important feature of log-log coordinates: namely, that logarithmic coordinates never actually reach zero. Thus the straight line with slope $+1$ extends indefinitely in both directions.

The corresponding phase-magnitude plot for a pole at $s = 0$ is equally simple. Let

$$H(s) = \frac{1}{s} \tag{13.41}$$

and refer to the diagram of Fig. 13.8. We must keep in mind in using Fig. 13.8 for a pole that the length of the arrow will occur in the denominator and that the contribution to the total angle of the system function will be $-\psi$.

From these facts, we obtain for the system function Eq. 13.41

$$|H(j\omega)| = \frac{1}{M} = \frac{1}{\omega} \tag{13.42}$$

$$\angle H(j\omega) = -\psi = -90°$$

The phase-magnitude plot for this case is shown in Fig. 13.10. Once again, the constant phase angle presents no problem, but the magnitude plot warrants closer examination.

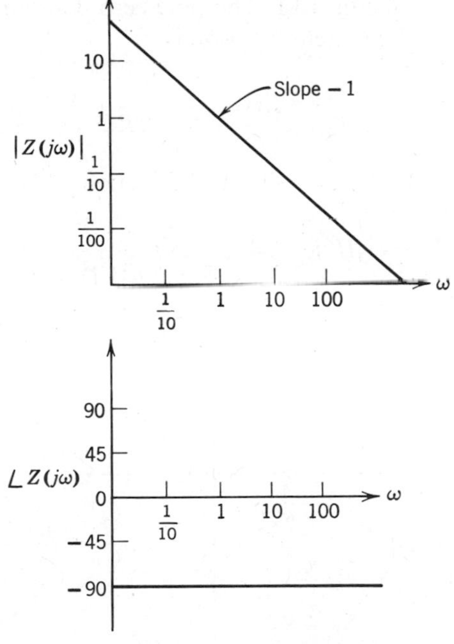

Figure 13.10
Phase-magnitude plot for a pole at $s = 0$.

Taking logarithms of the magnitude above, we obtain

$$\log |H(j\omega)| = \log \left(\frac{1}{\omega}\right) = (-1)\cdot(\log \omega) \tag{13.43}$$

It is seen here that while the log of the magnitude is proportional to log ω, the slope is -1.

Notice the reciprocal relations between the pole and zero. The magnitude for the zero is proportional to ω^{+1} and has slope $+1$ on log-log paper, while the magnitude for the pole is proportional to ω^{-1} and has a slope -1. Similarly, the phase angle for the pole is the negative of the phase angle for the zero.

13.2.5 Pole or Zero on the Real Axis

We shall work out in detail the phase-magnitude plots for a pole on the real axis. Then, we shall use the reciprocal relation between the pole and zero to deduce the corresponding results for the zero.

An example of a system with a pole on the real axis was already introduced as the low-pass filter of Fig. 13.1. The pole-zero diagram for this circuit is shown in Fig. 13.11. The system function is

$$\frac{V_2}{V_1} = H(s) = \frac{1}{1 + sRC} \tag{13.44}$$

which has a single pole at $s_a = -1/RC$. In Section 13.12, we showed that

$$|H(j\omega)| = \frac{1}{\sqrt{1 + (\omega RC)^2}} \tag{13.45}$$

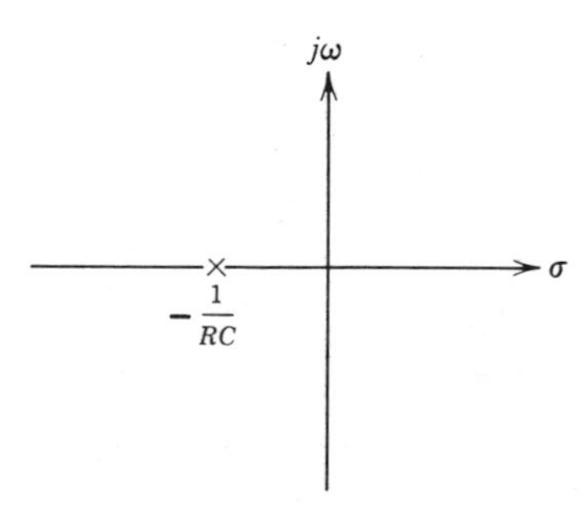

Figure 13.11
Pole-zero diagram for Fig. 13.1.

and

$$\angle H(j\omega) = -\tan^{-1} \omega RC \tag{13.46}$$

Let us now determine what these quantities look like when plotted on appropriate coordinates.

We shall use our graphical construction to visualize the answer before working with the analytical expressions. Figure 13.12 shows a sequence of

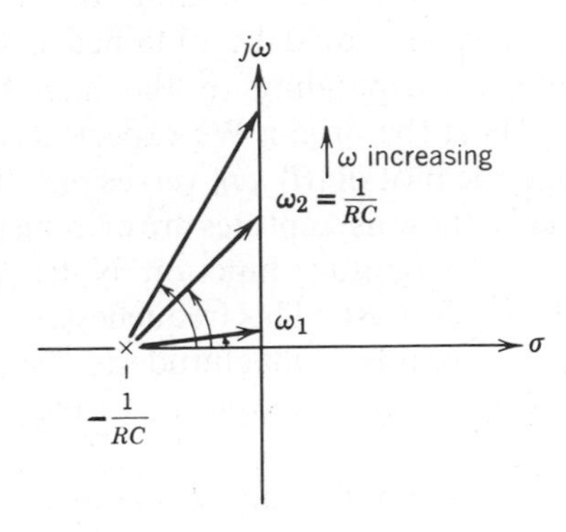

Figure 13.12
Graphical construction for a real axis pole.

three frequencies with the corresponding arrows drawn in. For very low frequencies (near ω_1), the magnitude M is essentially constant and the angle ψ is near zero. As the frequency increases, the magnitude begins to grow without limit while the angle increases toward a limit of $+90°$. An approximate dividing line between "low-frequency" and "high-frequency" regions occurs for $\omega = 1/RC$ (in the vicinity of ω_2 in Fig. 13.12). For frequencies below $1/RC$, the magnitude M is a weak function of frequency, while for frequencies much above $1/RC$, M becomes nearly proportional to frequency. We must keep in mind that the variation of M enters in the denominator of $|H(j\omega)|$ while the contribution to the angle is $-\psi$. Therefore we can conclude that for a single pole at $-1/RC$, the magnitude is roughly constant for $\omega \ll 1/RC$ and is proportional to $1/\omega$ for $\omega \gg 1/RC$. Furthermore, the angle is near zero for $\omega \ll 1/RC$ while the angle is about $-90°$ for $\omega \gg 1/RC$.

These same results, in more explicit form, can be obtained from the analytical expressions Eqs. 13.45 and 13.46. Considering the magnitude first, we note that when $\omega RC \ll 1$, the magnitude in Eq. 13.45 is approximately equal to unity. Furthermore, when $\omega RC \gg 1$, that term dominates the expression, so that the magnitude is approximately $1/\omega RC$. These two extremes of frequency and their corresponding approximate values for the magnitude comprise the *asymptotic* behavior of the magnitude, as summarized below;

$$\begin{aligned} &\text{Low-frequency asymptote: } \omega RC \ll 1 \qquad |H(j\omega)| \approx 1 \\ &\text{High-frequency asymptote: } \omega RC \gg 1 \qquad |H(j\omega)| \approx \frac{1}{\omega RC} \end{aligned} \tag{13.47}$$

The first of these asymptotes, corresponding to the low frequencies, has a constant magnitude, just as would be obtained in a resistive network. The second asymptote, corresponding to the high frequencies, behaves approximately like a pole at the origin. We expect, therefore, that for sufficiently high frequencies, the plot of $|H(j\omega)|$ versus ω will have a slope of -1 on log-log paper. Actually, these asymptotes are extremely useful in sketching approximate forms for the magnitude function. Notice that the two asymptotes become equal when $\omega = 1/RC$. This frequency at which the asymptotes intersect, which also corresponds in magnitude to the position of the pole, is called the *pole frequency*, the *corner frequency*, or the *cutoff frequency* of the system function.

Figure 13.13 shows both the high- and low-frequency asymptotes plotted on log-log coordinates, intersecting at the corner frequency. The exact curve is also plotted on the same coordinates. The deviation between the actual curve and the asymptotes is largest right at the corner frequency. Let us

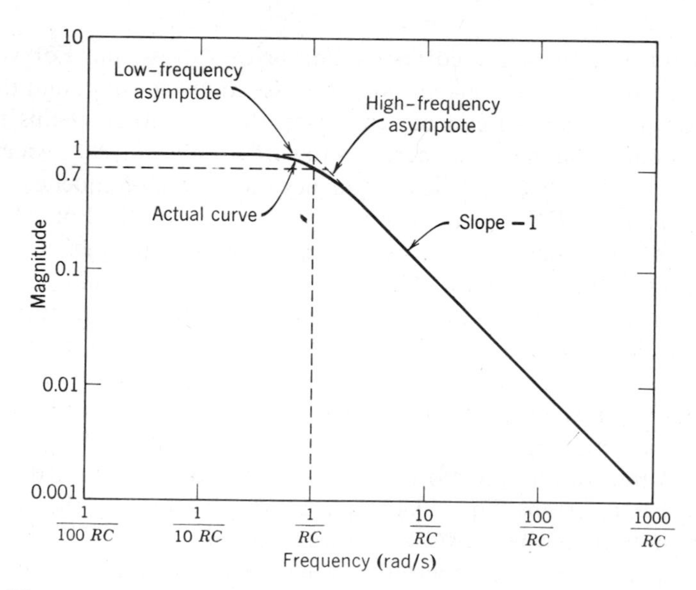

Figure 13.13
Magnitude versus frequency for a single pole at $s = -1/RC$.

estimate how far below the approximate asymptote curve the actual solution lies. When $\omega = 1/RC$, the magnitude has the value $1/\sqrt{2} = .707$. Expressed in decibels, at the corner frequency

$$|H(j\omega)| = 20 \log \frac{1}{\sqrt{2}} = -10 \log 2 = -3 \text{ dB} \qquad (13.48)$$

For this reason, the pole frequency (or corner frequency) is said to be the *3 dB frequency* for this system. More generally, the 3 dB frequency of a system is that frequency at which the magnitude of the frequency response falls 3 dB below some identifiable maximum. One can make very rapid sketches of the magnitude function for a single pole by first sketching the asymptotes, remembering that the low-frequency asymptote is constant out to the pole frequency, at which point the high-frequency asymptote *rolls off*[3] with a slope of -1. Then, at the corner frequency, one plots a point 3 dB (or a factor of .707) below the low-frequency asymptote. Finally, one sketches in a smooth curve that joins the asymptotes within a factor of 3 to either side of the pole frequency. Sketches of this sort, which are accurate to about 10%, can be made with great speed, and are very useful for estimating the behavior of a network.

The variation of the phase angle with frequency can be approached in a similar fashion. For very low frequencies, the magnitude of the frequency response is constant. Since a constant magnitude corresponds to the behavior of resistive networks, we expect that the phase angle should be approximately zero for low frequencies. Also, since the magnitude approaches that of a pole at the origin for high frequencies, we expect the angle to be at about $-90°$ for sufficiently high frequencies. The actual curve for the angle versus frequency is plotted in Fig. 13.14 along with some useful approximate curves that can be used for quick estimates. Indeed, as expected, the phase angle is approximately zero for frequencies less than 1/10 of the pole frequency, while the angle is approximately $-90°$ for frequencies more than about 10 times the pole frequency. At the pole frequency, where ωRC is exactly equal to unity, the angle is given by $-\tan^{-1}(1)$, which is equal to $-45°$. The piecewise linear approximate phase curve shown in Fig. 13.14 incorporates all these features. It is zero until 1/10 of the pole frequency, and descends smoothly through $-45°$ at the pole frequency to $-90°$ at 10 times the pole frequency, and then is constant thereafter. This approximation to the actual curve is accurate to within about seven degrees of angle throughout the entire range of frequencies, and is extremely useful for rapid estimates of network behavior.

---

[3] In the vocabulary of electronics, a slope of -1 is also expressed as a rolloff of 3 dB/octave or 10 dB/decade for power gain, and 6 dB/octave or 20 dB/decade for voltage and current gain.

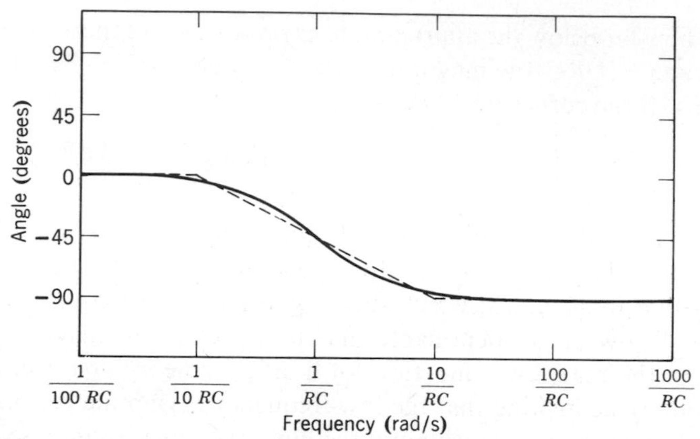

Figure 13.14
Angle versus frequency for a single pole at $s = -1/RC$.

The corresponding curves for the magnitude and phase for a zero on the negative real axis are shown in Fig. 13.15. The position of the zero is presumed to be at $s_i = -\omega_1$. Notice that the magnitude and phase curves are identical to those for the pole at the negative real axis, except that each curve

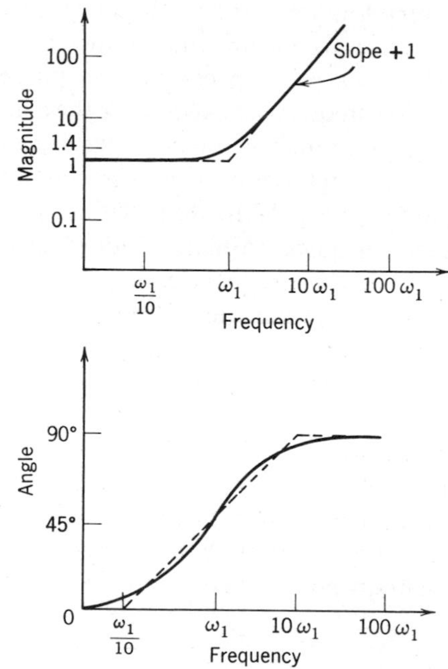

Figure 13.15
Phase-magnitude plot for a zero at $s_1 = -\omega_1$.

is inverted. The magnitude is constant out to the position of the zero, but then increases with a slope of $+1$, while the phase angle is zero at low frequencies, passes through $+45°$ at the corner frequency, and becomes a constant $+90°$ at high frequencies.

13.2.6 The Complex Pair

A complex pole pair is shown in Fig. 13.16. The poles have real parts equal to $-\omega_1$, and imaginary parts of $\pm j\omega_0$. The arrows for our graphical construction are also shown, with the magnitudes M_1 and M_2 and the angles ψ_1 and ψ_2 labeled for a typical value of frequency.

The overall magnitude is $1/M_1M_2$ and the overall phase is $-(\psi_1 + \psi_2)$.

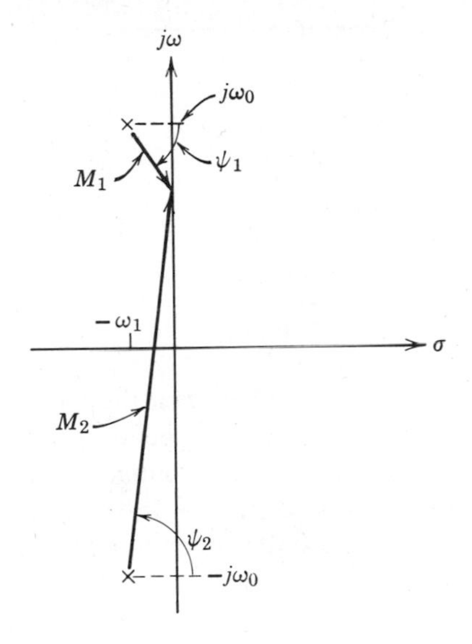

Figure 13.16
Complex pair of poles.

Near $\omega = 0$, the magnitudes of M_1 and M_2 both equal $\sqrt{\omega_1^2 + \omega_0^2}$. Therefore, at low frequencies, the magnitude of the response is $1/(\omega_1^2 + \omega_0^2)$, as shown in Fig. 13.17. The phase angle at low frequencies is zero, since for ω near zero, ψ_1 and ψ_2 are equal but opposite. As ω approaches ω_0, the magnitude of M_2 increases to about $2\omega_0$, but the magnitude M_1 drops sharply to a minimum value of ω_1. This produces a peak in the frequency response (Fig. 13.17), the height of the peak being inversely proportional to ω_1. That is, as the complex pair of poles get closer to the imaginary axis, the peak in

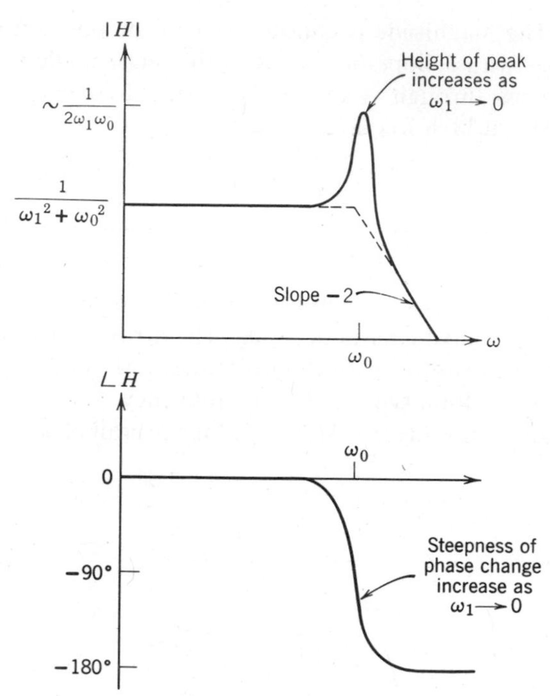

Figure 13.17
Phase-magnitude plot for a complex pole pair.

the response gets higher. At the same time, the width of the peak narrows, because ω must be within about ω_1 of ω_0 before substantial peaking occurs. The phase angle also undergoes sharp changes near ω_0, dropping steeply from angles near zero to angles near $-180°$. At frequencies well above ω_0, both M_1 and M_2 are increasing proportional to ω, so the high-frequency asymptote has a slope of -2, and the phase angle is a constant at $-180°$.

Because of the peak in the frequency response, systems with complex pole pairs are used for frequency-selective filters, passing a narrow band of frequencies near ω_0 and attenuating all other frequencies relative to the peak amplitude of the transfer function at ω_0.

13.3 BAND-PASS SYSTEMS

13.3.1 An Active Band-Pass Filter

A *filter* is a network that selectively passes certain ranges of frequencies. The single-pole RC network encountered in Fig. 13.1, with the frequency response shown in Fig. 13.13, is a *low-pass filter* because it allows frequencies below the corner frequency to be transmitted without attentuation, while it attenuates frequencies above the corner frequency, the amount of attenuation increasing with increasing frequency. This simple RC filter, being made entirely of passive components, is called a *passive filter*. In the example below, we shall present another filter circuit. It is a *band-pass filter*, since it allows unattenuated passage of frequencies between two limits and attenuates frequencies outside these limits. The filter is an *active filter*, because it incorporates an active element, in this case an op-amp.

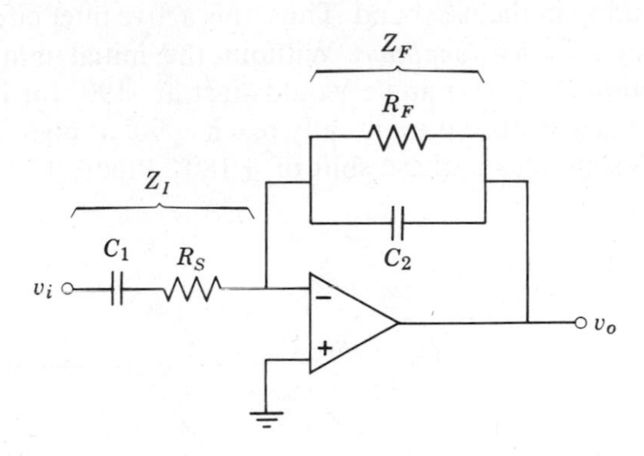

Figure 13.18
An active band-pass filter.

Figure 13.18 shows an op-amp circuit with RC networks both as the input impedance and feedback impedance. Let us assume that the op-amp can be considered ideal at all frequencies of interest (an assumption that will be examined further in Chapter 14). The transfer function can then be written in terms of these input and feedback impedances.

$$\frac{V_o}{V_i} = -\frac{Z_F(s)}{Z_I(s)} \tag{13.49}$$

By inspection of the circuit, we find that

$$Z_F = \frac{R_F}{1 + sC_2R_F} \tag{13.50}$$

and

$$Z_I = \frac{1 + sC_1R_S}{sC_1} \tag{13.51}$$

which yields

$$\frac{V_o}{V_i} = -\frac{sC}{(1 + sC_1R_S)(1 + sC_2R_F)} \tag{13.52}$$

Thus the transfer function has a single zero at $s = 0$ and two poles, one determined by the R_SC_1 product, the other by the R_FC_2 product. If we assume that $R_SC_1 \ll R_FC_2$, we obtain the pole-zero diagram and phase-magnitude plot shown in Fig. 13.19. The resistive (flat) region, between $1/R_SC_1$ and $1/R_FC_2$ is called the *passband.* Notice that if $R_F > R_S$, this filter has a gain greater than unity in the passband. Thus, this active filter circuit can be used as a *frequency-selective amplifier.* Without the initial minus sign on the transfer function the phase angle would start at $+90°$ for low frequencies, (from the zero at $s = 0$), and eventually reach $-90°$ at high frequencies. The overall minus sign adds a phase shift of $\pm 180°$. Figure 13.19 is drawn with the total phase shift.

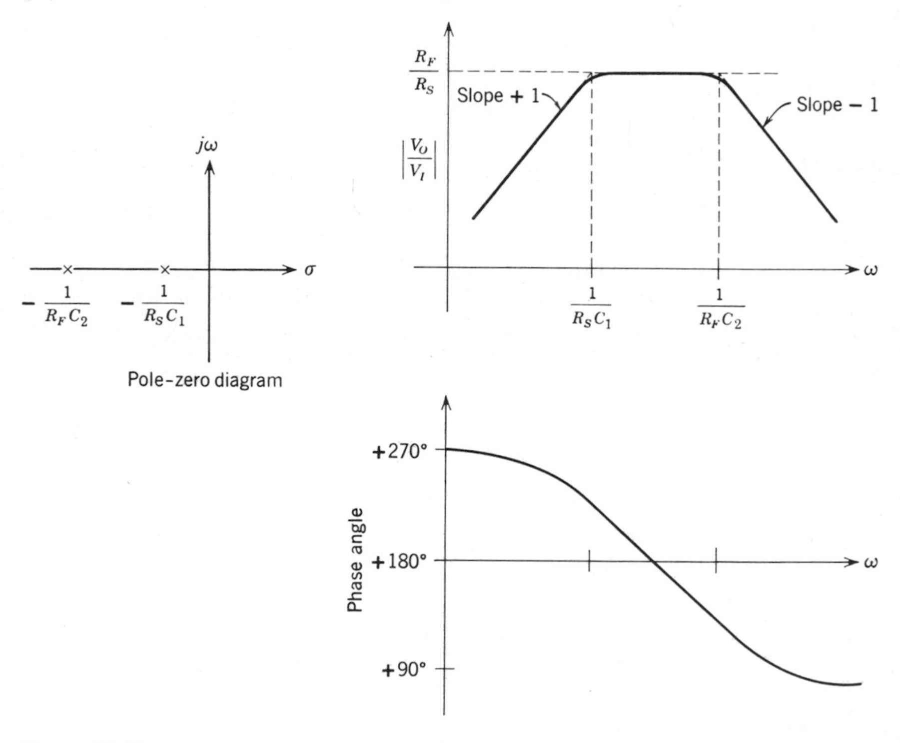

Figure 13.19
Frequency response of the band-pass filter.

Clearly, by adjusting the values of the capacitors and resistors in this network, one can in this ideal case match any desired passband characteristic, so long as it is consistent with one zero at $s = 0$ and two poles on the negative real axis. It is this design flexibility that accounts for increasingly widespread use of op-amps and of active filters using op-amps.

Active filters need not be restricted to transfer functions with poles on the negative real axis. By using more sophisticated amplifier-feedback combinations, active filters with all varieties of transfer functions can be built. Such filters are incorporated in many different types of communications and instrumentation equipment.

13.3.2 Transformer Frequency Response

Figure 13.20 shows a transformer used to couple a source v_S to a load resistor R_L. In this section, we shall examine the frequency response of this circuit using the transformer model introduced in Chapter 6. Figure 13.20*b* shows this model, in which an ideal transformer is used together with a leakage inductance L_e and a magnetizing inductance L_m. This model, which neglects all capacitive effects, is generally useful for audio-frequency transformers.

With the transformer model inserted into the circuit, the transformer relations (see Section 6.2.6) immediately yield the equivalent circuit of Fig. 13.20*c*. The resistance in the secondary R_L appears in the primary circuit as an equivalent resistor of magnitude R_L/n^2, and the voltage across L_m is v_L/n. Using this circuit, the transfer function between the complex amplitudes of v_S and v_L can be written

$$\frac{V_l/n}{V_s} = \frac{(sL_m \| R_L/n^2)}{R_S + sL_e + (sL_m \| R_L/n_2^2)} \tag{13.53}$$

This equation can be manipulated to the following form:

$$\frac{V_l}{V_s} = \frac{(nL_m R_L)s}{R_L R_S + [L_m(R_L + n^2 R_S) + L_e R_L]s + (L_e L_m n^2)s^2} \tag{13.54}$$

Equation 13.54 shows that the transfer function has a zero at $s = 0$ and two poles. The locations of these poles, however, must be obtained from the roots of an algebraically complicated quadratic equation. It is perfectly possible, indeed it is occasionally necessary, to press ahead with brute force to solve such equations. The physical interpretation of the resulting solution,

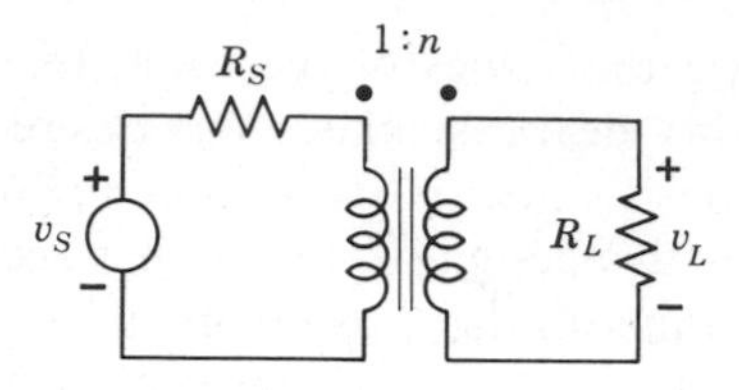

(*a*) Transformer coupling R_S to R_L

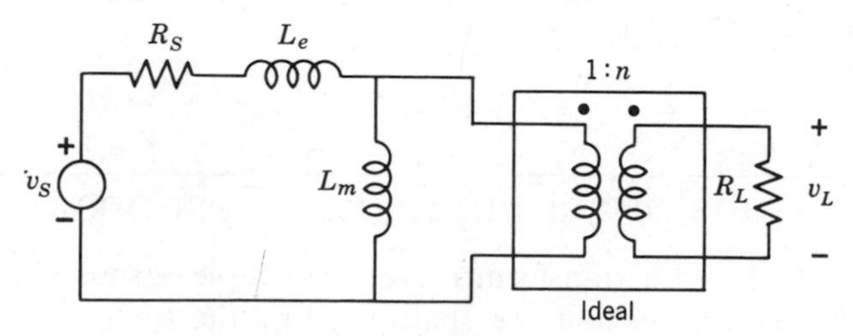

(*b*) Circuit model showing the leakage and magnetizing inductance

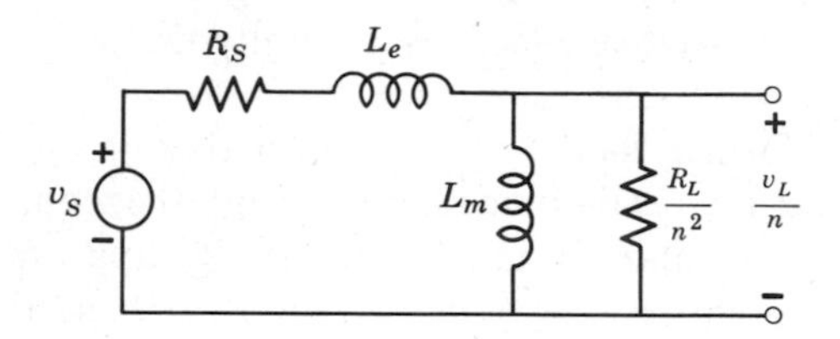

(*c*) Equivalent circuit

Figure 13.20
Transformer equivalent circuit.

however, might not be obvious. We prefer, therefore, to think of clever ways to get good approximations to the final answer and to sharpen our insight to the circuit's behavior.

The circuit has two poles, one for each of the two independent energy storage elements. In real transformers, the magnetizing inductance L_m is much larger than the leakage inductance L_e. Therefore, unless R_L/n^2 and R_S are many orders of magnitude different from one another, it becomes possible to separate the frequency domain into three regions, a low-frequency region where both L_e and L_m have small impedances compared to the resistive elements, a mid-frequency region where L_e is a small impedance but L_m is a large impedance compared to the resistive elements, and a high-frequency region where both L_e and L_m have large impedances compared to the resistive elements.

It becomes possible, then, to develop two much simpler circuits, each with only one energy storage element. The low-frequency circuit, Fig. 13.21*a*, covers the low- and mid-frequency regions. It includes L_m, but has replaced L_e with an effective short circuit. The high-frequency circuit, Fig. 13.21*b*, which covers the high- and mid-frequency regions, includes L_e, but has replaced L_m by an open circuit. In the mid-frequency region, where L_e is an effective short circuit and L_m is an effective open circuit, both Fig. 13.21*a* and 13.21*b* reduce to the same mid-frequency equivalent circuit, Fig. 13.21*c*. Notice that the mid-frequency region has a totally resistive transfer function, just as did the band-pass active filter discussed previously.

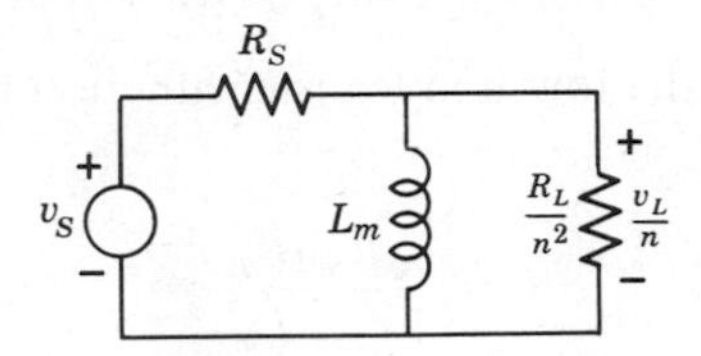

(*a*) Low-frequency circuit. L_e is a short circuit at low and mid frequencies

(*b*) High-frequency circuit. L_m is an open circuit at mid and high frequencies.

(*c*) Mid-frequency circuit. L_e is a short and L_m is open.

Figure 13.21
Equivalent circuits for different frequency ranges.

With this separation of the total circuit into two much simpler circuits, it is now possible to write down by inspection the transfer functions for the different frequency regions.

$$\text{Low-mid: (Fig. 13.21}a\text{)} \qquad \frac{V_l}{V_s} = \frac{n(sL_m \| R_L/n^2)}{R_S + (sL_m \| R_L/n^2)} \tag{13.55}$$

$$\text{Mid-high: (Fig. 13.22}b\text{)} \qquad \frac{V_l}{V_S} = \frac{R_L/n}{R_S + R_L/n^2 + sL_e} \tag{13.56}$$

There are many features to note from these transfer functions. First, if $s \to \infty$ in the low-mid transfer function, and if $s \to 0$ in the mid-high transfer function, both functions reduce to the correct mid-band transfer function

$$\text{Mid-band:} \quad \frac{V_l}{V_s} = \frac{nR_L}{R_L + n^2 R_S} \tag{13.57}$$

Next, we observe that the low-mid transfer function has a zero at $s = 0$ and a single pole at

$$s_l = -\frac{R_S \| (R_L/n^2)}{L_m} \tag{13.58}$$

a result that can also be obtained from the time constant of the network. This value of s_l represents an estimate for the location of the lower of the two poles (i.e., closer to the origin) in the total transfer function of Eq. 13.54 and for the companion value of the lower cutoff frequency ω_l. Finally, the mid-high transfer function has a single pole at

$$s_h = -\frac{R_S + R_L/n^2}{L_e} \tag{13.59}$$

This represents an estimate for the location of the upper of the two poles in the total transfer function of Eq. 13.54 and for the upper cutoff frequency ω_h.

This entire process of separating into two equivalent circuits is justified provided that the two poles obtained in this way are more than two decades apart (factor of 100). Assuming that the estimated low- and high-frequency poles are far enough apart, the total pole-zero diagram is that shown in Fig. 13.22. This pole-zero diagram is identical in form to the band-pass active filter of the previous section (Fig. 13.19). The magnitude of the frequency response is shown in Fig. 13.23, along with schematic indications as to which portion of the overall transfer function is estimated from which partial circuit.

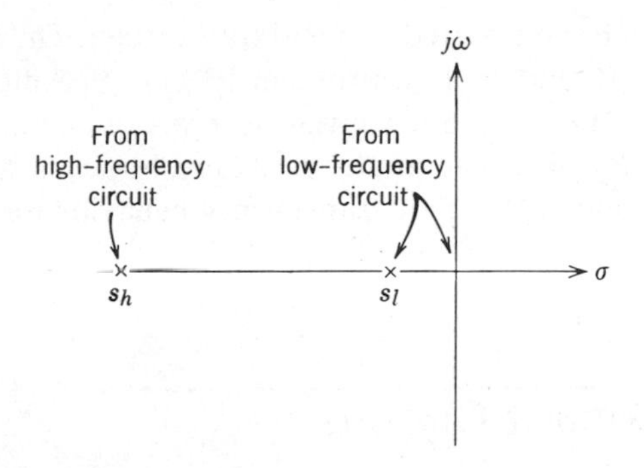

Figure 13.22
Pole-zero diagram for transformer transfer function.

Had the estimates of the low-frequency and high-frequency poles (s_l and s_h) been closer together than the desired two decades, it would be necessary to return to the full equation (Eq. 13.54) and solve explicitly for the two poles. In that case it turns out that both pole locations depend numerically on the values of both inductances. The overall pole-zero diagram is still similar to Fig. 13.22, but the algebraic expressions for s_l and s_h differ somewhat from Eqs. 13.58 and 13.59.

As a final remark, we note that in the passband, the transfer function is the same as that obtained by simply replacing the actual transformer by an ideal transformer. Thus ω_l and ω_h bound the frequency range within which

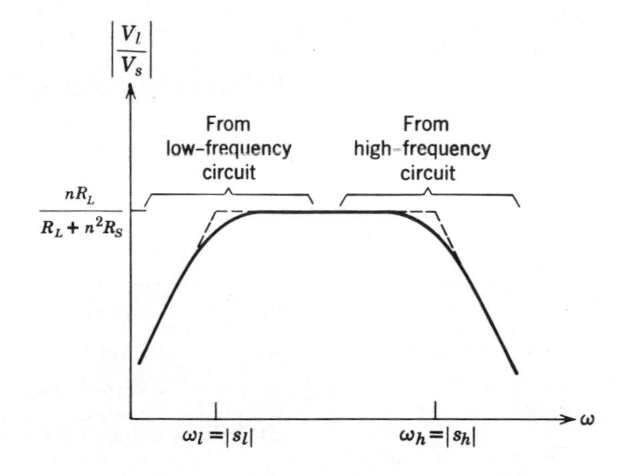

Figure 13.23
Transfer-function magnitude for the circuit of Fig. 13.20.

a transformer can be considered an ideal transformer. The limits of the ideal range depend on R_S and R_L (see Problem P13.11). Specifications for audio-frequency transformers are often given in terms of ω_l and ω_h for specified values of R_S and R_L. With the model used here, however, it is easy to obtain estimates for L_e and L_m so that transformer behavior for other choices of R_S and R_L can be successfully predicted.

13.3.3 Resonant Circuits

The subject of *resonant circuits*, or tuned circuits, is extensive. In this section we shall analyze one example of a resonant circuit (Fig. 13.24*a*) the so-called parallel-resonant *RLC* circuit. The transfer function of Fig. 13.24*a* and the impedance function of Fig. 13.24*b* are equivalent. We shall work with the impedance example of Fig. 13.24*b*.

The system function for this circuit has a form identical to the band-pass systems just encountered; that is, the parallel *RLC* circuit has a zero at $s = 0$ and two poles. Therefore, the system function can be written in the

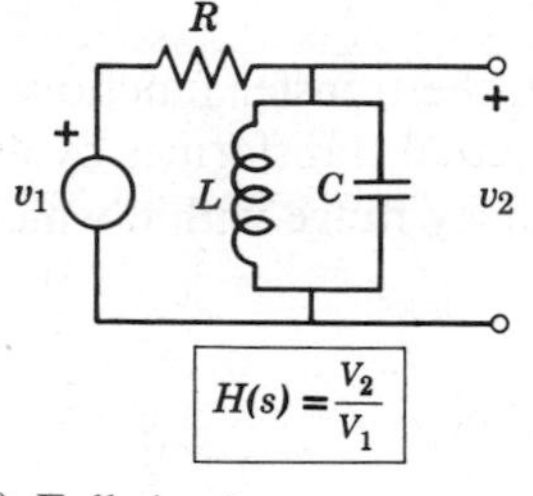

(*a*) Full circuit

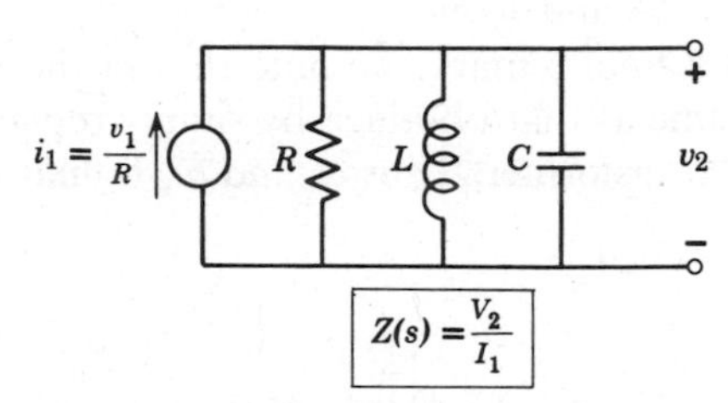

(*b*) Norton equivalent circuit

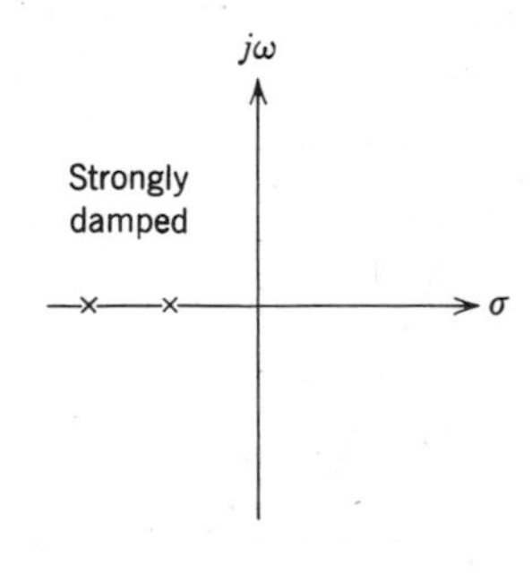

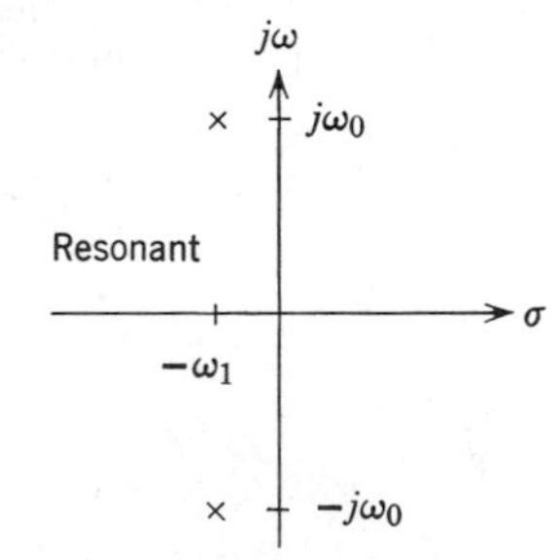

(*d*) Pole-zero diagram for $R \gg \sqrt{L/4C}$

Figure 13.24
The parallel-resonant *RLC* circuit

form

$$Z(s) = \frac{Ks}{1 + a_1 s + a_2 s^2} \tag{13.60}$$

where in this specific example

$$K = L$$
$$a_1 = \frac{L}{R} \tag{13.61}$$
$$a_2 = LC$$

The two poles are located at the roots of the denominator,

$$s_a, s_b = -\left(\frac{a_1}{2a_2}\right) \pm \sqrt{\left(\frac{a_1}{2a_2}\right)^2 - \frac{1}{a_2}} \tag{13.62}$$

Clearly, if $a_1^2 > 4a_2$, the poles are on the negative real axis (Fig. 13.24*c*), while if $a_1^2 < 4a_2$, the poles form a complex pair (Fig. 13.24*d*). We have already examined two band-pass systems with poles on the negative real axis. In terms of the vocabulary used to describe resonant circuits, the case of real-axis poles is called *strongly damped.* In Chapter 14, when we examine the step response of these circuits, the reason for this name will become clear.

The complex pole pair, Fig. 13.24*d*, is the characteristic resonant behavior. The phase-magnitude plot is sketched in Fig. 13.25. It differs from Fig. 13.17 (the complex pole pair) because of the presence of a zero at $s = 0$. At low frequencies, the asymptotic slope of $+1$ and the phase angle of $+90°$ are characteristic of the zero at $s = 0$. At high frequencies, the asymptotic slope of -1 and the phase angle of $-90°$ arise from the combined effect of the one zero and the two poles, leaving a net asymptotic behavior at high frequencies equivalent to a single pole. Near the center frequency ω_0, the complex pair produces a sharp peak in the response. We shall now examine the behavior near that peak more closely.

The general system function of Eq. 13.60 can be rewritten in terms of two new parameters ω_0 and Q, as follows:

$$Z(s) = \frac{sL}{1 + (1/Q)(s/\omega_0) + (s/\omega_0)^2} \tag{13.63}$$

where

$$\omega_0 = \frac{1}{\sqrt{a_2}} = \frac{1}{\sqrt{LC}} \tag{13.64}$$

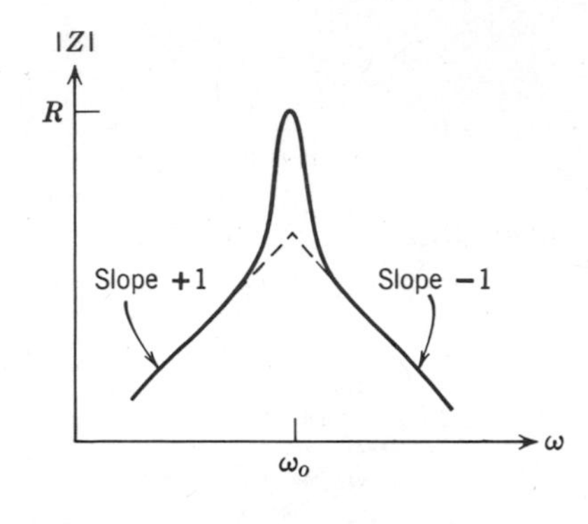

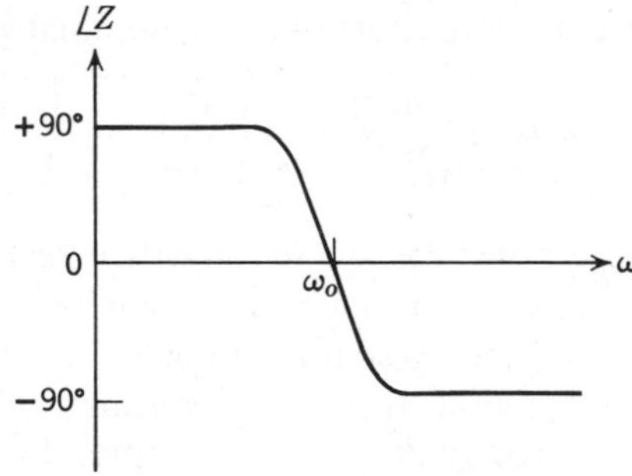

Figure 13.25
Phase-magnitude plot for the resonant circuit.

and

$$Q = \frac{1}{\omega_0 a_1} = \frac{R}{\omega_0 L} \tag{13.65}$$

We call ω_0 the *resonant frequency* while Q is called the *quality factor* and is a measure of the sharpness of the peak. (Notice that the distinction between resonant and damped behavior occurs at a Q of $\frac{1}{2}$.) If the Q is large, that is, greater than about 10, the positions of the poles can be expressed simply in terms of ω_0 and Q.

$$s_a, s_b = -\omega_1 \pm j\omega_0 \tag{13.66}$$

where

$$\omega_1 = \frac{\omega_0^2 L}{2R} = \frac{1}{2RC} = \frac{\omega_0}{2Q}$$

The impedance then becomes

$$Z(s) = \frac{\omega_0^2 Ls}{(s + \omega_1 + j\omega_0)(s + \omega_1 - j\omega_0)} \tag{13.67}$$

If we consider the frequency response at frequencies near ω_0, which is where the peak occurs, and substitute

$$s = j\omega = j(\omega_0 + \delta) \tag{13.68}$$

where δ is presumed small compared to ω_0, we obtain from Eq. 13.67

$$Z(j\omega) = (\omega_0^2 L)\frac{j(\omega_0 + \delta)}{[\omega_1 + j(2\omega_0 + \delta)][\omega_1 + j\delta]} \tag{13.69}$$

The magnitude of both the numerator and the first factor in the denominator do not change appreciably as δ varies in a narrow range about zero. Furthermore, under the assumption that $Q > 10$, it follows that $\omega_0 \gg \omega_1$. Therefore $Z(j\omega)$ becomes

$$Z(j\omega) \simeq (\omega_0^2 L)\left(\frac{j\omega_0}{2j\omega_0}\right)\left(\frac{1}{\omega_1 + j\delta}\right) = \left(\frac{\omega_0^2 L}{2}\right)\left(\frac{1}{\omega_1 + j\delta}\right) \tag{13.70}$$

from which we obtain the magnitude and angle

$$|Z(j\omega)| = \frac{\omega_0^2 L}{2}\frac{1}{\sqrt{\omega_1^2 + \delta^2}} \tag{13.71}$$

$$\angle Z(j\omega) = -\tan^{-1}\frac{\delta}{\omega_1}$$

Because δ can take on both positive and negative values (corresponding to frequencies just above and just below ω_0), it is customary to plot the magnitude and phase of $Z(j\omega)$ on a linear scale instead of on the logarithmic scales we have used in all other plots. The resulting curves are often plotted in parametric fashion (Fig. 13.26). The magnitude of the impedance is plotted relative to its maximum value, which occurs at the resonant frequency, $\delta = 0$. For the circuit used in our example,

$$Z_{max} = \frac{\omega_0^2 L}{2\omega_1} = R \tag{13.72}$$

The frequency axis is plotted in terms of the deviation from ω_0 relative to ω_1. We note from the plot that when δ is equal to ω_1, the magnitude of the response is down from the maximum by a factor $1/\sqrt{2}$, or 3 dB. Thus, ω_1 is often called the 3 dB frequency or *half-power point*. In this simple resonant circuit, the parameter $Q(=\omega_0/2\omega_1)$ can be interpreted as the ratio of the center frequency ω_0 to the full width of the peak between the upper and lower 3 dB points. Thus the Q is a measure of the sharpness of the peak of the response function.

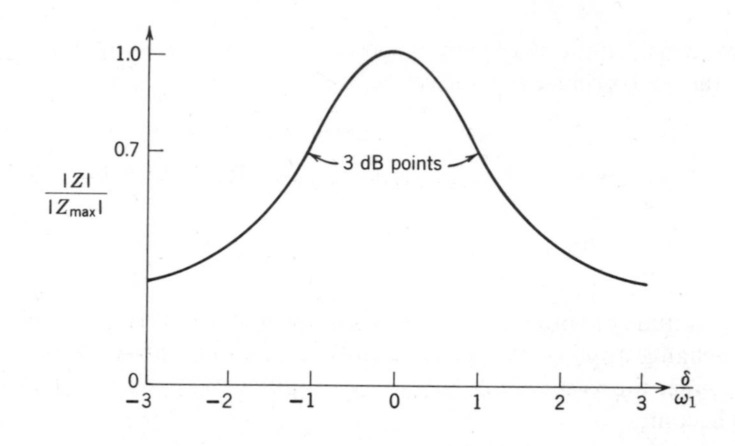

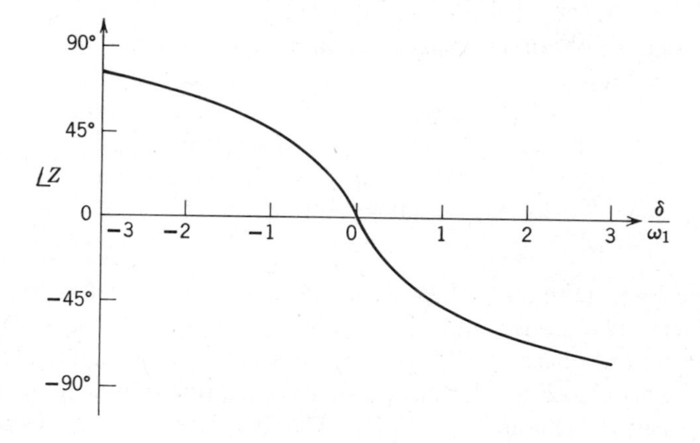

Figure 13.26
Phase-magnitude plot for a parallel resonant circuit near ω_0.

Notice that at the resonant frequency ω_0, the impedance reduces to that of a simple resistance. The parallel *LC* combination is behaving like an open circuit. An equivalent statement is that the impedance of the parallel *LC* combination has a complex pole pair *on* the imaginary axis at $\pm j\omega_0$. The *LC* impedance, therefore, is infinite at ω_0, leaving only the resistance R.

Figure 13.27 contains another *RLC* circuit, called a *series-resonant* circuit. This circuit has the property of having a minimum in impedance at its resonant frequency. For $R' \ll \sqrt{4L/C}$, the resonant frequency is equal

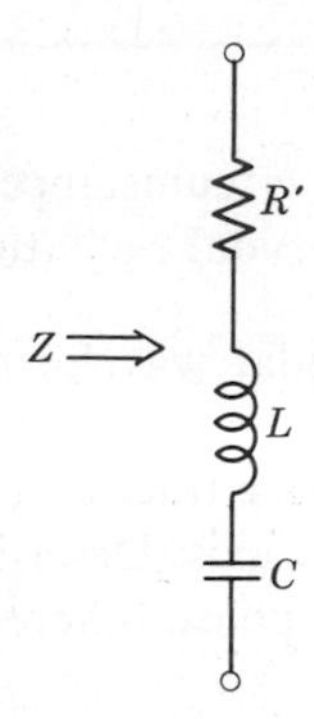

Figure 13.27
A series resonant circuit.

to $1/\sqrt{LC}$. The value of the minimum impedance at the resonant frequency is R'. The pole-zero diagram is the inverse of that shown in Fig. 13.24*d*, having a pole at $s = 0$ and a complex pair of zeros near the imaginary axis.

There are many other kinds of resonant circuits, all of which have in common a pole-zero diagram with one or more complex pairs of poles (or zeros) located near the imaginary axis. The more poles (or zeros) in the vicinity of a particular region of the imaginary axis, the more sharply peaked (or dipped) the response function, and the more rapidly the phase of the response varies from one side of the peak to the other.

The passbands of all the resonant circuits are narrow (compared for example, to the band-pass systems with poles on the negative real axis), and are used in applications where extremely precise frequency selectivity is required.

REFERENCES

R13.1 Amar G. Bose and Kenneth N. Stevens, *Introductory Network Theory* (New York: Harper and Row, 1965).

R13.2 Hugh H. Skilling, *Electrical Engineering Circuits* (New York: John Wiley, 1957).

R13.3 Bruce D. Wedlock and James K. Roberge, *Electronic Components and Measurements* (Englewood Cliffs, N.J.: Prentice-Hall, 1969), Chapters 8, 9, and 18.

QUESTIONS

Q13.1 Under what specific circumstances can the transients produced by turning on a sinusoidal excitation be ignored?

Q13.2 Suppose two sinusoidal waveforms, v_1 and v_2, have the same amplitude and frequency but different phase angles, ϕ_1 and ϕ_2. Describe an oscilloscope trace driven by v_1 on the horizontal axis and v_2 on the vertical axis. Describe how the appearance of this trace changes as the phase difference $\phi_1 - \phi_2$ varies from 0 to 360°.

Q13.3 Explain in simple terms the difference between impedance, resistance, reactance, admittance, conductance, and susceptance. Which are complex numbers and which are real numbers?

Q13.4 The shapes of the frequency response phase-magnitude plots are completely determined by the locations of the poles and zeros. What determines the position along the vertical axis for the magnitude plot? For the phase plot?

Q13.5 If you are told that "an amplifier has a gain of 40 dB," what additional information do you need to interpret that statement?

Q13.6 At what point in the analyses of the three band-pass examples of Section 13.3 is the common theme clearly visible? (*Answer.* Examination of the system function for the locations of poles of zero.)

EXERCISES

E13.1 For each of the pole-zero diagrams in Figure E13.1, sketch and dimension the magnitude of the system function and the angle of the system function versus frequency. Use log-log coordinates for the magnitude plots and semilog coordinates for the phase-angle plots.

E13.2 Match up the transfer functions V_2/V_1 for the networks in Fig. E13.2 with the pole-zero diagrams of Fig. E13.1. Be sure you

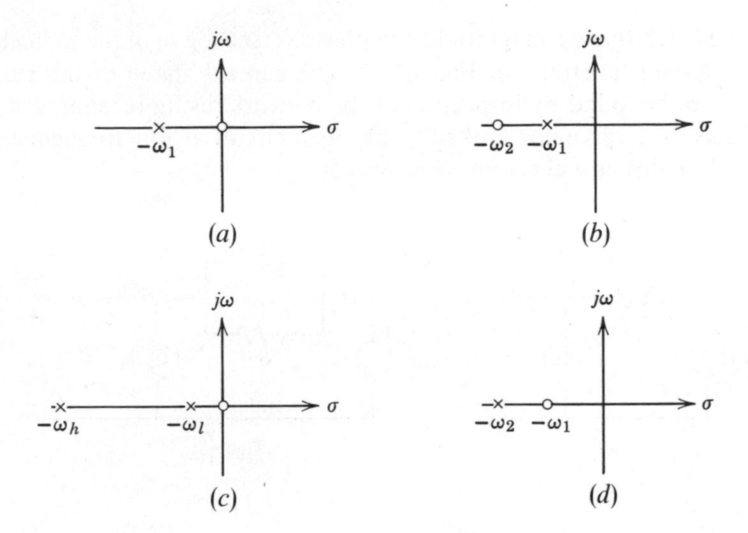

Figure E13.1

can relate the behavior of the magnitude versus frequency plot to the behavior of the capacitors in the network.

Answers. i—b; $\omega_1 = 1/C(R_1 + R_2)$; $\omega_2 = 1/CR_1$
ii—d; $\omega_1 = 1/CR_1$; $\omega_2 = 1/C(R_1 \| R_2)$
iii—a; $\omega_1 = 1/(C_1 + C_2)R$
iv—c; $\omega_l \approx 1/2C_1R$; $\omega_h \approx 2/C_2R$

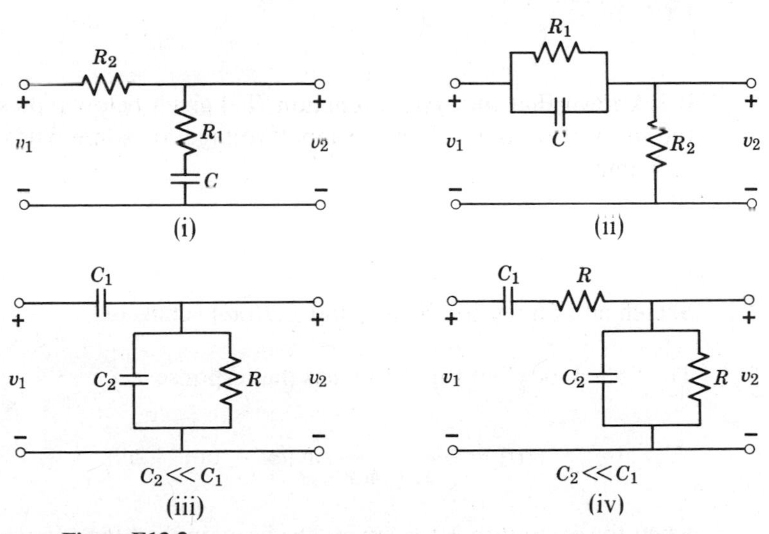

Figure E13.2

E13.3 Sketch the log-magnitude and phase versus log ω of the indicated system functions in Fig. E13.3. The general shape of the curve can be found by inspection of the network, using reasoning such as "the capacitor looks like an open circuit at low frequencies." Use this as a check on your results.

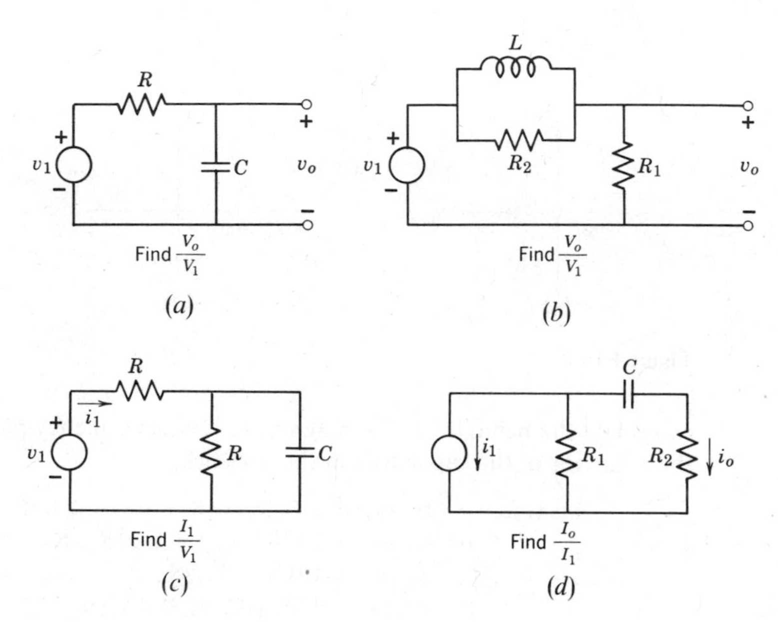

Figure E13.3

E13.4 It is known that the system function $H(s)$ given below represents the ratio of output voltage to input voltage for some particular hi-fi amplifier.

$$H(s) = 10^7 \frac{s(s + 10^2)}{(s + 10)(s + 10^3)(s + 10^5)}$$

Sketch and dimension $|H(j\omega)|$ and $\angle H(j\omega)$ versus ω.

E13.5 The "black box" in Fig. E13.5 has the response

$$v(t) = \frac{4}{\sqrt{1 + 4\omega^2}} \sin(\omega t - \tan^{-1} 2\omega)$$

when the excitation $i(t) = \sin \omega t$ has been applied for a long time.

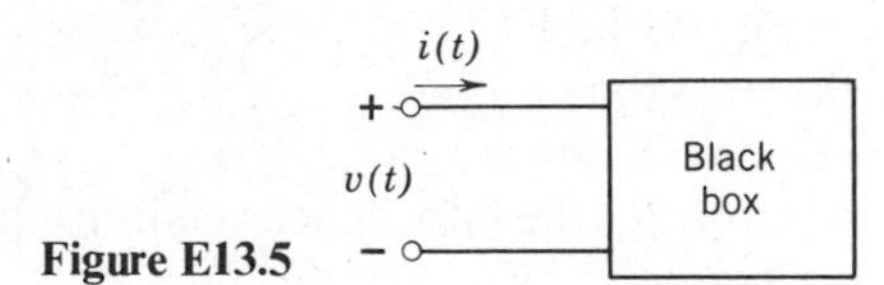

Figure E13.5

(a) What is the impedance system function for the box?
(b) Suppose $i(t) = u_{-1}(t)$. Sketch $v(t)$. (*Hint.* It may be easier to do part *c* first.)
(c) Devise a circuit containing one capacitor and one resistor that has the same system function as the black box. Find the required value of R and C and draw the circuit.

E13.6 A system with a power gain of 20 dB is driven by 10 mW of input power. What is the output power?

E13.7 A system with a voltage gain of 20 dB and an input resistance of 1 kΩ is driven by 10 mW of input power. What is the output voltage? Can the output power be determined?

E13.8 At a particular frequency a network element has a resistance of 6 Ω and a reactance of 8 Ω. Find the conductance and susceptance of the element at that frequency.

E13.9 A real 1 mH inductor is modeled by a 6 Ω resistor in series with a 1 mH inductance. At what frequency does the impedance of the inductor correspond to the network element of E13.8?

E13.10 Starting with the impedance of Fig. 13.24*b*, verify Eqs. 13.60, 13.62, and 13.63.

E13.11 It is often said that below its resonant frequency, a parallel resonant circuit "looks inductive," while above the resonant frequency it "looks capacitive." Refer to Fig. 13.25 and explain these statements.

E13.12 Real inductors have stray capacitance between the turns that produce an approximation to a parallel resonant circuit. That is, the real coil will be self-resonant at a particular frequency. Referring to E13.11, how would you expect an inductor to behave if used above its self-resonant frequency?

E13.13 (a) Consider the circuit in Fig. E13.13*a*,

(i) Calculate the impedance $Z(s)$.

(ii) Draw a pole-zero diagram.

(iii) Plot the magnitude of the impedance versus frequency on log-log coordinates.

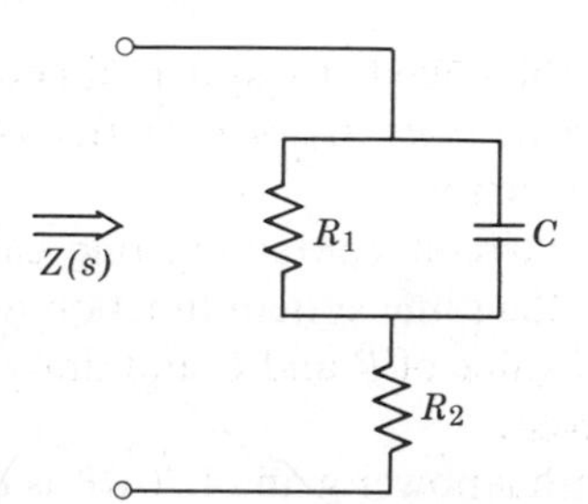

Figure E13.13*a*

(b) The network of Fig. E13.13*a* is used in the operation amplifier circuit of Fig. E13.13*b*.

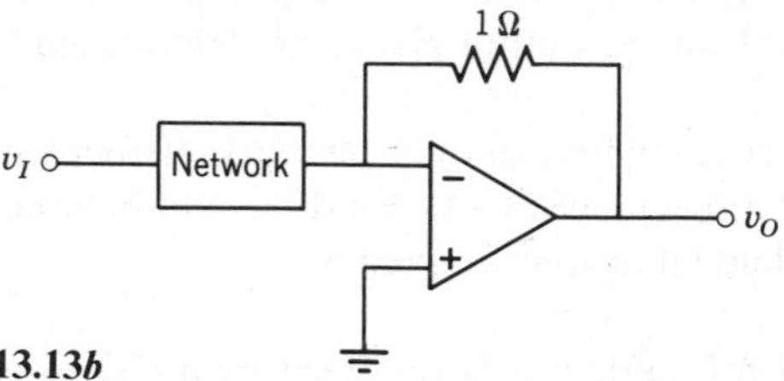

Figure E13.13*b*

(i) Find the transfer ratio V_o/V_i.

(ii) A required plot of the amplitude versus frequency for V_o/V_i is shown in Fig. E13.13*c*. Find R_1, R_2, and C to meet the required plot. (*Hint*. Consider what happens to the poles and zeros in part (i) when the network is connected into Fig. 13.13*b*. Also consider the gain at dc and high frequencies.)

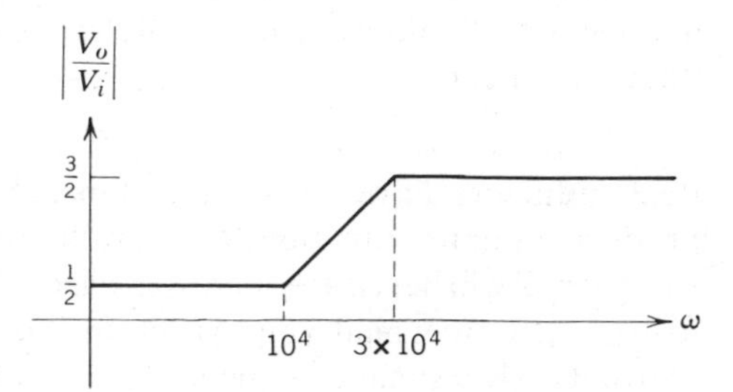

Figure E13.13*c*

PROBLEMS

P13.1 The network in Fig. P13.1 is called a *phase-shifting network.*

(a) Determine the system function, V_o/V_i, and plot the corresponding pole-zero diagram.

(b) Sketch and dimension the phase-magnitude plot for this system function.

(c) For a fixed-amplitude, fixed-frequency sinusoidal excitation, how does the response change as R is varied?

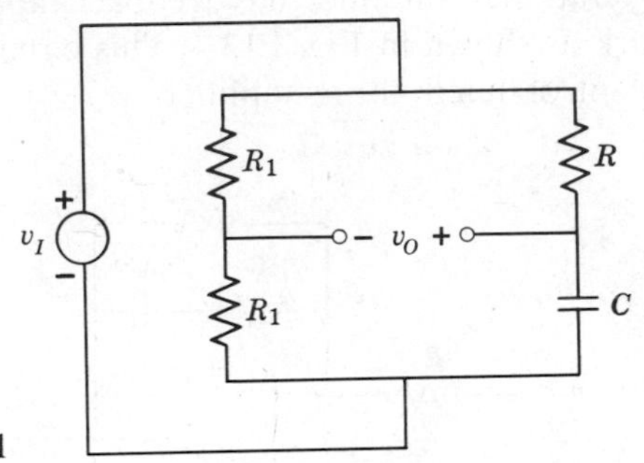

Figure P13.1

P13.2 For the circuit in Fig. P13.2 sketch and dimension $|H(j\omega)|$ and $\angle H(j\omega)$ versus ω.

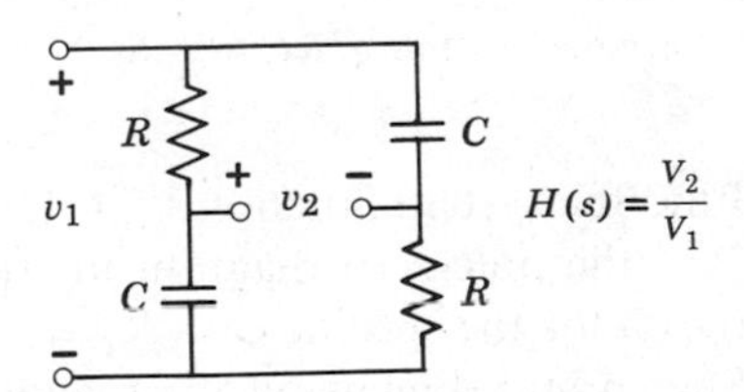

Figure P13.2

P13.3 Given an impedance $Z(s)$ (see Fig. P13.3*a*).

(i) At low frequencies, the impedance $Z(s)$ is equivalent to a 10 kΩ resistance.

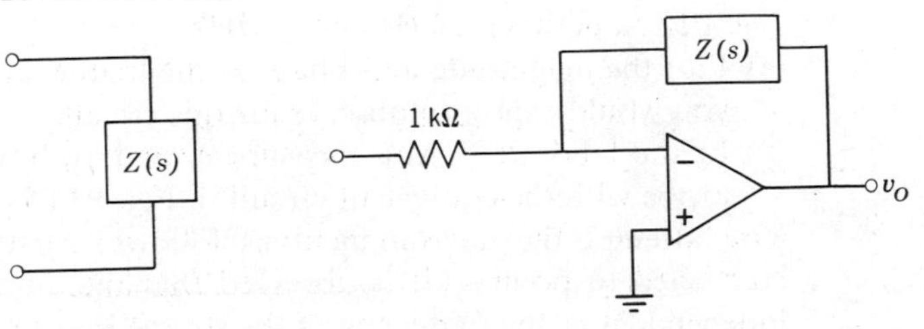

Figure P13.3

(ii) The impedance has a single pole as $s = -10^3$ rad/s and a single zero at $s = -10^4$ rad/s.

Assume, as usual, that the amplifier has large gain and input impedance, and small output impedance.

(a) Plot the frequency response of the *overall* transfer function V_o/V_i in Fig. P13.3*b*.

(b) If $v_I(t) = A\cos[(3 \times 10^3)t + (\pi/4)]$, estimate $v_O(t)$ from your plot.

P13.4 An operational amplifier has feedback applied through an *RC* network as shown in Fig. P13.4. This circuit might be used as a bass control in a hi-fi preamplifier.

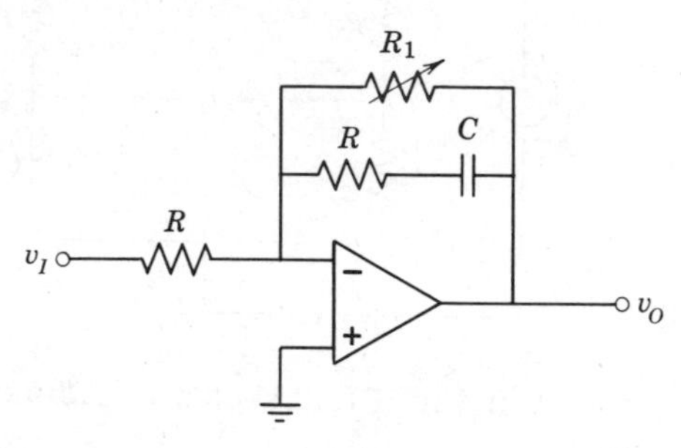

Figure P13.4
R_1 = variable resistance. $RC = 1$ ms.

(a) Find the system function V_o/V_i for the amplifier.

(b) Plot the pole-zero diagram for the system function found in part *a* for the specific case $R_1 = 100R$.

(c) Sketch and dimension the magnitude of the system function of part *a* for $R_1 = 100R$.

(d) Indicate on your plot in part *c* the effect of changing R_1.

P13.5 You wish to measure the frequency response of the circuit in Fig. P13.5*a* between 10 Hz and 1 MHz.

(a) Plot the magnitude and phase of the frequency response that you would expect to observe for this circuit.

(b) In the laboratory, you measure v_I and v_B by using a scope probe with the equivalent circuit in Fig. P13.5*b*.

You attempt the measurement as follows: First, the scope is connected to point *a*. It is observed that the amplitude of v_S is independent of the frequency of the source v_A between 10 Hz and 1 MHz. Next the scope is moved to point *b*, and the amplitude of

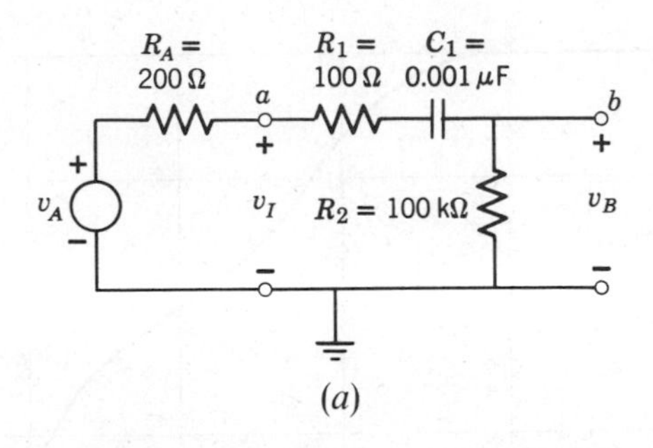

(*a*)

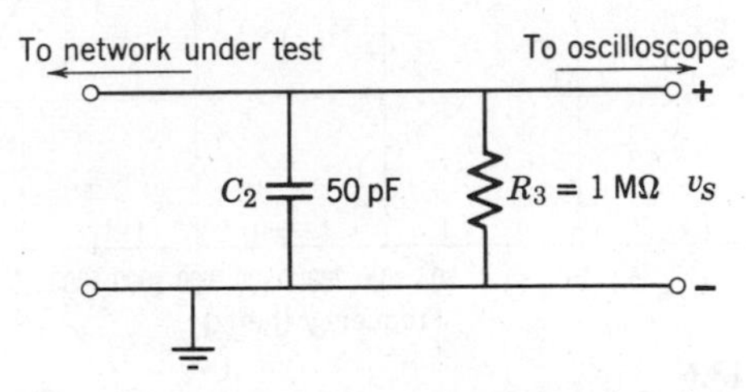

(*b*) Equivalent circuit of scope and probe.

Figure P13.5

v_S is measured as a function of source frequency over this same range. Show how the magnitude of the measured frequency response $|V_s$ (at b)/V_s (at a)$|$ will differ from the magnitude $|V_b/V_i|$ calculated in part (a).

Hint. C_1 and C_2 have very different magnitudes. This fact can be used to advantage when discussing either the high-frequency behavior or the low-frequency behavior of the network.

P13.6 Prior to being fed to the recording head that cuts the master for a conventional phonograph record, an audio signal is passed through a filter that attenuates the lower frequencies (to avoid overdriving) and boosts the high frequencies (to help combat the scratch noise). A compensating filter must then be inserted into the reproducing equipment. The standard RIAA equalizer characteristic is shown in Fig. P13.6.

You wish to construct this equalizer from a simple pole-zero diagram. Find a simple pole-zero diagram that corresponds to the magnitude response shown in the illustration. (*Hint.* More than one pole will be required.) Design a filter circuit with the correct transfer function.

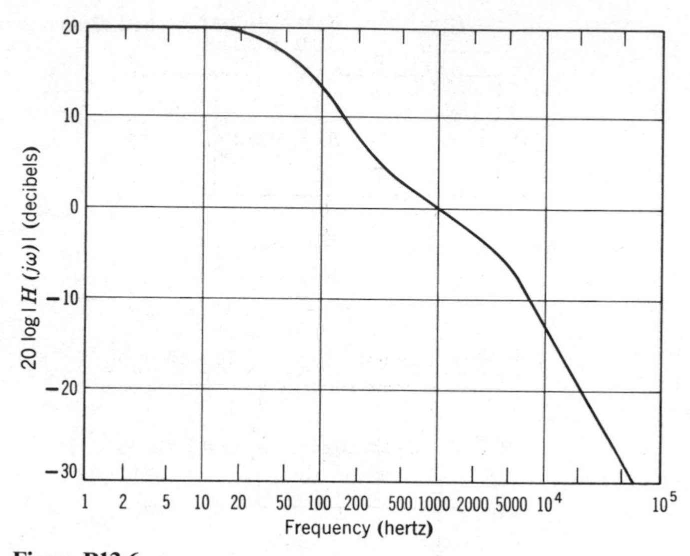

Figure P13.6

RIAA equalizer characteristic.

P13.7 A physicist desires to produce a 60 Hz magnetic field in the air gap of the structure shown in Fig. P13.7*a*.

When driven from the 120 Vac (170 V peak) 60 Hz power line, the magnet current i and, hence, the resulting magnetic field in

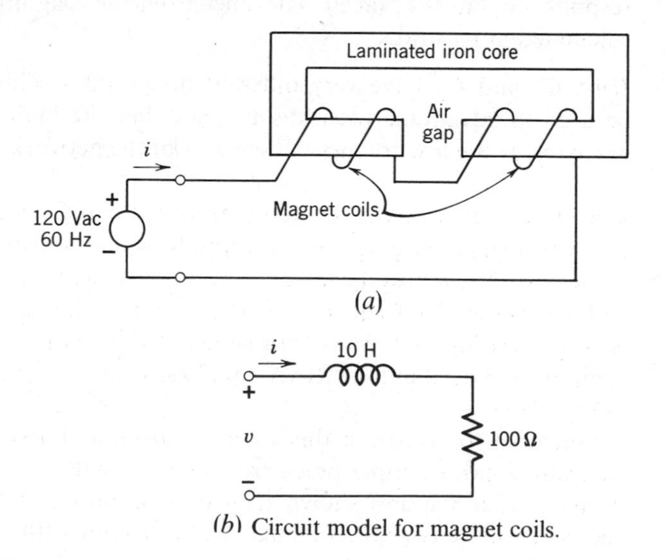

(*b*) Circuit model for magnet coils.

Figure P13.7

the air gap, is only one-tenth of that required for the desired experiment. Assume the magnet coils can be modeled as the series RL circuit in Fig. P13.7*b*.

(a) Find the rms magnitude of the magnet current i when driven from the 120 Vac 60 Hz power line.

(b) Design a simple passive network which, when connected between the power source and magnet coils, will produce at least a tenfold increase in magnetic current. Specify voltage, current, or power ratings for each component in your design.

P13.8 A $10\times$ attenuation probe is often used with an oscilloscope to extend its frequency response, as in the circuit in Fig. P13.8*a*.

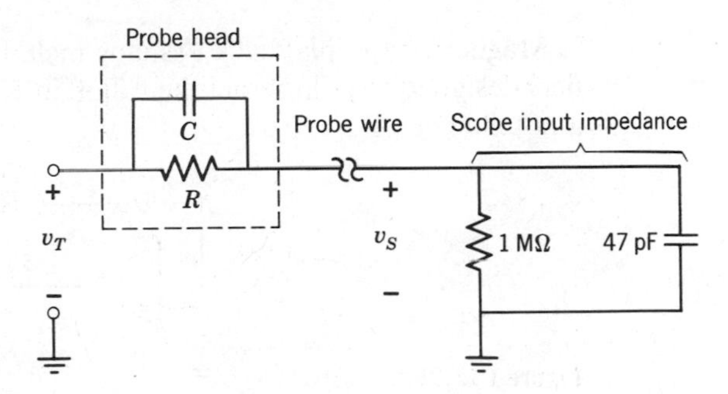

Figure P13.8*a*

v_T is the voltage we want to measure, and v_S is the voltage actually displayed on the scope's face (which we hope can be simply related to v_T). For simplicity the circuit can be redrawn as in Fig. P13.8*b*.

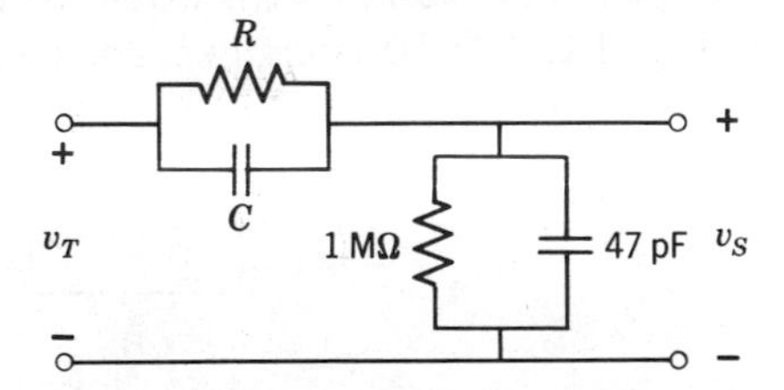

Figure P13.8*b*

(a) Find $H(s) = V_s/V_t$.

(b) Plot the pole-zero diagram, labeling all points.

(c) Find R such that the dc gain is 1/10.

(d) Find the value of C such that the pole and zero cancel.

(e) Plot $\log|H(j\omega)|$ versus $\log\omega$ for the value of C found in part *d*.

P13.9 In magnetic tape recordings the recording and playback process introduces a non-flat frequency response. That is, by the time the signal has been recorded on tape and then played back, the overall effect on the signal can be represented by a transfer function H_1 with the magnitude plot shown in Fig. P13.9*a*.

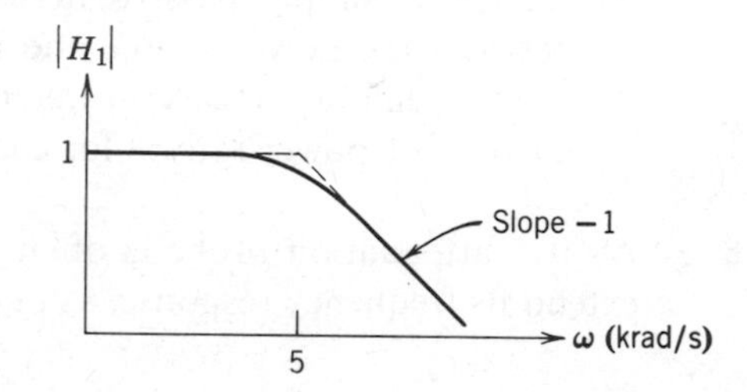

Magnetic tape playback systems include compensation amplifiers designed to counteract the rolloff in H_1. Figure P13.9*b* shows a possible circuit.

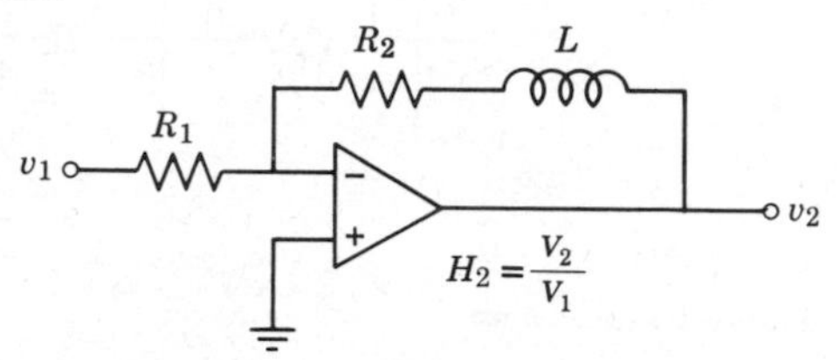

Figure P13.9*b*

Choose R_1, R_2, and L such that the dc gain is -5 and such that the total system frequency response determined by the product of H_1 and H_2 is flat.

P13.10 This problem examines the extent to which the simple *RC* circuit in Fig. P13.10*b* behaves like an ideal integrator. For the ideal

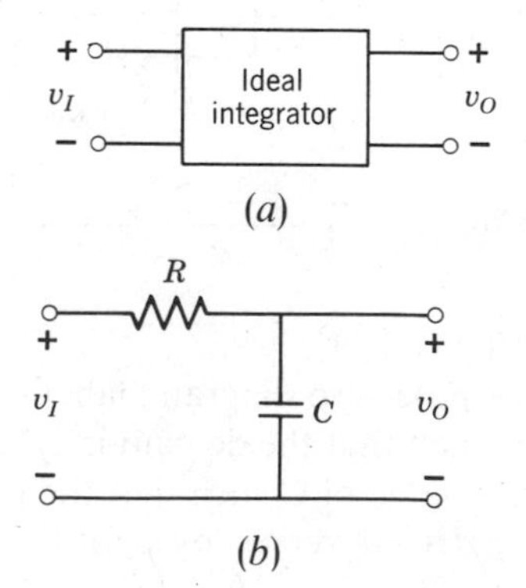

Figure P13.10

integrator, of course,

$$v_O = \int v_I \, dt$$

You are to use both the step and frequency responses in making your analysis. Specifically, for what kinds of signals does the RC network behave most nearly like the ideal integrator?

P13.11 For the bandpass transformer of Fig. 13.20

(a) Show that the ratio of the upper to lower cutoff frequencies ω_h/ω_l depends primarily on the ratio of the magnetizing inductance to the leakage inductance.

The ratio L_m/L_e depends only on the construction of the transformer. In practice its value can rarely be made larger than 10^3, which in turn places a limit on the bandwidth a transformer can achieve.

(b) Find ω_h/ω_l for the case where $L_m = 100$ mH, $L_e = 120$ μH and n is chosen to maximize the power delivered by v_S to R_L at mid-band frequencies.

(c) For $R_S = 100\ \Omega$, find ω_h and ω_l under the conditions of part *b*.

CHAPTER FOURTEEN
SIGNAL PROCESSING IN LINEAR SYSTEMS

14.0 INTRODUCTION

Most signal-processing applications require the use of active elements—amplifiers—in addition to the passive network elements discussed in most of the examples of the last chapter. In this chapter, we shall first examine the frequency response of transistor amplifiers, second, see how the use of feedback around the amplifiers can modify the frequency response and, finally, introduce some advanced concepts on the use of superposition to determine the response of linear systems to arbitrarily complex waveforms.

14.1 FREQUENCY RESPONSE OF TRANSISTOR AMPLIFIERS

In Chapter 11, a small-signal transistor model was introduced that contained two energy storage elements, one arising from the capacitance of the collector-base junction, the other arising from the combined effect of the capacitance of the emitter-base junction and the storage of excess injected minority charge in the base region. Also in Chapter 11, while discussing the biasing of transistor amplifiers, we concluded that in many situations the signals should be coupled into or out of the amplifiers through capacitors, and that capacitors could be used to bypass some of the resistors used for dc biasing. All of these capacitors have profound effects on the frequency response of amplifiers incorporating these elements. Other active devices, such as vacuum tubes and field-effect transistors, have similar internal capacitances, and the amplifiers built with them often use coupling and bypass capacitors. In the following discussion we shall focus entirely on the bipolar transistor, but the issues raised are much more general, and apply to amplifiers built with all types of devices.

14.1.1 Evaluation of the Hybrid-π Parameters

Figure 14.1*a* shows a bipolar junction transistor together with its small-signal model in Fig. 14.1*b*. The small-signal model is the hybrid-π model introduced in Section 11.4.1. We shall now outline a step-by-step method

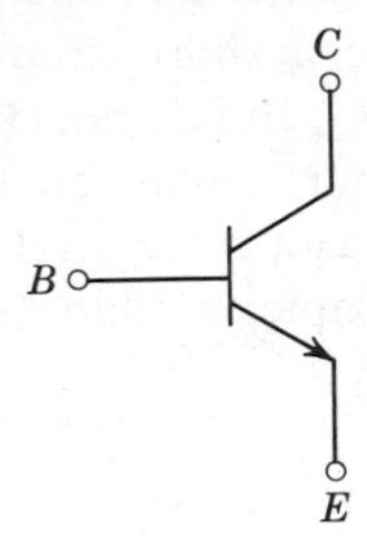

(*a*) Bipolar junction transistor

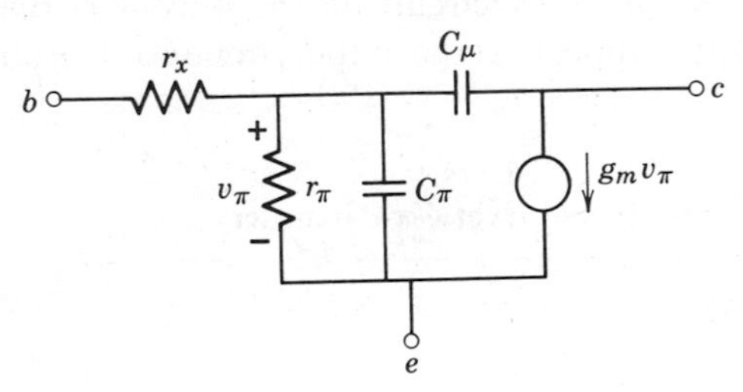

(*b*) Hybrid-π model

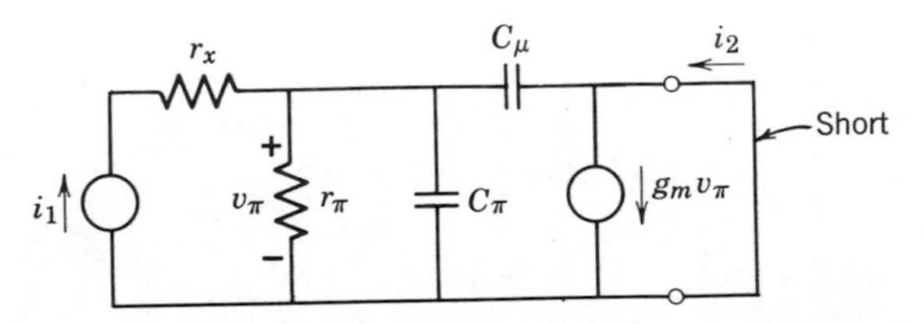

(*c*) Hypothetical common-emitter circuit to measure short-circuit current gain

Figure 14.1
The hybrid-π model.

for obtaining values for the various hybrid-π parameters from the transistor data sheet assuming the transistor's operating point (dc collector current) has already been determined. We shall then examine the frequency response of actual amplifier circuits.

Table 14.1 summarizes the procedure. First, one obtains g_m from the dc collector current as with the purely resistive model (remember that $kT/q \approx$ 25 mV at room temperature). Next, one reads the small-signal current gain, β_0, from the data sheet. It is usually listed as the equivalent h-parameter, h_{fe}. Then, one finds r_π as in the purely resistive model. The C_μ capacitance can be read directly from the data sheet. It is usually listed as the parameter C_{ob} (which means "output capacitance in the common-base configuration").

The relation used to obtain C_π requires a more detailed explanation. Figure 14.1c shows a common-emitter amplifier driven by an incremental current source, and loaded with a short circuit. This is not a practical circuit, but it does illustrate how C_π and C_μ can effect the frequency response of the device. Because of the short circuit on the output, C_μ appears in parallel with C_π. Therefore, one can solve for V_π directly using sinusoidal steady-state impedances. The complex amplitudes V_π and I_1 are related by

$$V_\pi = I_1\left[\frac{r_\pi}{1 + sr_\pi(C_\pi + C_\mu)}\right] \tag{14.1}$$

A further result of the short circuit on the output is that all the current from the g_m generator appears at the output; none goes through C_μ. Therefore,

TABLE 14.1
Determination of Hybrid-π Parameters

1. $g_m = \left(\dfrac{q}{kT}\right)|I_C|$
2. $\beta_0 = h_{fe}$ (data sheet)
3. $r_\pi = \beta_0/g_m$
4. $C_\mu = C_{ob}$ (data sheet)
5. $\dfrac{\beta_0}{2\pi r_\pi(C_\pi + C_\mu)} = f_T$ (data sheet)
6. $C_\pi = \dfrac{\beta_0}{2\pi f_T r_\pi} - C_\mu$
7. r_x (low frequency): data sheet, or estimate 50 to 100 Ω
   (high frequency): $\mathrm{Re}\left\{\dfrac{1}{y_{ie}}\right\}_{\omega > \omega_a}$, or estimate $\approx 25\ \Omega$.

one can write a transfer function $\beta(s)$ for the short-circuit current gain as

$$\beta(s) = \frac{I_2}{I_1} = \frac{g_m V_\pi}{I_1} = \frac{\beta_0}{1 + sr_\pi(C_\pi + C_\mu)} \tag{14.2}$$

Notice that the system function $\beta(s)$ has a single pole at

$$s_\beta = -\frac{1}{r_\pi(C_\pi + C_\mu)} \tag{14.3}$$

The magnitude of this pole frequency is called ω_β.

$$\omega_\beta = \frac{1}{r_\pi(C_\pi + C_\mu)} \tag{14.4}$$

At frequencies above ω_β, the magnitude of the short-circuit current gain rolls off at -6 dB per octave. (We shall see in Section 14.1.3 that in amplifiers with nonzero load resistors, the frequency at which the gain begins to roll off is generally lower than ω_β.)

The log magnitude of $\beta(j\omega)$ is plotted against frequency in Fig. 14.2. The asymptotes are shown as dotted lines, while the actual short-circuit gain is shown as the solid curve. The asymptotes intersect at the frequency ω_β. At high frequencies, the current-gain asymptote has a slope of -1. (The actual curve at very high frequency deviates somewhat from the hybrid-π asymptote. This inaccuracy in the model is not important for transistor operation in its ordinary useful range.) The frequency where the *extrapolated* asymptotic behavior passes through unity is called ω_T. The corresponding

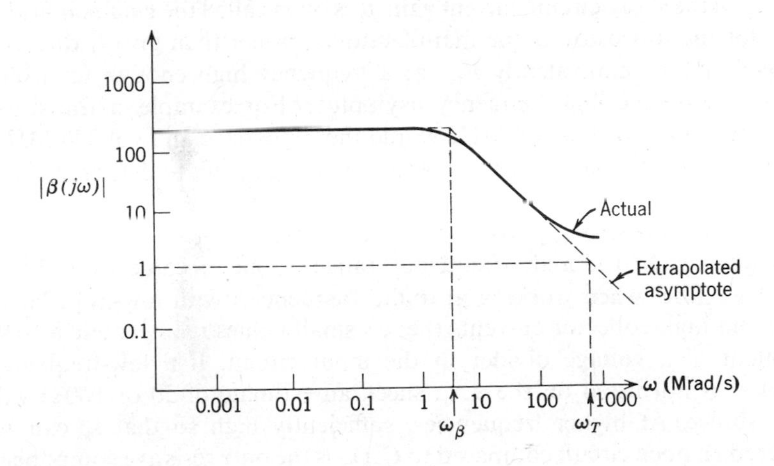

Figure 14.2
Typical short-circuit current gain for a transistor with $\beta_0 = 150, f_T = 100$ MHz.

frequency in hertz is called $f_T = \omega_T/2\pi$ and is a parameter given on the transistor data sheets. Along the high-frequency asymptote, the relation between $|\beta|$ and ω is simply

$$|\beta| = \frac{\beta_0}{(\omega/\omega_\beta)} \tag{14.5}$$

It follows, therefore, since $|\beta| = 1$ at $\omega = \omega_T$, that

$$\omega_T = \beta_0 \omega_\beta \tag{14.6}$$

or

$$\frac{1}{\omega_\beta} = \frac{\beta_0}{\omega_T} = \frac{\beta_0}{2\pi f_T} \tag{14.7}$$

From this expression, and from Eq. 14.4 it follows that

$$r_\pi(C_\pi + C_\mu) = \frac{\beta_0}{2\pi f_T} \tag{14.8}$$

which yields the relation in Table 14.1:

$$C_\pi = \frac{\beta_0}{2\pi f_T r_\pi} - C_\mu \tag{14.9}$$

Since from Eq. 14.6 we see that f_T is the product of β_0 and f_β, the corner frequency of the short-circuit current gain, f_T is often called the *gain-bandwidth product* for the transistor. Some manufacturers, rather than give f_T directly, will specify $|\beta|$, or equivalently h_{fe}, at a frequency high enough to insure operation along the high-frequency asymptote. For example, a transistor with a value of $h_{fe} = 3$ at 100 MHz would therefore have an f_T of 300 MHz.

The final hybrid-π parameter is r_x. It is the most difficult to determine and, fortunately, of least importance in general applications. The value of r_x does not vary too strongly from transistor to transistor, but it does vary with frequency. In the audio-frequency range, r_x has a value from 50 to 100 Ω. Therefore, when working at audio frequencies with low-impedance sources and high-collector currents (hence, small values r_π), r_x enters as one component of a voltage divider in the input circuit. If a low-frequency value of r_x is not listed on the data sheet, an estimate of 50 or 100 Ω will usually suffice. At higher frequencies, sufficiently high so that r_π can be considered an open circuit compared to C_π, r_x is the only resistive component of the input circuit. The input y-parameter y_{ie}, defined as the common-emitter input admittance with the output short-circuited, has this form at these high frequencies:

$$y_{ie} = \frac{s(C_\pi + C_\mu)}{1 + sr_x(C_\pi + C_\mu)} \tag{14.10}$$

This high-frequency y-parameter has a single pole at a frequency

$$\omega_a = \frac{1}{r_x(C_\pi + C_\mu)} \tag{14.11}$$

At frequencies above ω_a, an estimate of r_x can be obtained from the real part of y_{ie}.

$$r_x = \text{Re}\left\{\frac{1}{y_{ie}}\right\}_{\omega > \omega_a} \tag{14.12}$$

If data on y_{ie} at high frequencies are not available, a rough estimate at these high frequencies is $r_x \approx 25\ \Omega$.

14.1.2 Separation of High- and Low-Frequency Circuits

In most transistor amplifiers, there are some elements (such as coupling and bypass capacitors) whose dominant effects are at relatively low frequencies, and other elements (such as the internal capacitances C_π and C_μ) whose dominant effects are at relatively high frequencies. In this section, we will examine how to analyze an amplifier by splitting the frequency domain into low, middle, and high ranges, just as the frequency domain was divided in the analysis of transformer frequency response (Section 13.3.2).

The common-emitter amplifier of Fig. 14.3 has a coupling capacitor, C_S, which is inserted to keep the source resistance R_S and the source v_s from upsetting the transistor bias. It is intended to be a short circuit at the principal signal frequencies of interest. The emitter capacitor C_E, is also intended to be a short circuit at signal frequencies of interest. Its purpose is to increase the gain by shorting out the emitter resistor R_E at audio frequencies. If the emitter resistor were not bypassed in this fashion, the voltage gain would be limited to a value of about $-R_C/R_E$ (Section 11.3.2).

The full incremental circuit for the amplifier of Fig. 14.3 is shown in Fig. 14.4. Both C_S and C_E are shown, as well as the hybrid-π capacitances, C_π and C_μ. This circuit contains four energy storage elements. One anticipates, therefore, that the system function for the voltage gain, V_o/V_s, will contain

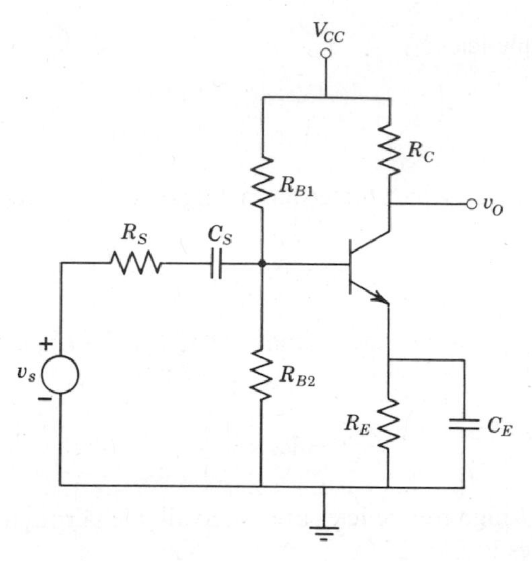

Figure 14.3
A common-emitter amplifier.

a fourth-order polynomial in s, leading to four poles in the response function. One could, if one wished, solve for the four poles directly, and obtain the frequency response using the methods of preceeding sections. But it would be foolish to solve such a cumbersome problem when one can easily separate it into two relatively simple problems.

The basis for this separation is that the coupling and bypass capacitors typically have values in the $>1\ \mu$F range, while the corresponding values for C_π and C_μ are usually less than $0.001\ \mu$F. Therefore, there will be three identifiably distinct frequency ranges: a low-frequency range, in which C_S and C_E are important, a mid-frequency range which is resistive, and a high-frequency range, in which C_π and C_μ are important.

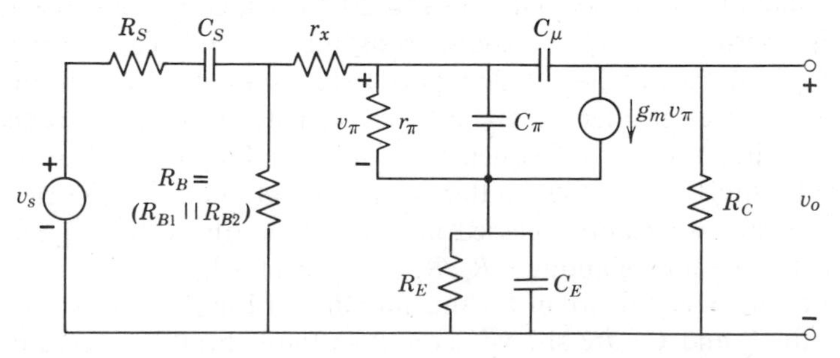

Figure 14.4
Small-signal model for Fig. 14.3.

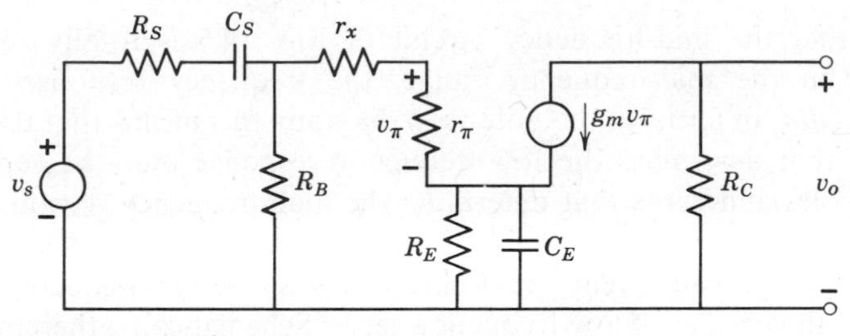

(*a*) Low-frequency circuit

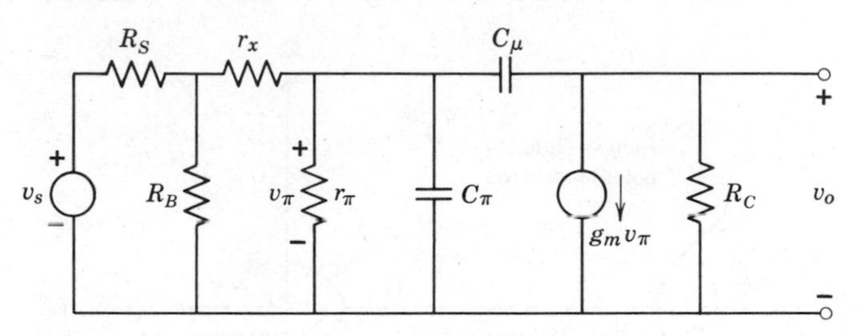

(*b*) High-frequency circuit

Figure 14.5
Separation of the circuit of Fig. 14.4 into low- and high-frequency circuits.

In both the low- and mid-frequency ranges, where C_π and C_μ are large impedances, one can use a circuit (Fig. 14.5*a*) in which C_π and C_μ are replaced by open circuits. Similarly, in both the mid- and high-frequency ranges, where C_S and C_E are small impedances, one can use the circuit of Fig. 14.5*b* in which both C_S and C_E are replaced by short circuits. Both circuits approach the same limit in the mid-frequency region. In that region, the coupling and bypass capacitors are effectively short circuits while the hybrid-π capacitors are effectively open circuits. The resulting mid-frequency circuit is shown in Fig. 14.6.

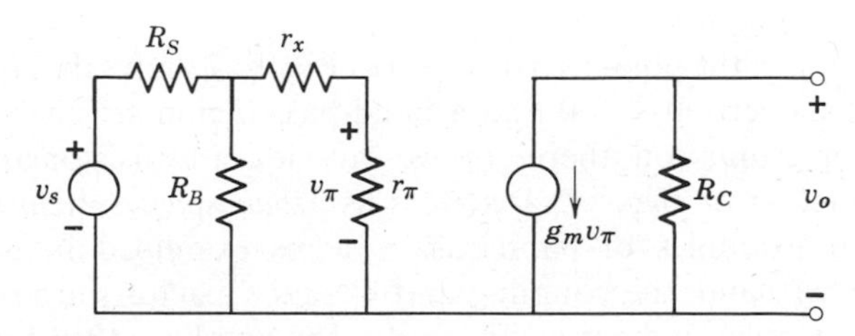

Figure 14.6
Mid-frequency circuit for Fig. 14.4.

Notice that the mid-frequency circuit of Fig. 14.6 is totally resistive. Therefore, in the mid-frequency range, the frequency response of the amplifier is *flat*. In terms of the pole-zero diagram, this means that the poles and zeros that determine the low-frequency response must be separated from the poles and zeros that determine the high-frequency response by a region that has no poles or zeros. Furthermore, in order to get resistive behavior in the mid-frequency region, the *number* of low-frequency poles must equal the *number* of low-frequency zeros. Schematically, therefore, the pole-zero diagram for the amplifier of Fig. 14.4 must look like the diagram in Fig. 14.7. There are two poles and two zeros in the low-frequency region.

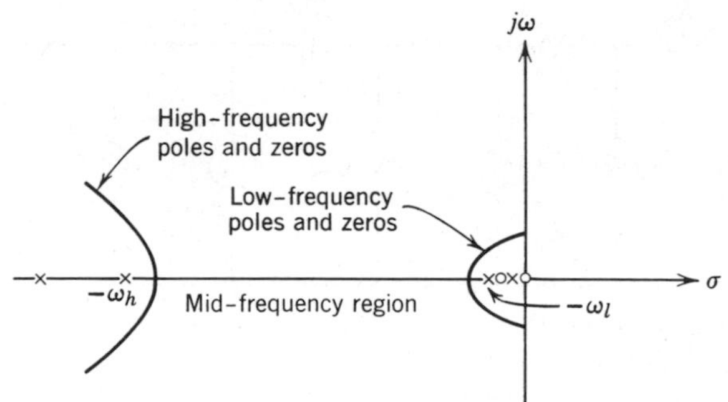

Figure 14.7
Schematic pole-zero diagram for the transistor amplifier of Fig. 14.4.

The highest of the low-frequency poles, at $-\omega_l$, determines the low-frequency limit of the resistive region. Two high-frequency poles are shown, although in practice the more distant of these two poles is very far out on the negative real axis. The lower (i.e., closer) of the high-frequency poles at $-\omega_h$ determines the upper limit of the resistive region. As was true with the transformer example in Section 13.3.2, the separation into low- and high-frequency circuits to determine ω_l and ω_h separately is fully valid as long as ω_l and ω_h differ by more than a factor of about 100, a condition that is usually satisfied in transistor amplifiers.

Examination of the pole-zero diagram of Fig. 14.7 shows that the transistor amplifier has a zero at $s = 0$ and a band-pass region set off by two poles. As a first approximation, therefore, we can view a transistor amplifier that includes coupling or bypass capacitors as a *band-pass system* very similar to the other examples of band-pass systems examined in Section 13.3. If, on the other hand, the coupling and bypass capacitors are removed, the low-frequency poles and zeros disappear. The amplifier then functions as a *low-pass system*, with a bandwidth determined by ω_h. In either case, the upper-frequency limitation on the transistor amplifier response is an intrinsic property of the device, while the low-frequency behavior is entirely determined by external elements. Further analysis of the low-frequency circuit, Fig. 14.5*a*, is left as an exercise. The determination of the high-frequency response from Fig. 14.5*b* is discussed in the next section.

14.1.3 The C_T Approximation

Let us calculate the mid- and high-frequency system function for the circuit of Fig. 14.5*b*. We shall assume R_B is large and can be ignored. The network then reduces to a two-node network. We can write KCL at each node using the complex amplitudes for v_s, v_π, and v_o.

$$\frac{V_\pi - V_s}{R_S + r_x} + V_\pi\left(\frac{1}{r_\pi} + sC_\pi\right) + (V_\pi - V_o)(sC_\mu) = 0 \tag{14.13}$$

$$\frac{V_o}{R_C} + g_m V_\pi + (V_o - V_\pi)sC_\mu = 0 \tag{14.14}$$

Transposing and collecting terms, we obtain

$$\frac{V_s}{R_S + r_x} = \left[\frac{1}{r_\pi} + \frac{1}{R_S + r_x} + s(C_\pi + C_\mu)\right] V_\pi - sC_\mu V_o \tag{14.15}$$

$$0 = (g_m - sC_\mu)V_\pi + \left(\frac{1}{R_C} + sC_\mu\right) V_o \tag{14.16}$$

This pair of equations can be solved to yield a transfer function V_o/V_s that has two poles and one zero. The zero occurs at $s_1 = g_m/C_\mu$, which is in the right half-plane and so far out on the real axis that it does not contribute to the frequency response at frequencies where the amplifier has useful gain. Similarly, the more distant of the two poles is so far out on the negative real axis that its contribution is also negligible at useful amplifier frequencies. Within its useful frequency range, therefore, the amplifier circuit behaves as if it had a single pole at $-\omega_h$. Algebraically, an excellent estimate of this dominant high-frequency pole is obtained by neglecting in Eq. 14.16 the admittance sC_μ compared to g_m and $1/R_C$. Since C_μ is $\approx 10^{-12}$ F and $1/R_C$ might be on the order of 10^{-3} S (g_m is usually even larger), ω would have to have a magnitude of $\approx 10^9$ rad/s (or 160 MHz) before the approximation fails seriously. With this simplification, Eq. 14.16 becomes

$$V_o = -g_m R_C V_\pi \tag{14.17}$$

which, when substituted in Eq. 14.15, yields the single-pole transfer function

$$\frac{V_o}{V_s} = \left(\frac{-\beta_0 R_C}{R_S + r_x + r_\pi}\right)\left(\frac{1}{1 + s[r_\pi \| (r_x + R_S)][C_\pi + (g_m R_C + 1)C_\mu]}\right) \quad (14.18)$$

At low frequencies ($s = 0$), this transfer function has the same gain as that obtained in Chapter 11 for the resistive model of the common-emitter amplifier, excepting only the addition of r_x to the input voltage divider. At high frequencies, the gain has a slope of -1, characteristic of a single-pole at $-\omega_h$, accompanied by a $-90°$ phase shift. The pole frequency is

$$\omega_h = \frac{1}{[r_\pi \| (r_x + R_S)][C_\pi + (g_m R_C + 1)C_\mu]} \quad (14.19)$$

An equivalent circuit, which has the same transfer function as Eq. 14.18, is shown in Fig. 14.8. The combined effect of C_π and C_μ is contained in the effective capacitance C_T.

$$C_T = C_\pi + (g_m R_C + 1)C_\mu \quad (14.20)$$

Notice that as the gain term $g_m R_C$ is increased, the effective capacitance C_T increases because of the C_μ term, reducing the bandwidth of the amplifier. This is one example of how feedback (here from output to input through C_μ) can substantially modify the frequency response of an amplifier. For typical values, $C_\pi = 300$ pF, $C_\mu = 5$ pF, and $g_m R_C = 75$, the contribution of C_μ to C_T would actually be larger than the contribution of C_π. By using these values and assuming $r_\pi = 1$ kΩ and $R_S = r_x = 50$ Ω, we obtain

$$\omega_h = 1.6 \times 10^7 \text{ rad/s} \quad (14.21)$$

from which we find the cutoff frequency to be

$$f_h = \frac{\omega_h}{2\pi} = 2.6 \text{ MHz} \quad (14.22)$$

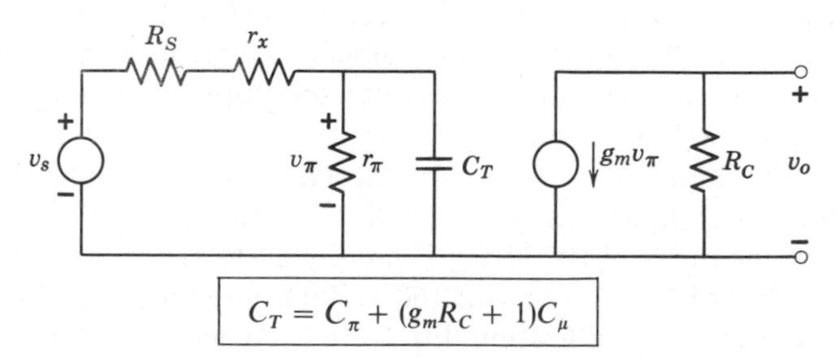

Figure 14.8
The C_T approximation for a single-stage common-emitter amplifier.

In addition to reducing the bandwidth of the amplifier, the feedback through C_μ substantially modifies the input impedance (or admittance) of the amplifier. In effect, the admittance sC_μ appears in parallel with the input terminals, but multiplied by the approximate voltage gain $g_m R_C$. The gain-dependent combination of a reduced bandwidth and increased input admittance because of capacitive feedback, called the *Miller effect*, is present in virtually every small-signal amplifier to some degree.

14.1.4 Multistage Amplifier

In amplifiers containing more than one transistor, the transfer functions are quite complex. The high-frequency circuits equivalent to Fig. 14.5*b* contain two capacitors for each transistor, and it is *not* legitimate in a multistage circuit to replace each transistor by its C_T equivalent (see Problem P14.8). A three-transistor amplifier, for example, will have a transfer function with six poles. Three of these poles are usually high enough in frequency to be ignored, leaving three poles or one per transistor that can be expected to contribute to the frequency response.

The lowest of the poles, called the *dominant pole*, defines the bandwidth of the amplifier, and corresponds to the pole at $-\omega_h$ in our previous discussion. There are excellent methods for estimating ω_h directly from the high-frequency circuit. The reader is referred to the advanced texts for these methods.

The poles beyond the dominant pole produce a more rapid rolloff of gain at higher frequencies, and also produce phase shift at high frequencies in addition to the $-90°$ phase shift from the dominant pole. In high-gain low-pass amplifiers, such as the kind used in op-amps, it is possible that at frequencies where the amplifier still has appreciable gain, the accumulated phase shift nears $-180°$. Because of this phase shift, what might be an inverting amplifier at dc can become a non-inverting amplifier at high frequencies. Furthermore, what is negative feedback at dc can become positive feedback at high frequencies, leading to instability and oscillation. The purpose of the *compensation networks* added to op-amps is to suppress these unwanted instabilities.

14.1.5 Frequency Response of an Op-Amp Inverting Amplifier

The open-loop frequency response of a commercial op-amp is usually compensated internally to produce a dominant pole at a very low frequency.

A typical open-loop characteristic is shown in Fig. 14.9. The corner frequency ω_o usually corresponds to a low-audio frequency ($\approx$tens of rad/s), while the open-loop gain at dc, A_o, will have a magnitude in excess of 10^4. At frequencies above ω_o the open-loop gain has a slope of -1. Accompanying this rolloff, there is a phase shift of $-90°$.

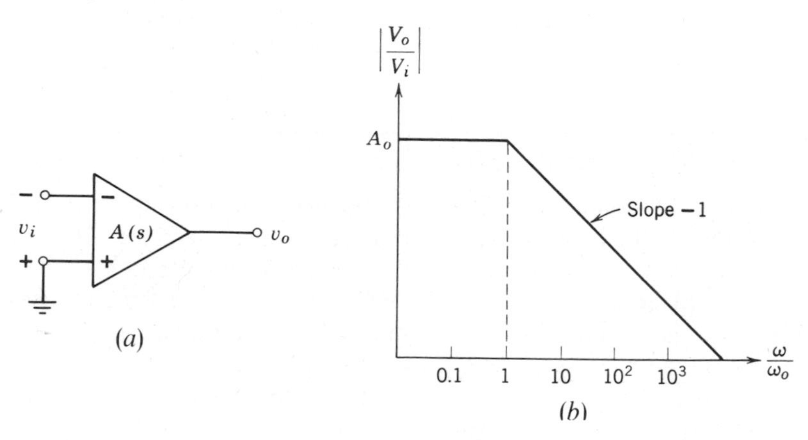

Figure 14.9
Open-loop op-amp frequency response.

If this single-pole behavior were to extend indefinitely, then the resulting transfer function could be written

$$\frac{V_o}{V_i} = A(s) = \frac{A_o}{1 + s/\omega_o} \tag{14.23}$$

In practice, every op-amp has additional poles at higher frequencies, but we shall assume for the moment that Eq. 14.23 correctly characterizes the op-amp transfer function.

Figure 14.10*a* shows a simple inverting amplifier. In analyzing the frequency response of this amplifier, we must be careful not to over-idealize the op-amp. If, for example, we were to assume the open-loop gain of the op-amp were infinite at all frequencies, we would obtain the simple transfer function

$$\frac{V_o}{V_a} = -\frac{R_F}{R_S} \tag{14.24}$$

Actually, the open-loop gain is not infinite at any frequency, and in particular, decreases rapidly as the frequency gets above ω_o. Therefore, we shall now analyze the circuit of Fig. 14.10*a* for finite op-amp gain. We shall continue

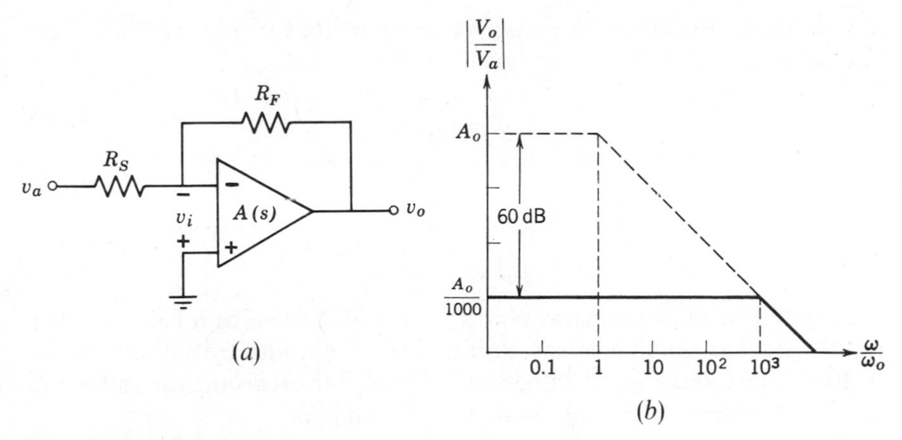

Figure 14.10
Closed-loop frequency response.

to assume, however, that the op-amp input current is zero. The relevant equations are given by KCL

$$\frac{V_a + V_i}{R_S} = -\frac{V_i + V_o}{R_F} \tag{14.25}$$

and by the op-amp transfer characteristic of Eq. 14.23. Substituting for V_i in terms of V_o we obtain for the transfer function

$$\frac{V_o}{V_a} = -\frac{A(s)}{1 + (R_S/R_F)[1 + A(s)]} \tag{14.26}$$

If we now substitute for $A(s)$, we find a single-pole overall transfer function

$$\frac{V_o}{V_a} = \frac{-A_o}{1 + (R_S/R_F)(1 + A_o) + [1 + (R_S/R_F)]s/\omega_o} \tag{14.27}$$

In all op-amps, the dc loop gain A_o is much larger than unity. If we also assume for algebraic convenience that $R_F > 10\,R_S$, we can approximate Eq. 14.27 as follows

$$\frac{V_o}{V_a} \cong \frac{-A_o}{[1 + (R_S/R_F)A_o] + s/\omega_o} \tag{14.28}$$

which has a pole at s_p given by

$$s_p = -\left(1 + \frac{R_S}{R_F}A_o\right)\omega_o \tag{14.29}$$

The resulting frequency response has been plotted in Fig. 14.10*b* for the specific case

$$\frac{R_F}{R_S} = \frac{A_o}{1000} \tag{14.30}$$

If, for example, $A_o = 10^5$, then this choice would correspond to a closed-loop dc gain of 100. Notice that the pole now occurs at a frequency

$$|s_p| \cong 1000\omega_o \tag{14.31}$$

That is, the negative feedback has decreased the dc gain by a factor of 1000 (60 dB) and has simultaneously increased the bandwidth by the same factor of 1000. Thus, there is the same kind of trade-off between gain and bandwidths as we encountered in the single-stage amplifier.

Capacitive elements could now be added to the circuit of Fig. 14.10*a* to build up the active band-pass filter discussed in Section 13.3.1. The simple analysis in that section for the pole and zero locations would continue to be correct, provided that the upper pole produced by the capacitor networks is *well below* the closed loop op-amp roll-off $A_o\omega_o/(R_F/R_S)$. If we take a typical example, with $A_o = 10^5$, $\omega_o = 2\pi \times 10$ Hz, and $R_F/R_S = 100$, then an active filter would work as predicted according to the simple op-amp analysis of Section 13.3.1 provided that its upper rolloff frequency is kept well below 10 kHz. If one wished to analyze the behavior at higher frequencies, it would be necessary to include the intrinsic op-amp frequency response and calculate the overall transfer function using the methods of this section.

If too much feedback is used around an op-amp, that is, if the ratio of the open-loop gain to the closed-loop gain is too large, the closed-loop bandwidth predicted by the simple model of Eq. 14.29 might be so large that the higher poles in the op-amp transfer function could no longer be neglected. In this case it becomes necessary to replace $A(s)$ in Eq. 14.23 with a more correct, multiple-pole op-amp transfer function and calculate the new resulting pole positions for the closed-loop system. Under closed-loop conditions, the poles form complex pairs, and even get close enough to the imaginary axis to produce instability and oscillations. It is often necessary, therefore, to provide additional compensation in applications involving large amounts of feedback.

Quantitative analysis of this fascinating subject is reserved for more advanced texts. The beginner, however, should recognize that increasing the amount of feedback in a system, either by increasing the open-loop gain or by decreasing the closed-loop gain compared to the open-loop gain, can lead to oscillations. When these oscillations occur, either the open-loop gain must be reduced or some form of frequency compensation must be added to the overall transfer function.

14.2 SUPERPOSITION IN THE FREQUENCY DOMAIN

We now introduce some advanced superposition methods for analyzing the response of linear systems to arbitrary waveforms. This section deals with superposition of sinusoids, while Section 14.3 deals with superposition of step and impulse functions. The mathematical skills required for these methods are more advanced than are required in the rest of the text. It is suggested, however, that the student who lacks the foundation in integral calculus necessary to follow the details can still benefit from reading the discussions of key issues.

14.2.1 Representation of Periodic Waveforms: Fourier Series

A *periodic waveform* is a waveform that repeats a basic pattern over and over. Figure 14.11 illustrates four different periodic waveforms. The time for one complete cycle through the basic pattern is called the *period*, and is denoted by T. The *fundamental frequency*, denoted here by f_o, is the number of complete cycles per unit time. The unit of frequency is the hertz (or Hz), which is the number of complete cycles in one second. Thus, if T_o is the period expressed in seconds, then

$$\begin{aligned} f_o &= 1/T_o \\ (\text{Hz}) &= (1/\text{s}) \end{aligned} \tag{14.32}$$

The four waveforms of Fig. 14.11 have different basic patterns, but they all have the same period (and therefore the same fundamental frequency). Each of these waveforms can be represented as a sum of sinusoids, whose frequencies are integral multiples of the fundamental frequency f_o. Such a sum is called a *Fourier series.*[1] The general form for a Fourier series is

$$v(t) \doteq A_o + \sum_{n=1}^{\infty} (A_n \cos n\omega_o t + B_n \sin n\omega_o t) \tag{14.33}$$

where A_o is a constant equal to the dc (or average) value of $v(t)$, ω_o is the fundamental frequency expressed in radians per second ($\omega_o = 2\pi f_o$), and A_n and B_n are constants that differ for different waveforms. The sum is an infinite sum, since there are an infinite number of multiples, or *harmonics*

[1] For non-periodic waveforms, the sum is replaced by a continuous integral called the Fourier integral, a topic that we will not discuss formally.

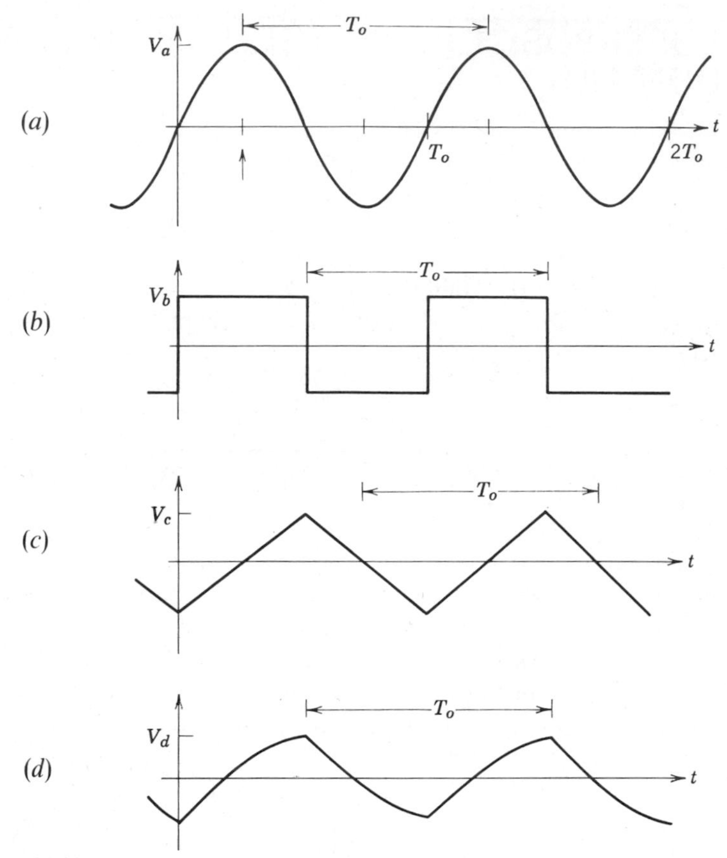

Figure 14.11
Periodic waveforms.

of the fundamental frequency ω_o. As a practical matter, most periodic waveforms can be represented with sufficient accuracy by using only the terms out to some maximum n.

Before examining examples based on the waveforms of Fig. 14.11, let us consider some of the general features of the Fourier series waveform of Eq. 14.33. The first question is whether Eq. 14.33 is in fact periodic. Consider each term. The constant A_o never changes, and thus repeats itself indefinitely. Each sinusoidal term at frequency $n\omega_o$ repeats its own basic pattern with a period T_o/n. Therefore, Eq. 14.33 consists of a sum of terms, one of which is constant, two of which are periodic with period T_o, two of which are periodic with period $T_o/2$ and, for the general term, two terms with period T_o/n. But a sinusoid that repeats itself every T_o/n seconds is also periodic with period T_o.

The basic pattern consists of n sinusoidal periods, this pattern being exactly repeated every T_o seconds. Thus every term in Eq. 14.33 is periodic with period T_o, so the entire waveform is periodic. Only the magnitudes of the various constants (A_o, A_n, B_n) distinguish one periodic waveform from another.

Let us now examine whether all terms are necessary for every periodic waveform. The constant A_o will clearly be nonzero only if the waveform has a dc component. In Fig. 14.11, all the waveforms have a zero average over one period and therefore have no dc component. Thus, the Fourier series for these waveforms will have no A_o term. Furthermore, we note that a cosine function is an *even* function of time (i.e., it has the same value at $+t$ as it has at $-t$), while the sine function is an *odd* function of time, having a value at $+t$ that is exactly the negative of its value at $-t$. One can reason, then, that periodic functions that are even functions of time will involve only cosine functions, while periodic functions that are odd functions of time will involve only sine functions. Looking once again to Fig. 14.11, waveforms a and b are odd functions of time, and therefore can be represented by a Fourier series involving only sine functions. Waveform c, on the other hand, is an even function of t. It therefore can be represented by a sum of cosines. Waveform d is neither even or odd. One expects, therefore, to need both sines and cosines in its Fourier series.

The question of evenness or oddness of a waveform is somewhat arbitrary, because the choice of origin ($t = 0$) is arbitrary. Indeed, a shift in the time origin does produce changes in the evenness or oddness of a waveform. Suppose, for example, that the origin of time were chosen at the time indicated by the vertical arrow in Fig. 14.11a. In that case, waveforms a and b would be even functions, c would be an odd function, and d would still be neither even or odd. If, instead, one were to choose the time origin somewhere else in the cycle, none of the waveforms would have any symmetry, and Fourier sums would require both sines and cosines. The principle to extract from this discussion is that shifts in the origin of time can change the relative sizes of A_n and B_n. However, in a fashion similar to complex numbers, there is a quantity involving A_n and B_n that is independent of the choice of time origin. This quantity is $(A_n^2 + B_n^2)^{1/2}$. Although we shall make no formal use of this fact in this text, it shows that although the Fourier series obtained with one particular choice of time origin may look very different from the Fourier series obtained with a different choice of time origin, there is a set of magnitudes that characterize the periodic waveform independent of the choice of time origin. For periodic functions with time-symmetry, it is possible to find a choice of the time origin for which either all A_n's or all B_n's are zero, while for waveforms without any symmetry, every choice of time origin will produce Fourier series with both sines and cosines in it.

The specific set of constants for any given waveform can be obtained directly from the waveform, using the integral expressions given below. Proofs of these results can be found in any intermediate calculus text.

$$A_o = \frac{1}{T}\int_0^T v(t)\,dt$$

$$A_n = \frac{2}{T}\int_0^T v(t)\cos n\omega_o t\,dt \qquad n \neq 0 \tag{14.34}$$

$$B_n = \frac{2}{T}\int_0^T v(t)\sin n\omega_o t\,dt \qquad n \neq 0$$

These relations have been used to obtain Fourier series for waveforms *a*, *b*, and *c* of Fig. 14.11.

$$\text{Sine wave:} \quad v_a = V_a \sin \omega_o t \tag{14.35a}$$

$$\text{Square wave:} \quad v_b = \sum_{n\,\text{odd}} \left(\frac{4V_b}{n\pi}\right) \sin n\omega_o t \tag{14.35b}$$

$$\text{Triangle wave:} \quad v_c = -\sum_{n\,\text{odd}} \frac{4V_c}{n^2\pi^2} \cos n\omega_o t \tag{14.35c}$$

The sine wave v_a requires only one term, while the square wave and the triangle wave require infinite sums.

Having obtained written forms for the Fourier series for three relatively simple periodic waveforms, let us illustrate in Fig. 14.12 how the superposition of more and more terms in the Fourier series gives an increasingly faithful representation of the actual waveform.

Figure 14.12*a* shows the first three terms of the Fourier series for the square wave plotted on the same axes. The sum of these terms is shown in Fig. 14.12*b*. It is seen that even with three terms, the sum is beginning to look square, but that the straight sides and flat top are imperfectly represented. In Fig. 14.12*c* is plotted a three-term representation of a triangle wave. In this case, the three terms give an excellent representation of most of the waveform, observable departures occurring only at the sharp peaks in the waveform.

There is an important lesson in these two examples. The more abrupt the changes in a waveform, the more terms one needs in the Fourier series to achieve any given level of approximation to the total waveform. The triangle wave is continuous, having only an abrupt change in slope at the peaks, and it is very well represented by a three-term Fourier series. The square wave, however, is itself an abruptly changing function. The same number of terms in the Fourier series represent the square wave less accurately than

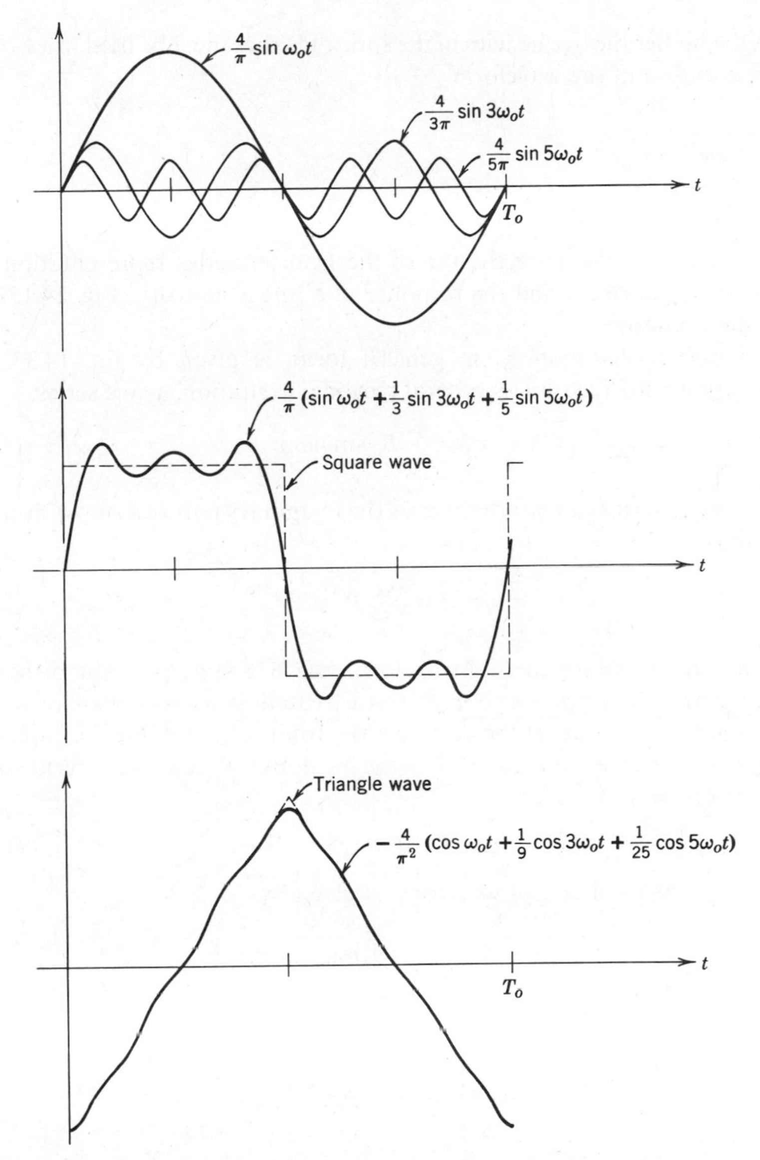

Figure 14.12
Fourier series representation of periodic waveforms.

the triangle wave. That is, one must include many more harmonics in the Fourier series for a square wave to get a representation that is as good as the three-term series for the triangle wave. This illustrates a general relation between time and frequency, that the more abruptly a waveform changes in time, the higher the frequencies of the sinusoids that must be used in a Fourier representation of the waveform.

14.2.2 Fourier Series and Linear Systems

We shall now illustrate the use of the Fourier series representation of a periodic waveform to find the response of a linear network (Fig. 14.13) to a periodic excitation.

A periodic excitation v_i, in general form, is given by Eq. 14.33. For simplicity we shall use a less general periodic excitation, a sine series.

$$v_i = \sum_n B_n \sin n\omega_o t \tag{14.36}$$

This Fourier series can be rewritten as the imaginary part of a sum of complex sinusoids.

$$v_i = \text{Im}\left\{\sum_n B_n\, e^{jn\omega_o t}\right\} \tag{14.37}$$

To find the response to this waveform, we use superposition principles. We first find the response to each term treated as an independent source. Then the responses are summed to get the total response. For example, in a system with transfer function $H(s)$ the response to a general term in the Fourier series is of the form

$$\text{Im}\,\{B_n H(jn\omega_o)\, e^{jn\omega_o t}\} \tag{14.38}$$

Therefore, the total output waveform is given by

$$v_o = \text{Im}\left\{\sum_n B_n H(jn\omega_o) e^{jn\omega_o t}\right\} \tag{14.39}$$

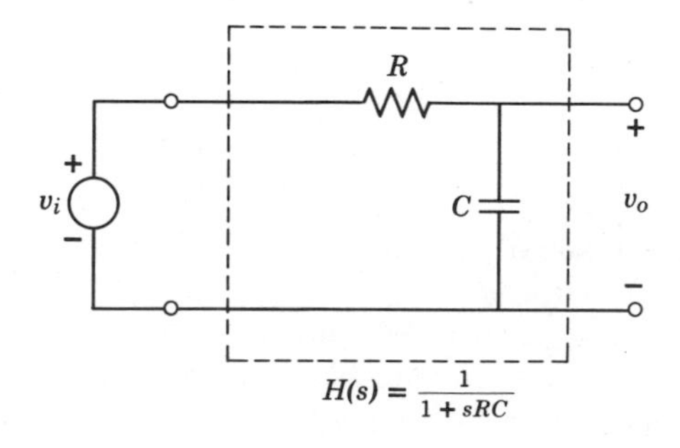

Figure 14.13

For the specific example in Fig. 14.13, in which $H(s) = 1/(1 + sRC)$ the output waveform is

$$v_o = \sum_n B_n \frac{1}{\sqrt{1 + (n\omega_o RC)^2}} \sin(n\omega_o t + \phi_n) \tag{14.40}$$

where

$$\phi_n = -\tan^{-1}(n\omega_o RC) \tag{14.41}$$

This can be brought into general Fourier series form by using the trigonometric identity

$$\sin(x + y) = \sin x \cos y + \cos x \sin y \tag{14.42}$$

which gives

$$v_o = \sum_n \left\{ \frac{B_n \sin \phi_n}{\sqrt{1 + (n\omega_o RC)^2}} \cos n\omega_o t + \frac{B_n \cos \phi_n}{\sqrt{1 + (n\omega_o RC)^2}} \sin n\omega_o t \right\} \tag{14.43}$$

Examination of the above result illustrates several important features of periodic waveforms. First, the output waveform is periodic, with the same period as the input waveform. Furthermore, the output waveform has harmonic content only for the harmonics present in the input waveform. But the output waveform differs from the input waveform in several important respects. In particular, the presence of phase shifts in the system function has converted a purely sine series at the input into a series that has both sine and cosine terms. That is, the fact that the phase shifts ϕ_n are different for different harmonics has destroyed the time symmetry of the total waveform. Thus both sine and cosine terms are necessary.

14.2.3 An Instructive Example

Let us consider a specific periodic excitation for the circuit of Fig. 14.13. Suppose that v_i is the square wave of Fig. 14.11*b*. From Eq. 14.35*b*, we note that all A_n's for the input waveform are zero and all B_n's for n even are also zero. The only nonzero coefficients are B_n for n odd, and these have the value $4V_i/n\pi$ where, V_i is the amplitude of the square wave.

We have already worked out in Eq. 14.43 the response to this waveform. Substituting for B_n, we find

$$v_i = \sum_{n\,\text{odd}} \left(\frac{4V_i}{n\pi}\right) \left\{ \frac{\sin \phi_n}{\sqrt{1 + (n\omega_o RC)^2}} \cos n\omega_o t + \frac{\cos \phi_n}{\sqrt{1 + (n\omega_o RC)^2}} \sin n\omega_o t \right\} \tag{14.44}$$

We shall now examine this output waveform for different choices of the RC time constant.

Case 1: $\omega_o RC \ll 1$

If $\omega_o RC \ll 1$, then for at least the first few terms, say up to n_k, the quantity $(n\omega_o RC)^2$ can be neglected compared to unity. Furthermore, if $n\omega_o RC \ll 1$ for $n < n_k$, then $\phi_n \approx -n\omega_o RC$, which yields

$$|\sin \phi_n| \approx |n\omega_o RC| \ll 1$$

and

$$\cos \phi_n \approx 1 \tag{14.45}$$

Therefore, the leading terms in the Fourier series are

$$v_i \cong \sum_{n\,\text{odd}}^{n_k} \frac{4V_i}{n\pi} \sin n\omega_o t - \sum_{n\,\text{odd}}^{n_k} \frac{4V_i}{n\pi}(n\omega_o RC) \cos n\omega_o t \tag{14.46}$$

This looks exactly like the leading terms in a square-wave series, except for the small additional cosine terms. Is this a reasonable result?

The square wave is a useful example here because we can solve for the exact periodic waveform rather easily in order to compare it with the Fourier series. If $\omega_o RC \ll 1$, the time constant for the RC circuit is very small compared to the period of the square wave. Therefore, we can look on the square wave as a series of step functions that are spaced sufficiently far apart that the transient excited by the step is able to die away completely before the next step occurs (Fig. 14.14*a*). The source square wave is shown in dashed lines with amplitude V_i. The response waveform simply has an exponential transient between $-V_i$ and $+V_i$, the transient starting whenever the square wave switches, and going to completion before the next switching of the source waveform.

Except for a few changes in the shape of the edges, the output waveform looks very much like a square wave. It has a flat top for times more than $3RC$ after v_o switches, it has steep sides, although not vertical, and the time between each crossing of zero is exactly $T_o/2$. But several things are different. First, the time symmetry of the input square wave is no longer present in the output square wave. No choice of origin will make v_o into an even or odd function. Therefore, its Fourier series will require both sine and cosine terms. Furthermore, the zero crossings do not occur at the same time as the zero crossings of the square wave. There is a time delay that depends on the RC time constant. This could be important if the zero crossings of a waveform were to be used for timing purposes. Finally, the output waveform is less abrupt than the input waveform. Therefore, we expect that the high-frequency

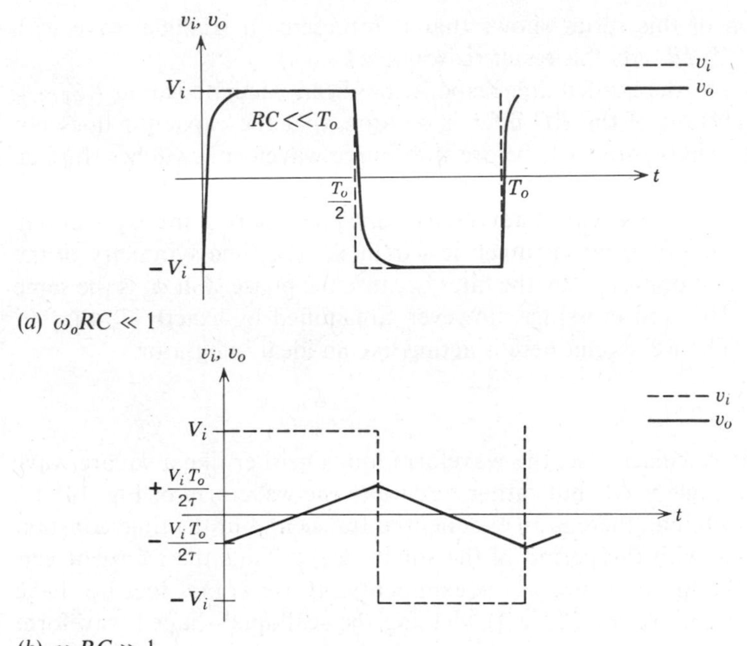

(a) $\omega_o RC \ll 1$

(b) $\omega_o RC \gg 1$

Figure 14.14
Response of Fig. 14.13 to a square wave.

terms in the Fourier series will decrease in importance more rapidly than the corresponding terms in the square-wave series. The RC circuit is a low-pass filter that attenuates the high-frequency components of the square wave. Since it is the high-frequency components that give the abrupt switching feature of the square wave, it is not surprising to find this abruptness somewhat smoothed out by the filter. Indeed, examination of the Fourier series for the output waveform shows that for high enough n, the terms $(n\omega_o RC)^2$ in the square root will become important. For these terms, the overall coefficient will then behave like $1/n^2$, instead of like $1/n$ for the square wave. Thus the high-frequency components are more strongly attenuated.

Case 2: $\omega_o RC \gg 1$

If $\omega_o RC \gg 1$, the $(n\omega_o RC)^2$ term dominates the square root in the denominator of every term in the sum. Furthermore, $\phi_n \approx -90°$ for every n, in which case $\sin \phi_n \approx -1$ and $\cos \phi_n \approx 0$. Therefore, the output series becomes

$$v_o = -\sum_{n \text{ odd}} \frac{2V_i T_o}{RC} \frac{1}{n^2 \pi^2} \cos n\omega_o t \qquad (14.47)$$

Examination of this series shows that it represents a triangle wave with amplitude $V_i T_o / 2RC$. Is this result reasonable?

In Fig. 14.14*b*, the explicit time response has been calculated for $\omega_o RC \gg 1$. The time constant of the RC filter is so long, that the capacitor does not have time to charge up to V_i before the source waveform switches. In fact, the requirement $\omega_o RC \gg 1$ is the same as the requirement $T_o \ll RC$. Therefore the resulting output waveform never leaves the linear part of the exponential, and has an amplitude that is much less than V_i. The time symmetry of the waveform is not destroyed by the filter because the phase shift ϕ_n is the same for every n. The zero crossings, however, are shifted by exactly $T_o/4$ (a 90° phase shift). The RC circuit here is acting like an ideal integrator.

Case 3: $\omega_o RC \approx 1$

In this intermediate case, the waveform looks neither like a square wave nor like a triangle wave, but rather resembles the waveform of Fig. 14.11*d*. After each switching there is an exponential transient, but the time constant is comparable with the period of the square wave. Thus the transient gets well out of the linear region of the exponential (Case 2) but does not have time to go to completion (Case 1), yielding the scalloped-shaped waveform shown.

Finally, as a useful exercise, consider what would happen to the square wave if it were put through a system with a sharply tuned transfer function, such as that encountered in Section 13.3.

14.3 SUPERPOSITION IN THE TIME DOMAIN

Just as it was possible in the frequency domain to build up complex waveforms from proper superpositions of sinusoids, it is possible in the time domain to build up waveforms from proper superpositions of simple time functions. Figure 14.15 shows a waveform containing a single positive-going pulse of amplitude A and width T. This waveform can be represented as a sum of two step functions with properly chosen amplitudes and time origins.

$$a(\tau) = Au_{-1}(\tau) - Au_{-1}(\tau - T) \tag{14.48}$$

Since this waveform is the sum of two terms, the response of a linear system to a source of this form would also contain two terms: one term being the response to the step $Au_{-1}(\tau)$, the other being the response to the step $-Au_{-1}(\tau - T)$.

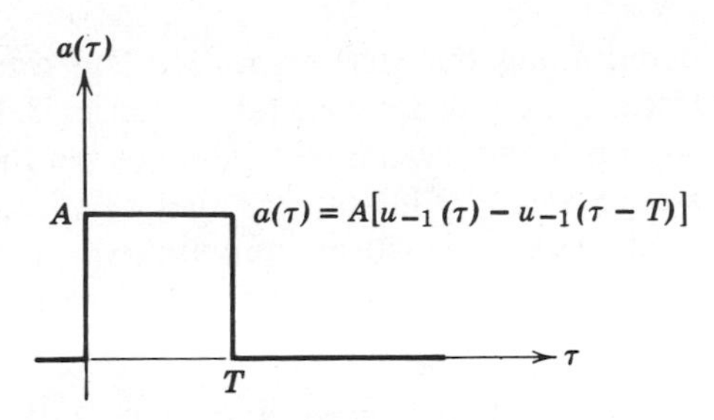

Figure 14.15
A single pulse represented by a superposition of step functions.

More complex waveforms based on the step response and waveforms derivable from the step response can be built up by extensions of this sum-of-terms superposition. We have already examined the step response of single-pole systems in great detail. In the next sections, we shall examine the step response of a band-pass system and use it to calculate the response to a single square pulse of the kind shown in Fig. 14.15. Finally, we shall present an advanced method for dealing with arbitrary waveforms.

14.3.1 Pulse Response of a Band-Pass System

Let us find the pulse response of a simple band-pass system, that is, a system with a zero at $s = 0$ and a pair of poles, either both on the negative real axis, or comprising a complex pair. We saw in Section 13.3.3 that by proper choice of R, L and C, in the parallel-resonant circuit of Fig. 14.16, it is possible to obtain either the broad passband of the untuned network or the narrow passband of the resonant circuit. Therefore, we shall use this network as our example of a general band-pass system.

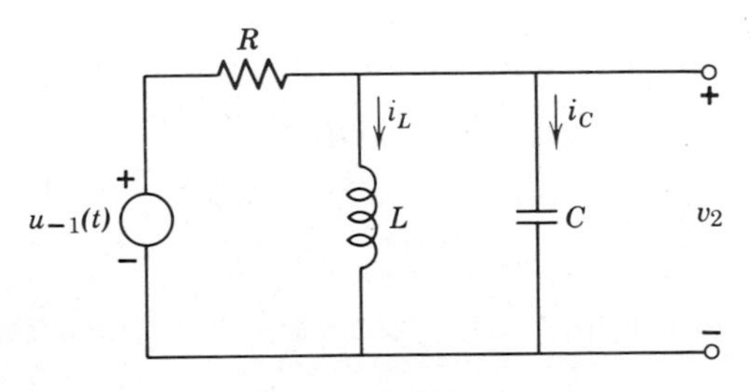

Figure 14.16
A band-pass system.

We begin by determining the step response. The total step response consists of a forced response plus the natural response. In this network, the forced response to a step is zero, because the inductor is a short circuit for dc. The presence of a zero at $s = 0$ guarantees that the forced response will *always* be zero in such systems. Because there are two poles, s_a and s_b, the natural response will contain two terms. The general form of the step response is

$$v_2(t) = A_1 e^{s_a t} + A_2 e^{s_b t} \tag{14.49}$$

where A_1 and A_2 are constants to be determined from initial conditions. The two initial conditions are obtained as follows:

Before the step, at $t = 0^-$, the voltage across the capacitor was zero. By the continuity condition for a capacitor, therefore,

$$v_2(0^+) = 0 \tag{14.50}$$

Similarly, before the step, at $t = 0^-$, the current in the inductor was zero. By the continuity condition for an inductor, therefore,

$$i_L(0^+) = 0 \tag{14.51}$$

But if $i_L(0^+) = 0$, we must have by KCL

$$i_C(0^+) = \frac{1}{R} \tag{14.52}$$

because the source has magnitude 1 volt at $t = 0^+$ and because v_2 is zero volts at $t = 0^+$. Equivalently, since

$$i_C = C\frac{dv_2}{dt} \tag{14.53}$$

this condition becomes

$$\left.\frac{dv_2}{dt}\right|_{t=0^+} = \frac{1}{RC} \tag{14.54}$$

Applying these initial conditions to Eq. 14.49 yields a pair of equations

$$\begin{aligned} A_1 + A_2 &= 0 \\ s_a A_1 + s_b A_2 &= \frac{1}{RC} \end{aligned} \tag{14.55}$$

which have as their solution

$$A_1 = -A_2 = \frac{1}{RC(s_a - s_b)} \tag{14.56}$$

Therefore, the general step response is

$$v_2(t) = \frac{1}{RC(s_a - s_b)}(e^{s_a t} - e^{s_b t}) \tag{14.57}$$

Let us consider two extreme cases of interest, the strongly damped system, where s_a and s_b are on the negative real axis and several decades apart, and the resonant system, where s_a and s_b form a high-Q complex pair.

For the strongly damped case ($R \ll \sqrt{L/4C}$). The difference $s_a - s_b$ has the value

$$s_a - s_b \approx \frac{1}{RC} \tag{14.58}$$

Therefore, the step response is shown in Fig. 14.17*b*, with a rapid rise to about unity with the time-constant characteristic of the high-frequency

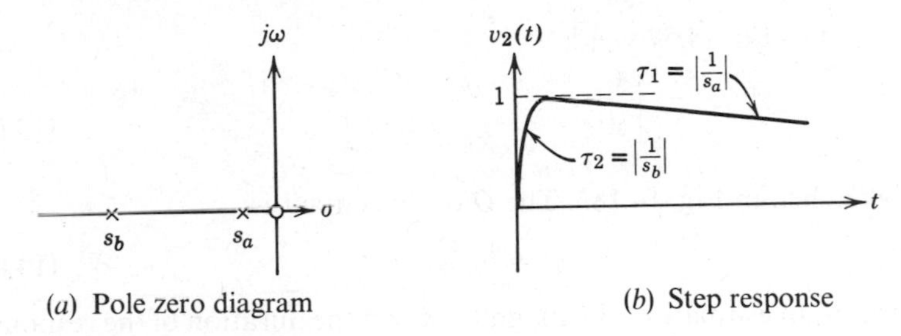

(*a*) Pole zero diagram (*b*) Step response

Figure 14.17
Step response of a band-pass system. For s_a, and s_b far apart, the rise and the decay would not be observable on the same time scale.

pole, followed by a much more gradual decay back to zero with the time constant characteristic of the low-frequency pole.

The resonant case ($R \gg \sqrt{L/4C}$) has poles at (see Fig. 14.18*a*).

$$s_a = -\omega_1 + j\omega_0 \tag{14.59}$$

and

$$s_b = -\omega_1 - j\omega_0 \tag{14.60}$$

where

$$\omega_0 = 1/\sqrt{LC}$$

and (14.61)

$$\omega_1 = 1/2RC$$

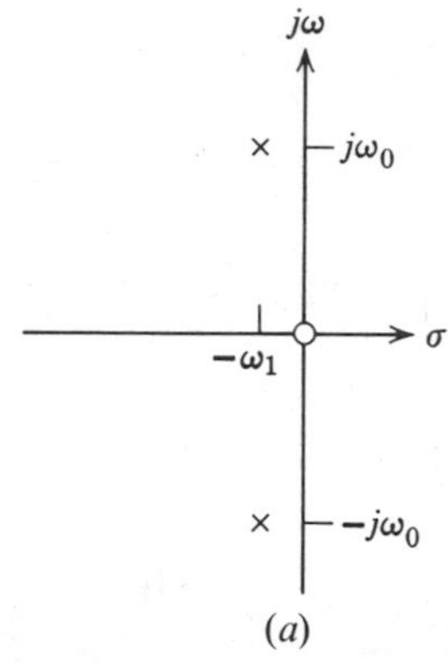

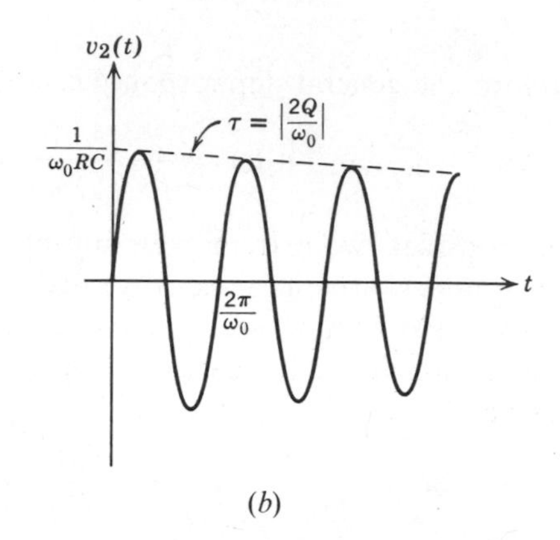

Figure 14.18
Step response of a resonant system.

substituting into Eq. 14.57 yields

$$v_2(t) = \frac{e^{-t/2RC}}{\omega_0 RC} \sin \omega_0 t \tag{14.62}$$

which is sketched in Fig. 14.18*b*. The Q of this circuit is

$$Q = \omega_0 RC \tag{14.63}$$

which permits an estimate of the magnitude and the duration of the response. The initial amplitude is $1/Q$, and the ringing at the frequency ω_0 decays to about 10% of its original amplitude in Q cycles. The higher the Q, the weaker the response, but the longer the ringing.

As an example of how the step response in either case might be used, let us consider the 1:1 transformer of Fig. 14.19 being used to couple a pulse

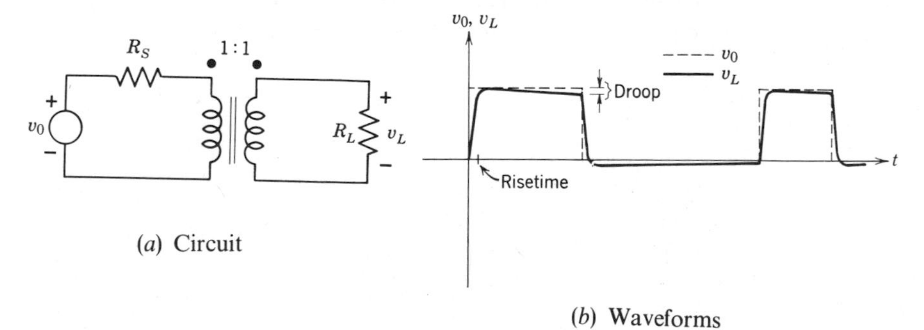

Figure 14.19
Pulse response of a transformer.

source to a load. We recall from Section 13.3.2 that the transformer has two real axis poles, and corresponds most closely to the "strongly damped" case of a band-pass system. The source pulse is perfectly square, but because of the band-pass nature of the transformer response, there is a finite risetime (determined by the upper cutoff frequency) and some decay or *droop* during the flat portion of the pulse (determined by the lower cutoff frequency). When the source pulse returns to zero, the output of the transformer actually overshoots zero. If another pulse comes along before this overshoot has died out, the amplitude of the second pulse is less than the first, and so on, until in the limiting case of a periodic pulse train the dc value of the output waveform becomes zero, as it must, since a transformer cannot pass dc.

Notice that the poles and zeros near the origin determine the long-time behavior of the response, while the poles at frequencies corresponding to the upper cutoff frequency of the pass band determine the short-time behavior of the response (i.e., the risetime). This is a general result of great importance in understanding the behavior and limitations of signal-processing systems.

14.3.2 The Convolution Integral

Any waveform could be represented as a sequence of little steps. Indeed, sums of the form of Eq. 14.48 could be extended to an infinity of infinitesimal steps, allowing any time function to be represented. An equivalent and more convenient generalization is based on the impulse function, $u_0(t)$. The basic impulse-function property of importance here is that for any amplitude function $a(t')$,

$$\int_{-\infty}^{\infty} a(t')u_0(t-t')\,dt' = a(t) \qquad (14.64)$$

Therefore, we could represent the pulse function of Fig. 14.15 as a superposition of impulse functions, as indicated schematically in Fig. 14.20.

The advantage of writing $a(t)$ as in Eq. 14.64, in what appears to be a redundant form, is that if the response of a network to the unit impulse $u_0(t-\tau)$ is known, the total response to any waveform can be found directly. Thus, if for a particular linear system

$$\begin{matrix} u_0(t-\tau) \rightarrow h(t-\tau) \\ \text{source} \rightarrow \text{response} \end{matrix} \qquad (14.65)$$

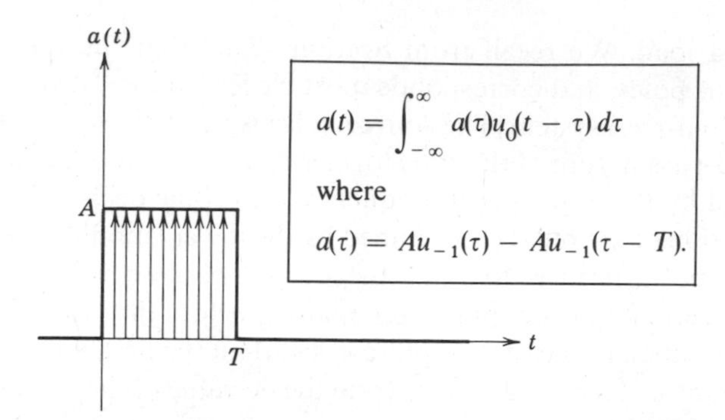

Figure 14.20
Time-domain superposition of impulses to represent the waveform of Fig. 14.15.

then the superposition of impulses represented by $a(t)$ in Eq. 14.64 yields this response $b(t)$:

$$b(t) = \int_{-\infty}^{\infty} a(\tau)h(t - \tau)\,d\tau \tag{14.66}$$

This integral is called the *convolution integral* or *superposition integral* and can be evaluated for any waveform and any linear system, provided that the impulse response of the network is known. Because the impulse response of a network $h(t)$ is always zero for $t < 0$ (causality demands it), the limits of integration of Eq. 14.66 can be changed to

$$b(t) = \int_{-\infty}^{t} a(\tau)h(t - \tau)\,d\tau \tag{14.67}$$

It is this form that we shall use in the example below.

14.3.3 A Convolution Example

Let us illustrate the use of the convolution integral with the same example that was solved in Section 12.2.3 using system function methods. In Fig. 14.21, a simple RC network is driven with a cosine wave that is turned on at $t = 0$. The source function, therefore, is

$$v_O = u_{-1}(t)V_o \cos \omega_o t \tag{14.68}$$

What is now needed is the impulse response of the RC network.

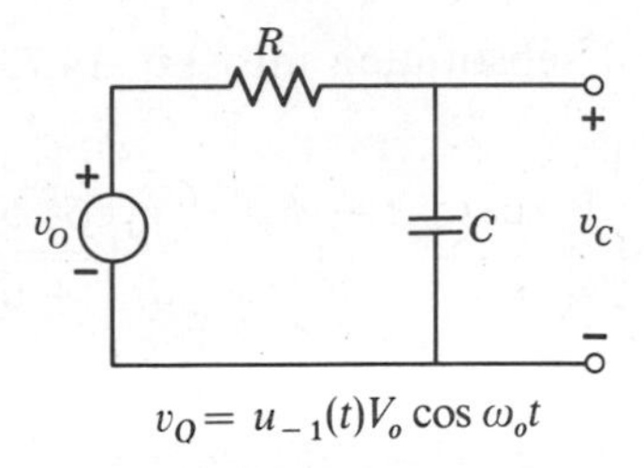

Figure 14.21
An example.

The easiest way to determine an impulse response of a network is to use the fact that differentiation is a linear operation. Since the unit impulse is the time derivative of the unit step, then the response to the unit impulse must be the time derivative of the response to the unit step. We know for this *RC* network that for a source $u_{-1}(t)$, the response is

$$u_{-1}(t)(1 - e^{-t/RC}) \tag{14.69}$$

The impulse response, $h(t)$ is therefore

$$h(t) = \frac{d}{dt}[u_{-1}(t)(1 - e^{-t/RC})] \tag{14.70}$$

or, since $u_0(t)f(t) = f(0)$,

$$h(t) = \frac{u_{-1}(t)}{RC} e^{-t/RC} \tag{14.71}$$

Substituting in Eq. 14.67 yields for the response v_c

$$v_C = \int_{-\infty}^{t} [u_{-1}(\tau)V_o \cos \omega_o \tau]\left[\frac{u_{-1}(t-\tau)\, e^{-(t-\tau)/RC}}{RC}\right] d\tau \tag{14.72}$$

This integral may be simplified as follows. The $u_{-1}(\tau)$ factor allows the lower limit to be moved from $-\infty$ to 0, and at the same time assures that $v_C = 0$ for $t < 0$. Furthermore, since $u_{-1}(t - \tau) = 1$ for $\tau < t$,

$$v_C = u_{-1}(t)\frac{V_o e^{-t/RC}}{RC}\int_0^t \cos \omega_o \tau \, e^{\tau/RC}\, d\tau \tag{14.73}$$

But

$$\begin{aligned} \int_0^t \cos \omega_o \tau \, e^{\tau/RC}\, d\tau &= \text{Re}\left\{\int_0^t e^{[j\omega_o + (1/RC)]\tau}\, d\tau\right\} \\ &= \text{Re}\left\{\frac{e^{[j\omega_o + (1/RC)]t}}{j\omega_o + 1/RC} - \frac{1}{j\omega_o + 1/RC}\right\} \end{aligned} \tag{14.74}$$

Taking the real part and substituting into Eq. 14.73 we find for the total response

$$v_C = u_{-1}(t)\left[\frac{V_o \cos(\omega_o t - \phi_o)}{\sqrt{1 + (\omega_o RC)^2}} - \frac{V_o \cos \phi_o \, e^{-t/RC}}{\sqrt{1 + (\omega_o RC)^2}}\right] \tag{14.75}$$

where

$$\phi_o = \tan^{-1} \omega_o RC \tag{14.76}$$

Notice that both the forced and natural parts of the response emerge from the one calculation.

14.4 A CONCLUDING REMARK

The concepts and methods introduced in the three-chapter segment just concluded, while restricted to linear systems, are not restricted to electronics. A single-pole system, with its characteristic frequency response and characteristic step response, remains a single-pole system whether it be built of electrical components, mechanical components, or the strictly mathematical components of the systems analyst. The notion that feedback modifies the frequency response (and also, therefore, the step response) of a linear system is true in any linear system, just as is the generalism that in systems with enough high-frequency poles, excess feedback will produce oscillations. The reciprocal relation between bandwidth and risetime is fundamental, and can be applied in any context.

The better one understands these basic issues, the better one can tackle problems in almost any area. With this exhortation to "think broadly," we now consider the applications of the material presented up to this point to problems in signal-processing and instrumentation.

REFERENCES

R14.1 Paul E. Gray and Campbell L. Searle, *Electronic Principles: Physics, Models, and Circuits* (New York: John Wiley, 1969), Chapters 12 to 20.

R14.2 Richard S. C. Cobbold, *Theory and Applications of Field-Effect Transistors* (New York: Wiley-Interscience, 1970).

R14.3 Charles L. Alley and Kenneth W. Atwood, *Semiconductor Devices and Circuits* (New York: John Wiley, 1971).

R14.4 Mischa Schwartz, *Information Transmission, Modulation, and Noise* (New York: McGraw-Hill, 1970), Chapter 1.

R14.5 Michael Kahn, *The Versatile Op-Amp* (New York: Holt, Rinehart and Winston, 1970).

R14.6 Jerald G. Graeme and Gene E. Tobey, eds., *Operational Amplifiers: Design and Applications* (New York: McGraw-Hill, 1971).

QUESTIONS

Q14.1 Will a transistor with f_T of 100 MHz be useful as an amplifier at 100 MHz?

Q14.2 Suppose you are in doubt on whether a particular circuit can be analyzed with separate low-frequency and high-frequency circuits. Suggest a practical way of evaluating the validity of the approach.

Q14.3 Why is it incorrect in a multistage transistor amplifier to replace each transistor by its C_T approximation equivalent?

Q14.4 Suppose that the value of one component in a network (such as a resistor or capacitor) is not constant with time. Is frequency-domain superposition valid in that network? Is time-domain superposition valid in that network?

Q14.5 Discuss how an op-amp's slew rate at unity gain is related to the op-amp's frequency response.

EXERCISES

E14.1 Find the transistor operating point in Fig. E.14.1 and draw the incremental circuit. Determine numerical values for the hybrid-π parameters.

E14.2 Assume in the circuit of Fig. E14.1 that $g_m = 0.1$ S, $r_\pi = 1$ kΩ, $r_x = 50\ \Omega$, $C_\mu = 3$ pF, and $C_\pi = 60$ pF. Determine the high-frequency response (magnitude and phase) using the C_T approximation.

E14.3 Determine the low-frequency response (magnitude and phase) of the circuit of E14.1. Above what frequency is the emitter resistor effectively bypassed by the 100 μF capacitor?

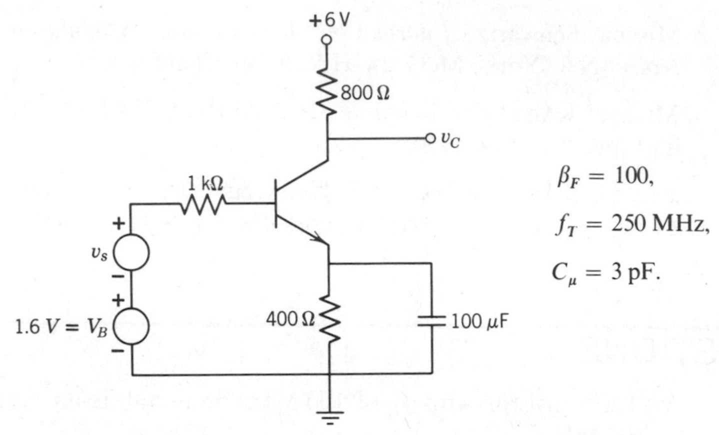

Figure E14.1

E14.4 (a) Draw the complete incremental circuit in Fig. E14.4. Determine numerical values for g_m, r_π, and C_π.

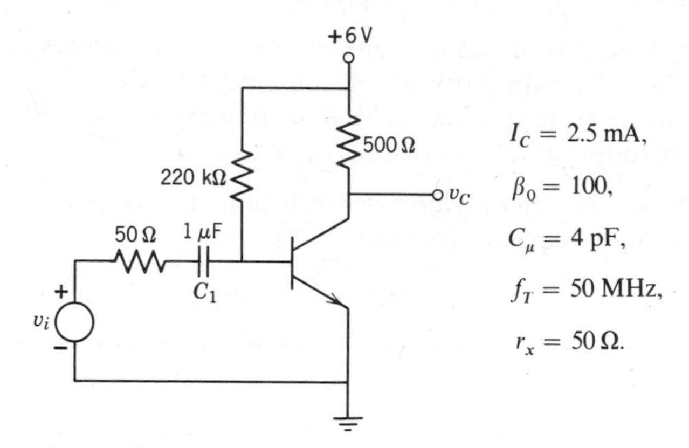

Figure E14.4

(b) Redraw the circuit using the C_T approximation for the high-frequency components.

(c) Determine the bandwidth of the amplifier. That is, determine the lower and upper cutoff frequencies, ω_l and ω_h.

Answer. $\omega_l = 900$ rad/s or 150 Hz
$\omega_h = 22$ Mrad/s or 3.7 MHz.

E14.5 An input voltage $v_i = 50u_{-1}(t)$ mV is applied to the transistor amplifier of E14.4.

(a) Assuming C_1 is a short circuit, find the response $v_C(t)$. How does the circuit time constant relate to ω_h?

(b) Assuming C_T is an open circuit, find the response $v_C(t)$. How does the circuit time constant relate to ω_l?

(c) If both C_1 and C_T are present, sketch the resulting response to v_i.

E14.6 The incremental model of Fig. E14.6*a* (with various choices of g_m and C's) can be used to represent most FETs throughout the audio and lower radio frequency ranges. Use this model to estimate the frequency response of the circuit in Fig. E14.6*b*.

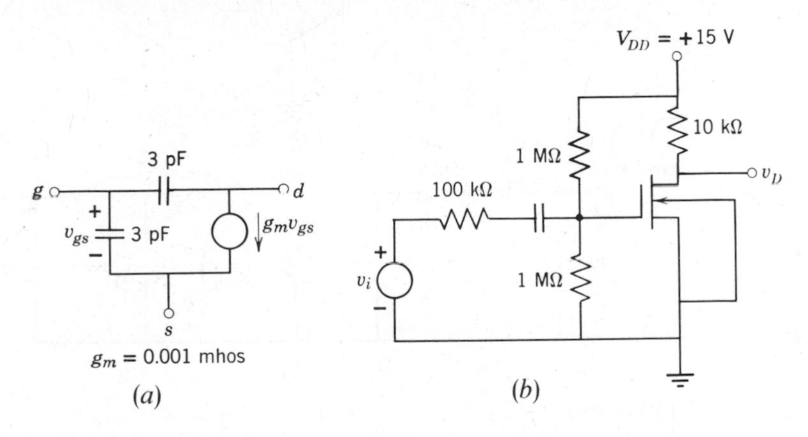

Figure E14.6

E14.7 Show that if an op-amp open-loop transfer function has two negative-real-axis poles, the closed-loop frequency response develops a complex pole pair as the amount of resistive feedback is increased (i.e., for smaller closed-loop gain). Comment on the effect of feedback on the step response of such a system.

E14.8 Show that if an op-amp open-loop transfer function has three poles, the poles of the closed-loop transfer function will actually lie on the imaginary axis when enough resistive feedback is used. Comment on the stability of such an amplifier system.

E14.9 Verify the Fourier series of Eq. 14.35.

PROBLEMS

P14.1 (a) Show that the correct low-frequency incremental circuit for the amplifier in Fig. P14.1*a* can be approximated as shown in Fig. P14.1*b* provided that $R_B \gg R_C, 1/g_m$.

(b) Determine the low- and intermediate-frequency response.

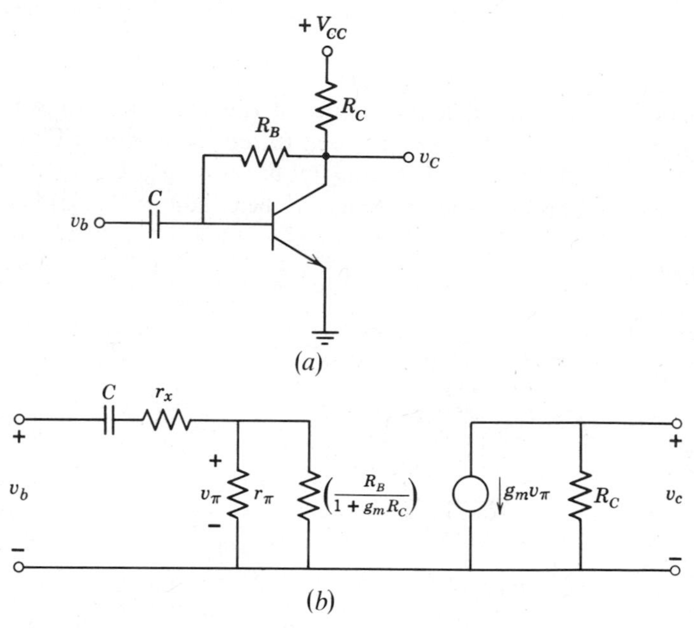

Figure P14.1

P14.2 Estimate the frequency response of the source follower using the model of Fig. E14.6*a*.

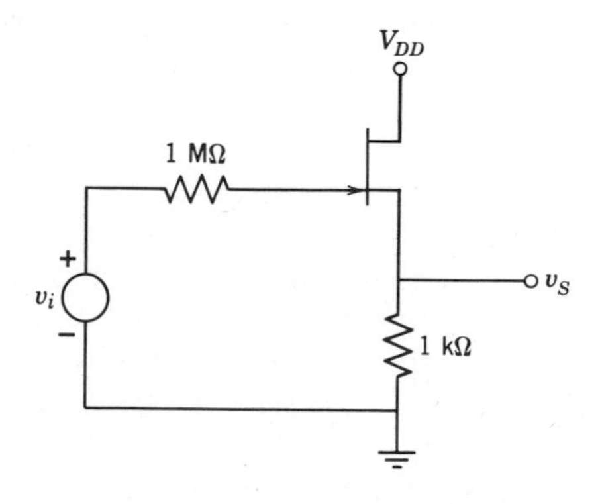

Figure P14.2
Source follower.

P14.3 The amplifier in Fig. P14.3 is *intended* to drive a loudspeaker so that the current i_O is proportional to the speech waveform v_s. Your job is to analyze the small-signal frequency response I_o/V_s. Use the simple r_π-g_m model for the transistor.

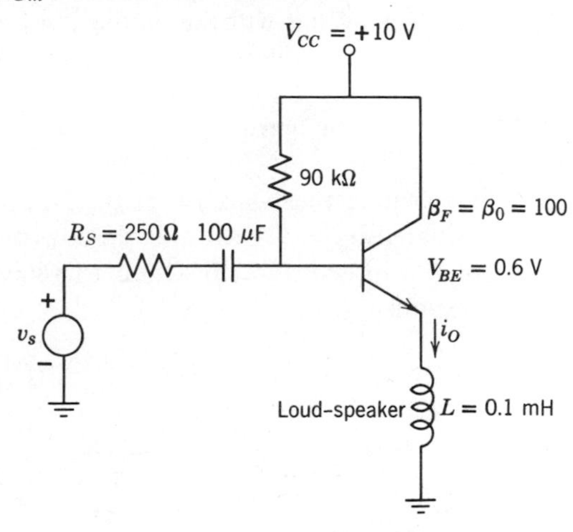

Figure P14.3

(a) Determine the values for r_π and g_m.
(b) Estimate the lower and upper cutoff frequencies for the response function I_o/V_s.

P14.4 The circuit in Fig. P14.4 is a schematic representation of a Colpitts oscillator, with the nominal dc bias current assumed to be known. All of the bias circuitry has been intentionally omitted. The purpose of this problem is to analyze this oscillator in steps to develop an understanding of the key issues.
(a) Calculate the impedance $Z(s)$ seen at the terminals of the inductancc. You can (1) neglect r_x and C_μ; (2) assume $r_\pi \gg 1/sC_\pi$; and (3) assume $C_\pi \ll C_1$.

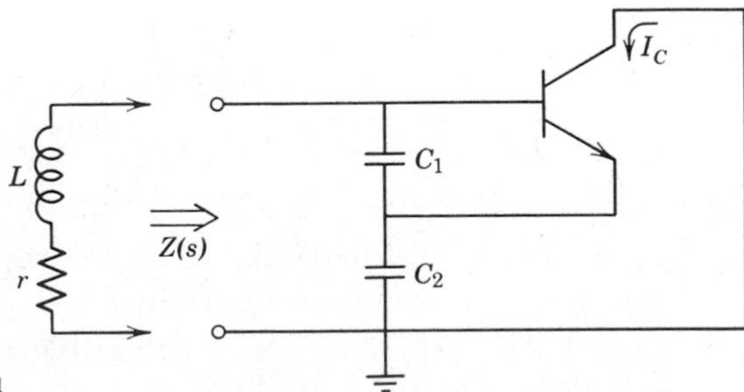

Figure P14.4

(b) The oscillator coil is modeled as an inductance in series with a small resistance. When it is connected to the appropriate terminals, the circuit *might* oscillate.
   (1) What condition must be met for oscillation?
   (2) What will the oscillation frequency be?
   (3) We know in practice that the oscillation amplitude is finite. Discuss what might determine the amplitude of the oscillator.

P14.5 Determine the response of the circuits in Fig. P14.5 to a ramp initiated at $t = 0$. Use either the convolution method or the linearity of the integration operation to relate this response to the step response.

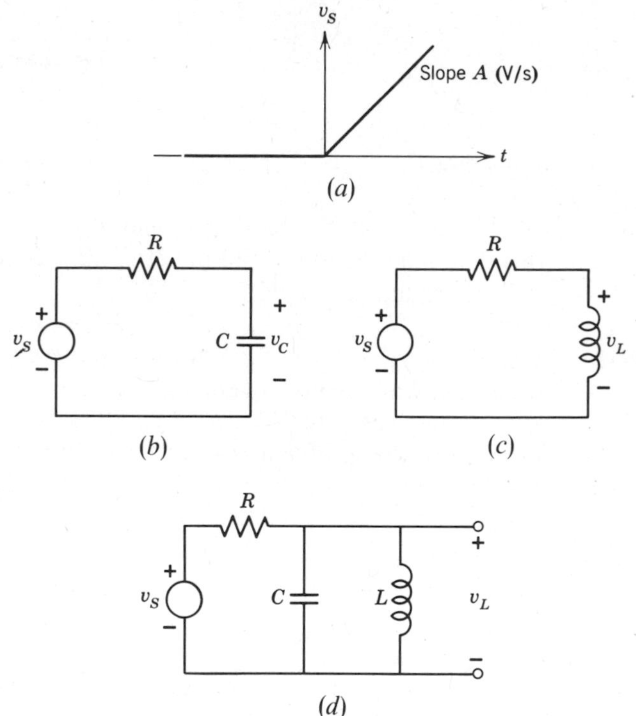

Figure P14.5

P14.6 A square wave with period T_o is used to drive a tuned filter. Plot the output amplitude of the filter as the resonant frequency is tuned slowly from below $1/T_o$ to a high frequency. Comment on the effect of varying the Q of the filter.

P14.7 A periodic sequence of positive pulses with period T_o and pulse width $T_o/2$ is passed through a bandpass filter with $\omega_l = 1/2T_o$ and $\omega_h = 5/T_o$. Plot the response.

P14.8 (a) Show that the equivalent high-frequency model of Fig. 14.8 is not valid if the load R_C is replaced by a reactive element or a complex impedance. See page 438 for figure.

(b) Consider the case where R_C in Figure 14.5*b* is replaced by a capacitor C_C. Derive a new high-frequency equivalent circuit for this case. See page 435 for figure.

PART D

APPLICATIONS

CHAPTER FIFTEEN

MODULATION AND DETECTION

15.0 INTRODUCTION

In this chapter, our emphasis shifts from the analysis of circuit behavior toward the understanding of useful applications of electronic circuits in signal processing. Let us begin by reviewing the kinds of signal-processing tasks we have already encountered. Resistive amplifiers produce faithful reproductions of any time-varying input waveform, and are used for increasing the strength of signals. Energy storage elements, by themselves or in combination with diodes, transistors, or amplifiers, can be used to filter signals or to perform a variety of wave-shaping, timing, or computational functions. This chapter explores the use of modulation and detection techniques, topics of fundamental importance in communications and experimental science.

15.1 BASIC CONCEPTS

15.1.1 Communication of Information

When we speak of "communications" and "experimental science," we are referring to endeavors in which the goal is the encoding, transmission, reception, and decoding of "information." Usually, the information to be communicated takes the form of a signal, or waveform. The signal might be a speech waveform to be transmitted by radio waves, or a series of pulses representing numbers to be sent over a data telephone line, or a waveform representing the absorption of light by a sample as a function of the wavelength of the light. In any case, one hopes to communicate a suitably faithful reproduction of the waveform to the proper detector, the loudspeaker of a radio receiver in the first case, a computer in the second case, and a chart recorder in the third case.

It is an interesting fact that a perfectly periodic waveform, such as a sine wave, cannot carry any information. A sinsusoid of constant amplitude, constant frequency, and constant phase persists indefinitely in a perfectly predictable fashion. If the amplitude frequency, or phase never change, the sinusoid cannot tell us how a speech waveform changes with time, or what sequence of pulses to put in the computer, or how the optical absorption of a sample changes with optical wavelength. For a sine wave to carry information, we must cause either the amplitude, the frequency, or the phase to vary according to the information-carrying waveform. This process is called *modulation*. The inverse process, the recovery of the information waveform from the modulated sinusoid, is called *demodulation*, or *detection*. The sinusoid carrying the information is called the *carrier*, while the waveform of information is often called the *modulation*.

15.1.2 Modulated Waves

Let us examine the three basic kinds of modulated waves in more detail. If v is a sinusoidal carrier waveform, we can write

$$v = A \cos (\omega t + \phi) \tag{15.1}$$

We will consider three cases: *amplitude modulation* (AM), in which the amplitude A varies with time; *frequency modulation* (FM), in which the frequency ω varies with time; or *phase modulation* (PM), in which the phase ϕ varies with time. These are illustrated in Fig. 15.1.

The modulation is the trapezoidal wave shown in Fig. 15.1*a* while the carrier is the cosine wave shown in Fig. 15.1*b*. In the amplitude modulated case, Fig. 15.1*c*, the modulation forms the *envelope* of the modulated carrier. That is, both the positive and negative excursions of the modulated carrier are determined by the magnitude of the modulation waveform. Notice that in the AM case neither the frequency nor the phase of the wave change with time.

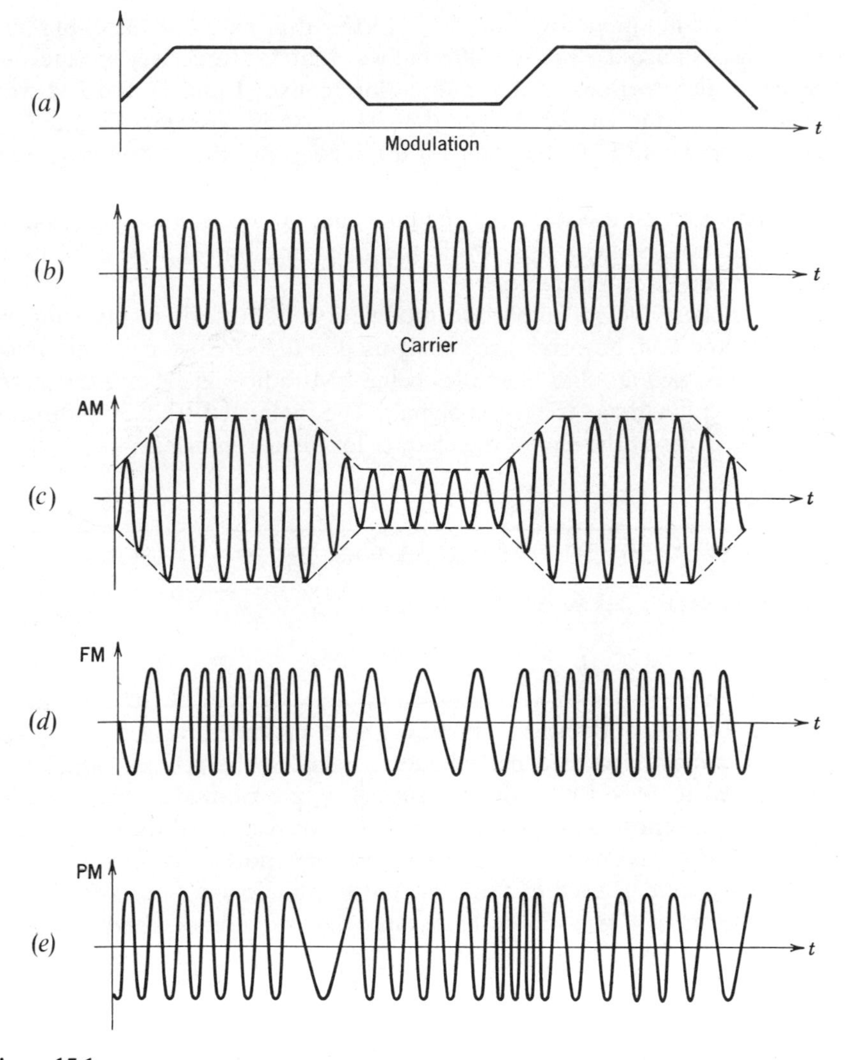

Figure 15.1

Modulated waves. The bunching and spreading in the FM and PM waves are exaggerated for clarity.

In the frequency-modulated wave, Fig. 15.1*d*, the amplitude is constant but the frequency varies. The frequency is lowest when the modulation is least positive. One can view the frequency-modulated wave as one in which the sinusoids bunch up or spread out according to the magnitude of the modulating wave.

The phase-modulated wave, Fig. 15.1*e*, also exhibits bunching and spreading of sinusoids, but in a different way than for frequency modulation. During the flat portions of the modulation (constant phase), the PM wave looks just like the carrier *except* that its phase is different. Figure 15.1*e* has been drawn so that the phase of the carrier and the PM wave are the same during the low portion of the modulation, and differ by 180° during the high portion of the modulation. The bunching up of sinusoids occurs when the modulation is increasing, while the spreading out of sinusoids occurs when the modulation is decreasing.

Our emphasis is on amplitude modulation because it is the simplest. Both FM and PM, however, are widely used in high-frequency communication systems, two familiar examples being FM radio signals and the sound portion of commercial television signals. The interested student is directed to the references at the end of the chapter for further information.

15.1.3 Amplitude Modulation by a Sinusoid

Let a_m be a sinusoidal modulation function,

$$a_m = 1 + m \cos \omega_m t \tag{15.2}$$

The constant m is called the *modulation index*. It specifies the fractional variation of the modulation function about its mean value. When this variation is expressed as a percentage, it is called the *percent modulation*, and is equal to $m \times 100\%$. Waveforms of a_m are plotted in Fig. 15.2 for three different choices of m. For $m < 1$, the waveform is always positive. For $m > 1$, the waveform is negative during some portion of a cycle.

If a wave like a_m is used to modulate the amplitude of a sinusoidal carrier, the waveforms of Fig. 15.3 result. Assume that the unmodulated carrier is

$$v_c = A \cos \omega_c t \tag{15.3}$$

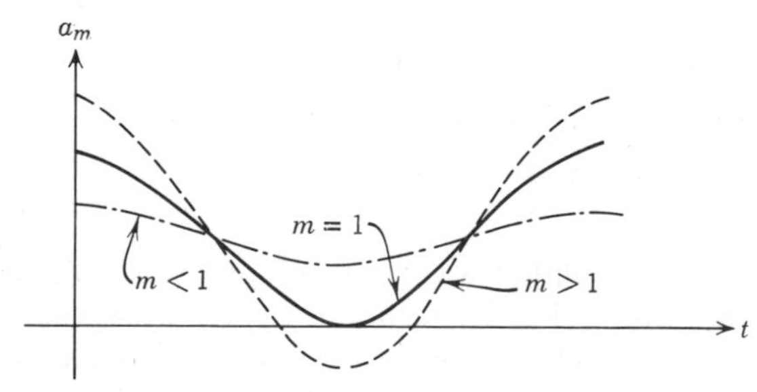

Figure 15.2
Sinusoidal modulation waveforms.

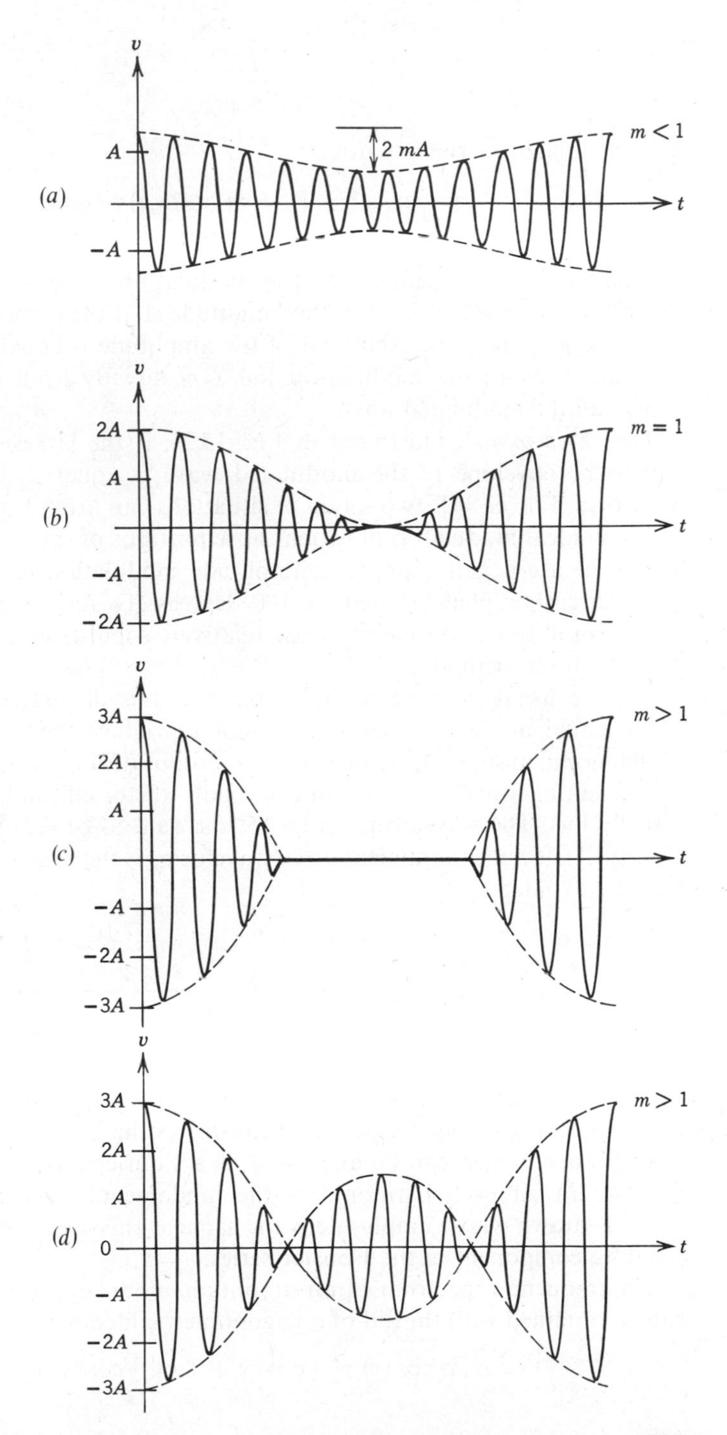

Figure 15.3
AM waveforms.

where

$$\omega_c > \omega_m \tag{15.4}$$

The modulated carrier is written

$$v = a_m v_c = A(1 + m \cos \omega_m t) \cos \omega_c t \tag{15.5}$$

Consider first $m < 1$, illustrated in Fig. 15.3*a*. The envelope of the modulated wave is identical to the modulation waveform. Furthermore, the mean amplitude is equal to the amplitude A of the unmodulated carrier, while the peak-to-peak excursion of the amplitude is equal to $2mA$. Thus, one can measure the modulation index m directly from an oscilloscope display of the modulated wave.

The value $m = 1$, illustrated in Fig. 15.3*b*, is the largest value of m for which the envelope of the modulated wave is equal to the modulation waveform. For $m > 1$, two kinds of situations can arise. Figure 15.3*d* illustrates an ideal modulator, in which those portions of the modulation waveform that are negative produce amplitude modulation of the carrier, but with the carrier phase shifted by 180 degrees. To recover the modulation waveform in this case, one must use relatively sophisticated phase-sensitive demodulation methods.

A more usual example for the case $m > 1$ is illustrated in Fig. 15.3*c*. Most actual modulation circuits cannot reproduce the negative portions of the modulation, yielding instead zero output during those time intervals. The resulting waveform envelope is badly distorted, and information is actually lost. The waveform of Fig. 15.3*c* is said to be *over-modulated*. Obviously, if faithful communication of waveforms is the goal, over-modulation must be avoided.

15.1.4 Frequency Spectrum

Whenever a waveform can be represented as a superposition of sinusoids, the amplitude of each sinusoid can be displayed graphically in a plot called the *frequency spectrum*. Figure 15.4 illustrates the frequency spectrum of a square wave, which can be expressed as a Fourier series of sinusoids (see Eq. 14.35*b*). At each harmonic of the fundamental frequency, a vertical arrow is drawn whose amplitude is the same as the amplitude of the corresponding component in the Fourier series.

The frequency spectrum of an amplitude-modulated wave like Eq. 15.5 can be obtained with the aid of a trigonometric identity:

$$\cos(x)\cos(y) = \tfrac{1}{2}\cos(x + y) + \tfrac{1}{2}\cos(x - y) \tag{15.6}$$

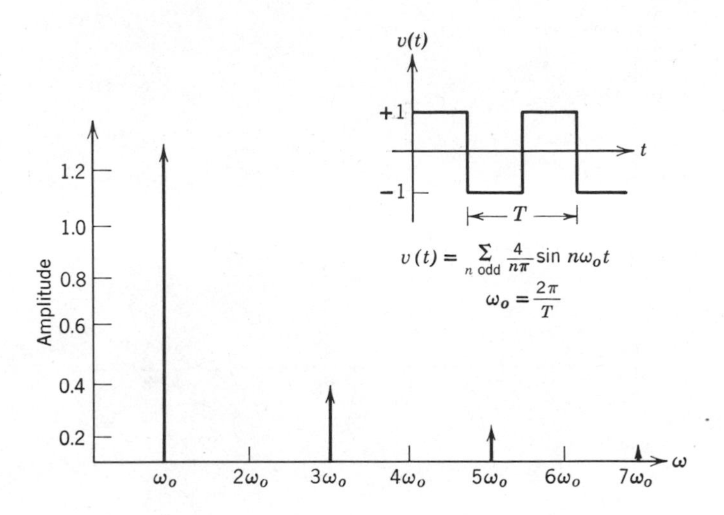

Figure 15.4
Partial frequency spectrum for a square wave.

If we use this relation, Eq. 15.5 becomes

$$v_c = A \cos \omega_c t + \frac{mA}{2} \cos [(\omega_c - \omega_m)t] + \frac{mA}{2} \cos [(\omega_c + \omega_m)t] \qquad (15.7)$$

We have now expressed the amplitude modulated wave as a superposition of three sinusoids, one at the carrier frequency ω_c, one at the frequency given by the sum of ω_c and ω_m, and one at the frequency given by the difference of ω_c and ω_m. The corresponding frequency spectrum is plotted in Fig. 15.5, using two different choices of m. Here are several important features of these spectra.

1. Each spectrum contains a central component at the carrier frequency. The amplitude of this component is independent of both the amplitude and frequency of the modulating signal.

2. There are two *sidebands* located at $\omega_c + \omega_m$ and $\omega_c - \omega_m$. The *amplitudes* of the sidebands are equal and are proportional to the modulation index. The symmetrical *position* of the sidebands relative to the carrier is determined only by the frequency of the modulating signal, not by its amplitude.

3. If over-modulation is to be avoided ($m \leq 1$), the maximum sideband amplitude is $A/2$.

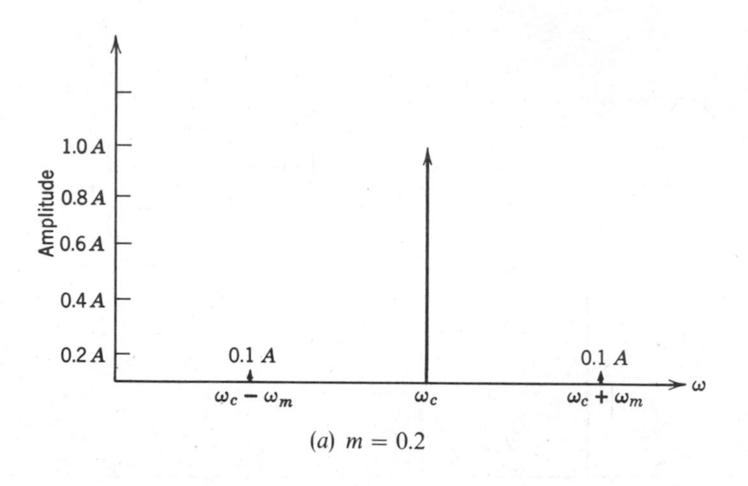

(a) $m = 0.2$

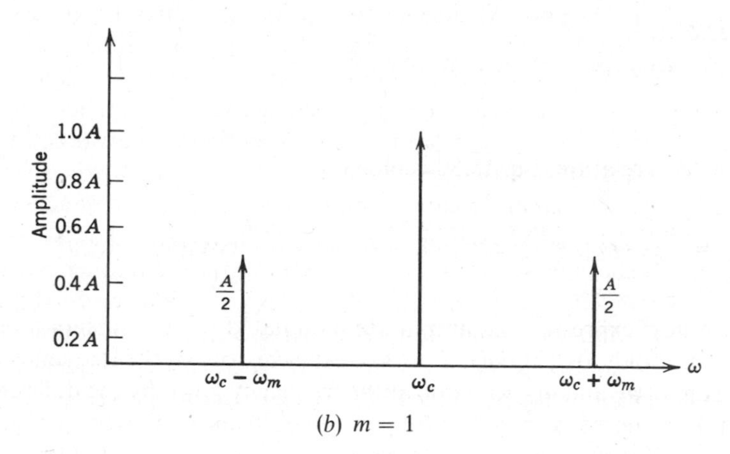

(b) $m = 1$

Figure 15.5
Frequency spectra of AM waves.

Let us extend this notion to a speech waveform. Normal speech can be represented as superpositions of sinusoids in the frequency range of approximately 200 Hz to 5 kHz. A schematic representation of the spectrum of a speech waveform is shown in Fig. 15.6*a*. For this plot, the frequency has been expressed in Hz (symbol f) instead of in radians per second (symbol ω). Figure 15.6*b* shows the spectrum of a carrier at frequency f_c, amplitude modulated by the same speech waveform. The sideband spectra are symmetrical about the carrier, and are faithful copies of the shape of the speech waveform spectrum.

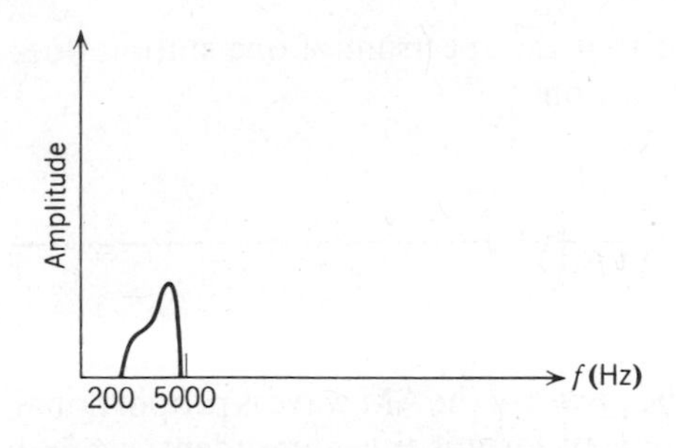

(*a*) Spectrum of a speech waveform

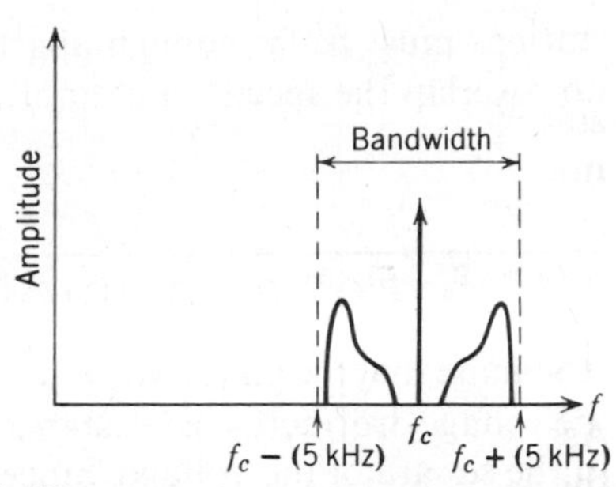

(*b*) Spectrum of a carrier at f_c modulated by the speech waveform

Figure 15.6

The spectrum of Fig. 15.6*b* illustrates a fundamental fact about the communication of information. We have already commented that a sinusoid with constant amplitude, frequency, and phase can carry no information. If we modulate the amplitude of a sinusoid with another signal, we *encode* some information onto the carrier. Figure 15.6*b* shows that the *spectrum of the modulated waveform has broadened* from a single line at the carrier frequency to a set of lines or, in the case of the speech waveform, to a whole range of frequencies around the carrier frequency. Thus, by modulating the sine wave, we cause its spectrum to occupy a finite *bandwidth*, or frequency range. The higher the frequency of the information waveform, the broader the bandwidth of the modulated wave. This determines the frequency spacing for radio or television stations. Each AM radio station needs 10 kHz of bandwidth to transmit amplitude-modulated speech and music waveforms. Each station is alotted a *channel*, with an assigned carrier frequency and a fixed 10 kHz bandwidth (see Fig. 15.7). The carrier frequencies of different

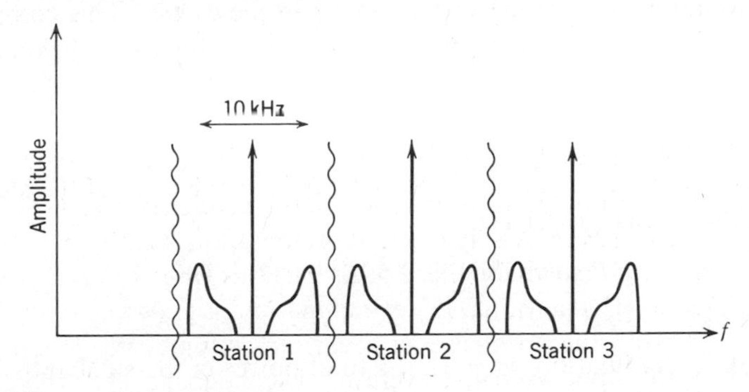

Figure 15.7
Frequency channels for different AM stations.

stations must be far enough apart so that the spectrum of one station does not overlap the spectrum of another station.

15.1.5 Power in an AM Wave

Assuming that the amplitude modulated wave, Eq. 15.5 or Eq. 15.7, represents the voltage drop across a resistance, the power in the AM wave is proportional to the square of the voltage. Since the waveform is time dependent, we find the *average power* by squaring Eq. 15.7 and averaging over time. (The bar above each term represents "time-average.")

$$\begin{aligned}\overline{v^2} = \overline{A^2 \cos^2 \omega_c t} + \left(\frac{A^2m^2}{4}\right)\overline{\cos^2 [(\omega_c + \omega_m)t]} + \left(\frac{A^2m^2}{4}\right)\overline{\cos^2 [(\omega_c - \omega_m)t]} \\ + A^2m\{\overline{\cos \omega_c t \cos [(\omega_c + \omega_m)t]} + \overline{\cos \omega_c t \cos [(\omega_c - \omega_m)t]}\} \\ + \frac{A^2m^2}{2}\overline{\{\cos [(\omega_c + \omega_m)t] \cos [(\omega_c - \omega_m)t]\}}\end{aligned} \qquad (15.8)$$

The time-average of $\cos^2\omega t$ is equal to 1/2 for any frequency ω. Furthermore, the time average of the product of two sinusoids at different frequencies is identically zero. Therefore, the average power in the AM wave is a sum of terms, one from each sinusoidal frequency component.

$$\overline{v^2} = \frac{A^2}{2} + \left(\frac{A^2m^2}{8} + \frac{A^2m^2}{8}\right) = \left(\frac{A^2}{2}\right)\left[1 + \frac{m^2}{2}\right] \qquad (15.9)$$

The first term represents the average power in the carrier. This component of the total power is not affected by the amount of modulation. The remaining term is proportional to the square of the amplitude of the modulation, and represents the average power in the sidebands.

It is interesting to compare the amount of power in the sidebands to the amount of power in the carrier. If we use Eq. 15.9, the ratio of the sideband power to the carrier power is

$$\frac{P(\text{sidebands})}{P(\text{carrier})} = \frac{2 \times (A^2m^2/8)}{A^2/2} = \frac{m^2}{2} \qquad (15.10)$$

Even at 100% modulation ($m = 1$), the total power in the sidebands is only 1/2 of the power in the carrier. That is, only 1/3 of the *total* power is in the sidebands, the remaining 2/3 of the power is in the carrier.

Since the carrier component of an AM wave does not carry any useful information, communication of information using AM is very wasteful of power. Nevertheless, amplitude modulation is widely used for commercial broadcasting. One reason is that compared to other forms of modulation, AM requires a relatively narrow bandwidth. The AM bandwidth needed is approximately twice the highest modulation frequency. The other forms of modulation, FM and PM, require much more bandwidth to carry the same modulation signal. A second reason for the widespread use of AM is that the *demodulation* process, the recovery of the modulation from the AM wave, is extremely simple (see Section 15.2.4 below).

There are several methods of amplitude modulation that have been devised to avoid the wasteful transmission of a carrier along with the information-containing sidebands. The best known of these is called single-sideband (SSB) transmission, in which only one sideband of the total spectrum is transmitted. Not only does this method concentrate the transmitted power in the sideband, but it reduces the total bandwidth needed to transmit a given signal. It is not surprising, however, that the price for this bandwidth and power economy is an increased complexity in both the modulation and demodulation processes (see the references at the end of the chapter.)

15.2 METHODS OF AMPLITUDE MODULATION

The key to understanding how to produce an amplitude-modulated wave lies in the form of the wave itself. The AM wave of Eq. 15.5 consists of a *product* between the carrier and the modulating waveform. Therefore, any circuit that can produce a product of two inputs can be used for producing amplitude modulation. We shall find that either a circuit with a time-varying transfer function or a nonlinear circuit can be used. Examples of each are examined below.

15.2.1 Emitter-Coupled-Pair Modulator

Figure 15.8 shows an emitter coupled pair (Q_1 and Q_2) biased by a transistor current source (Q_3). This circuit was analyzed as an amplifier in Chapter 10. We shall make use of the results of that analysis to show how the circuit can be used as a modulator.

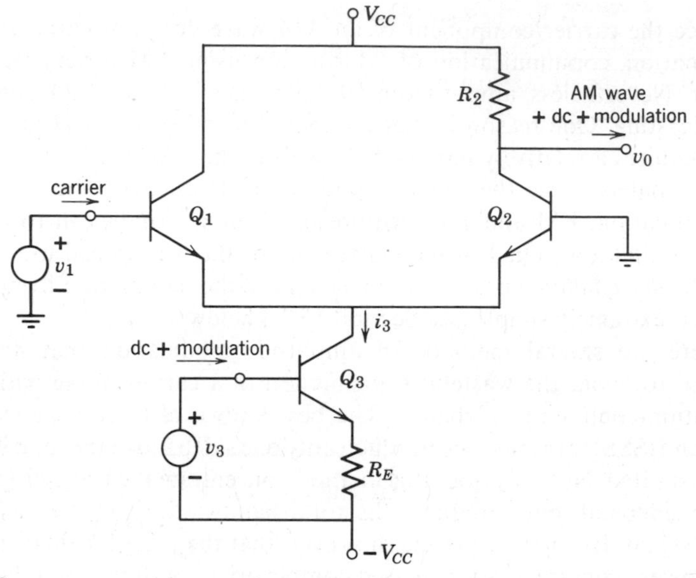

Figure 15.8
Emitter-coupled pair used to produce AM.

It is important in this discussion to keep track of all Fourier components of all signals. We assume that v_1 is a pure sinusoid (no dc component) at the carrier frequency:

$$v_1 = A \cos \omega_c t \tag{15.11}$$

We further assume that v_3 is a sum of dc bias component V_B plus the modulation v_m.

$$v_3 = V_B + v_m \tag{15.12}$$

For the present discussion, we assume that the carrier frequency ω_c is much higher than any frequency component in the modulation waveform v_m. (This restriction is not necessary; see Section 15.2.3 below.) Finally, we shall represent the output v_O as a sum of a dc component V_O and a small-signal component v_o.

$$v_O = V_O + v_o \tag{15.13}$$

We shall develop an expression for the output waveform and the output spectrum in terms of the waveforms and spectra at the two inputs.

If we use the large-signal model, the dc component of i_3 is

$$I_3 = \frac{V_B - V_{BE3}}{R_E} \tag{15.14}$$

where V_{BE3} is the base-to-emitter voltage of Q_3. The *variation* in i_3 due to v_m arises from emitter-follower action. The variation in the voltage drop across R_E follows v_m with nearly unity gain. Since the current through R_E equals i_3 to within a few percent, the total bias current is approximately

$$i_3 = I_3 + \frac{v_m}{R_E} \tag{15.15}$$

In the absence of a carrier ($v_1 = 0$), the bias current i_3 divides equally between Q_1 and Q_2, giving rise to a time-varying output

$$v_O = V_{CC} - \left(\frac{i_3}{2}\right)R_2 \tag{15.16}$$

which has a dc component

$$V_O = V_{CC} - \frac{V_B - V_{BE3}}{2R_E}R_2 \tag{15.17}$$

plus a component proportional to the modulation

$$-v_m\frac{R_2}{2R_E} \tag{15.18}$$

Since the variation of v_m is presumed to be slow compared to the carrier frequency, the response to a carrier ($v_1 \neq 0$) can be estimated from the time-varying transfer characteristic in Fig. 15.9. Three transfer characteristics are

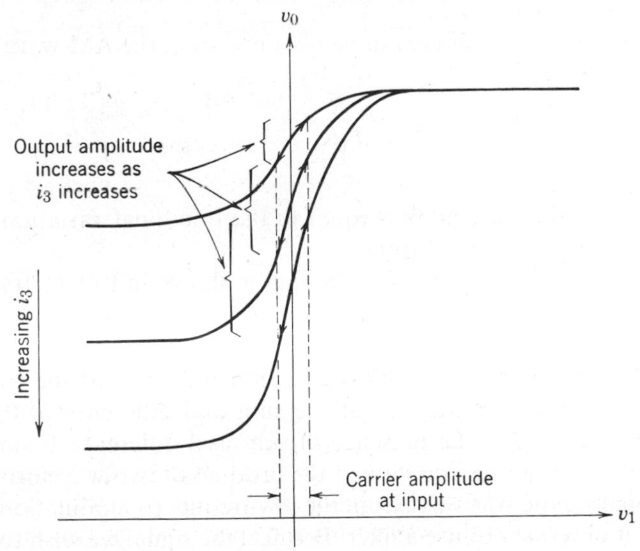

Figure 15.9
Emitter-coupled pair transfer characteristics for three different bias currents. If i_3 is time-varying the output amplitude is also time-varying.

shown, each one for a different value of i_3. As i_3 varies slowly, a fixed amplitude of carrier at the input will produce differing amplitudes at the output. That is, the output amplitude is *modulated* by the variation in i_3.

In quantitative terms, the slope of the v_1-to-v_0 transfer characteristic at the operating point ($v_1 = 0$) is, from Section 10.3.2,

$$\frac{v_o}{v_1} = \frac{qR_2}{4kT} i_3 \tag{15.19}$$

Therefore, there is a component in the output waveform that contains the product of v_1 and i_3. It is this product term that contains the AM waveform. Combining all results, we have the total output waveform

$$v_0 = \underbrace{\left(V_{CC} - \frac{I_3R_2}{2}\right)}_{\text{dc}} - \underbrace{\left(\frac{R_2}{2R_E}\right)v_m}_{+\ \text{modulation}} + \underbrace{\frac{qR_2}{4kT}\left(I_3 + \frac{v_m}{R_E}\right)v_1}_{+\ \text{AM wave}} \tag{15.20}$$

If we assume sinusoidal modulation

$$v_m = B \cos \omega_m t \tag{15.21}$$

Then the modulation index can be obtained from the AM wave of Eq. 15.20

$$m = \frac{B}{I_3R_E} = \frac{B}{V_B - V_{BE3}} \tag{15.22}$$

Thus the modulation index is equal to the fractional variation in the bias-plus-modulation supplied to Q_3.

The frequency spectrum of the output is shown in Fig. 15.10. The dc bias component and the component at the modulation frequency are shown together with the carrier-plus-two-sidebands characteristic of an AM wave. Altogether there are four small-signal sinusoids, two at the two input frequencies (ω_c and ω_m) and two at the sum and difference of the input frequencies ($\omega_c \pm \omega_m$). The presence of sum and difference frequencies is an automatic result of having formed the product of two waveforms.

To separate the AM wave from the low-frequency modulation component of v_0, it is necessary to use a filter. Because the signal we wish to keep is at a higher frequency than the signal we wish to reject, we can use a *high-pass filter*, such as the one shown in Fig. 15.11. At low frequencies, the capacitor

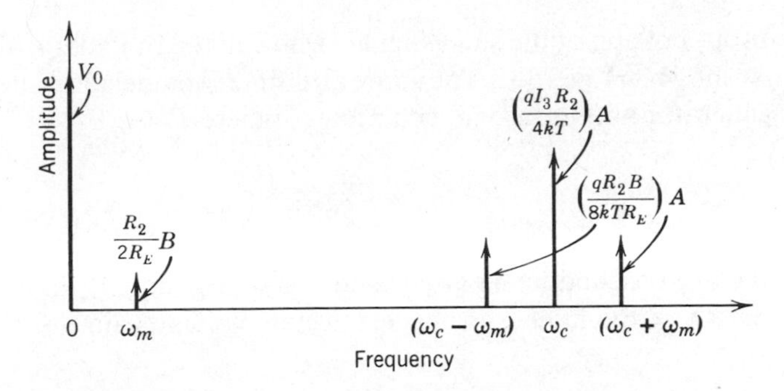

Figure 15.10
Spectrum of the output waveform from the circuit of Fig. 15.8.

looks like an open circuit, so no low-frequency current flows through R_3. At sufficiently high frequencies, however, the capacitor looks like a short circuit. It is simple to determine design restrictions on R_3 and C so that the capacitor will effectively filter out the low-frequency components of v_0.

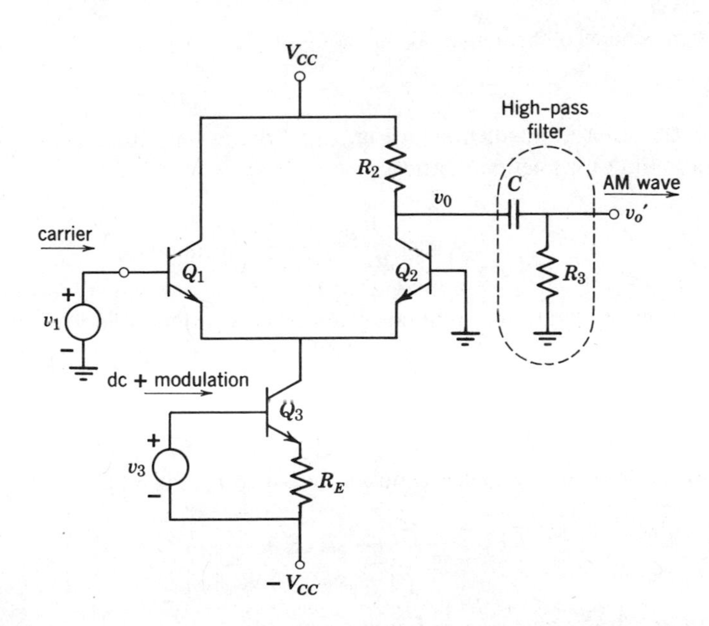

Figure 15.11
Use of a high-pass filter to remove the low-frequency components of v_0.

The output portion of the small-signal circuit of the modulator-filter combination is shown in Fig. 15.12. The entire amplifier is modeled as a dependent-current generator with transconductance g_m, where

$$g_m = \frac{qi_3}{4kT} \tag{15.23}$$

This choice of transconductance makes the voltage gain $g_m R_2$ in the absence of the filter. Once the filter is connected, a new voltage gain must be determined.

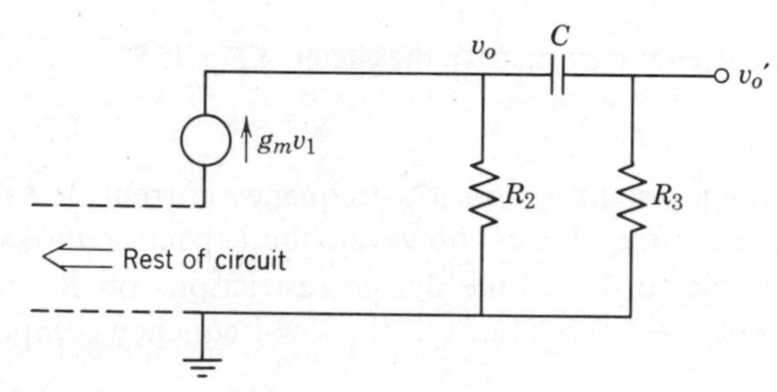

Figure 15.12
Small-signal model of the output portion of Fig. 15.11.

With the filter connected as in Fig. 15.12 the system function between the small-signal components, v_o and v_1, is

$$\frac{V_o}{V_1} = g_m \left[\frac{R_2(R_3 + 1/sC)}{R_2 + R_3 + 1/sC} \right] \tag{15.24}$$

Furthermore, v_o and v_o' are related by a simple voltage divider

$$\frac{V_o'}{V_o} = \frac{R_3}{R_3 + 1/sC} \tag{15.25}$$

Therefore, the overall transfer function between v_1 and v_o' is

$$\frac{V_o'}{V_1} = g_m \left[\frac{sR_2R_3C}{1 + s(R_2 + R_3)C} \right] \tag{15.26}$$

This relation has a zero at $s = 0$ and a pole at

$$s_a = -\frac{1}{(R_2 + R_3)C} \tag{15.27}$$

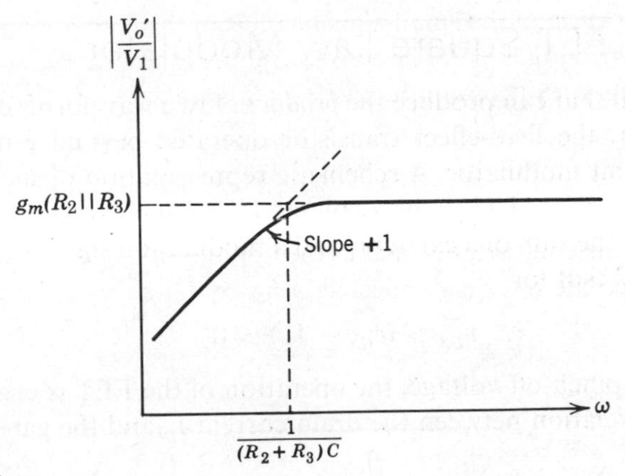

Figure 15.13
Frequency response for Eq. 15.26.

The magnitude versus frequency plot for the transfer function of Eq. 15.26 is shown in Fig. 15.13. If one chooses C and R_3 to placc the pole between ω_m and ω_c, the low-frequency component at ω_m is attenuated while the high-frequency components that constitute that AM wave are passed with an overall effective gain

$$g_m(R_2 \| R_3) \tag{15.28}$$

The maximum attenuation of the low-frequency components will occur when the pole location is as high as possible. Therefore, one should choose

$$\omega_m \ll \frac{1}{(R_2 + R_3)C} < \omega_c \tag{15.29}$$

The resultant spectrum of v_o' is shown in Fig. 15.14. The dc component is blocked by the capacitor, while the low-frequency component at ω_m is strongly attcnuated.

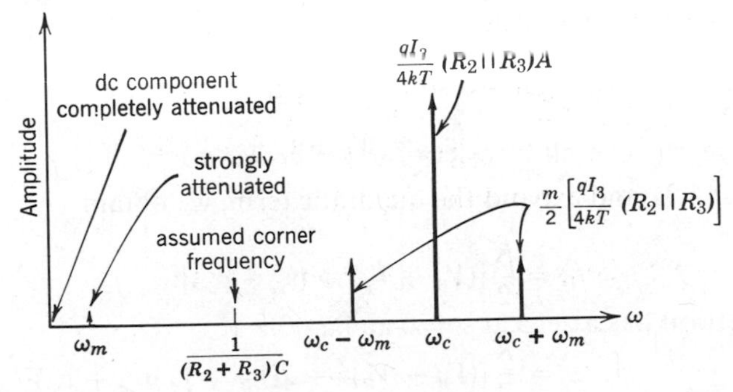

Figure 15.14
Output spectrum for the circuit of Fig. 15.11.

15.2.2 An FET Square Law Modulator

Since any circuit that can produce the *product* of two waveforms can be used as a modulator, the field-effect transistor operated beyond pinch-off can make an excellent modulator. A schematic representation of such a modulator is shown in Fig. 15.15. A bias voltage V_B is supplied to the gate of an FET along with the sum of a carrier v_c and a modulation signal v_m. We recall from Section 9.2 that for

$$v_{DS} > (v_{GS} - V_P) > 0 \tag{15.30}$$

where V_P is the pinch-off voltage, the operation of the FET is characterized by a quadratic relation between the drain current i_D and the gate-to-source voltage v_{GS}.

$$i_D = \frac{K}{2}(v_{GS} - V_P)^2 \tag{15.31}$$

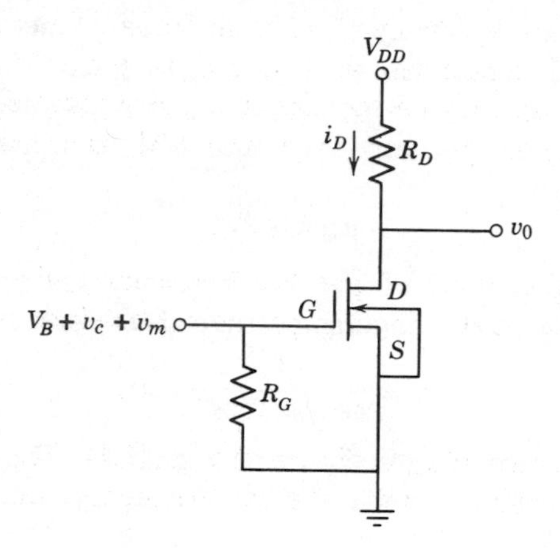

Figure 15.15
An FET square-law modulator.

If we insert the total gate-to-source voltage

$$v_{GS} = V_B + v_c + v_m \tag{15.32}$$

into Eq. 15.31 and expand the quadratic term, we obtain

$$\begin{aligned} i_D &= \frac{K}{2}[(V_B - V_P) + (v_c + v_m)]^2 \\ &= \frac{K}{2}[(V_B - V_P)^2 + 2(V_B - V_P)(v_c + v_m) \\ &\quad + v_c^2 + v_m^2 + 2v_cv_m] \end{aligned} \tag{15.33}$$

The last term in Eq. 15.33 above contains the *product* of v_c and v_m. Thus the output v_0 will contain an AM wave that can be separated from the other frequency components with a filter.

15.2.3 Mixers

The two circuits discussed above produced the modulation of one signal by another. Throughout the discussion, it was assumed that the carrier frequency ω_c was much greater than the modulation frequency ω_m. This restriction is not necessary. *Any* two frequencies that are combined in product form will produce components at the sum of the frequencies and at the difference of the frequencies (see Eq. 15.6). *Mixers* are circuits designed to combine two frequencies in this way to produce components at the sum and difference frequencies. Their principal use is in radio communications as discussed in Section 15.3.

In principle, either of the modulator circuits discussed previously will continue to work when the restriction $\omega_c \gg \omega_m$ is removed. Thus, either of the circuits can be used as mixers over a wide range of frequencies. In addition, many specialized mixer circuits have been developed for radio, television, and radar use. They fall beyond the scope of this text, but may be found in the references.

15.2.4 Demodulation of an AM Wave: The Peak Detector

In order to recover the modulation waveform from the AM wave, it is necessary to convert the amplitude variations of a wave at the carrier frequency into a signal at the modulation frequency. This demodulation process cannot be done by a linear circuit, because a linear circuit cannot change the frequency of a sinusoid.

Almost any kind of nonlinear circuit, however, will produce some degree of demodulation. Consider, for example, the FET circuit of Fig. 15.15. Suppose one were to feed an AM wave into the FET instead of $(v_c + v_m)$. The quadratic relation between output and input guarantees that the output contains the *product* between the carrier and a sideband. Since the product of two sinusoids contains both the sum and difference frequencies, and since the difference between the carrier and either sideband is equal to the modulation frequency, there would be an output component at the modulation

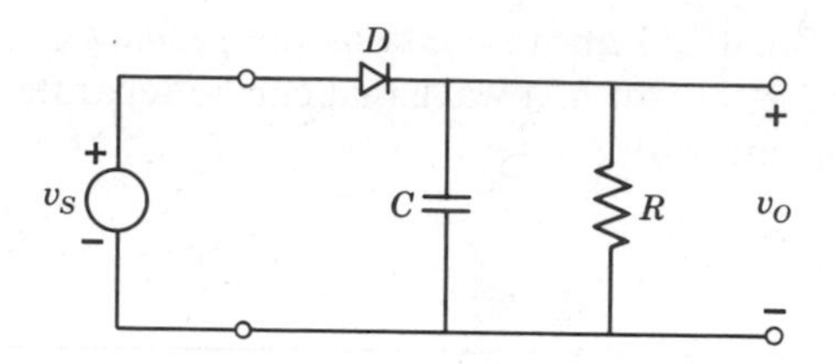

Figure 15.16
The peak detector.

frequency. But the FET demodulator is relatively inefficient, in that only a small fraction of the output represents the demodulated signal.

A far better demodulator is based on the peak sampler of Fig. 15.16. This circuit was originally presented in Section 8.4.1 for use in rectifiers. A diode (presumed ideal for this discussion) is connected between a source and a parallel RC network. Assuming for the moment that v_S is a sine wave, the analysis of the output waveform is very simple. Because of the ideal diode, the voltage v_O can never fall below the instantaneous value of v_S. If v_S tries to climb above v_O, the diode conducts (Fig. 15.17*a*), and enough current flows to charge the capacitor up to the value of v_S. Once v_S passes a

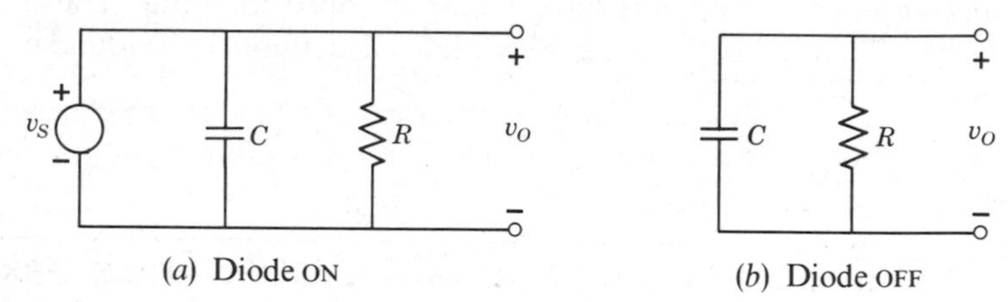

Figure 15.17

peak, however, the diode becomes back-biased, and the network of Fig. 15.17*b* results. The voltage on the capacitor can only discharge with a time constant of RC. Usually, one picks the RC time constant to be very long compared to the period of the sine wave, so the fractional decay is very small during one cycle. The waveforms for v_S and v_O are shown in Fig. 15.18. The charging of the capacitor through the ideal diode alternates with the gradual discharge of the capacitor through the resistance R. (When a real diode is used, the output waveform differs in two ways from that in Fig. 15.18. First, the peak capacitor voltage is about 0.6 V less than the peak source voltage because of the forward drop across a diode. Second, the charging of the capacitor is not infinitely fast because the diode has some small but finite forward resistance.)

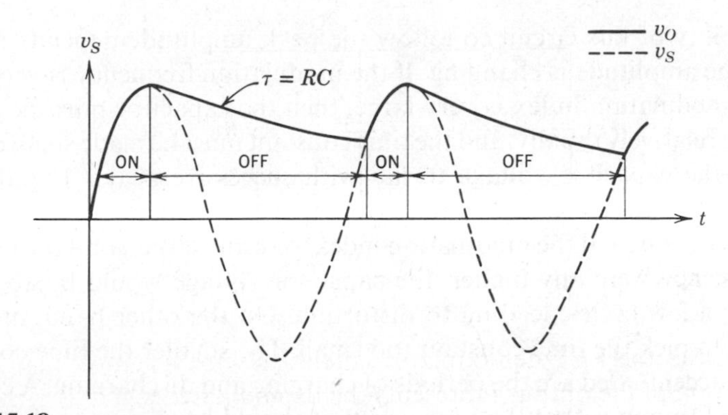

Figure 15.18
Peak detector waveforms.

The result of these alternating periods of charging and discharging is that for an *RC* time constant long compared to the sine-wave period, the output voltage remains very near the *peak* voltage of the sine wave. For this reason, the circuit of Fig. 15.16 is called a *peak detector*.

Let us examine the peak detector circuit when excited by an AM wave. Figure 15.19 illustrates the alternate charging and discharging of the capacitor in this case. The output voltage follows the peaks as they rise and fall with the amplitude modulation. Therefore, the capacitor voltage varies with time exactly as the original modulation, except for tiny peaks during the charging. Normally these peaks do not interfere with subsequent amplification of the low-frequency signal.

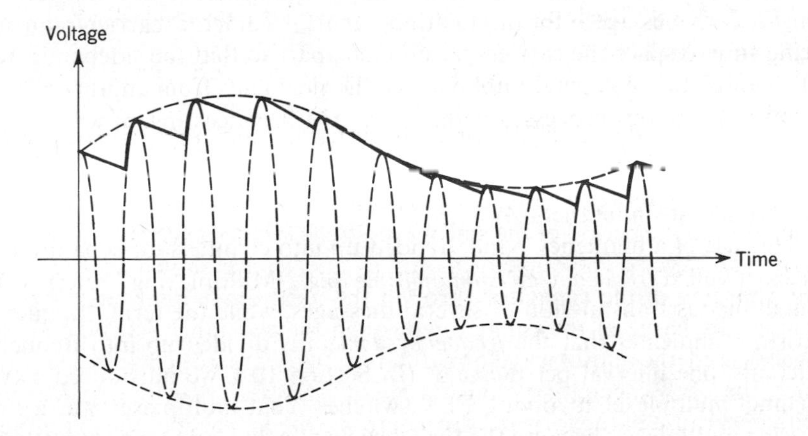

Figure 15.19
Output waveform for a peak detector driven by an AM wave.

The ability of this circuit to follow the peak amplitude depends on how rapidly the amplitude is changing. If the modulation frequency is very large, or if the modulation index is very large, then the capacitor must be able to discharge relatively rapidly, and the time constant must be made small enough to allow the capacitor voltage to fall with successive peaks. Figure 15.19 illustrates an extreme case in which the capacitor voltage is barely able to fall quickly enough. If the modulation index were any larger, or if the modulation frequency were any higher, the capacitor voltage would be above the peaks for a few cycles, leading to distortion. On the other hand, one does not want to pick the time constant too small. The smaller the time constant, the more accentuated are the periods of charging and discharging. A convenient rule of thumb is that the time constant should be chosen so that

$$RC < \frac{1}{m\omega_m} \tag{15.34}$$

15.3 THE SUPERHETERODYNE RECEIVER

15.3.1 Multiplexing in the Frequency Domain

In discussing the frequency spectrum of an amplitude-modulated wave, we introduced the notion of a *channel*, a band of frequencies within which a carrier plus its associated sidebands are intended to fit. If one wishes to transmit many messages simultaneously without mutual interference, one can use each message as the modulation signal for a different carrier frequency, being sure to space the carriers far enough apart so that the sidebands from one modulation pattern do not overlap the sidebands from another pattern. An example of this process is commercial AM broadcasting, in which many stations are broadcasting similar messages. Each station is assigned a carrier frequency and bandwidth in non-overlapping channels (Fig. 15.7), and all can broadcast simultaneously.

This use of a frequency band, divided up into channels for separate messages, is called *frequency-division multiplexing*. "Multiplexing" refers to the simultaneous transmission of several messages, while the term "frequency-division" indicates that the *frequency domain* is divided up into frequency intervals, one interval per message. (In Section 10.4, we introduced a two-channel multiplexer involving FET switches. That multiplexer was a *time-division* multiplexer, because the *time domain* was divided into time intervals, each time interval assigned for the transmission of a sample of a specific message.)

15.3.2 The Problem of Frequency Selectivity

Consider the waveform one might have on a radio antenna, with perhaps 20 stations broadcasting simultaneously, and with the electromagnetic radiation from each broadcaster producing a small component of current in the antenna. The spectrum of the current waveform in the antenna might be similar to the multiple-station spectrum of Fig. 15.7, with closely spaced channels each containing a carrier plus its sidebands. To receive the information in one channel without interference from the other simultaneous channels, it is necessary to use *frequency-selective circuits*. (As a contrast, the time-division multiplexing example of Section 10.4, required *time-selective circuits* with synchronized switches to separate one message from the other.)

One approach to the problem of frequency selectivity would be to build a sharply tuned filter that can pass one channel (carrier plus sidebands) while attenuating all other channels. The ease with which this can be done depends on the carrier frequency and the signal bandwidth. If the carriers are audio frequencies and the signal bandwidths are small (for slowly-varying signals), it is indeed possible to build filters that can select one channel at a time, and that can be tuned from channel to channel with relative ease. If the carriers are low radio frequencies (one the order of 1 MHz) and the bandwidths are typical AM bandwidths (10 kHz), it is more difficult, although still possible, to build a multiple-stage filter capable of separating one channel (or station) from the others. These filters, however, are difficult to tune from station to station.

The superheterodyne receiver, to be discussed in the next section, is a particularly clever solution to the frequency-selectivity problem for carrier frequencies in the radio-frequency range and above. It combines highly selective filtering with easy tuning. The key to its operation is the use of a mixer, as we will now explain.

15.3.3 Block Diagram of the Superheterodyne Receiver

Figure 15.20 shows the block diagram of a superheterodyne receiver. It contains an *rf* amplifier that strengthens the weak signals from the antenna, a local oscillator, mixer, and intermediate frequency (*if*) amplifier, whose functions are explained below, a peak detector for demodulation of the AM, and an audio amplifier for amplification of the audio-frequency signal.

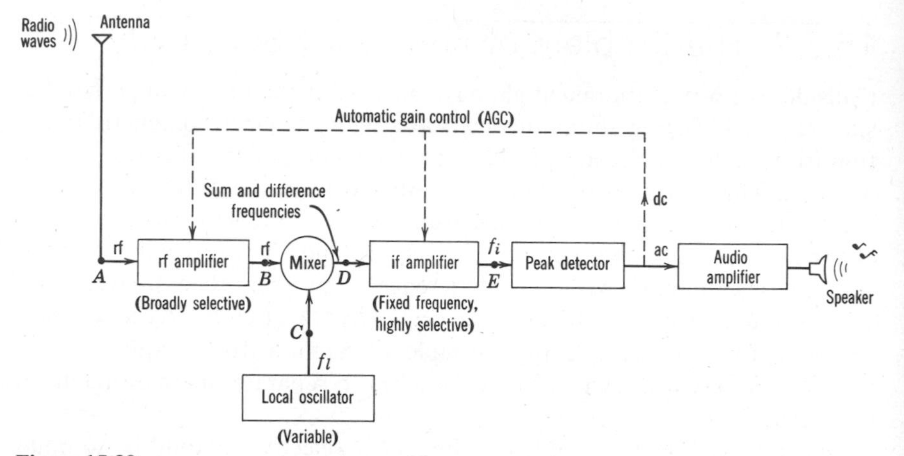

Figure 15.20
Block diagram of a superheterodyne receiver.

Figure 15.21 illustrates schematically how the *rf* amplifier, which is broadly tuned, selects several channels or stations from the complex spectrum of signals that impinge on the antenna. In the following discussion we shall refer to a carrier plus its sidebands by a single symbol, even though we recognize that the spectrum around f_1, for example, contains a continuous band of frequencies close to the carrier.

Figure 15.22 shows how the superheterodyne receiver works. The intermediate frequency amplifier, or *if amplifier*, is a highly selective amplifier, with a passband perfectly matched to the bandwidth of a single station and with a center frequency at some fixed intermediate frequency f_i. An "intermediate" frequency is a frequency below the *rf* frequencies coming in on the antenna and above the audio frequencies that constitute the modulation. For example, in an AM receiver with *rf* frequencies between 550 and 1600 kHz, the usual intermediate frequency is 455 kHz. In commercial FM receivers, which cover the band from 88 to 108 MHz, the intermediate frequency is usually chosen at 10.7 MHz.

A *local oscillator*, contained within the receiver, is tuned to supply a sinusoid at frequency f_l, located above the *rf* frequencies by an amount approximately equal to f_i. The *mixer* forms the product of the incoming *rf* with f_l, and produces both sum and difference frequencies. Whenever the local oscillator is adjusted so that the difference frequencies from a particular station fall within the passband of the *if* amplifier, the *if* amplifier will amplify that particular set of carrier-plus-sidebands while rejecting other nearby spectra. In the example of Fig. 15.22, f_l has been adjusted so that the difference frequencies in the vicinity of f_2 fall within the *if* passband. Therefore, at the

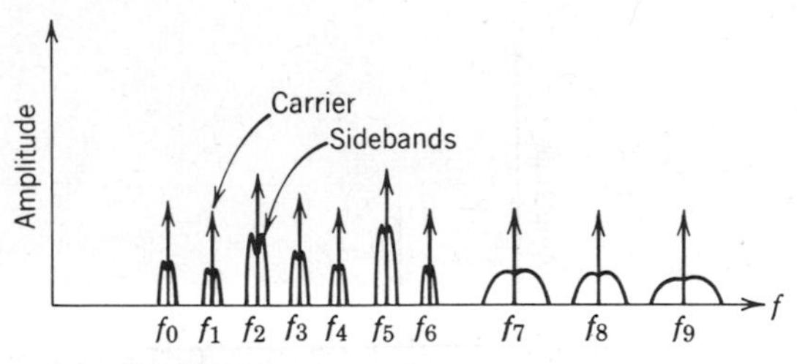

(a) Antenna spectrum (point A)

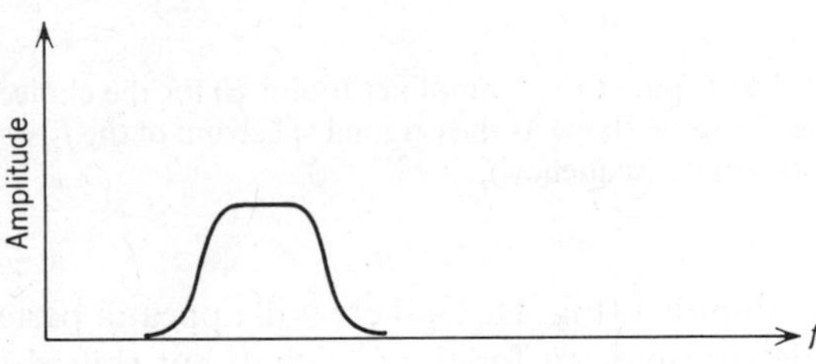

(b) rf amplifier passband

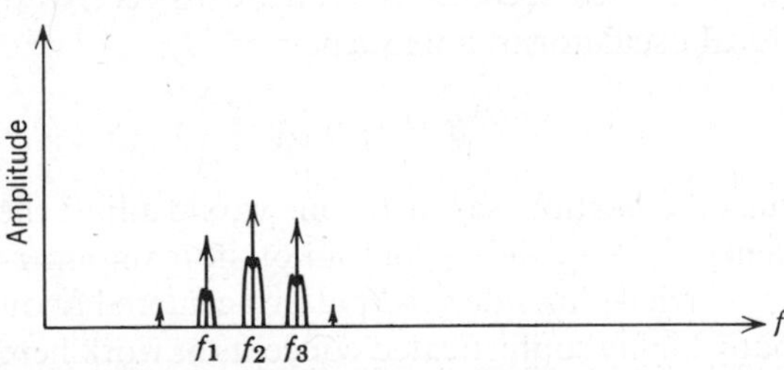

(c) rf amplifier output spectrum (point B)

Figure 15.21
The rf amplifier is broadly tuned, passing several stations at once, but rejecting frequencies far removed from the region of interest.

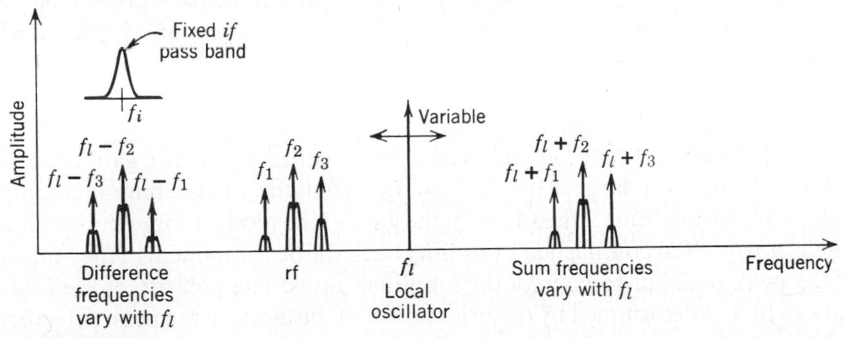

Figure 15.22
Spectrum at the output of the mixer (point D). In this illustration, f_l has been adjusted so that the difference frequency $f_l - f_2$ falls within the if passband.

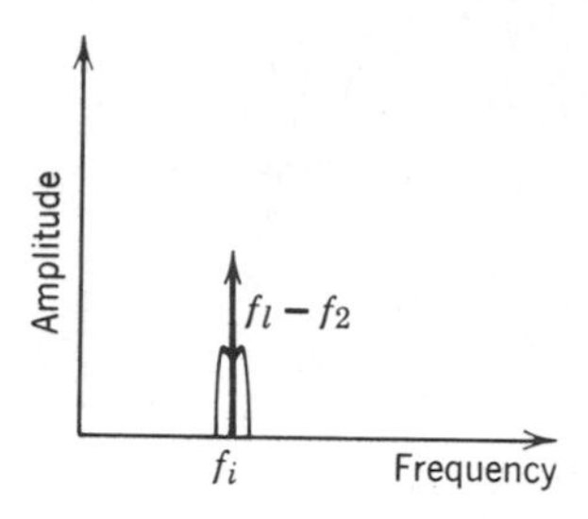

Figure 15.23
The spectrum at the output of the *if* amplifier (point E) for the choice of f_l in Fig. 15.22. The spectrum has the same shape as the original spectrum of the f_2 station, but is shifted down to the intermediate frequency f_i.

output of the *if* amplifier (Fig. 15.23) there will appear a perfect replica of the carrier-plus-sideband pattern for the f_2 signal, but shifted to a new center frequency at f_i. Thus, the receiver is "tuned" to receive the f_2 station by adjusting the local oscillator to a frequency

$$f_l = f_2 + f_i \tag{15.35}$$

To receive some other station, say at f_3, one would adjust the local oscillator to a new frequency, $f_3 + f_i$, and the output of the *if* amplifier would then be a replica of the f_3 carrier-plus-sideband pattern centered about f_i.

There are many highly sophisticated concepts at work here, and it is worth reviewing some of them. The use of a mixer in combination with a fixed frequency *if* amplifier permits both easy tuning (merely change f_l) and good selectivity (because the *if* amplifier response can be optimized). The local oscillator is chosen *above* the *rf* frequency (hence, the name *super*-heterodyne) because it requires a smaller fractional change in f_l to cover a given band if f_l is higher than the *rf*. The peak detector design can be genuinely optimized because it is always driven with an AM wave at the fixed intermediate frequency. Thus a single detector can be used to demodulate all stations regardless of the initial carrier frequency.

A useful feature, found on most receivers, is called automatic gain control (AGC). There is a large variation in the strengths of different incoming signals, requiring some adjustment in the gain of the receiver for every station. This is normally accomplished by feeding back the dc component of the output of the peak detector to control the amplifier gains. The polarity of the feedback (which is determined by the orientation of the diode in the peak detector) must be chosen so that the amplifier gains get smaller as the amplitude of the *if* output gets larger, thus stabilizing the signal level reaching the detector and audio amplifier.

Thus far, no mention has been made of the possibility of the receiver responding to an incoming signal located *above* f_l by an amount f_i. Indeed, if a signal at frequency $f_l + f_i$ is allowed to reach the mixer, the difference frequency when mixed with f_l would interfere with the desired station at $f_l - f_i$. Therefore, the "front-end" *rf* amplifier is broadly tuned to reject frequencies above f_l. The potentially interfering frequency at $f_l + f_i$ is called the *image frequency*. The degree of rejection of signals at the image frequency is one measure of the quality of a receiver.

A final note on commercial receivers: the block diagram of Fig. 15.20 was drawn as if each function had its own separate circuit elements. In modern design, particularly in widely sold inexpensive AM receivers, several functions are combined. It is not uncommon, for example, to find the *rf* amplifier, local oscillator, and mixer all combined in a rather complicated nonlinear one-transistor circuit. It takes some practice to recognize and analyze such circuits, but the functions performed are always the same.

REFERENCES

R15.1 Mischa Schwartz, *Information Transmission, Modulation, and Noise* (New York: McGraw-Hill, 1970).

R15.2 R. Ralph Benedict, *Electronics for Scientists and Engineers* (Englewood Cliffs, N.J.: Prentice-Hall, 1967).

R15.3 Frederick E. Terman, *Electronic and Radio Engineering* (New York: McGraw-Hill, 1955).

R15.4 Lawrence B. Arguimbau, *Frequency Modulation* (New York: John Wiley, 1956).

QUESTIONS

Q15.1 Describe the three kinds of modulated waveforms when digital signals (i.e., pulses) are used as the modulation.

Q15.2 Could a square wave be used as a carrier? How would the spectrum of an amplitude-modulated square wave compare with that of an amplitude-modulated sine wave of the same frequency?

Q15.3 Given a sine wave at ω_o, what kind of circuit might you use to produce a sine-wave signal at $3\omega_o$?

Q15.4 Compare the relative ease of tuning a superheterodyne receiver with tuning a receiver in which the local-oscillator frequency is below the carrier.

EXERCISES

E15.1 In troubleshooting signal-processing systems, it is sometimes important to differentiate between an amplitude-modulated signal and an ordinary linear superposition.

Given the two signals:

$$f(t) = c(t) + m(t)c(t)$$

and

$$g(t) = c(t) + m(t)$$

where $c(t) = \cos \omega_o t$ is the carrier and $m(t) = m_o \cos \omega_m t$ is the message, with $m_o < 1$ and $\omega_m \ll \omega_c$, sketch the waveforms over two cycles of the message signal, and indicate how the AM signal differs from the linear superposition.

E15.2 Plot these spectra and waveforms.

(a) $2 \cos 100\pi t + 0.1 \cos 90\pi t + 0.1 \cos 110\pi t$

(b) $0.1 \cos 90\pi t + 0.1 \cos 110\pi t$

E15.3 Plot these spectra and waveforms.

(a) $2(1 + 0.1 \cos 10\pi t) \cos 100\pi t$

(b) $0.2 \cos 10\pi t \cos 100\pi t$

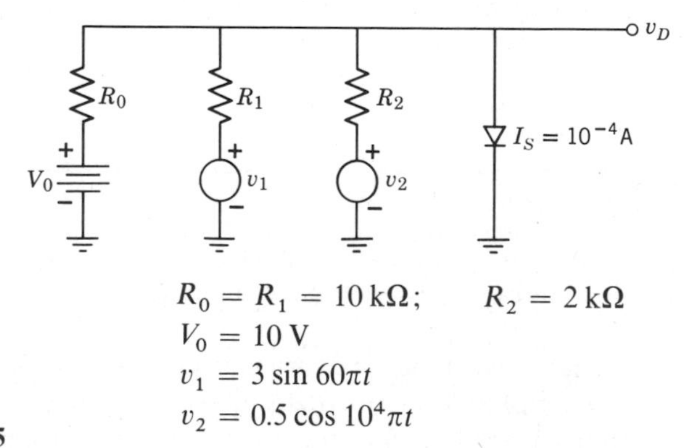

Figure E15.5

E15.4 The waveforms of E15.2b and E15.3b represent an AM wave from which the carrier has been removed. Can these waves be demodulated with a peak detector? (This illustrates with a two-sideband example the fact that single-sideband receivers must have provision for reinserting the carrier into the signal before demodulation).

E15.5 Find the output waveform v_D in Fig. E15.5 assuming an exponential diode. Express v_D in the form of an AM wave. Find the modulation index m.

E15.6 If, in the circuit of Fig. E15.6, $v_2 = -1 + \cos 20\pi t$ and $v_1 = 0.1 \cos 10^3\pi t$, find v_O. Express v_O as an AM wave. Find the modulation index m.

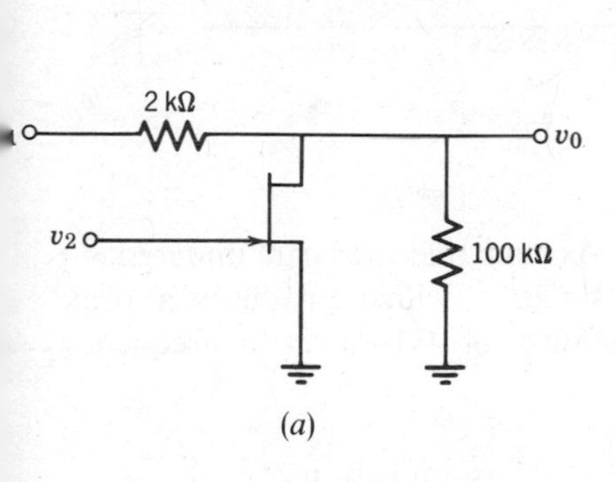

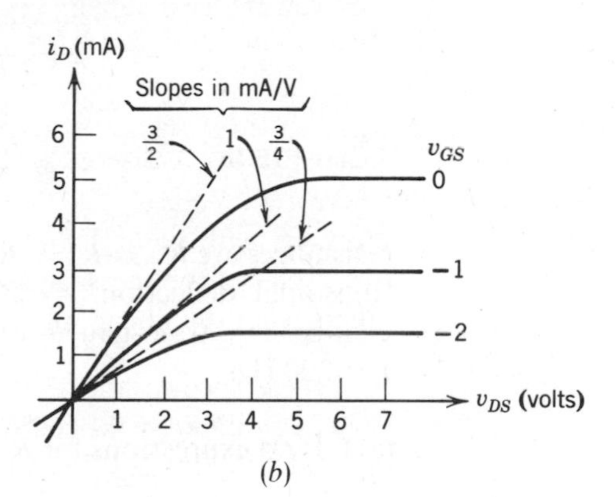

Figure E15.6

E15.7 A circuit has the following transfer characteristic:

$$v_O = Av_1 + Bv_2 + Cv_1v_2$$

where v_1 and v_2 are inputs and where A, B, and C are constants.

Assume v_1 is an AM wave centered at frequency f_1 and assume v_2 is a sine wave at frequency f_l. Plot the spectrum of v_O. Show that by adjusting f_l, the signal spectrum at f_1 can be made to appear at different center frequencies.

E15.8 Assume in E15.7, that f_l is above f_1 and is adjusted so that $f_l - f_1 = f_i$ (the intermediate frequency). Show that if v_1 contains any component at a frequency $f_l + f_i$ (the image frequency), the output will contain both the original signal at f_1 and the image signal at $f_1 + 2f_i$.

PROBLEMS

P15.1 Strain gauges are commonly made by cementing a fine wire to a carrier substrate. When the substrate is then cemented to a surface that undergoes strain, the elongation or compression of the wire produces a proportional increase or decrease in resistance, respectively. In this problem we will explore the use of strain gauges to investigate vibration in a cantilevered beam.

Two strain gauges are cemented to opposite sides of a beam as shown in Fig. P15.1*a*. With the beam at rest, the strain gauge

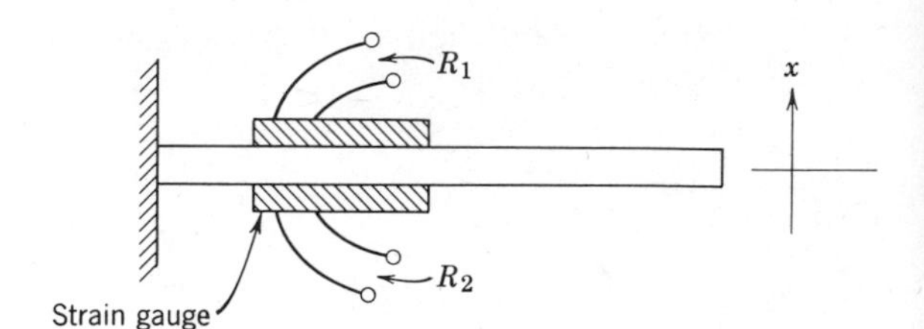

Figure P15.1*a*

resistances are $R_1 = R_2 = R_0$. Assume that the beam undergoes a sinusoidal deflection $x = X_0 \sin 2\pi f t$, which produces a peak change in strain-gauge resistance of $0.1\, R_0$ at a frequency $f = 200$ Hz.

(a) Find expressions for R_1 and R_2 as a function of time.

(b) If the strain gauges are connected in the bridge circuit shown in Fig. P15.1*b*, find and sketch $v_O(t)$.

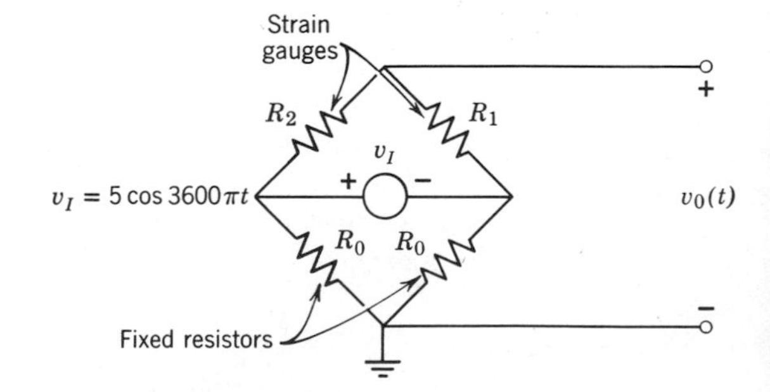

Figure P15.1*b*

(c) Plot the frequency spectrum of $v_O(t)$ showing the magnitude of each frequency component.

P15.2 A stable 200 kHz, 2 V peak-to-peak signal is needed for an experiment. However, only a stable 100 kHz, 2 V peak-to-peak oscillator is available.

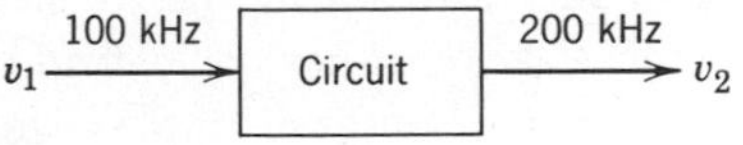

Figure P15.2

(a) Design a simple circuit (Fig. P15.2) that can do this frequency doubling. Estimate key component values.

(b) Suppose that along with the 100 kHz input there is some 60 Hz hum. That is, suppose

$$v_1 = v_{100\,\text{kHz}} + v_{60\,\text{Hz}}$$

Discuss how the hum will affect the output of your particular design and suggest modifications you might wish to make to reject the effect of the hum.

P15.3 Single-sideband transmission is very economical with transmitted power, but there is a price for this economy. To produce a single sideband, we must either (1) reject the unwanted sideband with filters or (2) null it out by using phase-shifting networks to produce the product $\sin \omega_m t \sin \omega_c t$ as well as $\cos \omega_m t \cos \omega_c t$ ($\omega_m =$ modulation frequency; $\omega_c =$ carrier frequency). This problem is concerned with method 2.

(a) Show that adding the waveform $A_1 \cos \omega_m t \cos \omega_c t$ to the waveform $A_2 \sin \omega_m t \sin \omega_c t$ produces just one sideband if $A_1 = A_2$. Assume for reference purposes that the carrier is $V_c \cos \omega_c t$ and that the modulation is $V_m \cos \omega_m t$.
How would you produce the 90° phase shifts in both the carrier and the modulation frequency needed to obtain $V_c \sin \omega_c t$ and $V_m \sin \omega_m t$?

(b) Discuss the additional problem that results when one wishes to use speech waveforms (instead of a single sinusoid) for the modulation. How would you suggest handling the problem?

P15.4 The transfer curve of the emitter coupled pair is of the form $\tanh q(v_1 - v_2)/2kT$, which has a nonlinearity to the lowest order of the form

$$v_0 = g v_I - \alpha v_I^3$$

where $v_I = (v_1 - v_2)$ and g and α are constants. Assume $v_I = \cos \omega_o t$. Find the spectrum of the output v_O. Could this circuit be used to produce $2\omega_o$ from ω_o? From $3\omega_o$? (The ratio of the amplitude of the third harmonic to the amplitude of the fundamental is called the *third-harmonic distortion.* The *second-harmonic distortion* of this particular example is zero.)

P15.5 Find the spectrum when the transfer characteristic of P15.4 is driven by

$$v_I = \cos \omega_o t + \cos \omega_1 t$$

What new frequencies appear that would not be present if v_I were driven by a single sinusoid? (These new frequencies are called *combination tones*. Their presence or absence can be used to test for linearity.)

P15.6 The small-signal junction capacitance of a back-biased diode depends on the bias voltage. Explain how such a device could be used for voltage-controlled tuning of a resonant circuit.

P15.7 Think of ways with which you could accomplish frequency modulation and phase modulation when the modulation signal is digital. Then try to develop methods for an analog modulation signal (see P15.6).

P15.8 An AM wave with carrier frequency ω_c and modulation frequency ω_m is passed through a resonant filter tuned to ω_c. Describe in qualitative terms what happens to the output spectrum and to the information content of the waveform as the Q of the resonant circuit is increased from a low value toward $+\infty$.

CHAPTER SIXTEEN
DIGITAL CIRCUITS AND APPLICATIONS

16.0 INTRODUCTION

Throughout this text we have made many passing references to digital signals and circuits. We shall now formalize our approach to this vast and important subject.

We recall from Chapter Two that digital signals are binary in nature, taking on values in one of two well-defined ranges. We shall see below that the set of basic operations that can be performed on digital signals is quite small, and can easily be mastered by the beginner. Furthermore, the behavior of *any* digital system, up to and including the most sophisticated digital computer, can be represented by appropriate combinations of digital variables and the digital operations from this small set. Finally, digital integrated circuits with input-output characteristics that correspond to each of the basic digital operations are inexpensive and easy to use. Thus, it is possible for the beginning student to design and successfully construct a wide variety of digital circuits, including circuits that perform logic and arithmetic functions, circuits for data storage and transmission, and circuits that interface between peripheral equipment and the small, commercial general-purpose digital computers known affectionately as "mini-computers".[1]

[1] The only thing "mini" about a minicomputer is the physical size. The minicomputers now on the market (for prices roughly equivalent to that of a high-quality oscilloscope) are capable of doing surprisingly large problems at surprisingly high speeds.

16.1 ALGEBRAIC REPRESENTATIONS OF DIGITAL VARIABLES

16.1.1 Boolean Algebra

This chapter is concerned with digital system variables that take on only two values (binary variables). We conventionally denote these values as "0" and "1," and then use a special set of rules called *Boolean algebra* to summarize the various ways in which digital variables can be combined. This algebra and much of the notation are adopted directly from mathematical logic. Thus, "logic variable" or "logic operation" are commonly used in place of "digital variable" or "digital operation."

Definition of the AND *operation.* Given two input variables, A and B, and an output variable C, the expression

$$C = A \quad \text{AND} \quad B \tag{16.1}$$

means

$$C = 1 \quad \text{if} \quad A = 1 \quad \text{AND} \quad B = 1 \tag{16.2a}$$

otherwise

$$C = 0 \tag{16.2b}$$

A circuit that performs the AND operation is called an AND *gate*. The logic symbol for a two-input AND gate is shown in Fig. 16.1*a*.

A	B	C
0	0	0
0	1	0
1	0	0
1	1	1

(*a*) AND

D	E	F
0	0	0
0	1	1
1	0	1
1	1	1

(*b*) OR

G	H
0	1
1	0

(*c*) NOT

Figure 16.1
Logic symbols and function tables for AND, OR, and NOT.

A dot is used as a shorthand for the AND operation, so that Eq. 16.1 may be written

$$C = A \cdot B \tag{16.3}$$

The dot is often omitted simplifying Eq. 16.3 further.

$$C = AB \tag{16.4}$$

One nice feature of digital operations is that the complete set of input-output variable values can be written down. Figure 16.1*a* shows such a *function table*, corresponding to Eq. 16.3, which lists all possible combinations of input variables A and B together with the corresponding output variable C. From this function table we see that in algebraic terms the AND operation is a form of multiplication, with these manipulation rules:

$$\begin{aligned} 0 \cdot 0 &= 0 \\ 0 \cdot 1 = 1 \cdot 0 &= 0 \\ 1 \cdot 1 &= 1 \end{aligned} \tag{16.5}$$

Definition of the OR *operation.* Given two input variables D and E, and an output variable F, the expression

$$F = D \quad \text{OR} \quad E \tag{16.6}$$

means

$$\begin{aligned} F = 1 \qquad &\text{if} \qquad D = 1 \\ &\text{OR} \qquad E = 1 \\ &\text{OR} \qquad \text{both} \quad D = 1 \quad \text{and} \quad E = 1 \end{aligned} \tag{16.7}$$

The + sign is used as a shorthand for OR, and is *never* omitted in algebraic expressions. Thus, Eq. 16.6 is written algebraically as

$$F = D + E \tag{16.8}$$

Figure 16.1*b* shows the logic symbol used for the two-input OR gate together with the corresponding function table. Algebraically, the OR operation is a special form of addition performed according to these rules:

$$\begin{aligned} 0 + 0 &= 0 \\ 0 + 1 = 1 + 0 &= 1 \\ 1 + 1 &= 1 \end{aligned} \tag{16.9}$$

Note that the last manipulation, $1 + 1 = 1$, differs from the ordinary arithmetic use of the + sign.

As in ordinary algebra, parentheses may be used in Boolean expressions to group terms and give precedence to operations. If there are no parentheses, the AND functions in an equation are evaluated first.

Definition of the NOT *operation.* In some situations, the opposite value of a particular variable is required. In Boolean algebra, the opposite value of a variable is called the *complement* of that variable, and is denoted by a bar drawn over the variable in question. The complement operation is summarized below using variable G as an example.

$$\begin{aligned} &\text{If } G = 1 \quad \text{then} \quad \overline{G} = 0 \\ &\text{If } G = 0 \quad \text{then} \quad \overline{G} = 1 \end{aligned} \tag{16.10}$$

The logic operation that produces the complement is called *inversion*, or the NOT *operation*. The logic symbol and function table for an inverter is shown in Fig. 16.1*c*.

DeMorgan's Theorem. DeMorgan's theorem is a Boolean algebra identity expression that states

$$\overline{(X + Y)} = \overline{X} \cdot \overline{Y} \tag{16.11}$$

or equivalently

$$\overline{X \cdot Y} = \overline{X} + \overline{Y} \tag{16.12}$$

(Note that the complete algebraic expression underneath the complement bar must first be evaluated, then the complement taken.) This theorem is easily verified by examining the function tables for the two sides of each equation. In summary, DeMorgan's theorem states that the complement of the OR operation is equivalent to performing the AND operation on the complement variables, and vice versa. DeMorgan's theorem is of great use in manipulating and simplifying Boolean algebraic expressions that contain more than one basic logic operation.

The composite operations NAND *and* NOR. Two combinations of basic operations arise so often that they are given individual names and logic symbols. The NOR operation is the complement of the OR operation (the name is simply a contraction of "NOT OR"), and is defined by

$$C = \overline{(A + B)} \tag{16.13}$$

or, by DeMorgan's theorem,

$$C = \overline{A} \cdot \overline{B} \tag{16.14}$$

Two equivalent symbols for the NOR gate, representing Eqs. 16.13 and 16.14 respectively, are shown in Fig. 16.2*a* along with the NOR function table. Note that the small circle adjacent to the input or output of the basic gate symbols produces the INVERSION of the variable in each case.

The complement of the AND operation is called the NAND operation (from "NOT AND"), and is defined by the two equivalent forms

$$F = \overline{D \cdot E} \tag{16.15}$$

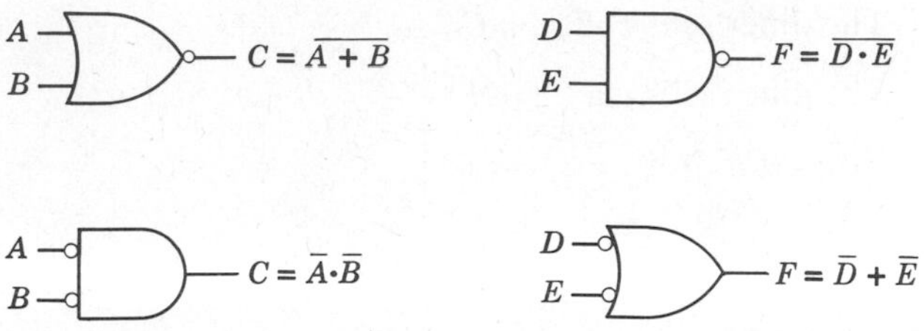

$C = \overline{A + B}$ $\qquad$ $F = \overline{D \cdot E}$

$C = \bar{A} \cdot \bar{B}$ $\qquad$ $F = \bar{D} + \bar{E}$

A	B	C
0	0	1
0	1	0
1	0	0
1	1	0

(*a*) NOR

D	E	F
0	0	1
0	1	1
1	0	1
1	1	0

(*b*) NAND

Figure 16.2
Equivalent symbols and function tables for NOR and NAND gates resulting from DeMorgan's theorem.

or

$$F = \bar{D} + \bar{E} \tag{16.16}$$

The two equivalent symbols for NAND gates and the function table are shown in Fig. 16.2*b*.

The principal importance of NOR and NAND is that they are the simplest logic functions to construct in integrated circuit form. Thus, while it may be easier for the beginner to learn to "think" with OR and AND, he should also practice thinking with NOR and NAND as these functions are likely to be used in the final circuit realization. Also, it is possible to synthesize all of the logic functions using only NOR gates or only NAND gates.

16.1.2 A Simple Example

Let us formulate a simple everyday situation in terms of digital variables and Boolean operations. Suppose you are driving home and become thirsty for a hot drink. You see a diner ahead and pull in. Let us develop a Boolean equation for whether or not you obtain a drink.

The first step is to assign variables for the problem:

The diner is open for business $\Rightarrow D = 1$, or simply D

($\Rightarrow$ means implies)

The diner sells coffee $\Rightarrow C$

The diner sells tea $\Rightarrow T$

You get a drink $\Rightarrow X$

The next step is to use AND and OR operations to construct this Boolean equation:

$$\underset{\text{Obtain drink}}{X} \Leftarrow \underset{\text{sells coffee}}{(C} \overset{\text{OR}}{+} \underset{\text{sells tea}}{T)} \overset{\text{AND}}{\cdot} \underset{\text{open for business}}{D} \tag{16.17}$$

or omitting the dot

$$X = (C + T)D \tag{16.18}$$

which is equivalent to

$$X = CD + TD \tag{16.19}$$

The final step is to construct a *logic flow diagram* using gate symbols. Two possible logic flow diagrams, corresponding to Eqs. 16.18 and 16.19, are shown in Fig. 16.3. As an illustration of how only NOR gates or only NAND gates can be used to synthesize any function, two additional implementations of this same example are shown in Fig. 16.4. Notice that when both inputs of the two-input NOR gate are connected together, as in Fig. 16.4*a*, the gate becomes an inverter.

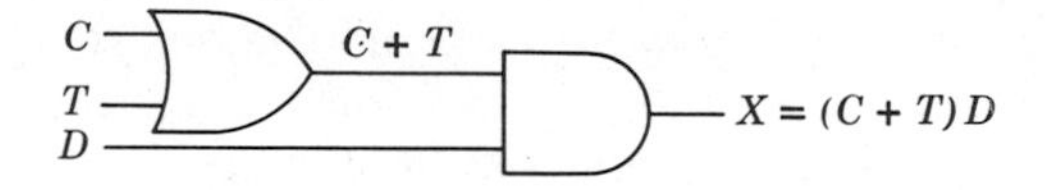

(*a*) Implementation of Eq. 16.18

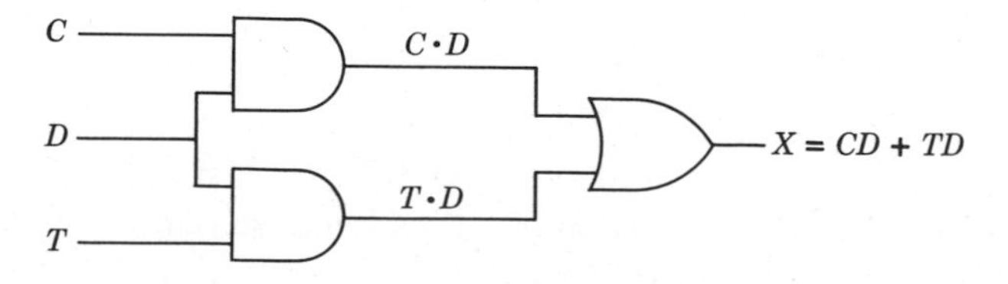

(*b*) Implementation Eq. 16.19

Figure 16.3
Logic flow diagrams for text example using AND and OR gates.

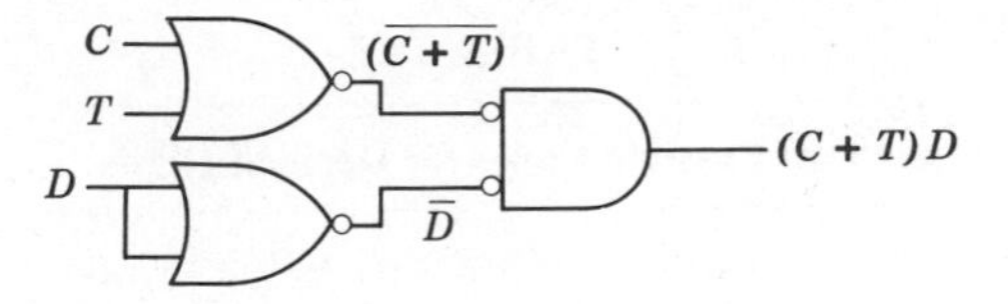

(a) NOR only

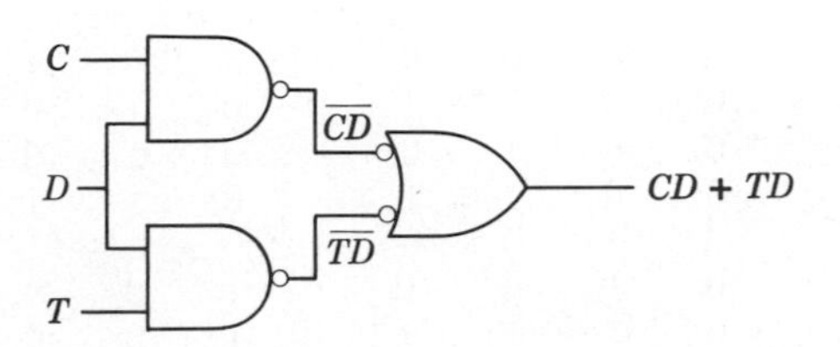

(b) NAND only

Figure 16.4
Implementation of text example using (a) only NOR gates and (b) only NAND gates.

16.1.3 Coding and Decoding; Binary Numbers

In the preceding example each of the quantities of interest was a binary variable. Either the diner was open or it was not; either coffee was served or it was not. In such a case, a single binary variable could be assigned to each separate quantity without any confusion. In many cases, however, such as those involving ordinary decimal numbers, the quantities of interest might take on several different discrete values. In these cases, more than one binary variable must be used to represent a given quantity. An example follows.

Suppose we wish to represent with binary digital variables the 10 possible values that an ordinary decimal digit D might assume. A single binary variable can represent two values. A pair of binary variables, each capable of taking on two values, can represent a total of $2 \times 2 = 4$ distinct values. A set of m binary variables can represent up to 2^m distinct values of a multivalued digital quantity. Thus, the representation of the 10 possible values of a decimal digit requires a minimum of *four* binary variables ($2^4 = 16 > 10$). Let us call the four binary variables in this set n_1, n_2, n_3, and n_4.

The process of assigning each of the 10 possible values of D to a particular set of 1's and 0's for $n_1 \ldots n_4$ is called *coding*. A *code* is simply a table showing for each discrete value of the quantity of interest the corresponding values, 0 or 1, of each member of the set of binary variables. Two examples of codes used for decimal digits are shown in Table 16.1.

TABLE 16.1

Two Possible Codes for Decimal Digits

Decimal Digit	Code A (1248 BCD)				Code B (Excess Three)			
D	n_4	n_3	n_2	n_1	n_4	n_3	n_2	n_1
0	0	0	0	0	0	0	1	1
1	0	0	0	1	0	1	0	0
2	0	0	1	0	0	1	0	1
3	0	0	1	1	0	1	1	0
4	0	1	0	0	0	1	1	1
5	0	1	0	1	1	0	0	0
6	0	1	1	0	1	0	0	1
7	0	1	1	1	1	0	1	0
8	1	0	0	0	1	0	1	1
9	1	0	0	1	1	1	0	0

There might be many practical reasons, usually connected with how a circuit is to be constructed, for selecting any one code in preference to another. However, one code is particularly useful for representing decimal numbers. It is called the *binary number system.* In this code, each binary variable in the set is assigned a place value in exactly the same way digits in a multiple-digit decimal number are assigned place values. The right-hand most place represents units (i.e., 2^0), the next place represents 2's (i.e., 2^1), the next represents 4's, and so on. The array of binary variables, written in order of place value, constitute a binary number with each binary variable playing the role of a binary digit or *bit*. Code A of Table 16.1 is actually a binary number representation of the numbers 0 to 9 with n_1 representing the unit's place, n_2 the 2's place, n_3 the 4's place, and n_4 the 8's place. Hence the code name 1248 Binary Coded Decimal. The binary number value in Code B of Table 16.1 is three greater than the corresponding decimal number; thus the name Excess Three Code.

One can do arithmetic with binary numbers by using the rules of carrying and borrowing just as with decimal numbers, and most digital computers use the binary number system to represent positive numbers.[2] However, one must remember that in the binary number system $1 + 1 = 10$ and $10 - 1 = 1$.

[2] There is more variability in the coding of negative numbers. Several different schemes have been devised for representing the sign of a number by a binary variable in such a way that subtraction and addition can be performed with easily implemented Boolean expressions. Further discussion can be found in the references.

Let us illustrate the process of binary addition by formulating Boolean expressions that compute the sum S of two binary digits B_1 and B_2, and determine whether there is a carry of 1 to the next higher digit. Such a circuit is called a *half-adder*. We denote the carry by C. Table 16.2 is the function table for the half-adder.

TABLE 16.2

Half-Adder Function Table

Inputs		Outputs	
B_1	B_2	S	C
0	0	0	0
0	1	1	0
1	0	1	0
1	1	0	1

Notice that the sum S is *not* equivalent to the OR operation. When doing binary arithmetic, $1 + 1$ yields a zero with a carry, but when combining logic variables, $1 + 1$ (OR) $= 1$.

By using the function tables in Table 16.2, one can write down Boolean expressions for each output function by noting each combination of inputs needed to produce a 1 in the output, and writing a "sum-of-products" for each case. For example, $C = 1$ only when $B_1 = 1$ AND $B_2 = 1$, while the sum S is 1 when $B_1 = 0$ AND $B_2 = 1$ OR when $B_1 = 1$ AND $B_2 = 0$. In Boolean terms, therefore,

$$S = B_1 \cdot \bar{B}_2 + \bar{B}_1 \cdot B_2 \tag{16.20a}$$

$$C = B_1 B_2 \tag{16.20b}$$

Two implementations of the half-adder are shown in Fig. 16.5; the first uses AND, OR, and NOT operations, and the second, five identical two-input NAND gates. Proofs that these networks are correct are left as exercises for the student.

The logic function described by Eq. 16.20*a* occurs in practice with sufficient frequency that it has been given the name EXCLUSIVE OR, with the symbol and function table shown in Fig. 16.6. From this table we see that the EXCLUSIVE OR gate has output 1 whenever either input, but not both, is 1. A half-adder realization utilizing an EXCLUSIVE OR gate is shown in Fig. 16.7.

The process of interpreting a code or of converting from one code to another is called *decoding*. Decoding operations can be represented as Boolean algebraic equations and implemented with gates. For example,

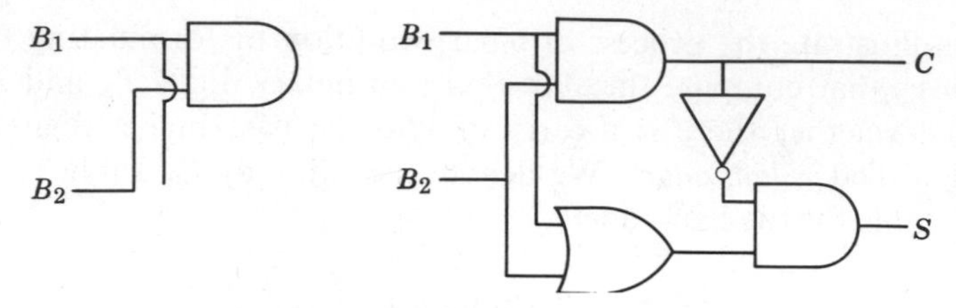

(*a*) Half-adder using AND, OR and NOT

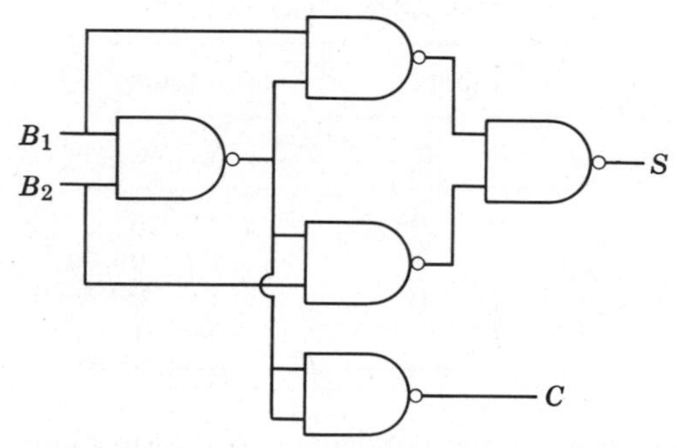

(*b*) Half-adder using five identical NAND gates

Figure 16.5
Half-adder logic flow diagrams.

A

B

$C = A \cdot \bar{B} + \bar{A} \cdot B$

A	B	C
0	0	0
0	1	1
1	0	1
1	1	0

Figure 16.6
Logic symbol and function table for the EXCLUSIVE OR gate.

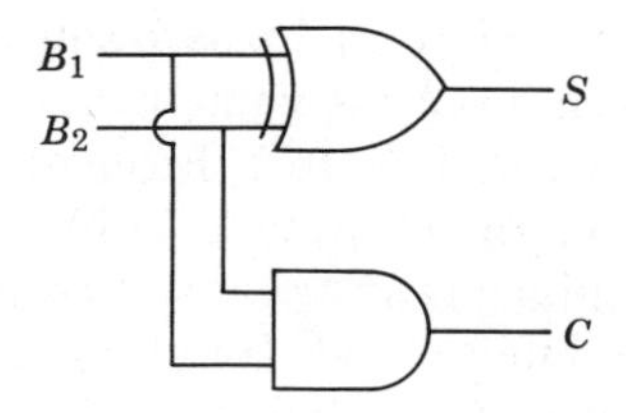

Figure 16.7
Half-adder implementation using an EXCLUSIVE OR and AND gate.

consider Code B of Table 16.1. Suppose we wish to detect whenever the decimal digits represented by $n_1 \ldots n_4$ is equal to 7. This will occur when n_2 and n_4 are 1's while n_1 and n_3 are 0's. If we define a new binary variable D_7 that equals 1 only when $D = 7$, the Boolean expression for D_7 is

$$D_7 = \bar{n}_1 \cdot n_2 \cdot \bar{n}_3 \cdot n_4 \qquad (16.21)$$

An implementation of D_7 using two inverters and a four-input NOR gate is shown in Fig. 16.8.

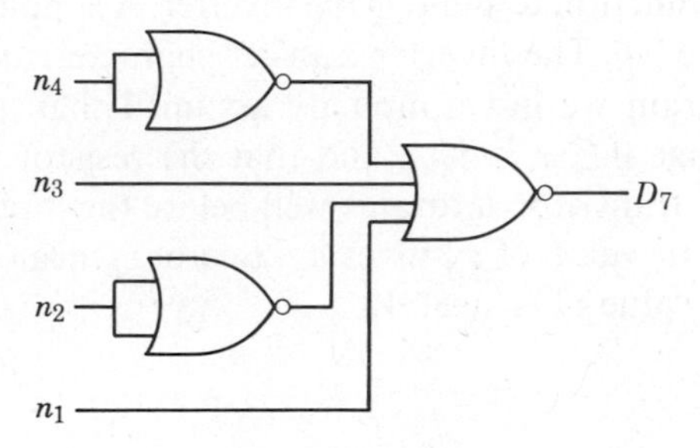

Figure 16.8
Implementation of Eq. 16.21.

16.2 CIRCUIT REALIZATION OF LOGIC FUNCTIONS

Up to this point we have treated digital signals and systems from a strictly mathematical viewpoint. We now turn our attention to the problem of circuit realization of the various logic functions presented in Section 16.1.

There are many different ways to build electronic circuits with input-output characteristics that correspond to the various digital operations. Several of these circuit types are manufactured in integrated circuit form. The collection of integrated circuit logic functions that share a common circuit type is called a *logic family*. Within each family, the wiring diagram for the logic inputs and outputs is identical to the logic flow diagram (only the power supply and ground connections must be added). Therefore, one generally selects a *single logic family* with which to implement all the digital circuits in a particular application. Occasionally, one must make connections between digital circuits constructed from logic families with input and output voltage ranges that are not compatible with one another. In these cases, it is necessary to construct additional circuits that *interface* one family with another.

The choice of one family over another for a particular application depends on such factors as circuit speed, cost, noise immunity, power dissipation, and available logic functions. In this section, we shall first survey some of the general problems of logic circuitry and then examine some properties of the more common families.

16.2.1 The Inverter as a Typical Logic Element

The simplest logic function to build is the inverter. A simple common emitter can be used (Fig. 16.9*a*). The inverter transfer characteristic is shown in Fig. 16.9*b*. In this diagram we have implicity assumed that all signal voltages must lie in the range $0 \le v \le V_{CC}$, and that the resistor values have been selected so that the transistor saturates well before the input voltage reaches V_{CC}. Notice that a HI value of v_A gives a value of v_B near $V_{CE,sat}$ while a LO value of v_A gives a value of v_B near V_{CC}.

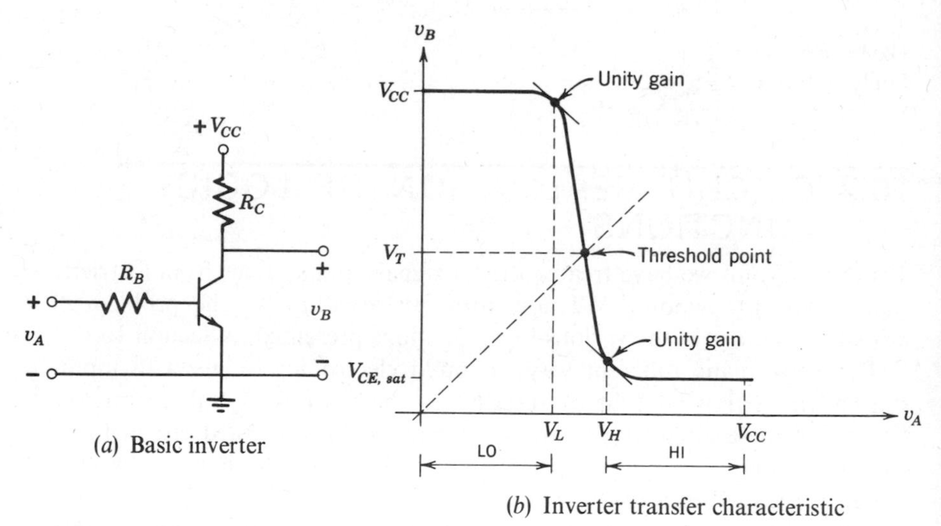

Figure 16.9
Basic inverter characteristics.

Logic Variables. The first question that must be resolved is which range of voltage is to correspond to the digital "1" and which to the digital "0." Throughout this text we shall assign the "1" to the HI or more positive voltage range and "0" to the LO or more negative voltage range. This convention is called *positive logic*. It is equally possible, although less common, to identify the HI range with "0," an assignment called *negative logic*.

Voltage Ranges. The second question to resolve is the selection of boundaries for the HI and LO voltage ranges. Since the digital variables "0" and "1" are a mutually exclusive set, so must be the corresponding ranges of voltages. In addition, because the output of one inverter can become the input to another identical circuit within a digital system, the voltage at the *threshold point* (defined as that input voltage that produces the same voltage at the output) represents an absolute boundary between the HI and LO voltage ranges. The threshold point is shown in Fig. 16.9*b* as the intersection of the transfer characteristic with the 45° line for which $v_B = v_A$, and can be easily measured in the laboratory by connecting the output of the inverter to the input. Choosing the threshold as the HI-LO boundary, however, would not produce satisfactory voltage ranges for several reasons. First, the transfer characteristic is very steep at the threshold, yielding a large change in output for a small variation in input. Second, there is a variation in the threshold between individual units that makes the specific voltage somewhat uncertain.

More conservative boundaries for the LO and HI voltage ranges are the *unity gain points* of the transfer characteristic. As shown in Fig. 16.9*b*, the unity gain points are those points on the transfer characteristic where the slope is -1. Thus, the ranges of voltage for LO and HI are given by

$$
\begin{aligned}
0 &\leq v_{\text{LO}} \leq V_L \\
V_H &\leq v_{\text{HI}} \leq V_{CC}
\end{aligned}
\tag{16.22}
$$

Note that there is a range of voltage between V_L and V_H that is indeterminate as to whether the state is LO or HI. Except during transitions, voltages between V_L and V_H should not be allowed to occur either as input or output voltages.

Noise Margin. For a digital system to operate reliably, it is essential that every precaution be taken to prevent false outputs due to spurious signals or noise that might be present along with an input signal. If an input signal plus noise should combine to cause an input voltage to cross a unity gain point, then that particular circuit may have a false output. A figure-of-merit that serves to measure the noise immunity of a gate is called the dc noise margin, and is specified as the difference between the signal voltage and the unity gain voltage. Thus, we have

$$
\begin{aligned}
\text{LO-state dc noise margin} &= V_L - v_A \\
\text{HI-state dc noise margin} &= v_A - V_H
\end{aligned}
\tag{16.23}
$$

where v_A is the actual LO- or HI-state input voltage. Clearly, the larger the noise margins, the more reliable the design. Acceptable noise margins will range from a few tenths of a volt to several volts, depending on the logic family and logic state.

Fan-out. In constructing the inverter transfer characteristic shown in Fig. 16.9*b*, we tacitly assumed that the output terminals were open-circuited. However, if the inverter or gate is part of a digital system, then the output will in general be connected to one or more subsequent inputs as directed by the logic flow diagram. Since an input will require some signal power to function properly, there is often a limit on the number of inputs a given output can drive. This number is called the *fan-out capability* of a gate. Because all inputs in a family of digital circuits do not require the same amount of power, the input drive requirements and output drive capabilities for a given family are often specified in terms of a *unit load*, which represents the lowest input drive requirement. The number of unit loads an input represents and an output can supply are often shown in parentheses on the circuit diagram beside the input and output terminals of a gate. Reliable system operation requires that output fan-out capability is not exceeded; that is, the output unit-load drive capability must equal or exceed the sum of the input unit loads being driven. Where a large fan-out is required, it may be necessary to add additional logic elements to achieve the required drive capability.

Speed. The speed of a digital system is governed by the transient response time of the gate circuits. In some applications, such as four-function calculators, speed is relatively unimportant since any logic family will perform much faster than the human reaction time. In other cases, where large quantities of data must be processed, system speed may be a critical factor in the choice of logic family. Since the transient waveforms encountered in digital signals are not normally exponential, their *rise* and *fall times* are specified as the time required for the waveform to change from 10 to 90% of the total logic signal swing (Fig. 16.10). The time required for the output of an inverter or gate to respond to a change in input is termed the *propagation delay* and is measured between the threshold points of the input and output waveforms. Propagation delay is not necessarily the same for a HI to LO transition and a LO to HI transition. Thus, a distinction is made in the respective symbols: t_{PLH} for LO to HI, and t_{PHL} for HI to LO. These definitions are summarized in Fig. 16.10.

Now that we have defined some of the terminology and presented some of the characteristics of digital circuits, we can turn our attention to several specific logic families.

16.2.2 Diode-Transistor Logic (DTL)

The first family of digital logic that we will examine is Diode-Transistor Logic or DTL for short. The basic configuration of a DTL gate is shown in Fig.

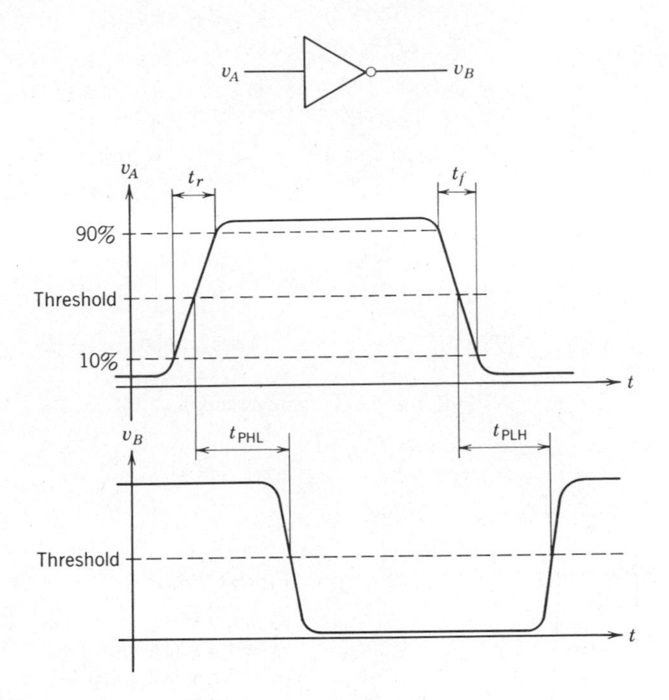

Figure 16.10
Definition of rise, fall, and propagation delay times.

16.11*a*. If we ignore the dashed portion of the input circuit for the moment, the circuit reduces to an inverter, and we can construct the transfer characteristic v_X versus v_A shown in Fig. 16.11*b*. If v_A is at zero volts, D_1 will be forward-biased and v_1 will be +0.6 V. This value of v_1 is not large enough to forward bias D_2, D_3, and the base-emitter junction of Q_1. Thus the current i_1 flows through D_1 and through the v_A voltage source to ground. The transistor Q_1 is in cutoff. With Q_1 in cutoff, v_X is +5 volts. When v_A increases, v_1 also increases until v_A reaches 1.2 volts and v_1 reaches 1.8 volts. At this point, D_2, D_3, and Q_1 turn on, and the current i_1 provides the base current for the transistor, driving it into saturation. Further increases in v_A reverse bias D_1, but do not affect the value of v_1 or the saturation state of Q_1. This relatively sharp transition between outputs of +5 volts and $V_{CE,sat}$ is illustrated in Fig. 16.11*b*. From the figure we see that the voltage ranges for the LO and HI logic states are approximately

$$0 \le v_{\mathrm{LO}} \le 1.2 \text{ V}$$
$$1.5 \le v_{\mathrm{HI}} \le 5 \text{ V} \tag{16.24}$$

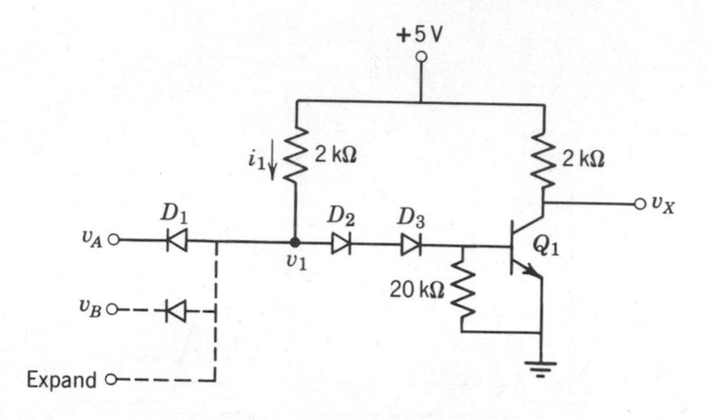

(*a*) Basic DTL gate circuit

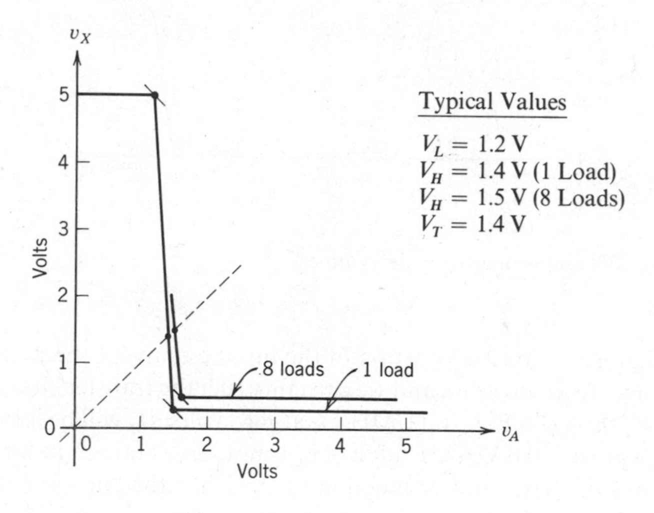

(*b*) DTL gate transfer characteristic

Figure 16.11
DTL gate and transfer characteristic.

In practice, v_{LO} is usually less than 0.4 volts and v_{HI} is very close to 5 volts, giving good dc noise margins.

When the input v_A is HI, diode D_1 is reverse biased, and hence requires a miniscule amount of power from the output of the driving source. However, when a gate input is held LO by the output of a preceding gate, as in

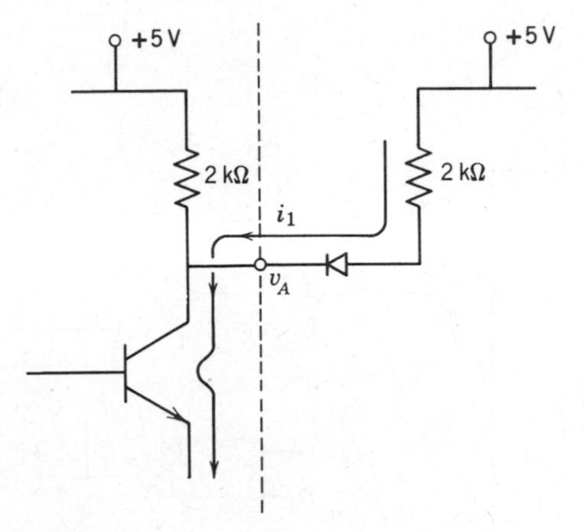

Figure 16.12
Loading effect in a DTL gate with v_A held LO.

Fig. 16.12, the current i_1 must flow out the gate input terminal and through the saturated transistor to ground. The magnitude of this current is approximately

$$i_1 = \frac{(5 - 0.6 - 0.2)\text{ V}}{2\text{ k}\Omega} = 2.1\text{ mA} \qquad (16.25)$$

and this represents one unit load. If n inputs are connected to one output, then the saturated transistor must conduct or *sink* n times i_1 to ground. As n increases, so will the voltage v_A since it is equal to $V_{CE,sat}$ for the output transistor. This effect is illustrated in Fig. 16.11*b*, where the transfer characteristic is shown for the case of one output load and eight output loads, the maximum allowed for the basic DTL gate.

If a second input diode is added to the circuit of Fig. 16.11*a* to provide an input v_B, then the output v_X will be HI if either input is LO, and the output will be LO when both inputs are HI. Thus, the logic operation produced by this circuit is

$$X = \overline{A \cdot B} \qquad (16.26)$$

which is the NAND operation. By adding additional diodes, whether discrete or in an integrated circuit array, to the EXPAND input, the number of inputs to the basic DTL NAND gate can be increased to as high as 20.

If the outputs of two (or more) DTL NAND gates are directly wired together as shown in Fig. 16.13*a*, the net result is to perform the AND operation on

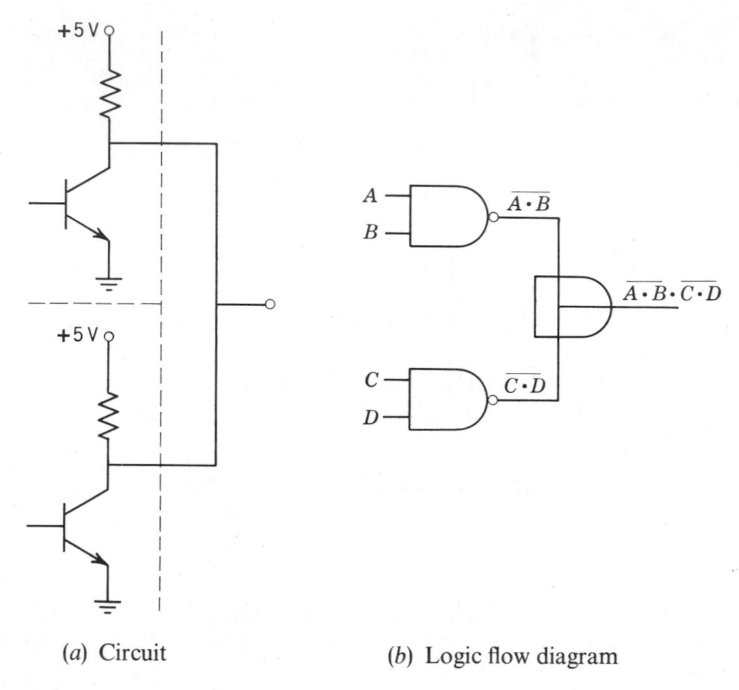

Figure 16.13
Wired AND gate connection.

the outputs of each NAND gate. This can be seen from the circuit since if either NAND gate output is LO, the output is LO. If both NAND gate outputs are HI, then the output is HI. Thus, this interconnection is called the *wired* AND connection and is shown on a logic diagram as in Fig. 16.13*b*. The fan-out capability of the total wired output must be reduced by one load for each additional output in the wired connection, since these outputs require sinking the current from their collector resistors when they are in the HI state.

The propagation delay time for a typical DTL gate is 30 nanoseconds, which is moderately slow, but adequate for many applications.

The DTL family includes AND, OR, NAND, NOR, and EXCLUSIVE OR gates. This is convenient for the logic designer, since he has every possible gate available, in contrast to some logic families, which do not provide the complete variety. Most designs involve several unused gate inputs. It is good practice to tie unused inputs to ground or +5 volts rather than to leave them open. This improves noise immunity and prevents degradation of the propagation delay time.

16.2.3 Transistor-Transistor Logic (TTL)

Transistor-Transistor Logic (TTL for short) evolved from the DTL integrated circuit family, but is unique in that it is not realizable in discrete component form. The basic TTL gate is shown in Fig. 16.14*a*.

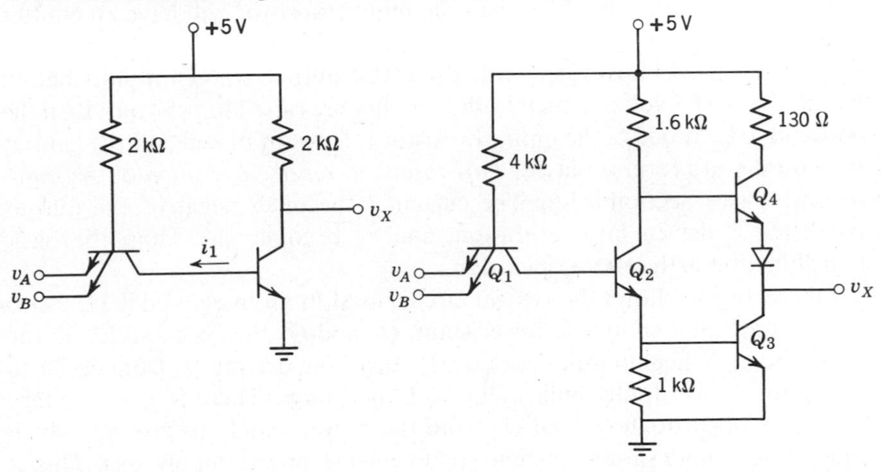

(*a*) Basic TTL gate circuit

(*b*) Practical TTL gate circuit with active pull-up

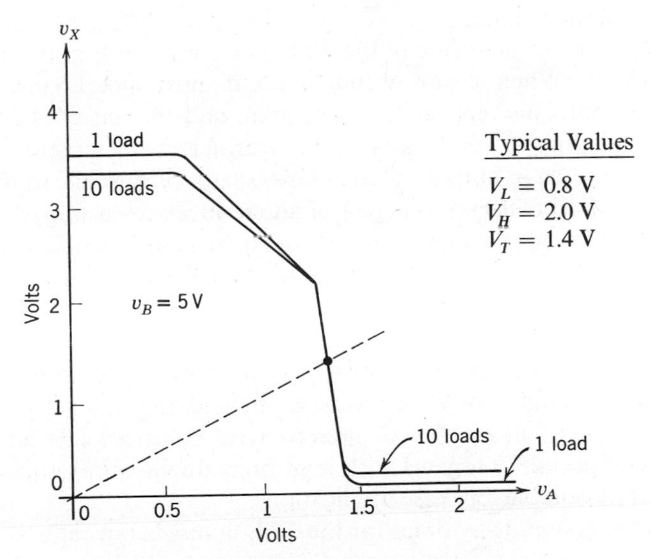

Typical Values

$V_L = 0.8$ V
$V_H = 2.0$ V
$V_T = 1.4$ V

(*c*) TTL gate transfer characteristic

Figure 16.14
TTL gate

The operation of the multiple-emitter input transistor (which has no discrete counterpart) differs from the normal single-emitter transistor, since its collector current is controlled by the *sum* of the emitter currents, with the sum replacing the single emitter current for the normal transistor. With this substitution, its operation is identical to the normal transistor.

If either v_A or v_B is held LO, then the input transistor will have an emitter current flowing to ground that will produce a positive collector current i_1. However, since a positive i_1 will drive the output transistor into cutoff, v_X will rise to +5 volts, or HI. If both v_A and v_B are raised to +5 volts, then the collector-base diode of the input transistor is forward biased and the emitter base diodes are reverse biased. This results in *reverse operation* of the input transistor (see Section 9.1.6). The current i_1 becomes negative, the output transistor is driven into saturation, and v_X becomes LO. Thus, the basic logic function is the NAND operation.

Figure 16.14*b* shows the typical circuit used in an integrated TTL NAND gate. A push-pull emitter-follower stage, Q_3 and Q_4, has been added to the basic circuit. When an input goes LO, Q_2 turns off, driving Q_4 from cutoff to saturation. This rapidly pulls v_X up to a HI voltage. There is a momentary current path through Q_3 and Q_4 from the power supply to ground, which can produce undesirable current spikes on the power supply bus. This is the price paid to increase the operating speed by use of the *active pull-up* provided by Q_4. Judicious use of *RC* filtering on the power bus minimizes the effects of the spiking.

The transfer characteristic of the TTL gate circuit of Fig. 16.14*b* is shown in Fig. 16.14*c*. When a gate output is LO, it must sink 1.6 mA from each input it drives. This represents a unit load, and the basic TTL gate has a fan-out capability of 10. As seen in the transfer characteristic, loading has some affect on the HI output voltage. This occurs because the HI output must supply "inverted collector" current of about 40 μA to each driven input.

TTL gates are also available in the "open collector" configuration shown in Fig. 16.15. This permits the "wired AND" interconnection by the addition of an external collector resistor to the 5 V supply. (Active pull-up gates such as in Fig. 16.14*b* cannot be interconnected in the "wired AND" configuration because there is the possibility of a very low resistance path between the supply and ground should one gate be HI and the other LO). The open-collector TTL circuit can also be used to drive external loads such as lamps and relays. Special TTL gates with high-breakdown-voltage open-collector outputs are available for these applications.

The propagation delay times for the TTL family is typically

$$\begin{aligned} t_{\mathrm{PHL}} &= 7 \text{ ns} \\ t_{\mathrm{PLH}} &= 11 \text{ ns} \end{aligned} \tag{16.27}$$

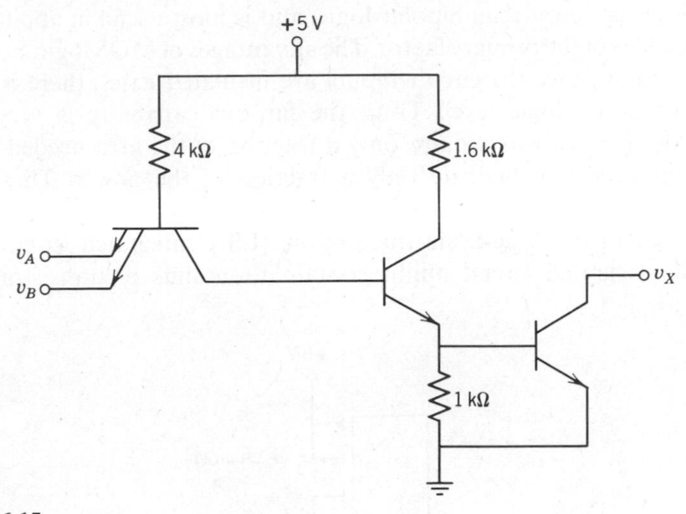

Figure 16.15
"Open collector" TTL gate.

Thus, TTL is considerably faster than DTL.

TTL and DTL logic families are generally compatible, as the transfer characteristics and drive requirements are similar. Thus, it is possible to intermix these families. This permits maximum economy when TTL is used for those portions of the design where speed is important and the slower DTL in those portions where speed is not a critical factor. TTL gates are available in the AND, OR, NAND, and NOR logic operations. As with DTL gates, any unused inputs should be connected to either +5 V or ground.

16.2.4 MOS Logic

A relatively new logic family uses enhancement-mode MOS transistors instead of bipolar transistors. We recall from Section 9.2 that MOS transistors can be used as ON-OFF switches. Furthermore, in the low-current part of the characteristics, an MOS transistor can be used as a resistor. Thus it is possible, using only MOS transistors, to fabricate all of the resistors and switches in a logic circuit with fabrication steps no more complex than for the transistors alone. One can fabricate MOS logic using *p*-channel devices, *n*-channel devices, or both together. Complementary-Symmetry MOS logic (COS/MOS), discussed in more detail below, uses both *p*- and *n*-channel transistors, and is of particular utility where minimum power consumption is a prime consideration.

MOS logic is slower than bipolar logic, and is most useful in applications where speed is not the primary factor. The advantages of MOS logic, however, are many. First, since the circuit inputs are insulated gates, there is no dc loading for either logic level. Thus, the fan-out capability is very high. Second, MOS transistors require only a fraction of the area needed for bipolar transistors and dissipate only a fraction of the power. These facts permit dense packing of components in a given area of silicon wafer, leading to what is called Large-Scale-Integration (LSI) integrated circuits. An MOS-LSI integrated circuit might contain thousands of transistors, and

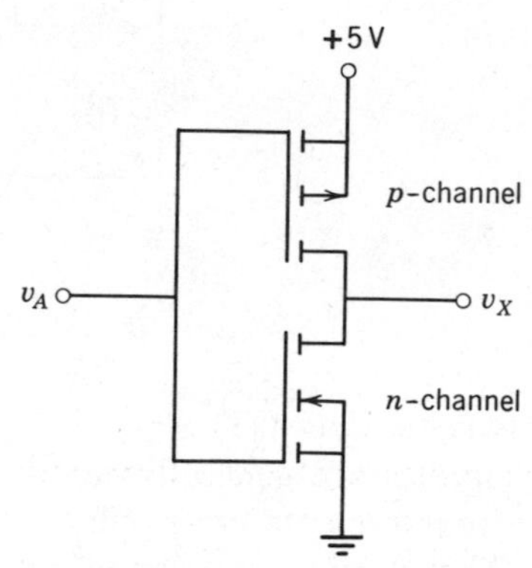

(*a*) Basic COS/MOS inverter circuit

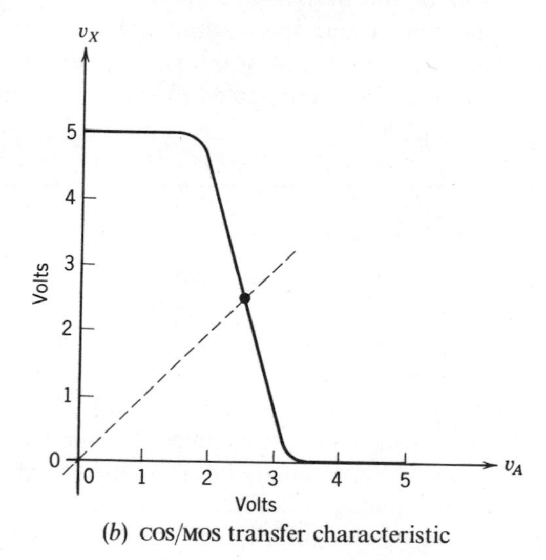

(*b*) COS/MOS transfer characteristic

Figure 16.16
COS/MOS inverter.

be equivalent in complexity to a small computer. Pocket electronic calculators, for example, require only one *p*-channel MOS-LSI circuit for all of the computational, logic, and control functions.

The basic operation of COS/MOS logic is illustrated by the inverter circuit shown in Fig. 16.15*a*. The *n*-channel MOSFET typically has a threshold voltage of +1.5 V, while the *p*-channel unit has a threshold voltage of −1.5 V. Thus, with v_A LO, the *n*-channel unit is cutoff and the *p*-channel unit has an induced channel, providing approximately a 500 Ω conducting path between the output terminal and the +5 V supply. As v_A increases, the *n*-channel unit will begin to conduct when v_A exceeds +1.5 V. The two MOSFETs are then both on, and v_X is determined by the voltage divider action between the two channel resistances. When v_A reaches +3.5 V, the *p*-channel unit is cutoff and the *n*-channel unit has an induced channel providing approximately a 500 Ω path from the output to ground. This behavior is summarized in Fig. 16.16*b*. Notice that there is no loading effect. Also, since there is no loading, the operating voltages are very close to 0 or +5 V. In this case, one of the MOSFETs is always cutoff, and the current drain on the +5 V supply is extremely small. This is the reason for the miniscule power consumption of this logic family.

The extension of the inverter circuit to a two-input NOR gate is shown in Fig. 16.17. If both v_A and v_B are LO, then both *p*-channel units are conducting and both *n*-channel units are cutoff. Thus, v_X will be HI. If either or both of v_A and v_B are HI, then v_X will be LO, resulting in the NOR operation. By reversing

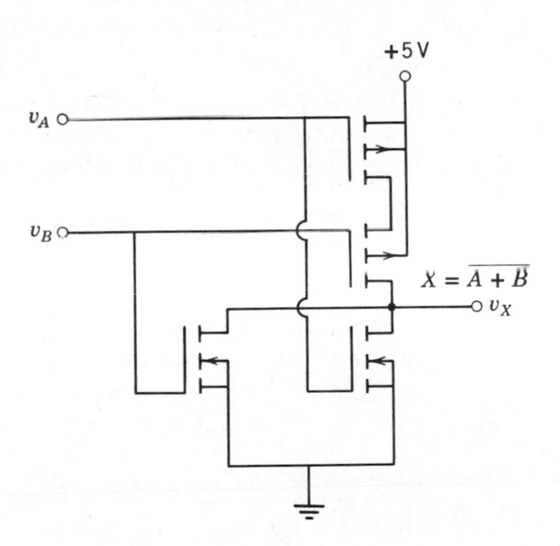

Figure 16.17
COS/MOS NOR gate.

the series-parallel arrangement of the MOSFETs (Fig. 16.18), a COS/MOS NAND gate results. Both the NOR gate and NAND gate can be expanded to three, four, or more inputs by simply increasing the number of series-parallel MOSFETs.

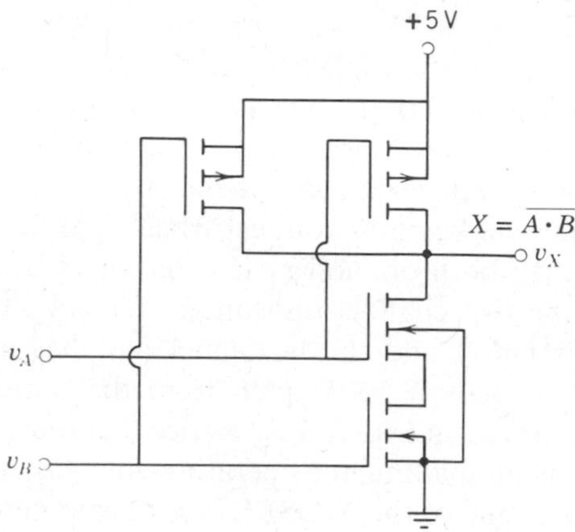

Figure 16.18
COS/MOS NAND gate.

The typical propagation delay time for a COS/MOS gate is 25 ns when driving a 15 pF load. Each gate input is approximately 5 pF. It is particularly important to check for unexpected transient outputs caused by variation in propagation delay through different portions of the circuit. A timing diagram, which plots the successive transitions through the circuit as the input change, is a helpful design tool in eliminating these errors.

COS/MOS logic is available in the NAND, NOR, and EXCLUSIVE OR operations, as well as several more complex gate configurations. Again, any unused circuit inputs should be connected to the positive or negative supply.

16.3 FLIP-FLOPS AND MONOSTABLES

Having examined how basic integrated circuit logic works, we now consider more complex digital circuits and their applications. To simplify the discussion, only DTL and TTL circuits will be shown. Other logic families will have implementations very similar to those shown here.

16.3.1 The *S-R* Flip-Flop

Figure 16.19*a* shows a pair of NAND gates connected in a loop. One input of each gate is connected through a pull-up resistor (typically 33 kΩ for

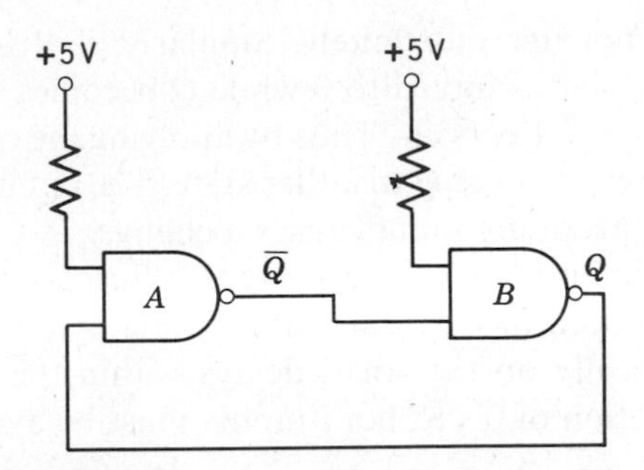

(*a*) Two inverters with feedback

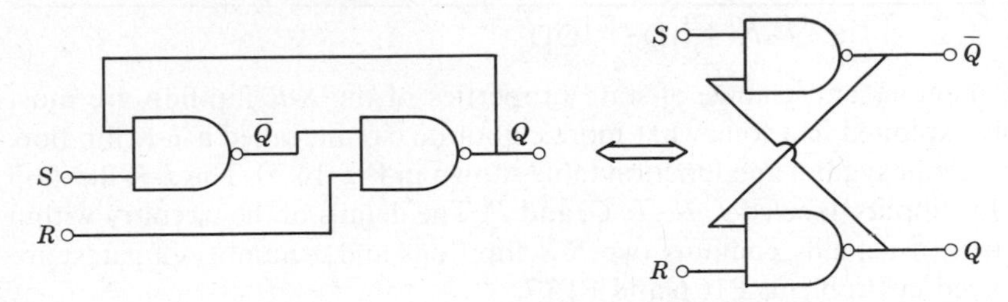

(*b*) Two equivalent schematics for the *S-R* flip-flop

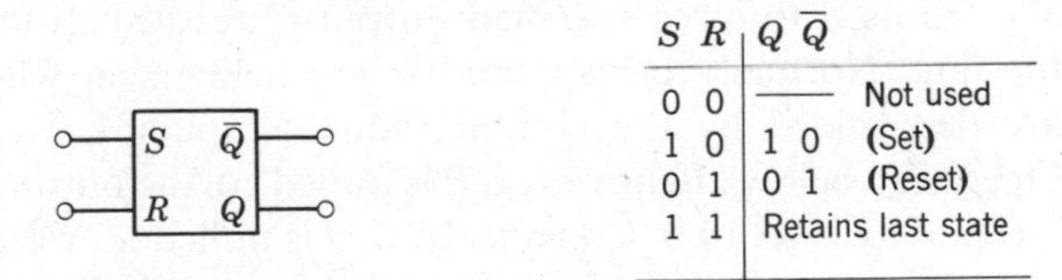

S	*R*	*Q*	$\bar{Q}$	
0	0	—	—	Not used
1	0	1	0	(Set)
0	1	0	1	(Reset)
1	1			Retains last state

(*c*) Symbol and function table for the *S-R* flip-flop

Figure 16.19
The set-reset flip-flop (*S-R*).

TTL) to the +5 V supply, signifying a HI or "1" input, and reducing each NAND gate to a simple inverter. This cyclical connection of two inverters is stable regardless of whether $Q = 1$ or $Q = 0$. The outputs Q and $\bar{Q}$ persist indefinitely. Thus, if we can manage to force Q to a desired value, that value will be stored or remembered by the circuit of Fig. 16.19*a*.

Figure 16.19*b* illustrates how one forces the outputs to either of the two states. The gate inputs that had been connected HI in Fig. 16.19*a* are removed from the supply and function as circuit inputs. The name given this configuration is the *set-reset* or *S-R flip-flop.* If both S and R are held HI, the circuit reverts to that of Fig. 16.19*a*. But if S is kept at 1 while R is set to 0, $\bar{Q}$ will be forced to 0 and Q will be forced to 1. If then R is returned to 1, the state $Q = 1$, $\bar{Q} = 0$ remains indefinitely. Similarly, if R is kept at 1 while S is set to 0, the Q output is forced to 0 while $\bar{Q}$ becomes 1. If S is returned to 1, the state $Q = 0$, $\bar{Q} = 1$ persists. Thus by applying the correct 1-0 pattern to the S-R inputs, one can force Q to either state. The circuit then maintains that state until an appropriate input causes a change.

One other combination of possible input values is the simultaneous application of 0's to both inputs. The output state is indeterminate in this case, depending critically on the small delays within the gates. Therefore, simultaneous application of 0's to both inputs must be avoided in practice.

16.3.2 The *J*-*K* Flip-Flop

The memory and change-of-state properties of the *S*-*R* flip-flop are most easily exploited in a somewhat more elaborate circuit, called a *J*-*K* flip-flop, with circuit symbol and function table shown in Fig. 16.20. The *J*-*K* flip-flop has five inputs, labeled J, K, T, C, and P. The details of the circuitry within the box (it actually contains two *S*-*R* flip-flops and a number of gates) are explored in Problems P16.6 and P16.7.

Asynchronous Operation. The C and P inputs, called *clear* and *preset* respectively, are used to force specified output states just as in the case of the *S*-*R* flip-flop. Normally both C and P are held HI, in which case the outputs are determined by the combination of inputs J, K, and T (see Clocked Operation below). If, however, P is pulled LO, the output Q is set to 1 ("setting" Q) and $\bar{Q}$ is set to 0. Conversely, if C is pulled LO while P remains HI, the output Q is set to 0 ("clearing" Q) and $\bar{Q}$ is set to 1. The simultaneous application of 0's to both P and C is avoided. Operationally, the P input of a *J*-*K* flip-flop is equivalent to the R input of an *S*-*R*, while the C input of a *J*-*K* flip-flop is equivalent to the S input of an *S*-*R*. The alphabetic confusion here is unfortunate. To make matters worse, some flip-flops have only one of the preset or clear inputs instead of both, and sometimes this single asynchronous input is denoted by P. In such cases it is necessary to consult the flip-flop data sheet to determine which function the P input performs. All unused P and C inputs should be connected to the +5 V supply.

Clocked or Synchronous Operation. With both preset and clear inputs held HI, the *J*-*K* flip-flop has the property that changes of output occur only when the clock line T falls from 1 to 0. The output state (Q and $\bar{Q}$) *after* T falls depend on the inputs J and K *before* T falls.[3] The function table relating

[3] If the circle at the T input is omitted from the *J*-*K* symbol, this signifies a flip-flop that changes state when T rises from 0 to 1. Both kinds are available commercially. This text uses only falling-edge flip-flops in its examples.

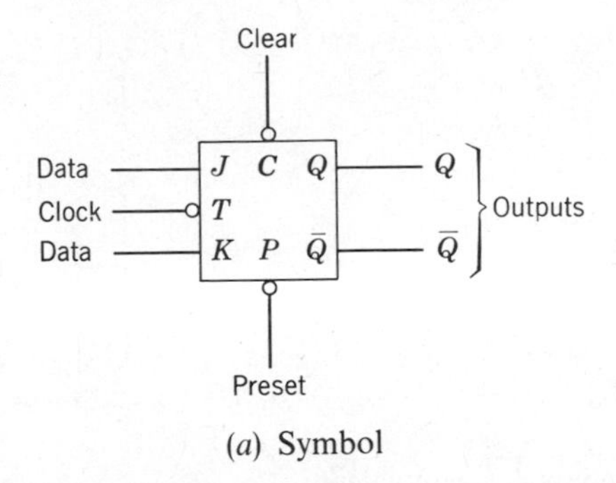

(*a*) Symbol

Asynchronous Operation using Preset and Clear Inputs

P	*C*	$Q \; \bar{Q}$
0	0	Not used
0	1	1 0
1	0	0 1
1	1	Clocked Operation

Clocked Operation ($P = C = 1$)

Before Clock Falls		After Clock Falls	Application
J	*K*	$Q \; \bar{Q}$	
0	0	No change	Memory
0	1	0 1	Data Transfer
1	0	1 0	Data Transfer
1	1	Complements	Counting

(*b*) Function table

Figure 16.20
The *J-K* flip-flop.

J and *K* before *T* falls to *Q* and $\bar{Q}$ after *T* falls is given in Fig. 16.20*b*. When both *J* and *K* are 0, a fall in *T* produces no change in the output, while when *J* and *K* are both 1, a fall in *T* causes a change of state at the output. If one of the inputs is a 1 while the other is 0, the flip-flop stores the pattern after *T* falls, with *Q* taking on the value of *J*, $\bar{Q}$ taking on the value of *K*. Because all changes of state occur when *T* falls, the changes of state can be synchronized with a master pulse generator or *clock*.

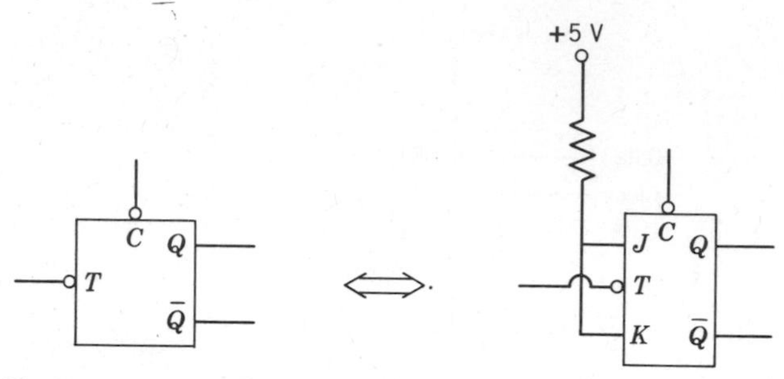

Figure 16.21
The type T flip-flop and its J-K equivalent.

There are many applications of the J-K flip-flop. The complement or toggle mode is used for counting, the no-change mode for data storage, and the remaining modes for data transfer. When flip-flops are designed for restricted use either in the toggle mode or the data transfer mode, they are given special names and notations. The type T flip-flop (Fig. 16.21) is equivalent to a J-K flip-flop with both J and K connected HI. Every time T falls, the output state changes. The type D flip-flop (Fig. 16.22) is equivalent

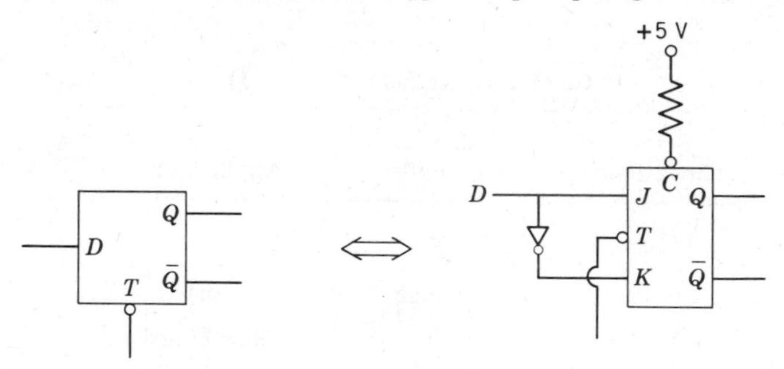

Figure 16.22
The type D flip-flop and its J-K equivalent.

to a J-K flip-flop with an inverter connected between J and K, thus insuring that K always is the complement of J. The input here has a new name, D. The data on D before T falls becomes the output Q after T falls. Selected applications of these flip-flops are presented in Section 16.4.

16.3.3 Monostables

Logic circuits can be used in conjunction with RC networks to generate a variety of pulse waveforms. Figure 16.23 illustrates a particularly simple

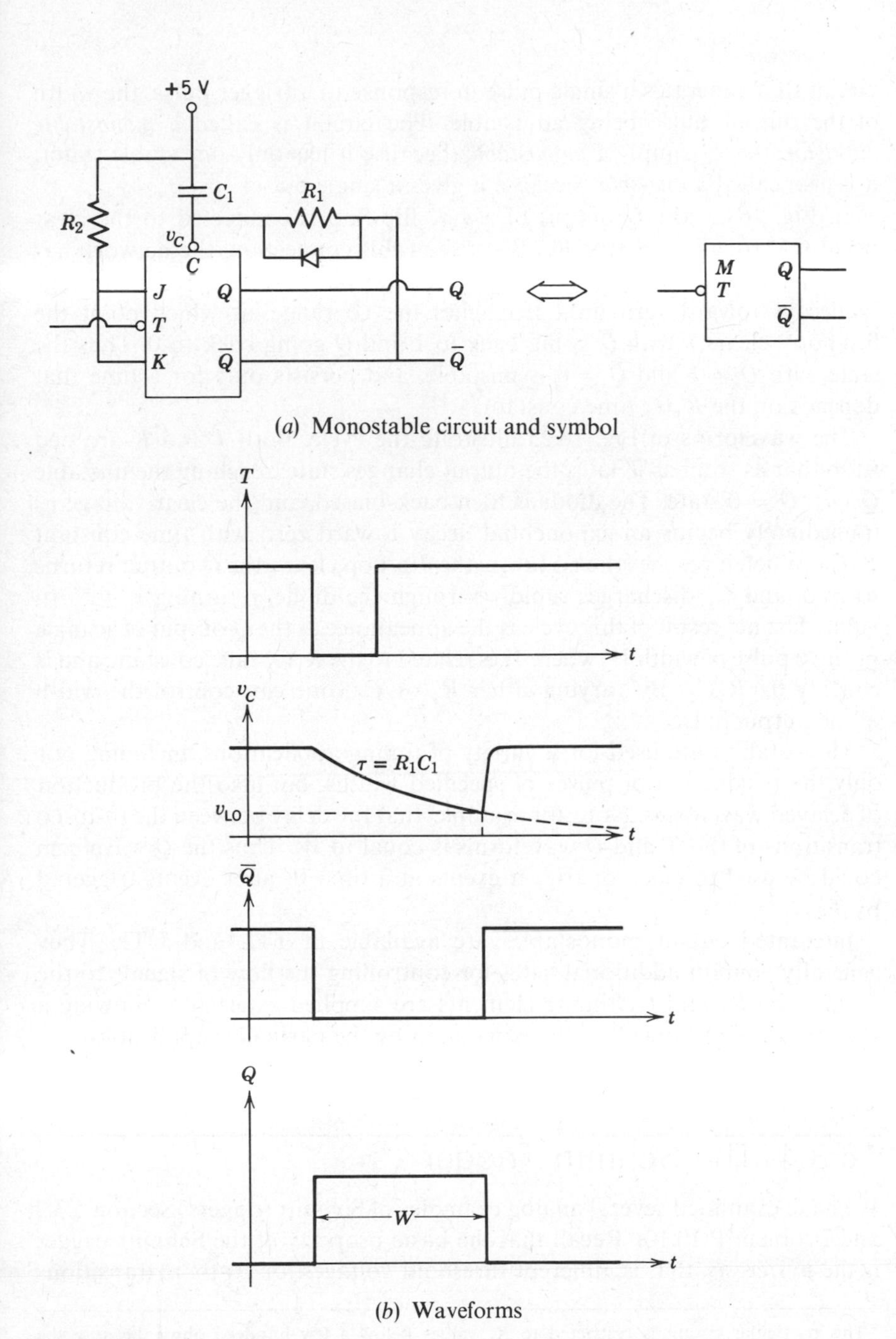

Figure 16.23
The basic monostable.

circuit that generates a single pulse in response to a trigger pulse, the width of the output pulse being adjustable. The circuit is called a *monostable multivibrator*, or simply a *monostable* (because it has only one stable state); it is also called a *one-shot* (because it gives a single pulse).

In Fig. 16.23, the $\bar{Q}$ output of a J-K flip-flop is connected to the clear input C through a resistor R_1. Because of this connection, the network has only one stable state: $\bar{Q} = 0$ and $Q = 1$. If $\bar{Q}$ does become zero, the voltage v_C decays toward zero until it reaches the LO range, at which point the flip-flop "clears," with $\bar{Q}$ going back to 1 and Q going back to 0. Thus the state with $Q = 1$ and $\bar{Q} = 0$ is unstable, and persists only for a time that depends on the R_1C_1 time constant.

The waveforms of Fig. 16.23 illustrate the cycle. Both J and K are tied HI so that as soon as T falls, the output changes state, reaching the unstable $Q = 1$, $\bar{Q} = 0$ state. The diode is then back-biased, and the clear voltage v_C immediately begins an exponential decay toward zero with time constant R_1C_1. When it reaches the LO range, the flip-flop clears, the Q output returns to zero, and C_1 discharges rapidly through the diode, returning v_C to a HI value. The net result of this cycle is the appearance at the Q output of a single positive pulse of width W, where W is related to the R_1C_1 time constant, and is roughly $0.6\,R_1C_1$. By varying either R_1 or C_1, one can control the width of the output pulse.[4]

Monostables are used for a variety of timing applications, including not only the production of pulses of specified widths, but also the production of delayed waveforms. Note, for example, that the delay between the HI-to-LO transitions of the T and Q waveforms is equal to W. Thus the Q waveform could be used to clock or trigger events at a time W after events triggered by T.

Integrated circuit monostables are available in TTL and DTL. They generally contain additional gates for controlling the flow of signals to the T line; the R_1 and C_1 timing elements are supplied externally, allowing a wide range of pulse widths W, as required by the particular application.

16.3.4 The Schmitt Trigger

We have examined several analog examples of Schmitt triggers (Section 5.3.2 and Problem P.10.10). Recall that the basic property of the Schmitt trigger is the *hysteresis*, that is, different threshold voltages for LO-to-HI transitions

[4] This particular circuit is restricted to R_1 values below a few hundred ohms, because the LO-state current from the C input must flow through R_1. Commercially available integrated-circuit monostables are designed to permit wide ranges of R_1 and C_1.

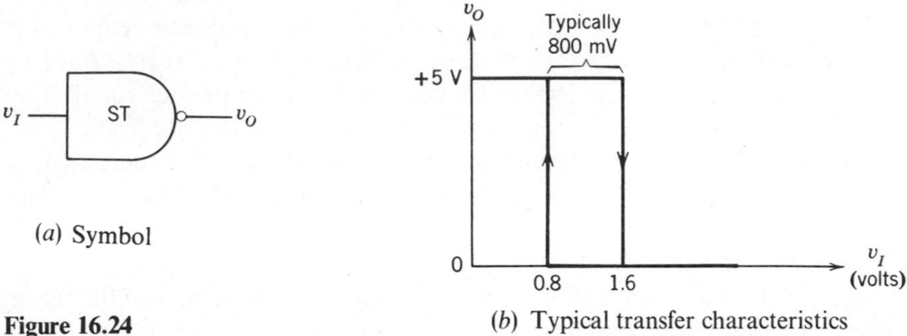

Figure 16.24
TTL Schmitt trigger.

and HI-to-LO transitions. A digital integrated-circuit version of the Schmitt trigger is available in TTL (e.g., Texas Instruments' SN7413). The SN7413 is available in NAND gate form, and has the symbol and transfer characteristic shown in Fig. 16.24. Note that the HI-to-LO transition occurs about 800 mV above LO-to-HI transition. Schmitt triggers are used for pulse shaping (producing sharp transitions from slowly varying waveforms), for threshold detection, and for the production of continuous series of pulses, as illustrated in the following examples.

Figure 16.25*a* shows the digital equivalent of the square-wave oscillator discussed in Section 6.4.3. Negative feedback around a Schmitt trigger

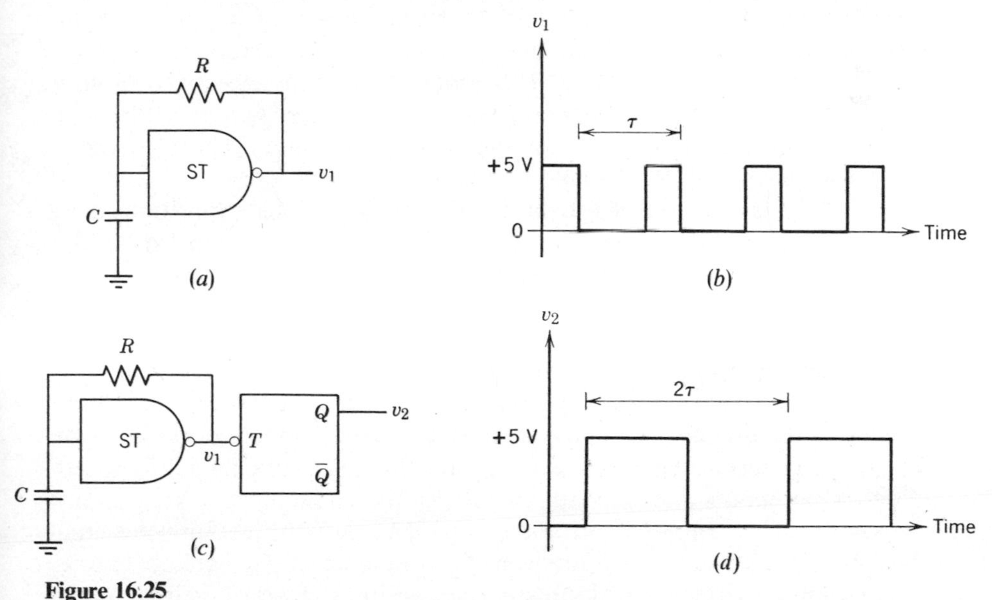

Figure 16.25
The use of a Schmitt trigger in a square-wave oscillator.

produces alternate charging and discharging of the capacitor, with time constant RC. When v_1 is HI, C charges up until the HI-to-LO threshold is reached, at which point v_1 becomes LO, and C discharges down to the LO-to-HI threshold. The output waveform is a periodic pulse train (Fig. 16.25*b*). However, because the threshold voltages are not symmetrically located between the HI and LO voltages, the pulse train is not symmetrical, having unequal HI and LO time intervals. If it is desired to have a symmetrical square wave, the circuit of Fig. 16.25*c* can be used. A Type-T flip-flop changes state on every falling edge of the v_1 waveform, producing a symmetrical square wave of period 2τ (Fig. 16.25*d*).

16.4 APPLICATION EXAMPLES

16.4.1 A Ripple Counter

Figure 16.26 shows three Type-T flip-flops connected as a counter. The preset and clear inputs (not shown) are presumed connected to HI, so that the flip-flops are in the toggle mode. Every time the signal on a T input undergoes a HI-to-LO transition, the corresponding flip-flop changes state. In the figure, the output of one flip-flop is connected to the T terminal of the next. Thus Q_0 changes state on every falling edge of the input waveform A. On every other Q_0 state change, Q_1 changes state. On every other Q_1 state change, Q_2 changes state. This sequence of events effectively counts the number of pulses in the input train, giving as an output a three-bit binary number representation of the number of pulses. The Q_2 output represents the 4's, Q_1 the 2's, and Q_0 the unit's.

Because there are switching delays in the flip-flops, not all transitions take place at the same time. For example, after the falling edge labeled "4" in Fig. 16.26*b*, the Q_0 state change occurs slightly after A falls. The Q_1 and Q_2 transitions then follow in sequence, and are further delayed. Thus the carry from place to place "ripples" down the line of flip-flops, giving rise to the name *ripple counter* for this configuration.

Notice that this counter cycles through its eight output states and, thus, counts events modulo 8. If one is interested in determining a count between 0 and 7, it is necessary, first, to clear the counter to zero with a LO pulse applied to all clear inputs and, second, to restrict the counting time to prevent overflow. One example of the use of this counter is to measure the time interval required to reach a predetermined number of data pulses. Figure 16.27 shows a circuit that, following the receipt of a START pulse, counts until five counts are accumulated, then resets itself. The output is HI during

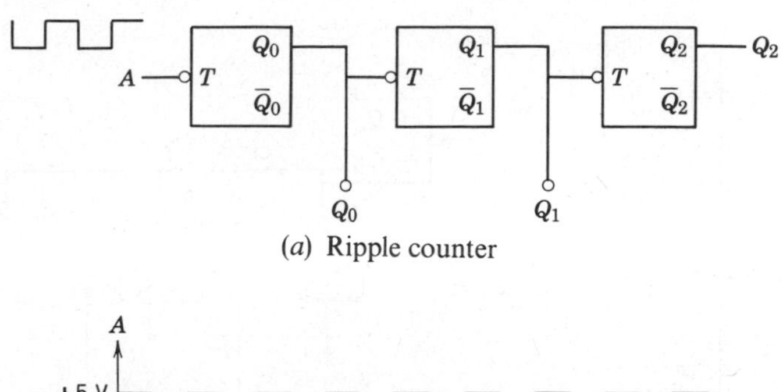

(*a*) Ripple counter

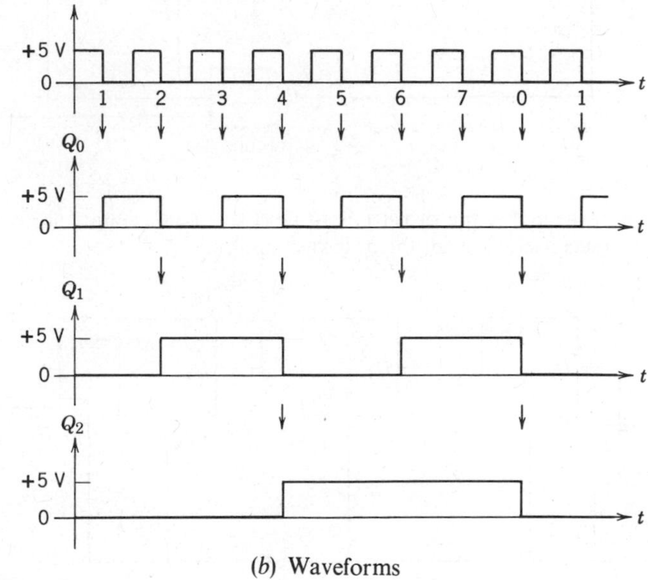

(*b*) Waveforms

Figure 16.26
A three-bit ripple counter. The arrows show how a falling edge at the input produces sequential state changes.

the interval when these counts are being accumulated. Notice the use of the $\overline{Q}_M$ output of a monostable to produce a LO pulse to clear the counter, and the use of the three-input NAND gate to decode the counter output.

16.4.2 Shift Registers

Figure 16.28 shows three Type-*D* flip-flops connected as a shift register. The output of the Q_0 flip-flop is connected directly to the input of the Q_1 flip-flop, and so on. All the *T* inputs are tied together, and are driven in

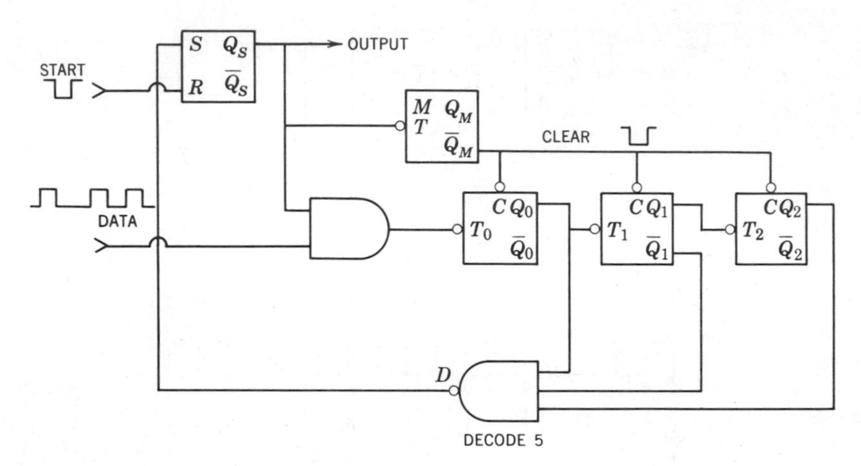

Figure 16.27
Following a START pulse, the OUTPUT is HI until five DATA counts are complete. The system then resets itself to await the next START pulse.

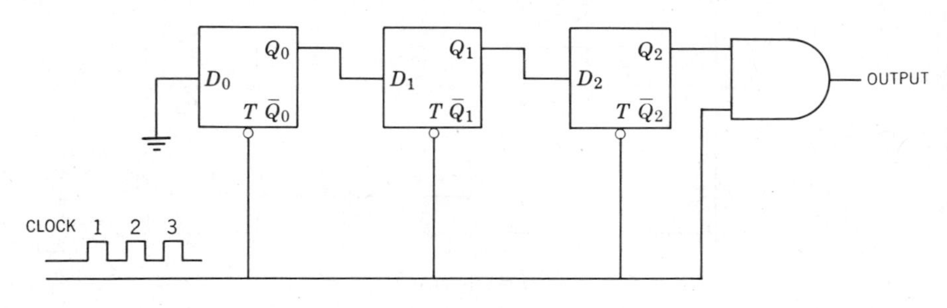

(*a*) Circuit

Time	Function	Q_0	Q_1	Q_2	Output
Before Interval 1	Store	A	B	C	0
During 1	Read Q_2	A	B	C	C
Between 1 and 2	Shift	0	A	B	0
During 2	Read Q_2	0	A	B	B
Between 2 and 3	Shift	0	0	A	0
During 3	Read Q_2	0	0	A	A
After 3	Store	0	0	0	0

(*b*) Function table

Figure 16.28
A three-bit shift register and function table. The initial values *A*, *B*, and *C* in the flip flops are read out in a sequence of pulses.

this case with a sequence of three positive pulses. The output of the Q_2 flip-flop is gated with the clock line to form the OUTPUT. During each positive pulse, the value of Q_2 appears at the OUTPUT. On each falling edge of the clock, the content of each flip-flop shifts one space to the right (with the content of Q_2 being lost). Thus if A, B, and C are the values previously loaded into Q_0, Q_1, and Q_2 before pulse 1 arrives (see the next example for a discussion of loading), the sequence of outputs and the values in each flip-flop follow the table in Fig. 16.28. Notice that the D input of Q_0 is tied LO. This causes Q_0 to be low after each fall on the clock line, and has the effect of clearing the register to 0 after three bits are read out of the OUTPUT line.

Shift registers can be used to convert data from parallel form to a series of pulses (as in the preceding example) or in the conversion of a series of pulses to parallel form. They also find numerous applications in the arithmetic sections of computers, permitting for example bit-by-bit addition of two numbers and the shift-and-add operations required in the multiplication of two numbers. Shift registers can also be used as counters, with a single 1 being loaded at one end, and then shifted down a line of flip-flops. The following example will illustrate some of the issues involved in the design of shift registers.

Suppose we wish to construct a 3-bit shift register that is controlled by two logic signals, A and B, and that can operate in one of three modes depending on the values of A and B:

$A = B = 0$	No shift. Data is stored as is.
$A = 1; B = 0$	Shift to the right, loading successive zeros in the left-most flip-flop.
$A = 0; B = 1$	Load a 3-bit number $X_0X_1X_2$ (externally available) in parallel form.

We assume that A and B will not both be 1 at the same time. To carry out this design, we must develop logic expressions for the way in which the flip-flop inputs depend on the control variables A and B. For example, assuming D_2 is the input of the rightmost of three flip-flops to be used in the shift register, and Q_0, Q_1, and Q_2 are the outputs of the three flip-flops just before T falls, then after T falls we require that

$$D_2 = Q_2 \text{ (if } A = B = 0) \quad \text{OR} \quad X_2 \text{ (if } B = 1) \quad \text{OR} \quad Q_1 \text{ (if } A = 1) \tag{16.28}$$

The corresponding Boolean expression is

$$D_2 = \bar{A}\bar{B}Q_2 + BX_2 + AQ_1 \tag{16.29}$$

Similarly, we obtain

$$D_0 = \overline{A}\overline{B}Q_0 + BX_0$$
$$D_1 = \overline{A}\overline{B}Q_1 + BX_1 + AQ_0 \qquad (16.30)$$

where the provision for loading zeros into D_0 when shifting has been included. An implementation of this shift register design is shown in Fig. 16.29.

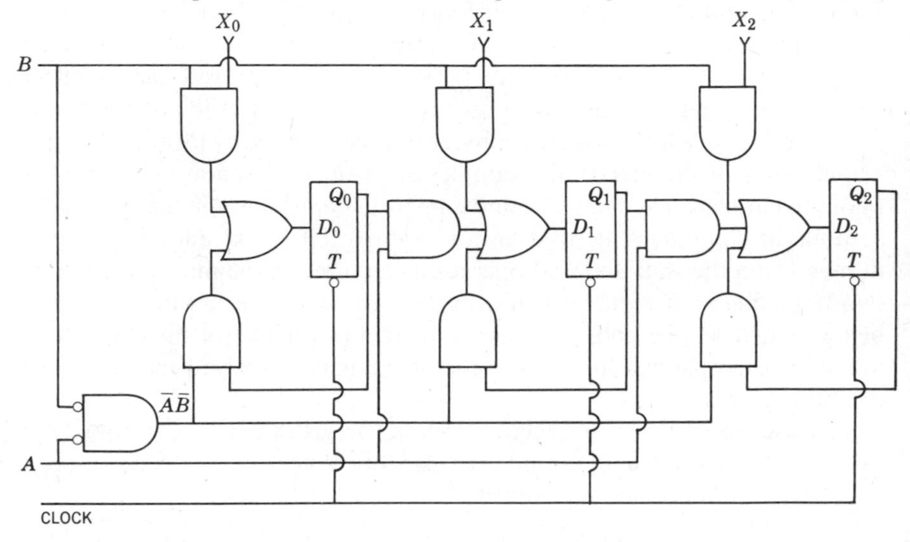

Figure 16.29
A parallel-in series-out three-bit shift register.

16.4.3 Digital-to-Analog Conversion

It is often necessary to convert from a digital representation of a number to an analog voltage of magnitude proportional to that number. A circuit which performs this conversion is called a *digital-to-analog converter*, abbreviated D-to-A converter. Two D-to-A converter circuits are explored in Problems P16.12 and P16.13.

The reverse conversion process, of an analog signal to a corresponding digital representation, is discussed in Section 18.3.

REFERENCES

R16.1 William E. Wickes, *Logic Design with Integrated Circuits* (New York: John Wiley, 1968).

R16.2 Taylor L. Booth, *Digital Networks and Computer Systems* (New York: John Wiley, 1971).

R16.3 Richard S. C. Cobbold, *Theory and Application of Field-Effect Transistors* (New York: Wiley-Interscience, 1970).

R16.4 Jacob Millman and Herbert Taub, *Pulse, Digital, and Switching Waveforms* (New York: McGraw-Hill, 1965).

R16.5 D. H. Sheingold, ed., *Analog-Digital Conversion Handbook* (Norwood, Mass.: Analog Devices, 1972).

R16.6 Bruce D. Wedlock and James K. Roberge, *Electronic Components and Measurements* (Englewood Cliffs, N.J.: Prentice-Hall, 1969), Chapter 17.

R16.7 Paul E. Gray and Campbell L. Searle, *Electronic Principles: Physics, Models, and Circuits* (New York: John Wiley, 1969), Chapters 21 to 24.

R16.8 Texas Instruments Incorporated, IC Applications Staff, *Designing with TTL Integrated Circuits*, Robert L. Morris and John R. Miller, eds. (New York: McGraw-Hill, 1971).

R16.9 Data sheets and applications notes from major manufacturers (Texas Instruments, Motorola, RCA, National Semiconductor, Fairchild, Signetics, and others).

QUESTIONS

Q16.1 Given a two-input gate that performs the AND operation for positive-logic identification of HI and LO. What operation would this same gate perform is negative logic were used? (This is, in effect, an application of DeMorgan's theorem.)

Q16.2 Why is it desirable to have a gap between HI and LO ranges?

Q16.3 Can resistor networks by themselves be used to perform AND, OR, or NOT operations?

Q16.4 Can resistors and diodes (and bias sources) be combined to perform an AND, OR, or NOT operation? Try it, then see Excercise E16.6.

EXERCISES

E16.1 Simplify the following Boolean algebraic expressions:

(a) $Z = A(AB + C) + \bar{A}(B + \bar{C})$

(b) $Z = (A + B + \bar{C})B + (\bar{A} + B)C$

(c) $Z = (\overline{A + B})C + (\overline{AC}) + AB$

(d) $Z = \overline{(\overline{A + B + C}) + (\overline{AB + AC})}$

E16.2 Determine whether the following statements are valid:

(a) $\bar{A}CD + \bar{B}CD + \bar{B}C\bar{D} = \bar{B}C + \bar{A}CD$

(*Hint*. Note that $D + \bar{D} = 1$.)

(b) $(\bar{Y} + Z)(\bar{Z} + W)(\bar{W} + Y) = (Y + \bar{Z})(Z + \bar{W})(W + \bar{Y})$

(c) $ABC + ACD + \overline{BC}D = ABC + A\bar{B}D + \overline{BC}D$

E16.3 Implement the following expressions using AND, OR, and NOT functions:

(a) $Z = B + AC + \bar{A}\bar{C}$

(b) $Z = B + \bar{A}C$

(c) $Z = A\bar{B}C$

(d) $Z = A(B + C)$

E16.4 Implement the expressions of E16.3 using only two-input NOR gates. Try to minimize the number of gates required.

E16.5 Implement the expressions of E16.3 using only two-input NAND gates. Try to minimize the number of gates required.

E16.6 *Diode logic*. Determine the logic operations performed in Fig. E16.6 and suggest practical HI and LO ranges for this logic.

E16.7 How should v_B in Fig. E16.6*a* be connected to obtain operation as an inverter?

E16.8 Plot the waveforms that appear at the outputs Q_S, Q_M, Q_0, Q_1, Q_2, and D in Fig. 16.27 during one complete cycle of operation.

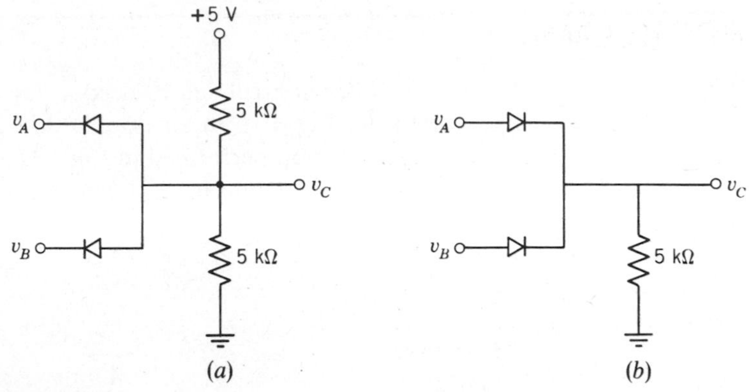

Figure E16.6

E16.9 Prove that the circuit of Fig. 16.5*b* is a correct implementation of Eq.16.20.

E16.10 Assume that when a traffic light appears other than red, green, yellow, or green and yellow, the light is malfunctioning. Design a NAND-gate logic circuit to detect improper operation.

E16.11 Develop the function table for the SET-RESET NOR-gate flip-flop in Fig. E16.11. Show how it differs from the *S*-*R* built with NAND gates.

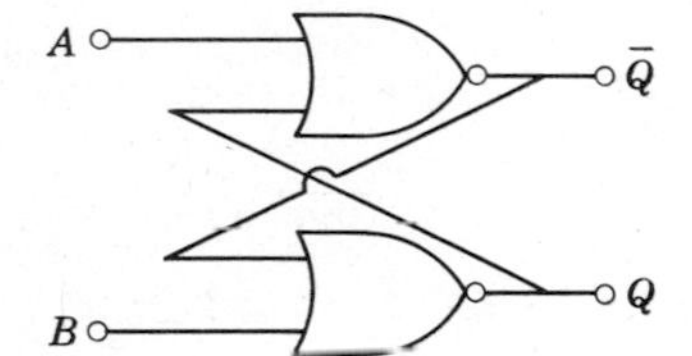

Figure E16.11

E16.12 Plot *C* versus time in Fig. E16.12. Notice how *B* can either *enable* or *inhibit* the passage of $\bar{A}$ through the gate.

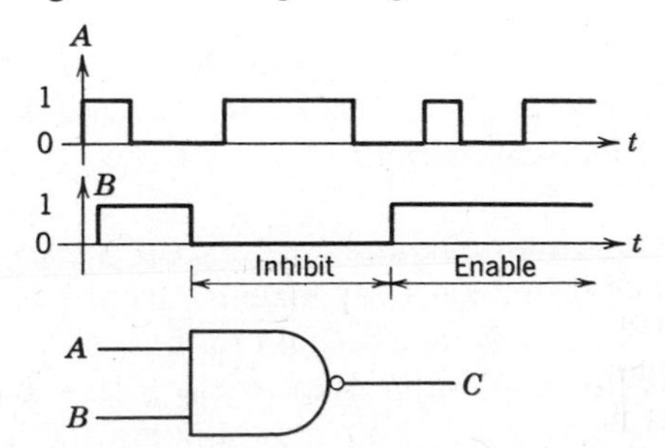

Figure E16.12

PROBLEMS

P16.1 Switches for ON-OFF power control can be used as logic elements, either single throw as in Fig. P16.1*a*, or double throw as in Fig. P16.1*b*. Find the logic function performed in Figs. P16.1*c* to *f*, if a completed path represents an output of 1 and an open circuit represents an output of 0.

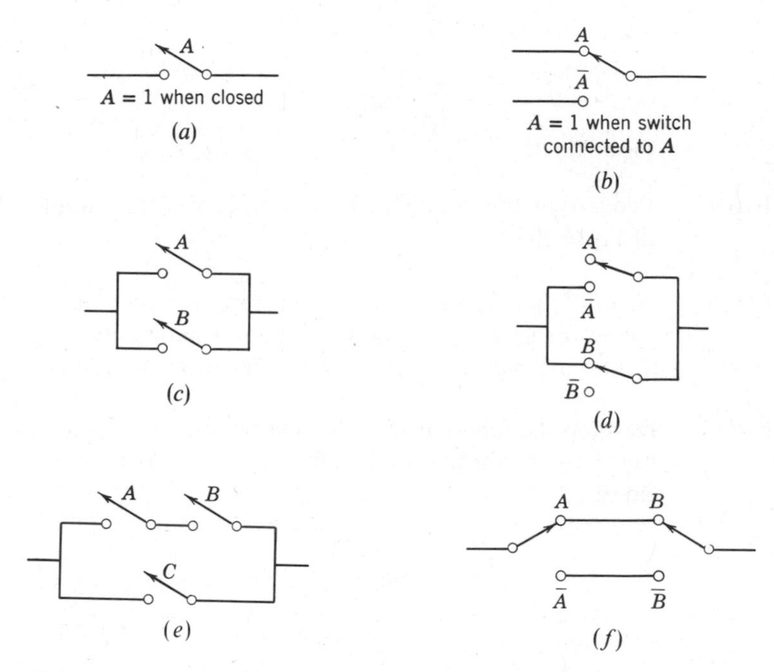

Figure P16.1
"Three-way switch"

P16.2 Design a switching network to turn a light ON or OFF by using any one of *three* switches. (The solution for *two* switches that is commonly found in home wiring is shown in Fig. P16.1*f*). You should construct the truth table and derive a Boolean expression to aid your design.

P16.3 Design an *interlock* (i.e., a set of switches) that prevents the high voltage on an X-ray machine from being turned on unless (1) the filaments F have been on for five minutes, (2) the cooling water W is flowing, (3) the lead shield S is in place, and (4) the exposure time E is set. You may assume the availability of transducers that drive single-pole double-throw relays (i.e., electrically operated switches) in response to the variables of interest. How sensitive is your design to the failure of one of the transducers or relays?

P16.4 Design the combinational logic network for a coin-operated vending machine containing items costing 15 cents. The machine must accept nickels, dimes, or quarters, and give correct change. Implement your design with NAND gates.

The machine should accept binary digital input signals N, D, and Q. N will be 1 momentarily when a nickel is inserted, D will be momentarily 1 when a dime is inserted, and Q will be momen-arily 1 when a quarter is inserted.

Your machine should have three outputs, N_0, D_0, and F. N_0 should be momentarily 1 if a nickel is to be returned, D_0 should be momentarily 1 if a dime is to be returned, and F should be momentarily 1 if food is to be dispensed.

P16.5 Four stockholders of a corporation meet. Stockholder A holds 40% of the stock, B holds 30%, C 20%, and D 10%. Design a logic network with which the YES-NO votes of the four individuals can be automatically tallied in terms of stock percentage. The output of your network should indicate either YES wins, NO wins, or a tie.

P16.6 This problem (Fig. P16.6) illustrates how the change in flip-flop states can be controlled. The C waveform is used to read data into the flip-flop at a particular time.

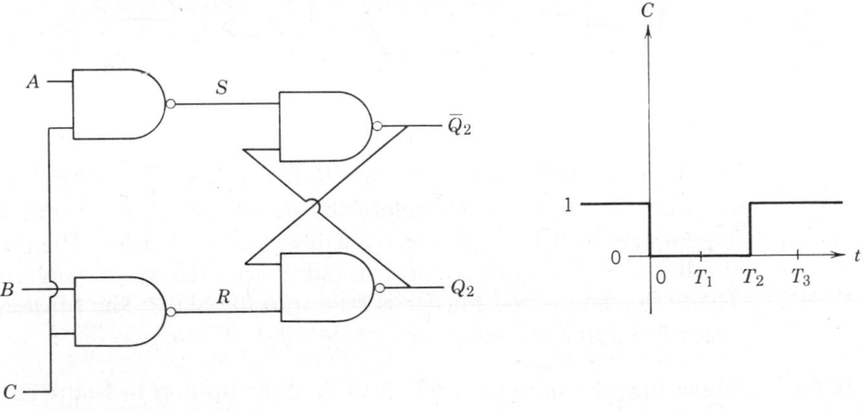

Figure P16.6

(a) Before $t = 0$, how do Q_2 and $\overline{Q}_2$ depend on A and B? (See Exercise E16.12).

(b) Plot S, R, Q_2, and $\overline{Q}_2$ as functions of time for $t > 0$, assuming that $A = 0$ and $B = 1$ for $0 \leq t \leq T_3$.

(c) What happens to your result in (b) if A and B were to change to 1 and 0, respectively, at time T_1? At time T_2? At T_3?

P16.7 *Master-Slave Flip-Flop*

Q_1 represents the input transistor of a TTL inverter. Suppose v_1 makes a LO to HI transition in time τ (Figs. P16.7*a* and *b*). Show that because of the 220 Ω resistor, v_2 must reach the HI range slightly before v_1 does. Use this result to show in Fig. P16.7*c* that

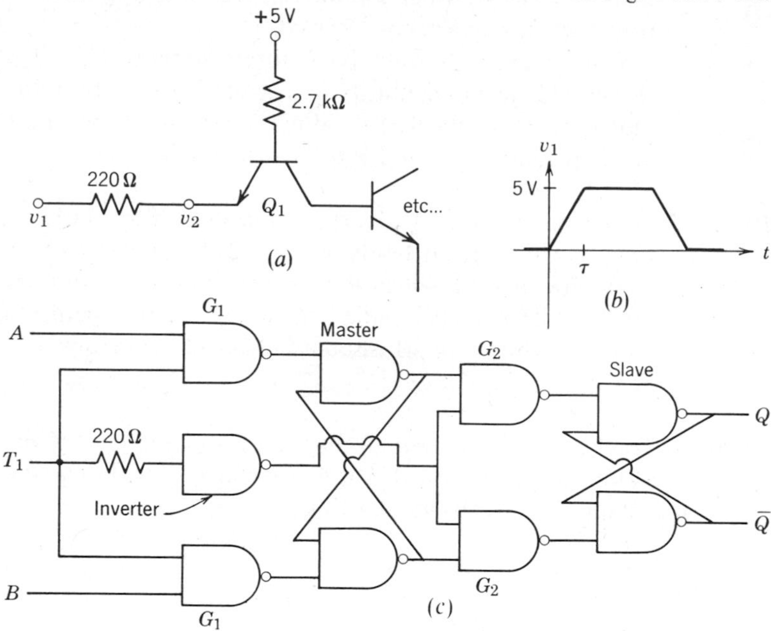

Figure P16.7

the G_2 gates stop reading the data from the Master flip-flop into the Slave flip-flop *slightly before* the new data on the A and B inputs gets read through the G_1 gates into the Master. Then as T falls, the G_1 gates disconnect A and B from the Master *slightly before* the data in the Master is read into the Slave. The Master-Slave flip-flop is the basic unit inside a J-K flip-flop.

P16.8 Show that the use of a DTL or TTL S-R flip-flop in Fig. P16.8 eliminates the effect of push-button contact bounce.

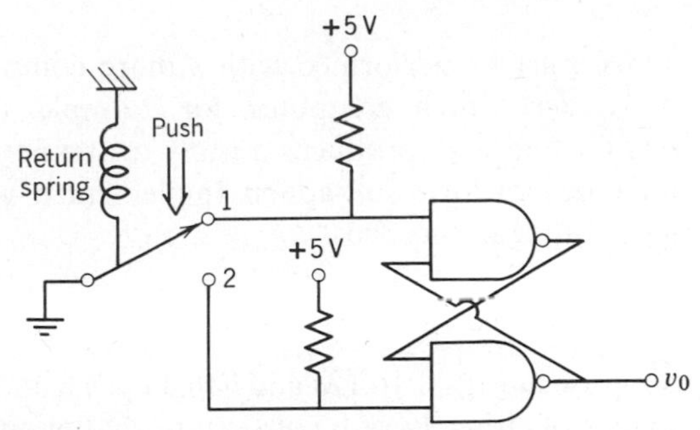

Figure 16.8
The switch will bounce on making contact when pushed to position 2 or when returned to position 1 by the spring.

P16.9 Devise a flip-flop circuit that in response to a push-button START gives a HI output until a positive pulse arrives on a DATA line. Show how such a circuit could be used in conjunction with a clock and counter to measure time intervals.

P16.10 Design a timer that, when started, produces two 10 μs pulses spaced by a time interval that can be varied between 0.1 and 10.0 s. (This problem can be solved using monostables. It can also be solved with counters if a clock frequency, such as the 60 Hz line, is available. You should attempt both kinds of design). The timer should reset itself so that it may be used repeatedly.

P16.11 Figure P16.11 is a diagram of a 4-bit adder. The least significant bits, A_0 and B_0 can be added with a half-adder, but the other

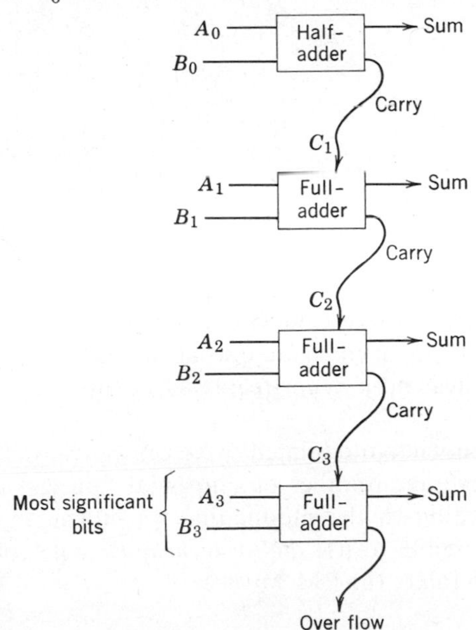

Figure P16.11

sums must be performed with a more complex network called a *full-adder*, which computes, for example, the sum of A_1, B_1, *and* C_1 and also produces a carry to the next place. Perform the logic design for a full-adder. Implement it with two-input NAND gates (nine are needed).

P16.12 *Digital-to-Analog Conversion*

Suppose in Figs. P16.12*a* and *b* that v_A, v_B, and v_C are binary digital signals of either V_0 or 0 volts. Show that in either circuit v_D will be an analog voltage proportional to the magnitude of the three-bit binary number ABC where $A = 1$ corresponds to $v_A = V_0$, etc.

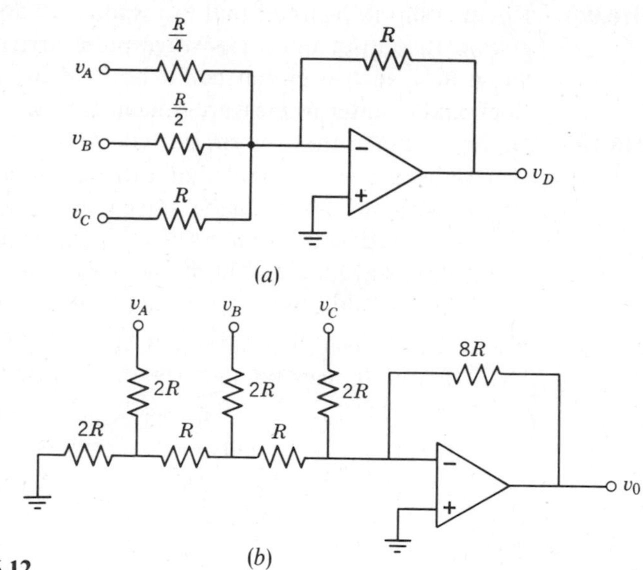

Figure P16.12

P16.13 Examine the circuits of P16.12 and show why v_A, v_B, and v_C must *not* be ordinary gate-output levels, with ranges of HI and LO voltages. Suggest how one might produce a 0 volt LO and a V_0 volt HI from gate-output levels (using perhaps transistor switches and a Zener-regulated V_0 supply).

P16.14 A frequency-modulated wave can be viewed as a wave in which the average number of sinusoidal "pulses" per unit time varies according to the modulating waveform. In this case, it should be possible to use *digital techniques* (e.g., counters and gates) to demodulate the FM wave.

You are to design, in at least block-diagram form, a digital FM demodulator. Your design should include conversion to a properly filtered analog output.

P16.15 Design a three-bit shift register that can perform the following operations according to the values of three control variables A, B, and C.

$A = B = C = 0$ No shift. Data is stored as it is.
$A = 1$ Shift to right with data Y_0 loaded into Q_0.
$B = 1$ Shift to left with end-around carry ($Q_2 \rightarrow Q_0$).
$C = 1$ Clear register on next clock pulse.

P16.16 Given two three-bit binary numbers, A and B, design a logic network that compares A to B and provides outputs to indicate when $A > B$, $A = B$, or $A < B$ (Fig. P16.16). If your solution

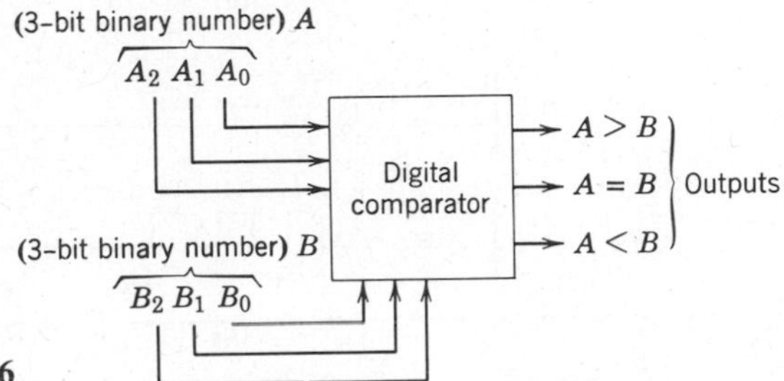

Figure P16.16

uses a clocked sequence of steps (a possible approach) you may assume that a line carrying clock pulses is available. In this case, however, the *number* of clock pulses needed to complete a comparison must be clearly indicated, and a procedure for resetting the comparator so that it can accept new input data must be provided.

P16.17 You are to design a digital combination lock for your door based on a five-digit sequence selected by push buttons. The START button sets up the lock to accept numbers. If the correct set of digits are then pushed in the correct order, a two-second output pulse is produced that can be used to control a door-opening mechanism.

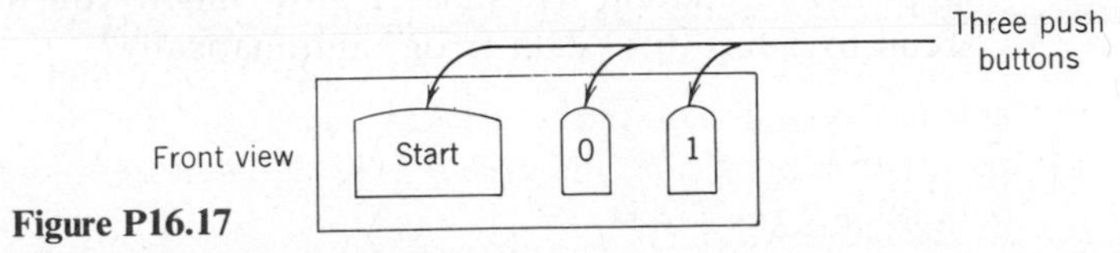

Figure P16.17

P16.18 (a) You are to design an interface between two inputs (A and B) and a set of lights (Fig. P16.18a).

The inputs are telemetry data and have the following form. Input A supplies pulses occurring at widely scattered random intervals. Immediately following each pulse from A, input B will supply a train of pulses with either 1, 2, or 3 pulses in the train. A and B are both LO in the interim periods. A sample set of inputs might appear as in Fig. P16.18b.

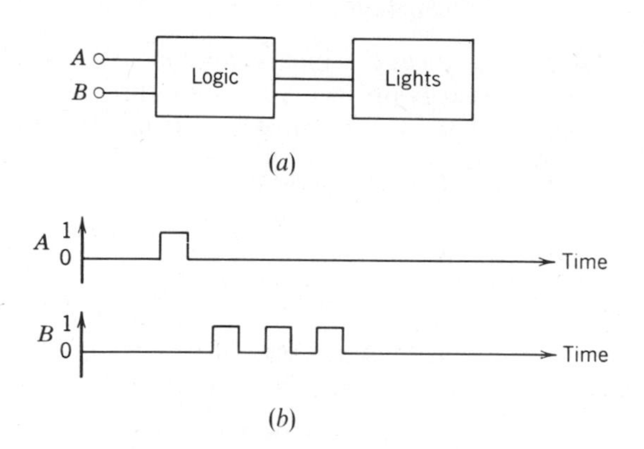

Figure P16.8

Your output wires go to a set of three lights labeled "1," "2," and "3." Your circuit should light the appropriately labeled light corresponding to the number of pulses that occur in the pulse train from input B. The pulse in A should turn off the lights. After the pulse train in B is finished, the appropriate light should be lit and remain lit. For example, as shown in Fig. P16.18b, the pulse in A would erase the previous number and light "3" would be lit after the pulse train in B had ended. Design the logical interface as described.

(b) For your design, what would happen if the pulse train in B happened to contain five pulses? How might you modify your circuit to detect this "data error" automatically?

CHAPTER SEVENTEEN
NOISE

17.1 SIGNALS AND NOISE

Too often in introductory texts, noise is considered to be one step down from the weather: hardly anyone even talks about it. Yet it is the noise level in a circuit that ultimately limits the ability of that circuit to transmit faithfully the information carried by the signals being processed. In this chapter we shall discuss noise not only from a theoretical point of view, but also with a view toward the proper practical design and use of electronic circuits and instruments.

What is noise? Generally, any signal that can interfere with or obscure the signal of interest can be called noise. We recall that the information carried by a signal might be coded in several ways: a speech waveform is an analog representation of sound intensity as a function of time; an amplitude-modulated wave carries information in the frequency, amplitude, and phase of its sidebands; a set of pulses of HI and LO voltage levels can carry digital information, provided that the code for interpreting the pulses or the pattern of HI and LO voltages is known. In each case, the recovery of the information depends on proper processing of the information-carrying waveform. Any other signal that gives nonzero output when similarly processed can interfere with the recovery of the information, and is thus a potential source of noise.

It is useful to group noise sources into the following three categories.

1. Interfering Signals. This category includes all waveforms that might under other circumstances be called "signals" but that happen, in a particular application, to interfere with the signal of interest. An example is the signal broadcast by one radio station when you might be interested in listening to a different station. To obtain an error-free replica of the messages from your station, it is necessary to filter out all other stations prior to demodulation of the carrier. The superheterodyne receiver discussed in Chapter 15 is designed to perform this frequency-selecting function.

As a less trivial example of an interfering signal, imagine that you are working in a biology laboratory. Every time a subway train starts in a nearby station, there is a large current surge in the subway power line. The magnetic field from this current surge couples inductively into your microprobe or other low-level measurement apparatus. The result: every time a subway leaves the station (an unpredictable event at best), your high-gain amplifier is driven into saturation. You observe only that occasionally your apparatus malfunctions, destroying the results of hours of careful preparation.

As a final example in this category, consider the interference at 60 Hz and harmonics of 60 Hz introduced by couplings between the power lines and your apparatus, or between heavy equipment (e.g., large transformers and motors) and your apparatus. This kind of interference is so widespread that it deserves special attention (see Section 17.1.1).

2. Drifts. Even the best-designed apparatus has limits of stability. Amplifiers, current sources, capacitors, the temperature of a sample, and so on *ad infinitum*, all tend to be "unstable" if measured over a sufficiently long time period. The instability might be random and unpredictable, or it might follow slow periodic fluctuations. In either case, if you wish to assume that you are measuring a certain quantity under certain specified conditions, and those conditions or the apparatus with which you are doing the measurement change, then your result will acquire an uncertainty because of the drifts and instabilities. We shall discover that it is possible to make a useful semi-quantitative model of drifts and, thus, to consider them as equivalent to a source of noise.

3. Device Noise. This category includes the noise that is an inherent adjunct to the functioning of any electronic device, whether resistor, diode, infrared detector, etc. It arises from the finiteness of the electronic charge. Voltages and currents in any signal are average quantities, representing in reality the average behavior of a statistical assembly of electrons. The statistical fluctuations in the assembly of electrons give rise to noise—random components in both voltage and current.

Device noise, precisely because it has well-definited statistical properties, gets most of the attention in elementary treatments of noise in electronic systems. This remains true in spite of the overwhelming agreement among experimenters of almost all persuasions that it is interfering signals and drifts that are responsible for limiting the ultimate achievable sensitivity in most experiments. In our discussion, we shall bow to custom and concentrate our theoretical discussion on device noise and its propagation in linear systems. At the same time, however, we shall incorporate into the discussion some ideas about interference and drifts that will be used in the next chapter for our discussion of practical instrumentation problems.

17.1.1 The Interference Problem

Figure 17.1 illustrates the interference problem as it applies to audio-frequency sources.[1] A source network is connected through a cable to an

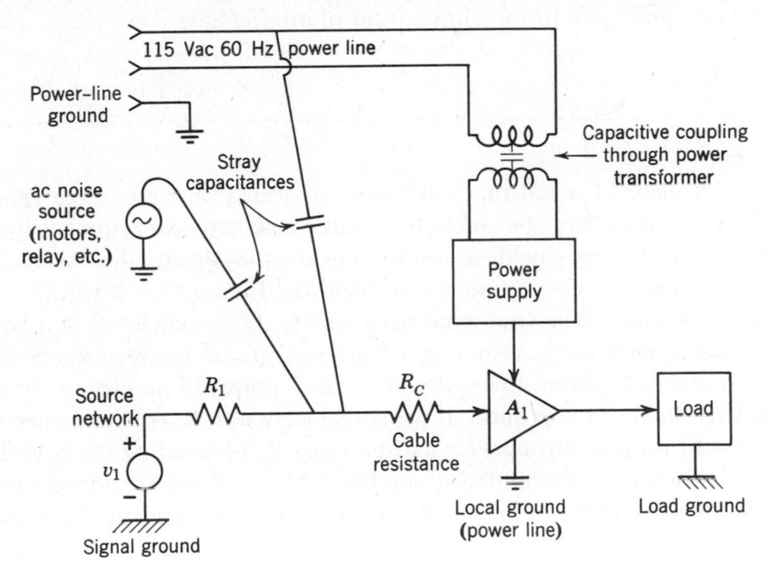

Figure 17.1
The interference problem.

amplifier A_1 that is connected to a load. The amplifier is driven from a power supply that is itself driven by the 60 Hz power line. Three stray capacitive couplings are shown, two from the power line (and lighting fixtures) and from electrically noisy machines such as motors and relays, and one directly through the power supply transformer. The magnitude of

[1] At radio frequencies, the stray inductance of conductors must be included, as well as possible transmission-line effects. These subjects will not be considered here.

the first two capacitances might be fractions of a picofarad, while the primary-to-secondary capacitance in a typical power transformer might be 1000 pF, and is modeled as connected from the center of one winding to the center of the other. Each of the components (source, amplifier, and load) is connected to a different nominal ground, and because of couplings among and current patterns in the various conductors that tie these nominal grounds together, it is necessary to allow for the grounds to be at different potentials, with differences as large as 1 to 10 Vac at 60 Hz being common. In the example in Fig. 17.1, the third "neutral" line that accompanies the 60 Hz power line is used as the ground for the amplifier.

All of the sources shown except for the v_1 signal source can produce interference. The following discussion will survey qualitatively the steps to be taken to reduce this interference. In any practical situation, not all steps may be needed. As a rule, the weaker the signal being processed, the more attention one must pay to the elimination of interference.

17.1.2 Shields

A *shield* is a piece of metal that surrounds or nearly surrounds the space being shielded. The effect of a shield is to reduce the stray capacitances from conductors inside the shield to those outside the shield, the reduction factor depending on how complete the shield is.[2] Figure 17.2 illustrates the operation of a shield. The source network of Fig. 17.1 is enclosed in a box, and shielded cable is used. Assuming a 1 pF capacitance between the power line and the shield, about 40 nA (i.e., 4×10^{-8} amps) of ac current flows through the shield to the ground. If the shield were not present, this current would flow to ground through R_1 and perhaps R_C (depending on how the capacitive coupling is distributed along the cable). For a source-plus-cable resistance of 1 kΩ, there would be 40 μV of 60 Hz ac present on the signal line in addition to v_1. A good shield can be expected to reduce this amplitude by a factor of between 100 and 1000.

Figure 17.3 illustrates a problem encountered whenever shields are used. A shielded transformer, which contains a shield layer between the primary and secondary windings,[3] is shown, with the shield connected to the power-line ground and to the amplifier shield. Thus both the source network and

[2] Such electrostatic shields also provide some reduction in the inductive pickup of interference from time-varying magnetic fields. However, complete exclusion of low-frequency magnetic fields from a region requires enclosures made of high-permeability magnetic materials, a subject not discussed further in this book.

[3] A single transformer shield, depending on the details of its fabrication, can reduce the primary-to-secondary capacitance to the order of 10 pF.

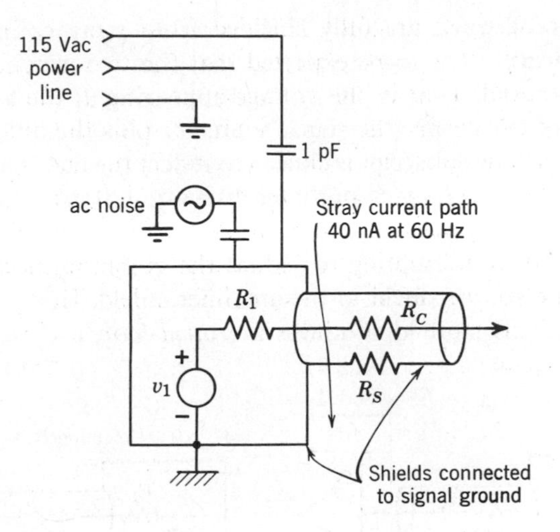

Figure 17.2
A shield connected to ground shunts stray currents away from signal source. The cable shield (with resistance R_S) is depicted as surrounding but not contacting the inner conductor of the cable (R_C).

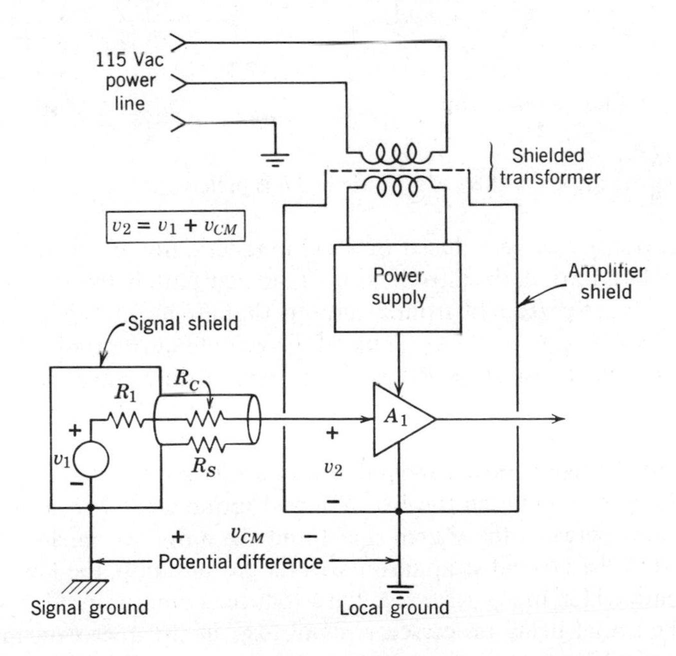

Figure 17.3
The signal and local grounds can be at different potentials.

the amplifier network are fully shielded from stray couplings to outside sources. However, it is to be expected that the two grounds will not be at the same potential. That is, the voltage appearing at the amplifier input v_2 will consist of two terms, the signal source v_1 plus the difference in ground potentials v_{CM}. The subscript is chosen to reflect the fact that v_{CM} will appear as a *common-mode voltage* in our later discussion. If v_{CM} is at all comparable to the magnitude of the signal, substantial errors can be introduced.

One method of attempting to reduce the common-mode voltage v_{CM} is to connect the source shield to the amplifier shield. However, as illustrated in Fig. 17.4*a*, this method produces a *ground loop*, a closed path in which

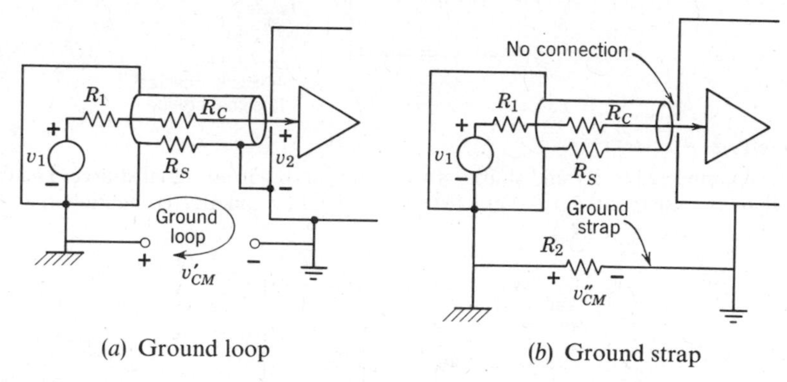

(*a*) Ground loop

(*b*) Ground strap

Figure 17.4
Attempts to reduce the effect of v_{CM}. Method *b* is preferred.

large currents can be induced by stray magnetic flux from transformers or motors. In addition, the introduction of the new path between ground points could alter the pattern of ground currents that are inevitably associated with power distribution networks. Some of the ground current may flow through the cable-shield resistance R_S. In both cases, a different noise voltage v'_{CM} appears, which often can be worse than the original voltage v_{CM}.

A somewhat better method for reducing v_{CM} noise is to connect a heavy conductor called a *ground strap* (braided cable is often used because of its low inductance) between the two nominal grounds, and to make no other connection between the source shield and the amplifier shield. Although it is true that the ground strap also creates a ground loop, the low impedance of the path will in many cases result in a reduced common-mode voltage v''_{CM}.

If the signal being processed is small (e.g., in the microvolt range), or if the electrical environment is particularly noisy, it is necessary to use more elaborate shielding schemes to reduce interference. These are described in the next section.

17.1.3 Guards

A *guard* is a shield that is connected to the common-mode voltage of the signal being processed. Figure 17.5 illustrates the use of a guard, together with other isolation devices, to achieve an excellent degree of noise isolation. The A_1 amplifier is driven by a power supply which is connected to the

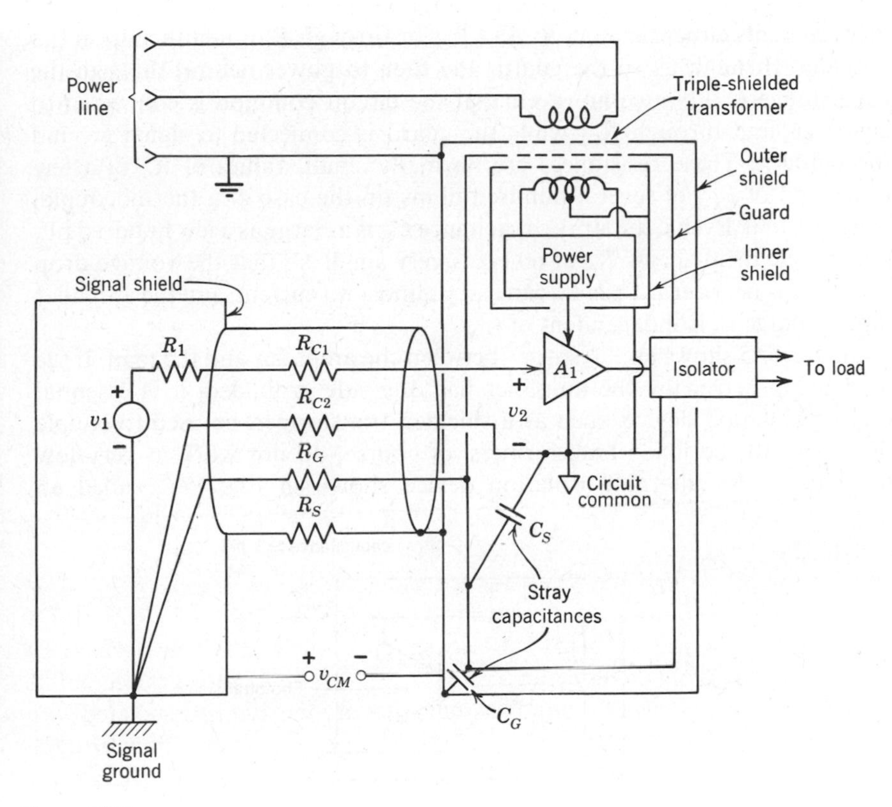

Figure 17.5
The use of a guard and a triple-shielded transformer.

60 Hz power line through a triple-shielded transformer, a transformer with three distinct layers of shielding between primary and secondary. The shield closest to the primary is tied to power neutral and to the amplifier outer shield. The middle shield of the transformer is connected to the guard, which is represented as a box-in-a-box, fully enclosing both amplifier and power supply. The power supply is enclosed in a third shield that is connected to the center tap of the transformer secondary and to a new "ground"

called the *circuit common.* This "ground" or "common" is actually connected to earth only through stray elements. With a properly constructed triple-shielded transformer the primary-to-secondary capacitance can be reduced to below 0.1 pF, thus effectively isolating the power supply from the power-line potentials. Let us examine how the guard works to reduce the effect of common-mode voltage v_{CM}.

The only way for v_{CM} to affect the amplifier input voltage v_2 is for v_{CM} to drive currents either through R_1 and R_{C1} or through R_{C2}, and then from the amplifier through C_S to the guard, and then to power neutral through the capacitance C_G. Notice however, that the circuit common is connected to signal ground through R_{C2} while the guard is connected to signal ground through R_G. These resistances are normally small, values of R_G of a few ohms and of R_{C2} of several hundred ohms (in the case of a thermocouple) being normal. Even if the stray capacitance C_G is as large as a few hundred pF, the voltage drop across R_G at 60 Hz is very small, so that the voltage drop across C_S is near zero. Thus C_S can carry almost no current, and the amplifier input voltage v_2 is independent of v_{CM}.

Figure 17.5 shows an "isolator" between the amplifier and the load. If the load being driven by the amplifier has one side grounded, it is essential that an isolation device, such as a shielded transformer, be used to couple the signal to the load. Transformers, of course, do not work at very low frequencies. An alternate isolation device shown in Fig. 17.6, called an

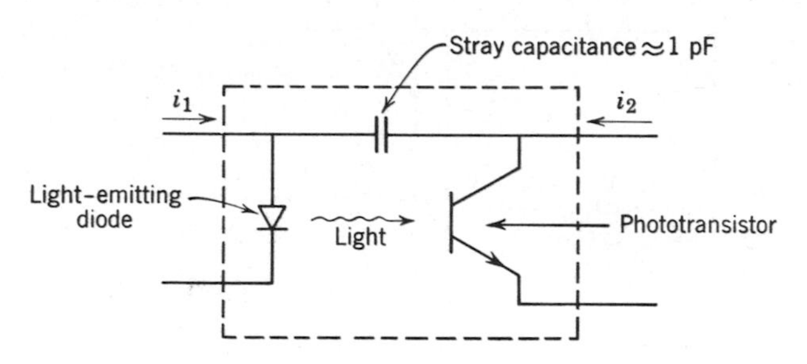

Figure 17.6
Optoelectronic isolator.

optoelectronic isolator, consists of a light-emitting diode and a phototransistor mounted close enough to one another so that the light from the diode carrying i_1 falls on the base of the transistor, produces hole-electron pairs, thereby injecting carriers into the base and giving rise to collector current i_2. Optoelectronic isolators operate at dc, but have the disadvantage that the current transfer characteristic, i_2 versus i_1 is not linear over large ranges. This subject is explored further in Chapter 18.

Having introduced the guard concept, we can now remove the restriction that the signal ground of Fig. 17.5 actually be connected to earth ground. *Any* potential difference between the amplifier ground and the signal common is effectively shielded by the guard connection. It is customary to express the quality of the guarding in terms of a common-mode-rejection ratio, which is the ratio of the amplitude of common-mode voltage v_{CM} to the amplitude of the component of v_2 produced by v_{CM}. With good design, it is possible to achieve common-mode-rejection ratios in the vicinity of 120 dB at 60 Hz (a factor of 10^6), which means that 10 V of common-mode signal produces only 10 μV of noise at the v_2 terminal.

17.2 CHARACTERIZATION OF NOISE

17.2.1 Time-Average

With the exception of interference from power-line signals and other extremely well-characterized sources, noise waveforms are characterized by a certain degree of unpredictability, or randomness. Information-carrying waveforms must also be somewhat unpredictable – otherwise they could carry no information. Nevertheless, one knows generally many more features of the signal of interest (such as principal Fourier components for analog signals or the HI and LO ranges for digital signals) than one does for noise. There is a tendency, therefore, not altogether rigorous, to consider the signal of interest to be "nearly predictable" and the noise to be "random" by comparison.

Because we must deal with random waveforms, we are restricted to making statements of a statistical nature about the waveforms. The first property is the time-average value. Given a waveform $v(t)$, we define the time-average value $\bar{v}$ by

$$\bar{v} = \lim_{T \to \infty} \frac{1}{T} \int_{-T/2}^{T/2} v(t)\, dt \qquad (17.1)$$

In the strict sense, if $v(t)$ has a time average that is not zero, the time-average value should be called a "signal" since it represents a predictable component of the waveform. Throughout this discussion, however, we will encounter waveforms that are random except perhaps for a dc component. We shall continue to call these waveforms "random."

17.2.2 Mean Square Amplitude

We can obtain a measure of how much $v(t)$ fluctuates about its time-average value as follows. Let $v_n(t)$ represent the departure of $v(t)$ from its average or mean value. That is,

$$v_n(t) = v(t) - \bar{v} \tag{17.2}$$

The time-average value of $v_n(t)$ is clearly zero.

One measure of the magnitude of v_n is called the *mean square voltage* (or *current*, if one is dealing with a current waveform). It is defined as

$$\overline{v_n^2} = \lim_{T \to \infty} \frac{1}{T} \int_{-T/2}^{T/2} [v_n(t)]^2 \, dt \tag{17.3}$$

The mean square voltage has several useful properties. Consider, for example, the circuit of Fig. 17.7. The time-average power dissipated in the 1 Ω resistance is

$$P_R = \overline{v^2} = (\bar{v})^2 + \overline{v_n^2} + \overline{2v_n\bar{v}} \tag{17.4}$$

or

$$P_R = \quad P_{dc} + P_n + 0 \tag{17.5}$$

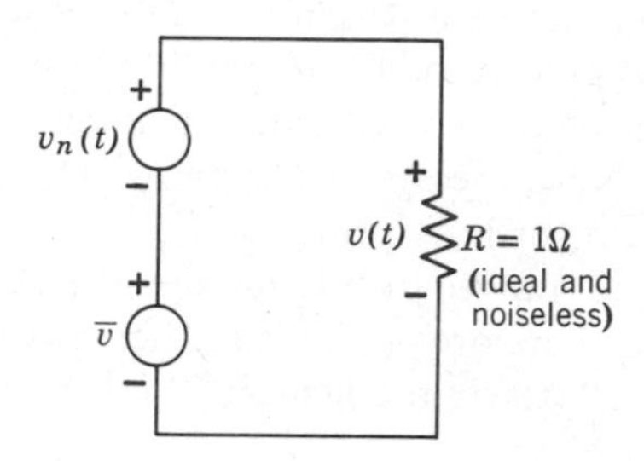

Figure 17.7
A noise source represented as its time average $\bar{v}$ plus fluctuations $v_n(t)$.

The time-average power P_R is equal to the sum of P_{dc}, the power supplied by the time-average value of $v(t)$, plus P_n, the "noise" power supplied by the mean square fluctuation, $\overline{v_n^2}$. The third term in Eq. 17.4, the time average of the product of $v_n(t)$ and $\bar{v}$, is zero because $\bar{v}$ is a constant and $v_n(t)$ has zero time average.

We can expand this argument to include any two series-connected voltage sources. Suppose we have two noise sources, $v_1(t)$ and $v_2(t)$, each with zero time average. If the total voltage $(v_1 + v_2)$ is applied to our ideal 1 ohm resistor, the time-average power dissipated in the resistor is given by

$$P_R = \overline{v_1^2} + \overline{v_2^2} + \overline{2v_1v_2} \tag{17.6}$$

The first two terms on the right-hand side of the above equation clearly represent the mean square voltage from each of the two sources. The third term, however, is the time average of the product of two random voltages. Whether this term has a nonzero time average or not depends on the statistical properties of the two voltage sources. That is, if v_1 goes positive whenever v_2 goes positive, and they both go negative together as well, the product $v_1 v_2$ will have a positive time average. Similarly, if v_1 goes negative whenever v_2 goes positive, and vice versa, the product $v_1 v_2$ will have a negative time average. Finally, if the variations of v_1 are completely independent of the variations of v_2, the product will have zero time average, and we say that v_1 and v_2 are *statistically independent.* Therefore, the third term in the equation would be zero, and we have the result:

> The total mean square value of the noise from several statistically independent noise sources is the sum of the mean square values from each noise source.

Expressed in terms of power, the noise power delivered to a load by statistically independent noise sources is the sum of the powers delivered by each source. Most of the noise sources we encounter in real system are statistically independent of one another. As a result, we can use a kind of superposition argument to determine the total mean square noise at a particular point in a circuit: we determine the mean square noise power from each statistically independent noise source, then sum the results.

(*Note.* A similar argument applies for current noise sources. Mean square currents replace mean square voltages—and the powers from statistically independent current sources add together.)

We have now a useful measure of how much noise is present, and we know how to combine noise from statistically independent sources. The final issue is the comparison between signal and noise, the crucial aspect of any measurement. We define *signal-to-noise* ratio, S/N, as the ratio of the mean square signal power P_S to the mean square noise power P_N.

$$S/N = \frac{P_S}{P_N} \tag{17.7}$$

Note that although we can discuss signals in terms of amplitude, the fact that we can only speak of mean square noise amplitude (or noise power) means that for comparison purposes, we must also talk about mean square signal amplitude (or signal power). Also, since signal-to-noise ratio is a power ratio, it is not uncommon to find decibel notation used.

$$S/N = 10 \log \frac{P_S}{P_N} \text{ (d}B\text{)} \tag{17.8}$$

The larger the signal-to-noise ratio, the easier it should be to detect a signal. As a very general rule of thumb, the S/N must exceed unity for detectability.

17.2.3 Spectral Density Function

Having addressed ourselves to the issue of how much noise is present, we must now consider the actual time dependence of the random voltage $v(t)$. Although we cannot use well-defined Fourier components to describe this time dependence, we certainly should be able to say something that differentiates between a random signal that changes its value very rapidly and a random signal that changes its value slowly. The detailed theory of the time variation of random processes goes well beyond the scope of this presentation. The student who wishes to pursue this theory is advised to consult a text on probability, random variables, or statistical communication theory. We shall approach the subject from a quasi-empirical point of view that is, at least, not inconsistent with the detailed theory.

We shall attack this issue by measuring the time-dependent properties of our random waveform with the aid of an idealized narrow-band filter, having the transfer function magnitude shown in Fig. 17.8*b*. (We are using

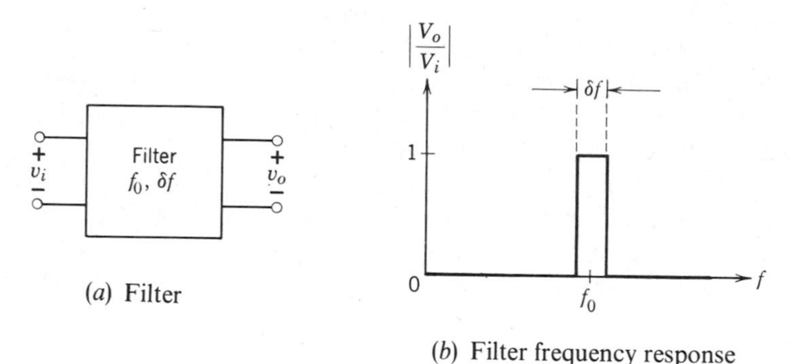

Figure 17.8
Idealized filter for noise characterization.

frequency in hertz instead of angular frequency in radians per second for the frequency variable. This is done to be consistent with other common usage.)

The filter transfer function has unity magnitude within a bandwidth δf about f_0, and is zero elsewhere. If we now connect this filter between a noise source and an ideal load resistor, we get the system shown in Fig. 17.9.

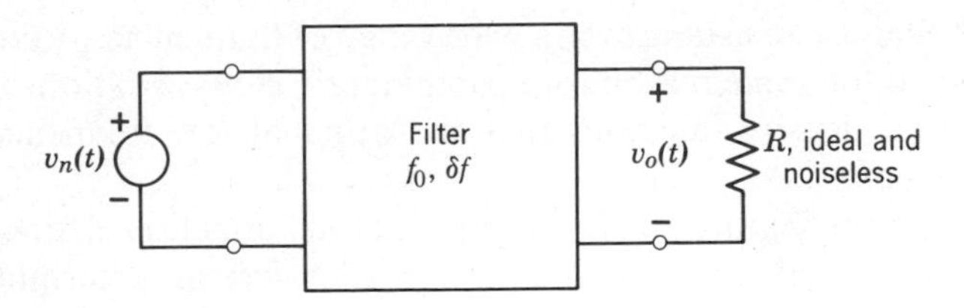

Figure 17.9
The ideal filter of Fig. 17.8 is used to determine the spectral density of the source $v_n(t)$.

Having constructed this system we can now measure the mean square voltage across the ideal load resistor. The value of the mean square voltage will, of course, depend on the choice of center frequency f_0 and on the bandwidth δf. Therefore, let us write the mean square output noise as a function of these two variables:

$$\overline{v_o^2} = \overline{v_o^2(f_0, \delta f)} \tag{17.9}$$

Let us try to understand what the dependence of $\overline{v_o^2(f_0, \delta f)}$ might be on these two variables. Imagine that the bandwidth is set rather small, and that we vary f_0 from low to high frequencies. The filter produces output only for that fraction of the input signal that "looks like" Fourier components near f_0 for a time on the order of $1/\delta f$. Even if the output is a random voltage, there will be identifiable Fourier components for short times that can produce filter output. What we measure is the long-time average of this filter output. If the random voltage changes its value only very slowly, the variations in its amplitude will be best approximated by Fourier components of predominantly lower frequencies. Thus the mean square output of the filter will be greatest at low frequencies. Similarly, if the random voltage is a completely random function of time, there will be no preference for rapid or slow variations, and we might expect the filter to produce much the same mean square output at all frequencies.

It is a characteristic property of the noise sources with which we shall deal that for a δf small enough, the dependence of $\overline{v_o^2(f_o, \delta f)}$ on δf is a proportional one. That is, one can write

$$\overline{v_o^2(f_0, \delta f)} = S_n(f_0)\, \delta f \tag{17.10}$$

where $S_n(f)$ is a function of frequency and is a characteristic of the noise source. This function is called the *spectral density function* (often referred to by other names: power density function, or power spectrum). Note that $S_n(f)$ is a characteristic of the noise source, *not* of the filter—we have merely used the filter to measure $S_n(f)$. Based on the preceding discussion, we expect that totally random functions of time will have spectral density functions that are constant over a wide range of frequencies. Functions that are restricted for some reason to predominantly slow variations in time will have spectral density functions that are larger at low frequencies than at high frequencies.

An important feature of the spectral density function defined above is that if we know this function at all frequencies, we can determine the total mean square voltage from the integral of the spectral density function over all frequencies.

$$\overline{v_n^2} = \int_0^\infty S_n(f)\, df \tag{17.11}$$

We shall use this result in the discussion to follow.

17.2.4 Noise in Linear Systems

We have already learned how *signals* pass through linear systems. A characteristic example is shown in Fig. 17.10. The signal source, v_s, produces a

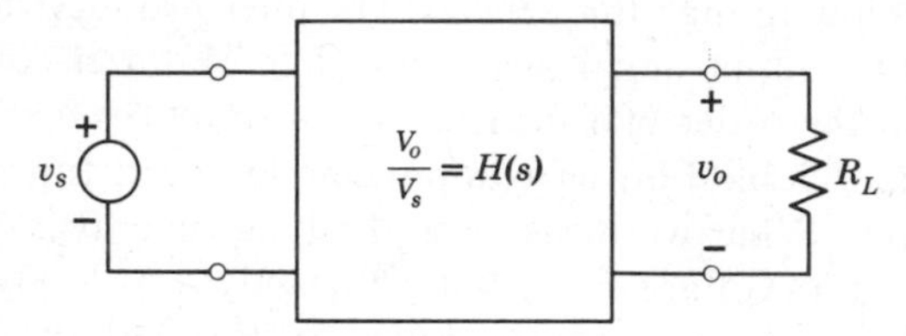

Figure 17.10
A signal source driving a linear system.

waveform of known Fourier composition. The linear system is characterized by the system function $H(s)$. The signal at the output can be found by (1) multiplying each Fourier component of v_s by the magnitude of $H(s)$ evaluated at $s = j2\pi f$, where f is the frequency of the Fourier component; (2) shifting the phase of each Fourier component by the angle of $H(s)$ evaluated at the Fourier component frequency; and (3) summing the results (superposition).

With noise voltages, we cannot follow the above recipe because we cannot break down the random voltage into its Fourier components. One can, however, determine the spectral density function at the output from the spectral density function at the input (see Fig. 17.11). Let $v_n(t)$ be a noise source with spectral density function $S_n(f)$. The spectral density function $S_o(f)$ of the output noise voltage, $v_o(t)$ is given by

$$S_o(f) = |H(j2\pi f)|^2 S_n(f) \tag{17.12}$$

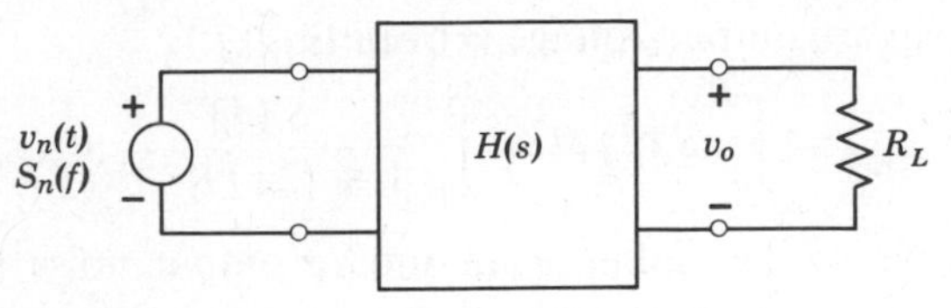

Figure 17.11
A noise source driving a linear system.

That is, the spectral density function at the output of a linear system equals the spectral density function at the input of the linear system multiplied by the square of the magnitude of the system function. For this particular example, the system function is a *voltage* transfer function, and the spectral density function characterizes the *square of the voltage.* Thus it is certainly plausible that the relation between input and output spectral density functions should involve the square of the system function. What is not obvious, and must be demonstrated by proof,[4] is that the relation has the particularly simple form of Eq. 17.12.

This relation lets us determine the mean square noise output from any linear system if we know (1) the spectral density function of the noise source and (2) the system function relating the output to the input noise source. We shall demonstrate this with Fig. 17.12. We assume that the spectral

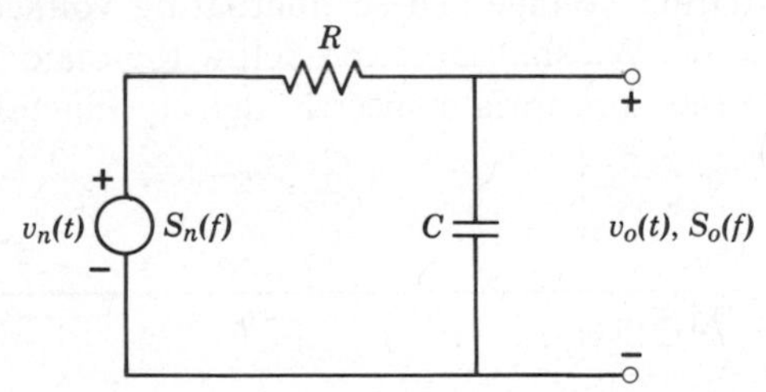

Figure 17.12
The determination of the mean square noise output from any linear system.

density function of the noise source, $S_n(f)$, is known. We must first determine the system function, V_o/V_n. By inspection, this function is

$$\frac{V_o}{V_n} = H(s) = \frac{1}{1 + sRC} \tag{17.13}$$

Therefore, substituting into Eq. 17.12, the spectral density function at the output is

$$S_o(f) = \frac{1}{1 + (2\pi f RC)^2} S_n(f) \tag{17.14}$$

[4] See Reference R17.2, Chapter 6, pp. 405–407.

and the mean square output voltage is from Eq. 17.12

$$\overline{v_o^2} = \int_0^\infty S_o(f)\,df = \int_0^\infty \frac{S_n(f)}{1 + (2\pi f RC)^2}\,df \tag{17.15}$$

We are thus able to determine mean square output noise so long as we know the spectral density function of the noise source. We shall now investigate what kinds of devices produce noise, and what are the spectral density functions for the various noise sources of importance.

17.3 NOISE SOURCES

Noise in electronic devices arises from the very basic fact that electric charges are discrete. Electrons in a conducting medium are in rapid thermal motion, somewhat like an ideal gas. When electrical conduction takes place, this electron gas develops a rather small average drift velocity in the appropriate direction. The electric current crossing any particular cross-section of the conductor is not constant, but fluctuates slightly because of the random thermal motions of the electrons. Even in a conductor in which there is no average current flow, the microscopic charge density is constantly fluctuating. Therefore, between any two points in a conductor, it is possible to observe a fluctuating voltage. These fluctuating voltages and currents constitute device noise. We shall describe below the major kinds of device noise, and indicate the appropriate spectral density function for each type of noise.

17.3.1 Shot Noise

Shot noise is the noise associated with the electrical current. Suppose we have an electrical current $I = nq$, where q is the electronic charge and n is the *average* number of carriers passing a certain cross-section per second. If we were to denote the full time-dependent fluctuating current by $i(t)$, then the time-average of $i(t)$ is simply

$$\overline{i(t)} = I \tag{17.16}$$

The noise current is the fluctuation in $i(t)$ about its mean value. Thus the noise current is

$$i_n(t) = i(t) - I \tag{17.17}$$

This noise current has zero time average. The mean square value of $i_n(t)$ depends on the precise way one models the charge distribution and the arrival pattern of charges.

The simplest model is to assume that each electron represents an impulse of charge, that the time required for each electron to cross the point of interest is infinitely short, and that the arrival of one electron is completely uncorrelated with the arrival of any other electron. With this very simple (and idealized) model, the spectral density function for the noise current is

$$S_i(f) = 2qI \tag{17.18}$$

Several features of this formula should be noted. First, the noise spectrum is constant at all frequencies. Spectra with this characteristic are called *white noise*. Second, the amplitude of the power density spectrum is proportional to the current I. Note, therefore, that while the "signal" power is proportional to I^2, the noise power is proportional to I, and the signal-to-noise ratio is, therefore, proportional to I.

This constant spectral density function is an idealization of the actual spectral density function. If we were to calculate the mean square current by integrating the spectral density function from zero to infinity, and if $S_i(f)$ were truly constant, the mean square current would be infinite. Since no real current can supply infinite noise power, there must be some effective upper cutoff frequency for this spectral density function.

In modeling shot noise, one should include (1) the finite size of the electron, thus requiring a finite time for the electron to pass the point of interest, and (2) the repulsive force between electrons, which prevents simultaneous arrival of two electrons at the same point. Both of these issues limit the rapidity of the fluctuations at the high frequency end of the spectral density function. Fortunately, this upper cutoff frequency in most applications is *much higher* than any frequency of interest. The additional external networks through which we pass the noise introduce upper cutoff frequencies much below the frequency at which the shot noise model fails. Therefore, we are justified in assigning to each electrical current a shot noise current generator with the spectral density function $2qI$ at all frequencies of interest.

The next question concerns which types of devices exhibit significant shot noise. The key issue is the source impedance one must associate with the shot noise current generator. If the Norton equivalent of the shot noise source is a current source in parallel with a short circuit, then none of the noise current can actually leave the device. On the other hand, if the Norton equivalent resistance is large, then almost all the noise current will flow in the external circuit. Consider, for example, the back-biased collector-base junction of a transistor. We normally model this as an incremental open circuit, the only connection to the collector being a dependent current generator between collector and emitter. Thus even if our transistor model is slightly modified and improved, the effective source impedance at the collector of a transistor is large, and therefore, the shot noise associated with the collector current will be significant.

This choice of example raises another important point. Normally, the collector current in a transistor amplifier is chosen for convenience, the only constraint being that the transistor be biased far enough into the linear region to handle the output swing of the signal being amplified. The introduction of noise can produce an additional constraint on the biasing if very-low-level signals are so amplified, because the amount of noise produced by the transistor depends on the magnitude of collector current. Therefore, the choice of bias conditions can critically influence the ability to amplify weak signals (see Section 17.3.5).

Other devices in which shot noise plays an important role in determining overall performance are vacuum tubes of all varieties (here the shot noise is associated with the plate current), and semiconductor diodes. We should note that the amount of shot noise in a forward-biased thermionic (vacuum-tube) diode can be easily adjusted over a wide range of levels. For this reason, thermionic diodes are often used as calibrated sources of white noise.

17.3.2 Johnson, or Thermal Noise

We have already remarked that even in a conductor carrying no current there can be fluctuating voltage drops across the conductor because of the random thermal motions of the electrons. This thermal noise, called *Johnson noise*, can be analyzed with a variety of models; those that assume that the voltage fluctuations are totally random yield a white noise spectral density function. This spectral density function is written with reference to the equivalent circuit for a noisy resistance shown in Fig. 17.13*b*. The Thévenin equivalent circuit of a noisy resistance is a noiseless resistance of magnitude R

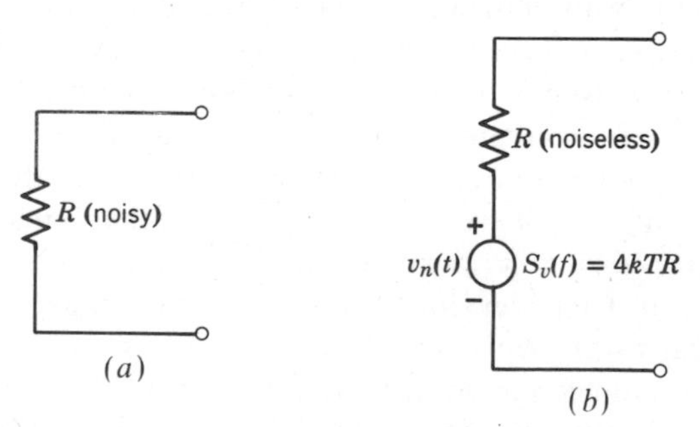

Figure 17.13
Equivalent circuit that represents thermal noise in a resistor.

in series with a voltage noise source $v_n(t)$. The statistical properties of $v_n(t)$ are that it has a zero time average and a spectral density function

$$S_v(f) = 4kTR \tag{17.19}$$

where k is Boltzmann's constant and T is the temperature in Kelvins. Notice once again that thermal noise is white noise, so that the frequency characteristics of the associated external circuitry will determine the total mean square noise power supplied by this noise source. The temperature T is ideally equal to the actual temperature of the resistance in Kelvins. In practice, because of additional sources of noise, such as the continuous generation and recombination of carriers, resistors are often noisier than indicated in Eq. 17.19. Their departure from ideality is often accounted for by ascribing to each resistor an equivalent *noise temperature* that may be much higher than the actual operating temperature of the resistor.

All resistors exhibit thermal noise. In an amplifier or instrumentation system, however, it is the input resistance to the first stage and the source resistance for the signal of interest that are most crucial, because it is the noise at the input that is most strongly amplified, and can most easily interfere with the signals of interest.

17.3.3 Low-Frequency Noise and Drifts

Transistors and, to a lesser degree, vacuum tubes exhibit a kind of noise that is much greater at low frequencies than the shot noise and Johnson noise characteristics of the device would predict. This excessive low-frequency noise has a spectral density function that can be approximated by a $1/f$ dependence in the range of interest. In typical triodes, $1/f$ noise is important at frequencies below about 1 kHz, while in junction transistors, $1/f$ noise persists to slightly higher frequencies.

Low-frequency noise from another type of source is often modeled as $1/f$ noise. We refer here to random drifts of amplifier gain, capacitor temperatures, power supply voltages, etc. It might be impossible to prove that apparatus drifts constitute a random process with a well-defined spectral density function. Nevertheless, the kinds of theoretical results one obtains by modeling drifts as $1/f$ noise are remarkably similar to experimental experience, so we shall include under the "$1/f$ umbrella" both low-frequency device noise *and* random system drifts.

A spectral density function proportional to $1/f$ produces several problems in analysis, and one must use caution together with common sense. We shall illustrate this problem with a simple example.

17.3.4 Noise Bandwidth

Consider the circuits in Fig. 17.14. The $v_1(t)$ source in Fig. 17.14*a* is a white noise source with spectral density function $S_1(f) = A$, constant at all frequencies. From Eq. 17.15, the mean square noise at the output $\overline{v_{o1}^2}$ is given by the following integral, where we have let $f_p = 1/2\pi RC$ be the pole frequency.

$$\overline{v_{o1}^2} = \int_0^\infty \frac{A\,df}{1 + (f/f_p)^2} = Af_p \int_0^\infty \frac{dx}{1 + x^2} = Af_p\frac{\pi}{2} \qquad (17.20)$$

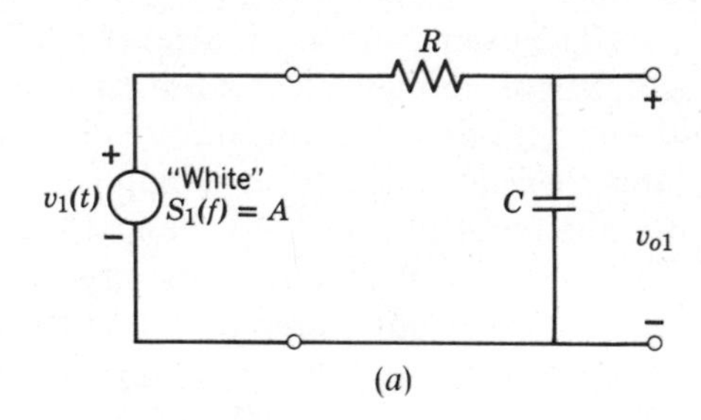

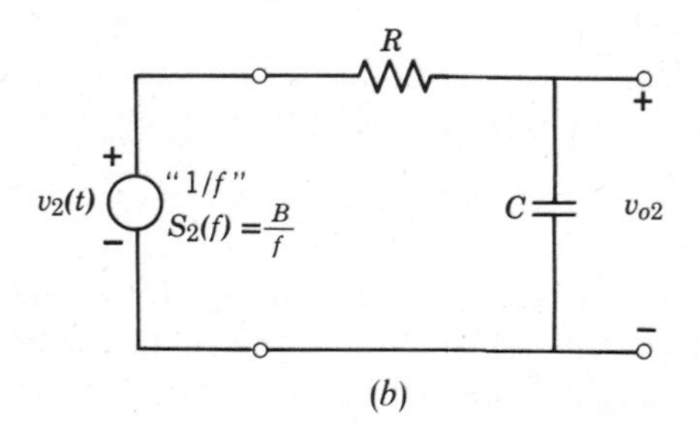

Figure 17.14
The noise bandwidth concept for white noise and $1/f$ noise.

That is, the amount of white noise that passes through a single-pole low-pass filter is the same as would be passed by an abrupt filter extending from $f = 0$ to $f = f_p(\pi/2)$. We call this bandwidth $f_p(\pi/2) = 1/4RC$ the *equivalent noise bandwidth* of the filter.

Now let us ask what happens to the source in Fig. 17.14*b*, assumed to have a spectral density function $S_2(f) = B/f$. If we proceed to substitute B/f into Eq. 17.15, we obtain a divergent integral:

$$\overline{v_{o2}^2} \stackrel{?}{=} \int_0^\infty \frac{B\,df}{f[1 + (f/f_p)^2]} \to \infty \qquad (17.21)$$

The problem here is that we have over-idealized our source. The spectral density function itself approaches infinity as f approaches zero. In practice, one does not observe the system for an infinite time. Let us suppose, for example, that we observe the voltage on the capacitor for a time T. Any variation in $v_2(t)$ that corresponds to frequencies much lower than $1/T$ would not change during the observation. Therefore, the total observation time for our *estimate* of $\overline{v_{o2}^2}$ introduces a lower cutoff frequency, and we should not integrate from zero frequency up to infinity, but rather from about $1/T$ up to infinity. Any fluctuation much slower than $1/T$ appears as a constant voltage on the capacitor, and as such contributes to the estimate of the time average instead of to the estimate of the noise. Therefore, a realistic way to evaluate this integral is

$$\overline{v_{o2}^2} = \int_{\frac{1}{T}}^{\infty} \frac{B\,df}{f[1 + (f/f_p)^2]} = \frac{B}{2} \ln\,[(f_pT)^2 + 1)] \qquad (17.22)$$

A similar modification of Eq. 17.20 yields for the white noise case

$$\overline{v_{o1}^2} = Af_p\left[\frac{\pi}{2} - \tan^{-1}\frac{1}{f_pT}\right] \qquad (17.23)$$

Normally the time T would be long compared to the time constant of the filter. Therefore in normal circumstances $f_pT \gg 1$, and we have

$$\overline{v_{o1}^2} = Af_p\frac{\pi}{2}\left[1 - \frac{1}{(\pi/2)T}\right] \rightarrow Af_p\frac{\pi}{2} \qquad \text{for long times} \qquad (17.24)$$

and

$$\overline{v_{o2}^2} \rightarrow B\ln(f_pT) \qquad \text{for long times} \qquad (17.25)$$

Note that for white noise, the noise bandwidth is $f_p(\pi/2)$, and is well defined after long observation times. For $1/f$ noise, however, the mean square noise keeps growing and growing. The longer one observes the system, the greater the mean square noise.

We shall examine the implications of these results in Chapter 18. For the moment, at least, one can see that $1/f$ noise has some properties that look pathological.

17.3.5 Amplifier Noise

We have already seen that associated with any signal source is a Johnson noise source with a magnitude determined by the source resistance and its noise temperature. What we seek to investigate here is how one describes the additional noise contributed by one's amplifying system because of the Johnson noise of the input resistance, the shot noise associated with the biasing currents, and any other noise sources in the amplifier itself. Figure 17.15 shows a signal source, consisting of an ideal signal source plus a Johnson noise source [spectral density function $S_n(f) = 4kT_SR_S$] in series with the source resistance R_S. The source network drives an amplifier that is modeled as having gain A over an effective noise bandwidth δf, and zero gain elsewhere.

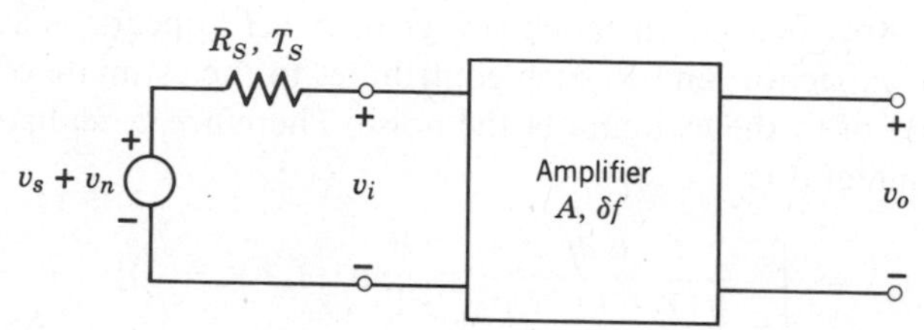

Figure 17.15
Circuit for illustrating amplifier noise.

If the amplifier were perfectly noiseless, the mean square noise at the output would be $\overline{v_o^2} = A^2 4kT_SR_S\,\delta f$. In practice, for any real amplifier, the mean square noise is *always* larger. A convenient way of describing the amount of additional noise contributed by the amplifier is to write the actual measured output noise in this form:

$$\overline{v_o^2} = A^2 4k(T_S + T_A)R_S\,\delta f \tag{17.26}$$

With a proper choice of the parameter T_A, one can make the mean square output as large as one wishes. Thus T_A, called the *amplifier noise temperature*, is a measure of the amount of noise contributed by the amplifier. It is written in this form to emphasize the idea that we get an equivalent result by assuming that the source resistance really has a temperature $T_S + T_A$ and then assuming the amplifier to be ideal.

Another measure of amplifier noise is called the noise figure, F. It is the ratio of the signal-to-noise ratio of the source to the signal-to-noise ratio at the output. Assuming that the signal is simply multiplied by A on passing through the amplifier, then this ratio is

$$F = 1 + \frac{T_A}{T_S} \tag{17.27}$$

Often one finds noise figures expressed in dB.

$$\text{N.F.} = 10 \log F \text{ (dB)} \tag{17.28}$$

Thus a 3 dB noise figure corresponds to $F = 2$ or $T_A/T_S = 1$.

Each component within an amplifier can produce noise. In a transistor, there is shot noise associated with the base current and with the collector current, and Johnson noise associated with the incremental resistances r_π and r_x. Bias resistors, particularly at the input stage of an amplifier, can also contribute Johnson noise. The contributions of each potential noise source to the total output noise can be estimated using a linear incremental model to which the noise sources have been added. The results exhibit a complex dependence both on the frequency at which one wishes to use the amplifier, and on the magnitude of the source resistance. Manufacturers often provide data on specific devices, usually expressed graphically as in Fig. 17.16.

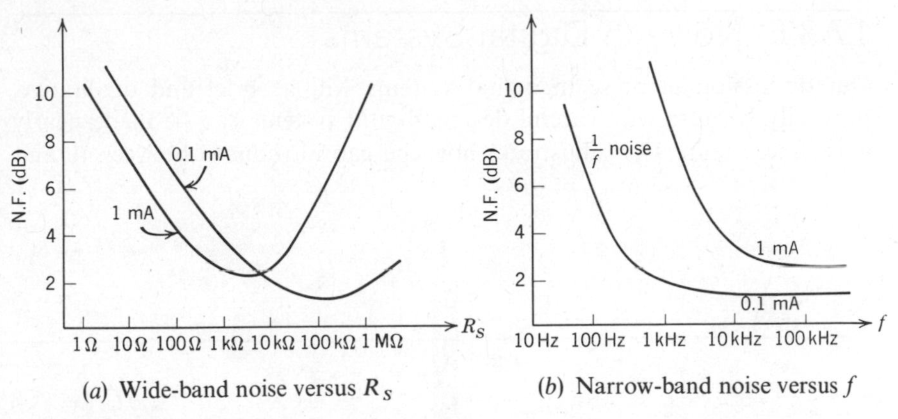

Figure 17.16
Typical transistor noise figure plots for two different collector currents.

Figure 17.16*a* illustrates a typical dependence of the common-emitter transistor noise figure on source resistance for two different values of collector current. The abscissa is labeled "wide-band noise" because the bandwidth δf used to measure the noise figure generally covers the entire audio range (perhaps 10 Hz to 15 kHz). Notice that for each specific operating point (collector current), there is an optimum choice of source resistance. In general, the reason for the existence of the optimum is that for low values of source resistance, the noise produced by the source is so small that Johnson noise in r_x and shot noise associated with the collector current overwhelm the noise from the source. As the source resistance and its associated noise increase, the ratio of transistor noise to source noise drops until the source resistance gets large enough so that Johnson noise in r_π is no longer shorted out by voltage divider action between R_S and r_π. More elaborate discussions can be found in the references.

The narrow-band noise data illustrated in Fig. 17.16*b* are plotted assuming the optimum choice of source resistance for each value of collector current. The term "narrow-band" is used because the noise bandwidth used to make these measurements must be only a few Hertz (otherwise the frequency dependence would be obscured). Notice that at low frequencies, the noise figure has a $1/f$ dependence, falling off with a slope of approximately -1. At sufficiently high frequencies, the noise figure becomes flat, characteristic of white noise sources. The overall magnitude of N.F. $\approx$ 2 to 3 dB is typical of so-called "low-noise" transistors.

17.3.6 Noise in Digital Systems

Our discussion of noise in digital systems will be brief and qualitative, primarily because with careful design, digital systems can be made nearly noise-free. Figure 17.17 illustrates how the gap introduced between HI and

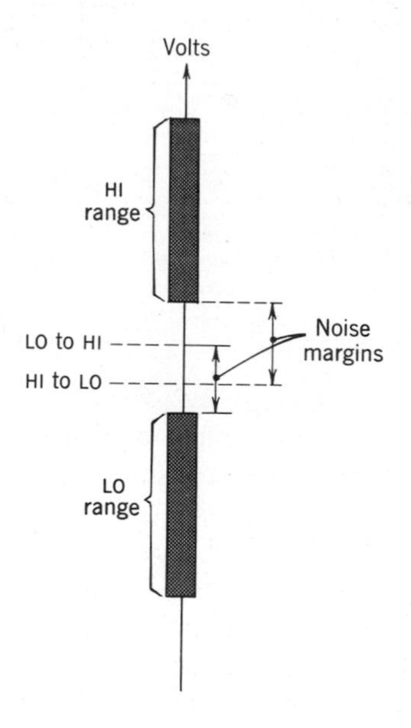

Figure 17.17
The digital noise margin.

LO voltage ranges provides a certain margin of error against fluctuations in voltages. The defined HI and LO states are shown in heavy shading. In practical digital circuits, however, the LO-to-HI transition can occur above the upper edge of the LO region. Therefore, if a LO voltage happened to be at the top of the LO range, and because of coupling to a noise source, it suddenly got driven more positive, there is still a margin of error, called the *noise margin*, before an erroneous LO-to-HI transition would occur. Similarly, a HI voltage would have to be driven below the HI-to-LO transition point before an error could occur. Few of the components in digital systems are intrinsically noisy enough to produce such erroneous transitions. The principal hazards, therefore, are capacitive couplings to other digital waveforms that might switch suddenly (a particular problem in closely packed circuits), improperly shielded power-line transients, or transmission-line effects in which pulses are "reflected" from the input of one gate and can return to the output of a previous gate to induce a transition. In TTL logic, where the current pulses during a gate transition can be large, it is possible for the small resistance in the bus wires used to distribute power to cause a momentary drop in the local power supply voltage, thereby causing an error. It is recommended, therefore, that capacitors between the supply and ground be distributed throughout digital circuits to provide local reservoirs of stored energy that can supply the transient switching currents.

REFERENCES

R17.1 Ralph Morrison, *Grounding and Shielding Techniques in Instrumentation* (New York: John Wiley, 1967).

R17.2 Mischa Schwartz, *Information Transmission, Modulation, and Noise* (New York: McGraw-Hill, 1970).

R17.3 Aldert van der Ziel, *Noise: Sources, Characterization, Measurement* (Englewood Cliffs, N.J.: Prentice-Hall, 1970).

QUESTIONS

Q17.1 What kinds of noise do you encounter in listening to a radio? Characterize, as well as you can, each kind of noise according to type of source, spectral density, and method of removal.

Q17.2 Repeat Q17.1 for a tape player or phonograph. Include mechanical noise sources as well as electrical noise sources.

Q17.3 Repeat Q17.1 for a television receiver. Here both audio and video noise can occur.

Q17.4 Suggest a method for shielding a circuit from inductive pickup of the stray magnetic flux produced by a nearby motor.

Q17.5 Suggest a way to use an optoelectronic isolator (with its nonlinear transfer characteristics) to transmit analog information linearly. (*Hint.* See Chapter 15 on frequency modulation.)

Q17.6 Explain carefully the difference between *ground* and *guard.*

Q17.7 Can you think of a kind of noise for which Eq. 17.10 is inappropriate? What about the subway train example of Section 17.1?

EXERCISES

E17.1 A circuit model for the measurement system in Fig. E17.1*a* is shown in Fig. E17.1*b*, including the *common mode voltage,* v_{CM}, which is the difference in potential between the signal and instrument grounds.

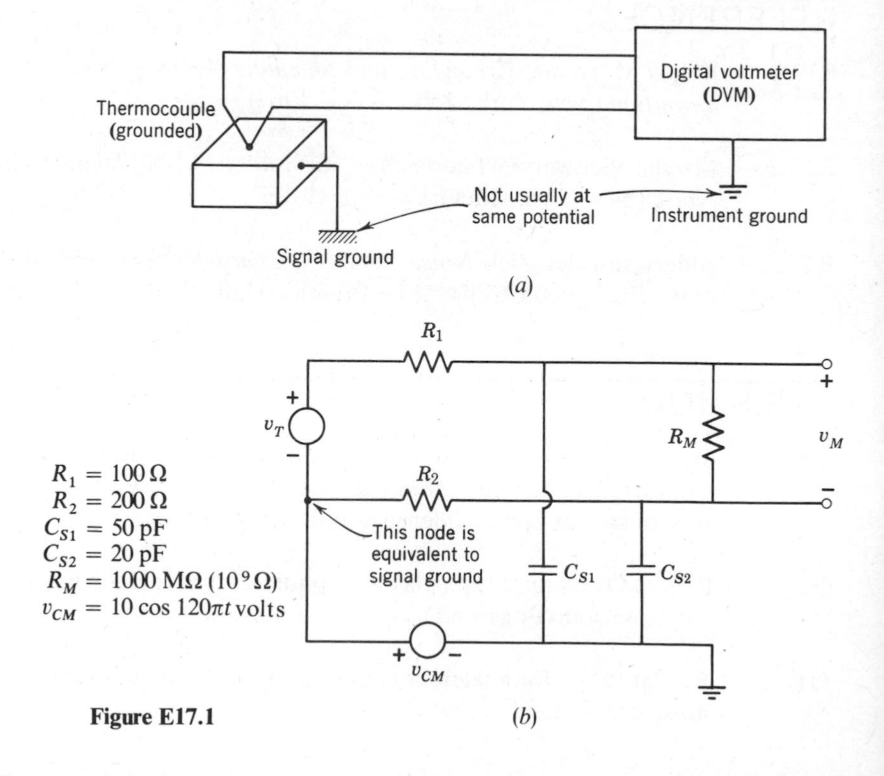

Figure E17.1

In the above circuit, v_T represents the signal from the thermocouple, R_1 and R_2 are the lead resistances (usually unequal), and v_M is the signal appearing across the DVM input resistance R_M. The capacitors C_1 and C_2 represent the capacitive couplings (usually unequal) between the thermocouple leads and instrument ground. (If these capacitors were absent, v_{CM} produces no interference.) Find v_M. Use reasonable approximations.

Answer. $v_T + 1200\pi(R_2C_{S2} - R_1C_{S1}) \sin 120\pi t$
$= v_T - (3.8\ \mu\text{V}) \sin 120\pi t.$

E17.2 The circuit in Fig. E17.2 models the effect of a guard (Fig. 17.5) on the circuit of E17.1. Find v_M in this case. Compare it with the result of E17.1. Show that the use of the guard reduces the 60 Hz component of v_M by more than 40 dB.

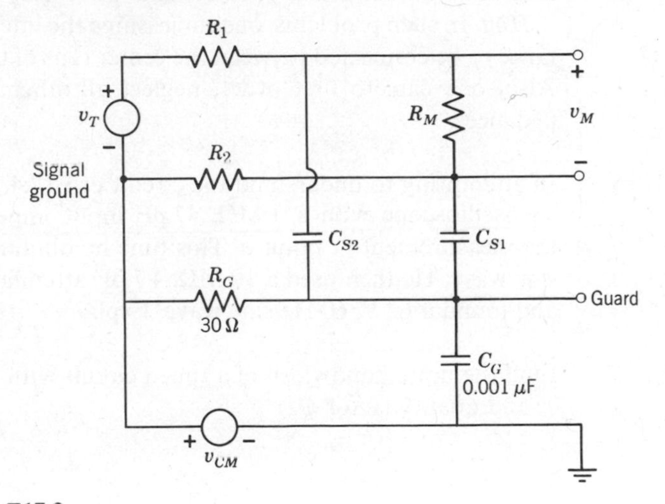

Figure E17.2

E17.3 Qualitatively, how would the following conditions affect the 60 Hz component of v_M in E17.2?

(a) R_1 and R_2, originally equal, become unequal.
(b) C_1 and C_2 originally equal, become unequal.
(c) Decrease R_G.
(d) Increase C_G.

E17.4 An unshielded transformer is used in an attempt to build an isolated supply V_0 (Fig. E17.4).

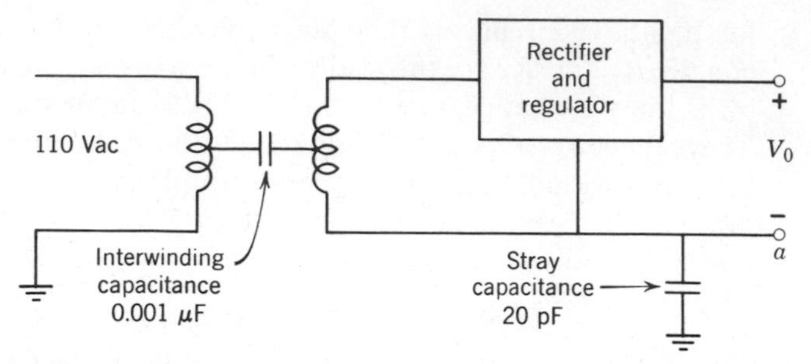

Figure E17.4

The stray capacitance between the LO side of this supply, point a, and ground is only 20 pF. Using an ac voltmeter with a 100 kΩ input resistance, a student measures the voltage at point a relative to ground and finds a 3 V, 60 Hz sine wave. Explain this result.

Hint. In such problems, one can assume the interwinding capacitance to be connected between the center taps of the two windings. Also, one can, to first order, neglect all other transformer impedances.

E17.5 In attempting to understand the circuit of E17.4, the student used an oscilloscope with a 1 MΩ, 47 pF input impedance to repeat the measurement at point a. This time he obtained a 22 V, 60 Hz sine wave. He then used a 10 MΩ, 4.7 pF attenuating scope probe and found a 62 V, 60 Hz sine wave. Explain.

E17.6 Find the noise bandwidth of a tuned circuit with center frequency f_o and quality factor Q.

Answer. $\dfrac{\pi f_o}{2Q}$.

PROBLEMS

P17.1 In Fig. P17.1, i_1 is a source of current noise with spectral density

$$S_1(f) = \begin{cases} A & \text{for } f_1 \le f \le 2f_1 \\ 0 & \text{elsewhere} \end{cases}$$

v_2 is a source of voltage noise (statistically independent of i_1) with spectral density

$$S_2(f) = \begin{cases} B & \text{for } 0 \le f \le f_1 \\ 0 & \text{elsewhere} \end{cases}$$

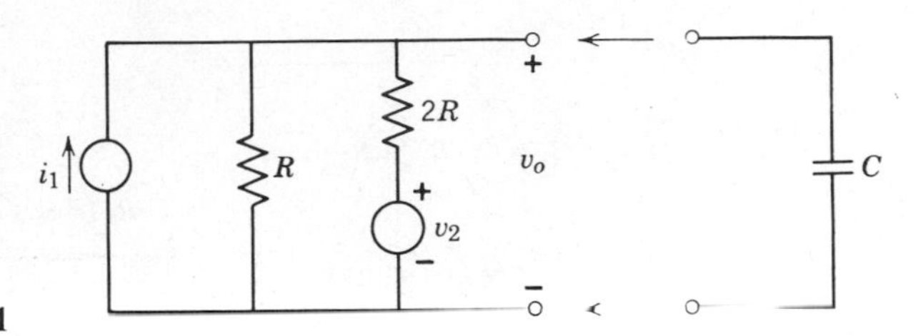

Figure P17.1

The resistors are noiseless.

(a) Find the total mean square value of the output voltage, v_o, when the capacitor is *not connected* (open circuit).

(b) Find the mean square output noise when C is connected. Explain in one sentence whether the noise is less or greater than that in (*a*), and why.

Hint. This is a linear circuit. Therefore the two system functions, relating V_o and I_1 and V_o and V_2 can be found using superposition.

P17.2 Estimate the small-signal ac gain of the amplifier in Fig. P17.2 and use Fig. 17.16 to estimate the wide-band noise figure and the narrow-band noise figure at 1 kHz. Assume $\beta_F = \beta_0 = 100$.

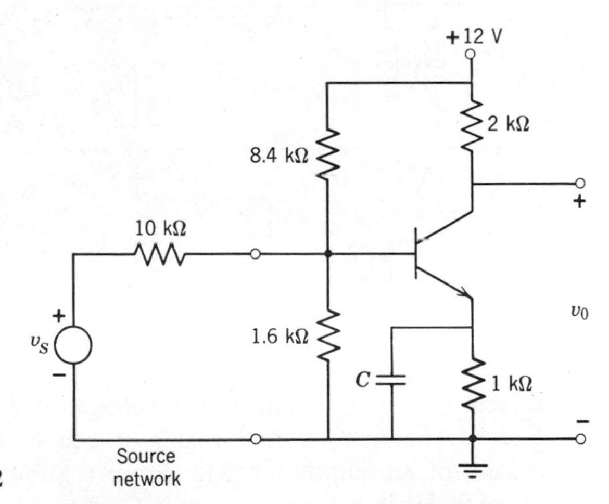

Figure P17.2

P17.3 Show, using the examples of series and parallel resistors in Fig. P17.3, that the noise source of a Thévenin equivalent network has spectral density $4kTR_T$ where R_T is the Thévenin resistance.

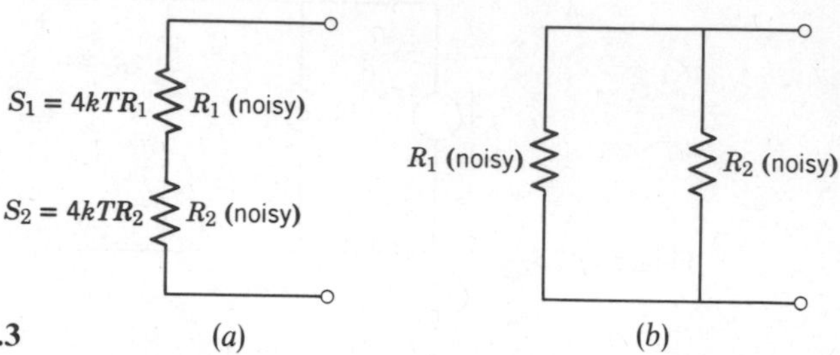

Figure P17.3 (*a*) (*b*)

P17.4 What is the Thévenin resistance driving the base of the transistor in P17.2? Assuming that the noise figure of the transistor is 3 dB, estimate the total output noise spectral density function (see P17.3). Show that the Johnson noise from the 2 kΩ load resistor can be ignored compared to the amplified Johnson noise from the source resistance.

P17.5 This problem illustrates how an improperly isolated power supply for a thermocouple compensator can introduce power-line interference (Fig. P17.5). The interwinding capacitance is, in effect,

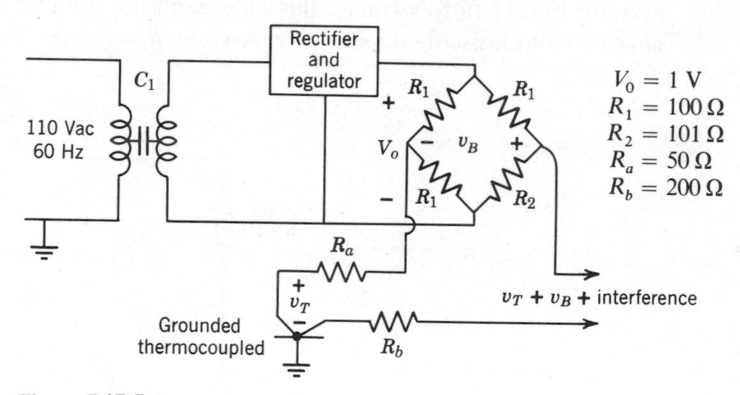

Figure P17.5

driven with a common-mode voltage of 55 Vac (see Exercise E17.4). The rectifier and regulator can be considered a short circuit for all common-mode currents. Show that the amplitude of the 60 Hz interference is $\omega C_1[(R_1/2) + R_a]$ times the power-line amplitude. Evaluate this interference amplitude for

(a) $C_1 = 0.001\ \mu$F (unshielded).
(b) $C_1 = 50$ pF (single shield).
(c) $C_1 = 0.1$ pF (properly guarded circuit with triple-shielded transformer).

CHAPTER EIGHTEEN

INSTRUMENTATION APPLICATIONS

18.0 THE INSTRUMENTATION PROBLEM

In general, the term *instrumentation* refers to the use of transducers and associated signal-processing devices to obtain information about a physical process either for the purpose of collecting data about that process or for the feedback control of some variable determining the process. Literally every topic we have encountered thus far in this text is relevant to some aspect of the instrumentation problem. Our purpose in this final chapter is to examine how these various topics interact with one another when one seeks the solution to a practical problem. The example we shall use is the data-acquisition system illustrated in Fig. 18.1. A transducer (or analog source) is

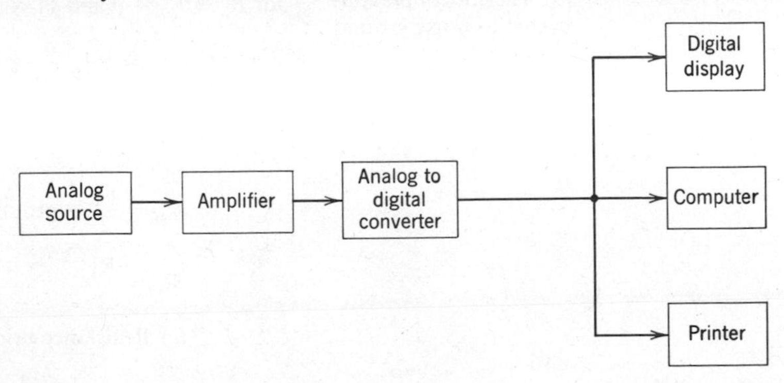

Figure 18.1
A single-channel data acquisition system.

connected to an amplifier, then to an analog-to-digital converter, and finally to a variety of digital components either for recording, display, or subsequent processing in digital form. We shall examine briefly each stage of this system.

18.1 SIGNAL CHARACTERIZATION

18.1.1 Equivalent Source Networks

Figure 18.2 illustrates the use of a rather general source network. The model, shown in Fig. 18.2*a*, includes a signal source v_S, resistances R_a and R_b in each signal lead, and a Thévenin equivalent source of common-mode signal referenced to a true ground. Two examples are given in Figs. 18.2*b* and *c*. The

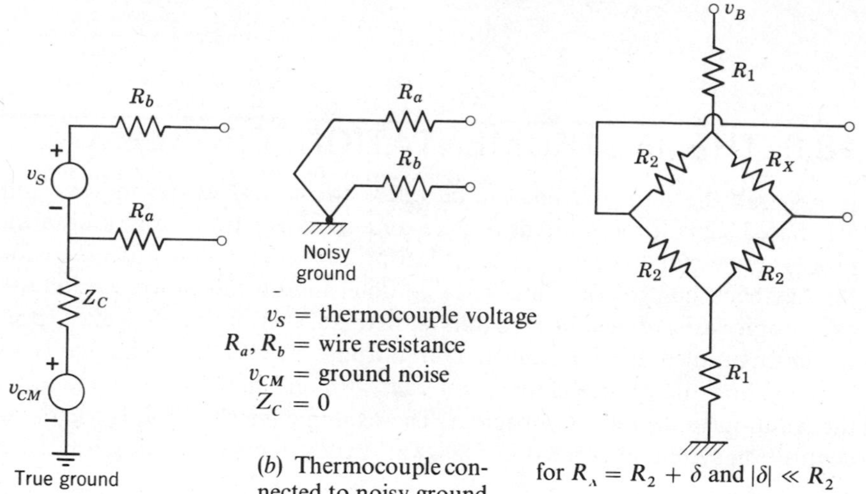

True ground

(*a*) Model

Noisy ground

v_S = thermocouple voltage

R_a, R_b = wire resistance

v_{CM} = ground noise

$Z_C = 0$

(*b*) Thermocouple connected to noisy ground

for $R_x = R_2 + \delta$ and $|\delta| \ll R_2$

$$v_S = \frac{\delta}{4(2R_1 + R_2)} v_B$$

$$R_a, R_b = \frac{R_2}{2}$$

$$v_{CM} = \frac{v_B}{2} + \text{ground noise}$$

$$Z_C = \frac{R_1}{2}$$

(*c*) Resistance bridge

Figure 18.2
The general source model.

thermocouple connected to a noisy ground has a signal source equal to the thermocouple voltage, and has lead resistances R_a and R_b arising from the thermocouple wires. Because the thermocouple is connected directly to ground, the common-mode impedance Z_C is zero, and the common-mode voltage is equal to the ground noise interference that may be present.

The second example, a resistance bridge, is used for temperature measurement with resistance thermometers, for resistive strain gage measurement, or with any other resistive transducer. The source v_B can be either dc or ac, and for the bridge near balance ($R_X \cong R_2$), the signal voltage and source resistances R_a and R_b are those given in the figure. Note that in this example, there is a common-mode signal present in addition to ground noise, and a nonzero common-mode impedance.

18.1.2 Frequency Content of Source Waveform

Figure 18.3*a* illustrates how a physical quantity of interest, such as temperature or resistance of a strain gage, might change with time. It is useful to attempt to represent this waveform as an approximate superposition of Fourier components, in order that the operation of subsequent signal-processing steps can be assessed. Clearly, one must include Fourier components at frequencies near f_1 (see Fig. 18.3) in order to represent the slow variations, and one must include Fourier components near f_2 in order to represent the "fine structure" or the bump in the curve. As a general rule, the more rapid the variation of a waveform, the higher the frequencies of the Fourier components that must be included in its description.

With these ideas in mind, we can now examine the spectral densities of the signal waveforms for the two examples of Fig. 18.2. The thermocouple, for example, is an intrinsically dc transducer, yielding a voltage that has a time dependence directly related to the time variation of the temperature. Thus the spectral density of the output of a thermocouple can be expected to look like that shown in Fig. 18.3*b*. The impulse at $f = 0$ represents the dc component of the thermocouple voltage, while the remainder represents the time-varying components, and has identifiable contributions around f_1 and f_2.

If the source v_B used to drive the bridge example of Fig. 18.2*c* were a dc source, then the output voltage would have a time dependence that, as in the thermocouple example, would have a spectral density of the form of Fig. 18.3*b*. If, however, the bridge were driven with a sinusoidal waveform, then the variation of the resistance R_X with time would produce amplitude modulation of the sinusoid, and would produce the sideband pattern shown in Fig. 18.3*c*.

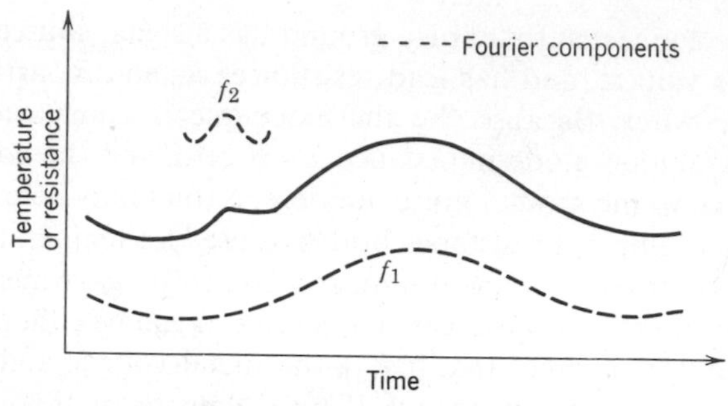

(*a*) Physical quantity versus time

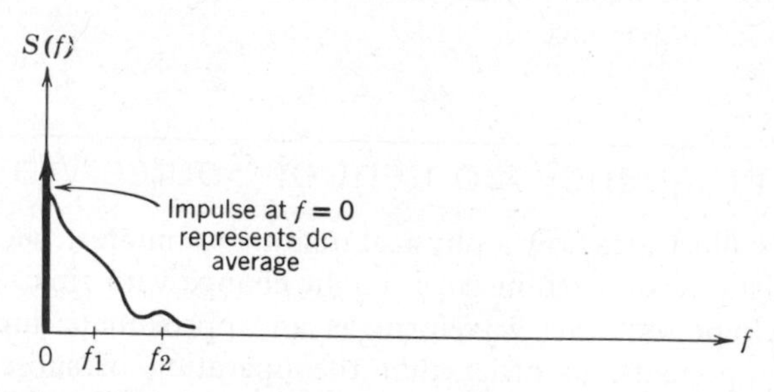

(*b*) Spectral density of thermocouple voltage

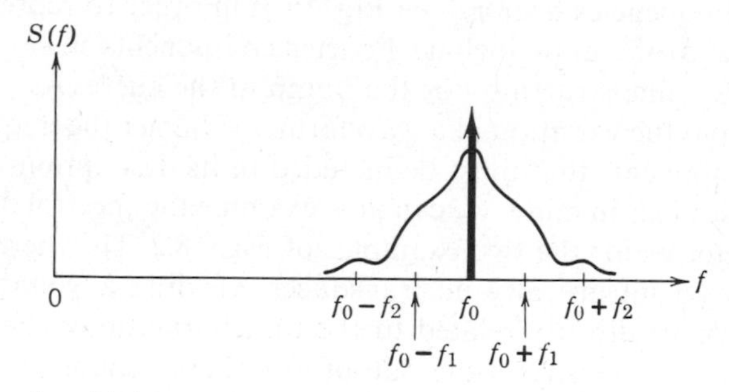

(*c*) Spectral density of bridge output when driven by sinusoid at f_0

Figure 18.3

Relation between (*a*) the time variation of a physical quantity, and (*b*), (*c*), the spectral densities of transducer output signals.

In each of these examples, it is necessary to consider first, the time-variation of the physical process of interest and, second, the kind of transducer and transducer excitation used to observe that variation. All natural physical processes have some upper limit to their speed. Therefore the spectral density of any physical phenomenon will be *band-limited*, that is, confined to frequencies below some identifiable maximum.[1] If one uses a dc transducer, such as the thermocouple, or the dc-excited bridge, the spectral density of the signal is a replica of the spectral density of the physical process. If the transducer is of the type that forms the product between the physical phenomenon and some external excitation (such as the bridge), then it is possible to produce amplitude modulation, with the average value of the physical process controlling the amplitude of the carrier, and the spectral density of the process defining the sidebands of the carrier. It is also possible to use transducers that produce frequency or phase modulation of a carrier. The spectral densities in these cases are more complicated.

18.1.3 Noise Components

Referring once again to the equivalent source network of Fig. 18.2*a*, one can represent noisy sources by adding appropriate thermal noise generators in series with R_a, R_b, and the resistive component of Z_C. If the transducer has additional noise, either because of other physical processes (such as shot noise in a photomultiplier tube) or because of random drifts (which can be modeled as $1/f$ noise), then an equivalent voltage source with an appropriate spectral density function can be added in series with v_S. In the following discussion, we shall tend to treat the noise produced by the source network in qualitative rather than quantitative fashion. We must be prepared, however, to discuss not only white noise produced by Johnson and shot noise sources, but also the possible presence of $1/f$ noise, and the presence of power-line interference, particularly in the common-mode voltage.

18.2 AMPLIFIERS

We have discussed a variety of amplifier circuits in this book. This section addresses itself to their proper use in instrumentation situations.

18.2.1 Difference-Mode Gain and Common-Mode Rejection

The term *difference-mode gain* (often called by an equivalent term *normal-mode gain*) refers to the transmission of difference-mode signals through some component or system. *Common-mode rejection* refers to the attenuation of the

[1] Remember, however, that the upper limit for noise processes can be very high.

common-mode signal relative to the difference-mode signal. What is confusing is that these terms can be applied either to a single component such as an op-amp, or to an entire instrumentation system, and that the identification of normal-mode and common-mode signals can differ greatly in the two cases. To clarify this situation, we shall discuss first an individual op-amp, and then address a complete source-amplifier combination.

Figure 18.4 illustrates the two basic kinds of op-amp amplifier circuits. We recall that in the linear region of operation, v_- and v_+ are almost exactly equal to one another. Therefore, any nonzero value of v_+ represents a common-mode voltage *as far as the op-amp concerned.* If we call the difference-mode gain A_D and the common-mode gain A_C, we can write for the open-loop linear-region transfer characteristic

$$v_O = A_D(v_+ - v_-) + A_C v_+ \tag{18.1}$$

where v_+ in the second term replaces the average of v_+ and v_-.

If we use this transfer characteristic to analyze the two circuits of Fig. 18.4, the gains and input resistances shown in the figure are obtained. Let us consider the inverting configuration first. The common-mode voltage in this

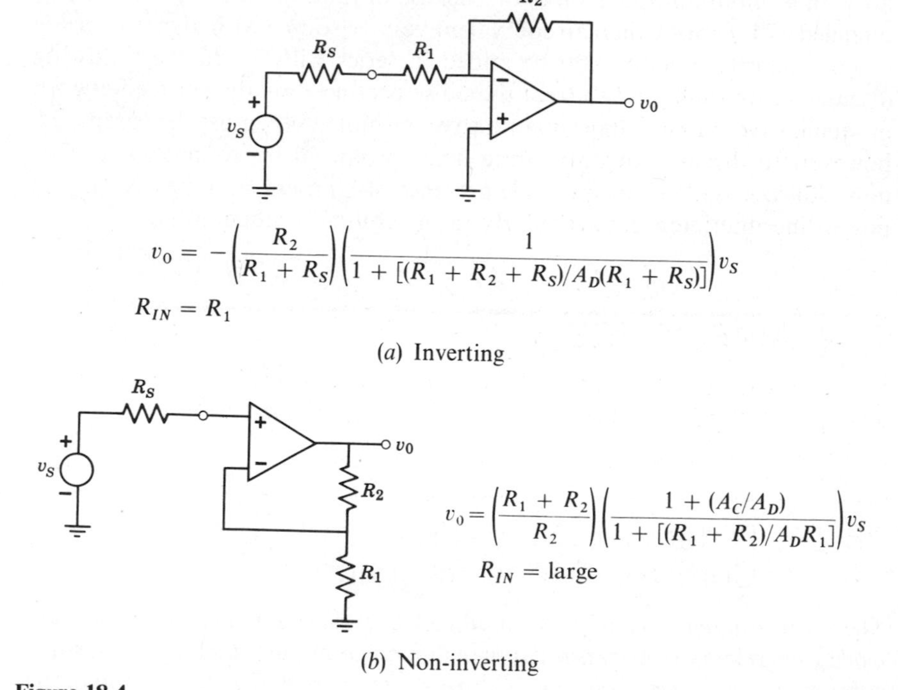

Figure 18.4
Two basic amplifier configurations.

case is zero (since v_+ is grounded), and so long as the difference-mode gain A_D is much greater than $(R_1 + R_2 + R_S)/(R_1 + R_S)$, the closed-loop gain is simply $-R_2/(R_1 + R_S)$. The "cost" of this absence of common-mode error is the relatively low input resistance R_1, which leads to the possibility of errors in predicting the gain unless R_1 is consistently much greater than R_S.

The non-inverting configuration eliminates the problem of low input resistance, but instead has a nonzero common-mode voltage (equal to v_S) applied to the op-amp. The closed-loop transfer function depends both on A_C and A_D, these quantities appearing in ratio form A_C/A_D (which is simply the inverse of the common-mode rejection ratio for the op-amp.) Therefore, in order for the closed-loop gain to equal $1 + R_2/R_1$, not only must this closed-loop gain be much less than A_D, but the common-mode gain A_C must be less than A_D. That is, an op-amp intended for use as a non-inverting amplifier must have high common-mode rejection, not because of ground noise or other common-mode sources external to the op-amp, but because the non-inverting configuration itself establishes a nonzero common-mode voltage at the op-amp's input.

We shall now examine a source-amplifier combination in which the source network itself has a common-mode voltage. Figure 18.5 shows a *differential amplifier*, a configuration that is an extension of the inverting amplifier to the case where the external source has a common-mode voltage. If the op-amp itself has perfect common-mode rejection, the output voltage is given by the

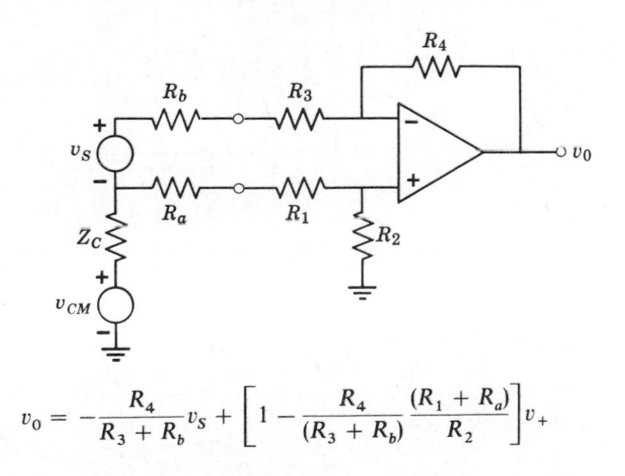

$$v_O = -\frac{R_4}{R_3 + R_b} v_S + \left[1 - \frac{R_4}{(R_3 + R_b)} \frac{(R_1 + R_a)}{R_2}\right] v_+$$

where

$$v_+ = \frac{R_2[(R_b + R_3)v_{CM} - Z_C v_S]}{(R_b + R_3)(R_1 + R_2 + R_a) + Z_C(R_1 + R_3 + R_a + R_b)}$$

Figure 18.5
The differential amplifier.

expression in the figure. Notice that in addition to the ordinary difference-mode gain of $-R_4/(R_3 + R_b)$, there is a term proportional to v_+. This term will vanish for the particular choice of resistors

$$\frac{R_4}{R_3 + R_b} = \frac{R_1 + R_a}{R_2} \tag{18.2}$$

With Eq. 18.2 satisfied, and with a perfect op-amp, this source-amplifier system exhibits perfect rejection of the v_{CM} signal. As was the case for the inverting amplifier, however, the closed-loop gain of this system depends on the network source resistance R_b. In addition, the network source resistance R_a enters the balance condition (Eq. 18.2) for perfect common-mode rejection. Therefore, unless R_a and R_b are known and stable, the gain and common-mode rejection of the differential amplifier become somewhat unpredictable. Beyond these problems, it must be recalled that the common-mode voltage applied to the op-amp is nonzero in this circuit configuration, v_+ depending both on v_S and on the external common-mode signal v_{CM}. Therefore, when an actual op-amp with finite common-mode rejection characteristics is used, there will be a gain-error due to the v_S term in v_+, and some of the v_{CM} signal will appear in the output.

An alternate approach to the rejection of external common-mode signals is the use of a guard. Figure 18.6 shows in schematic form a non-inverting

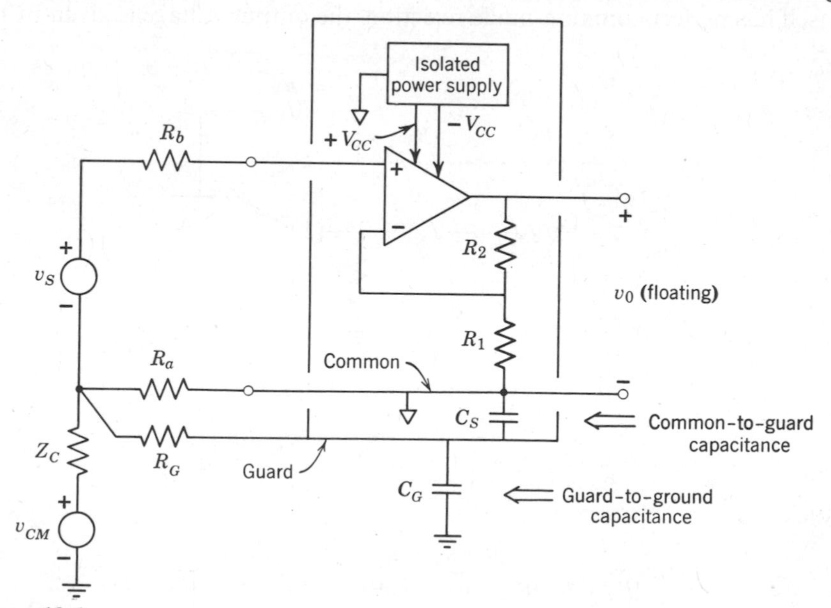

Figure 18.6
A guarded non-inverting amplifier. $v_O = [(R_2/R_1) + 1]v_S$.

amplifier driven from an isolated power supply, and fully enclosed in a guard (a shield driven at the common-mode potential). As discussed in Chapter 17, all common-mode currents are shunted away from the input resistance R_a and R_b, so that neither the difference-mode signal nor the common-mode signal applied to the op-amp depend on the external common-mode voltage v_{CM}. The one disadvantage of this method of rejecting common-mode signals (aside from the mechanical problems of surrounding all circuits with shields) is that the output of this circuit is "floating," with neither output terminal at ground potential. If this amplifier is to drive a grounded load, some kind of isolation device (transformer or optoelectronic coupler) must be inserted between the amplifier and the load. In many applications, such as analog-to-digital converters, the insertion of isolation devices is easily accomplished. In such systems, therefore, guarding is the method usually employed to provide overall common-mode rejection for a system.

Numerical values of common-mode rejection ratios are usually expressed in decibels. The common-mode rejection of individual op-amps varies from about 60 dB (factor of 10^3) to in excess of 100 dB (factor of 10^5) for amplifiers intended for low-level instrumentation use. In differential amplifier systems such as Fig. 18.5, where resistance ratios must be precisely maintained, common-mode rejection rarely exceeds 80 to 100 dB. In fully guarded systems with proper isolation of both the power supplies and the load, common-mode rejection ratios in excess of 120 dB (factor of 10^6) can be obtained.

18.2.2 DC Amplifiers

In this section we shall examine some of the problems specifically associated with the amplification of dc and low-frequency signals, such as the thermocouple waveform of Fig. 18.3*a*. The main problem is that op-amps have nonzero input offset voltages, and nonzero bias currents, and that all of these quantities drift. Figure 18.7 shows how the output of a simple noninverting amplifier depends on the input offset voltage v_I and on the two bias currents, i_+ and i_-. It is obvious from the transfer function in the figure that if v_S is at all comparable to v_I, there will be an intolerable error in the output. It is true that most commercial op-amps provide means for nulling v_I, usually with the adjustment of an external potentiometer. Even with this nulling feature, however, if the *drift* in v_I becomes comparable to v_S, unacceptable errors can result. The bias currents pose a similar problem. It is possible for a given R_S to make choices of R_1, R_2, R_3, and R_4 so that for a specified gain, the effect of the two bias currents will cancel. However, bias

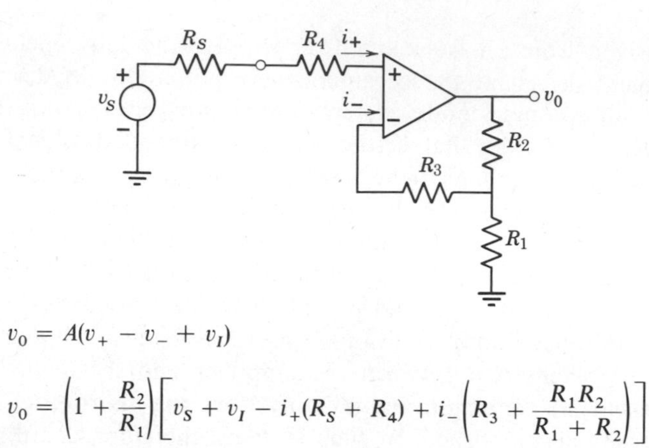

$$v_O = A(v_+ - v_- + v_I)$$

$$v_O = \left(1 + \frac{R_2}{R_1}\right)\left[v_S + v_I - i_+(R_S + R_4) + i_-\left(R_3 + \frac{R_1R_2}{R_1 + R_2}\right)\right]$$

Figure 18.7
The effects of the input offset voltage v_I, and the bias currents i_+ and i_-.

currents also drift, which means that a null under one set of conditions cannot be expected to persist indefinitely.

It is convenient to think of the offset and drift problem in terms of an equivalent noise spectral density function for an amplifier (Fig. 18.8). At low frequencies, the spectral density has a $1/f$ characteristic, representing both the drifts of these offset voltage and bias currents and also the $1/f$ noise inherent in the amplifier's transistors. The problem with dc signals is that the signal spectral density tends to be located within this $1/f$ region of the spectrum. It is much more difficult, therefore, to achieve high signal-to-noise ratios at the output of the amplifier when amplifying dc signals than when amplifying ac signals. Assuming there is a small signal that one wishes to amplify and measure, there are two general approaches one can take to

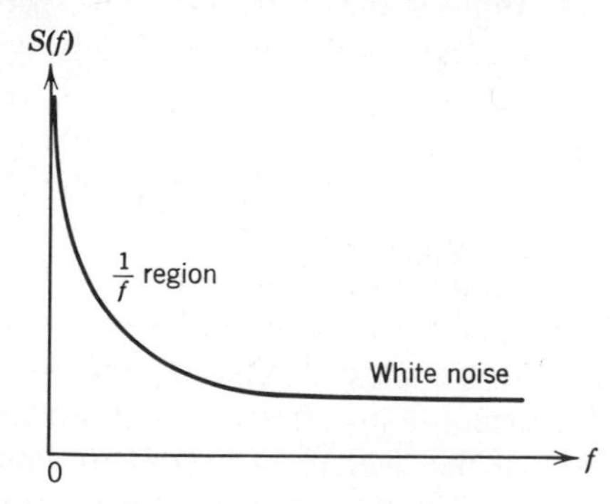

Figure 18.8
Noise spectrum of an amplifier. Linear scales have been used to emphasize the divergence of $S(f)$ as f approaches zero.

improve the acquisition of the data. One approach is to improve the drift characteristics of the amplifier; the other is to modify the use of one's transducer to make the signal into an ac signal. Both approaches have in common the use of methods of amplitude modulation, as discussed in the following section.

18.2.3 Modulation and Chopping Methods

Consider the resistance bridge of Fig. 18.2*c* when driven by a sinusoidal source at frequency f_0. The spectral density in this case (Fig. 18.3*c*) is not located at $f = 0$ but, instead, is centered at f_0 with sidebands covering a bandwidth of roughly $2f_2$ about f_0. If f_0 could be chosen high enough, this spectral density would lie above the $1/f$ region of one's amplifier, permitting a much higher signal-to-noise ratio than could be obtained using the same bridge and amplifier but with the bridge driven with a dc source. Thus, by operating the transducer so that the physical variable of interest produces an amplitude-modulated carrier, the signal spectrum can be shifted away from dc to a more favorable frequency interval. This principle is widely used in data acquisition problems.

It is necessary when using modulation methods, of course, to demodulate the carrier after amplification, a process that can be carried out either with an ordinary peak detector, or with a synchronous demodulator driven at the modulation frequency f_0. These two types of demodulators are illustrated in Fig. 18.9. In the case of the synchronous demodulator, the FET

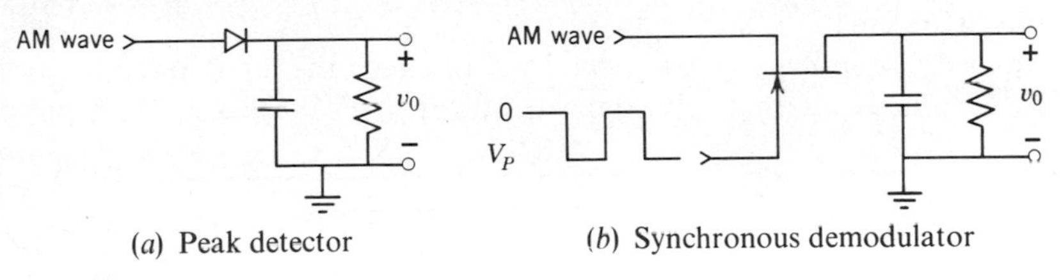

Figure 18.9
Half-wave demodulators. The gate-drive waveform in (*b*) is a square wave at f_0 with its phase adjusted for synchronization with the carrier.

switch must be driven with a square wave at the carrier frequency, and this square wave must have the correct phase so that the FET is conducting during the positive half-cycle of the carrier waveform and not conducting during the negative half-cycle. We note in passing that full-wave demodulators, which use both the positive and negative half-cycles of the AM wave, can be built by a simple extension of these examples. There are commercial instruments available called *lock-in amplifiers*, which contain an ac amplifier, a synchronous demodulator, and the necessary phase control circuits for use in conjunction with this modulation method.

The modulation method outlined above is not limited to the specific noise and drift problems associated with dc amplifiers. Some transducers, such as photomultiplier tubes and photodiode detectors, are themselves sources of large $1/f$ noise components. It is common practice in optical measurements, therefore, periodically to interrupt or *chop* the light beam, in order that the change in detector output between the beam-off and beam-on states carries the information about the light intensity in the beam. By chopping the beam at an audio frequency, the signal information can be made to occur in a spectral region away from the $1/f$ noise of the optical detector. The output of the detector is then amplified with a band-pass amplifier that rejects the low-frequency noise, passes sufficient bandwidth about the chopping frequency to allow for amplitude variations of the beam intensity, and rejects frequencies above this range. After synchronous demodulation, the resulting low frequency signal is free of the $1/f$ noise that would have been present if straight dc amplification were used.

Having now given the impression that modulation methods can solve all problems, we must force our attention back to the situations for which direct modulation methods provide no assistance. The best example is the thermocouple, which as stated earlier, is inherently a dc transducer. It is not possible to operate a thermocouple so that the temperature information appears as modulation on an ac carrier. Therefore, if one wishes to amplify thermocouple voltages, one must use modulation methods on the amplifier rather than on the transducer to reduce the effect of offsets and drifts in the amplifier system.

Figure 18.10 shows a *chopper amplifier*, in which the input and a fraction of the output are alternatively connected through a capacitor to the input of

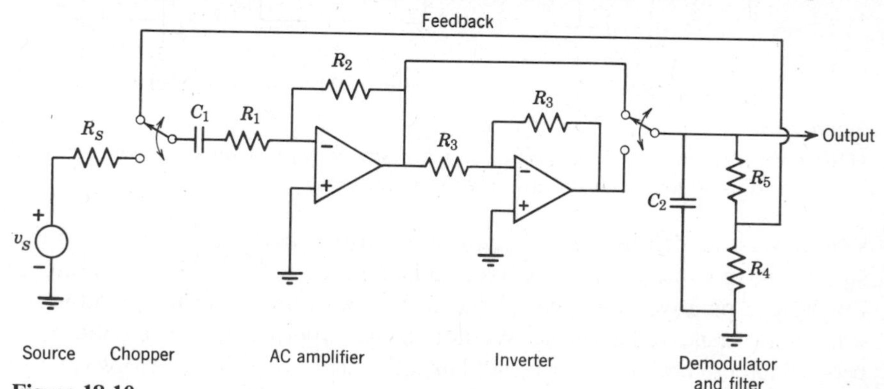

Figure 18.10
A chopper amplifier with gain $(1 + R_5/R_4)$.

an inverting amplifier. It is presumed that the capacitor C_1 is sufficiently large to be considered a short circuit at the chopping frequency, and that R_2/R_1 is much larger than the final system gain of $1 + R_5/R_4$. The demodulator must be driven synchronously with the chopper, both in frequency and phase. Notice that any offset errors appearing at the output of the ac amplifier are averaged out by the synchronous demodulation process. The overall effect, therefore, is an amplifier system with greatly reduced offset errors. The choppers can be made with FETs or with relays. With careful design, overall drifts can be reduced to an equivalent voltage offset drift on the order of 1 μV/°C, which is low enough for most instrumentation applications.

18.2.4 Bandwidth Considerations

The spectral density of the signal of interest places restrictions on the amplifier bandwidths that can be used. A dc amplifier must have a pass band that extends to high enough frequencies to pass the highest Fourier component of the signal. Ac amplifiers designed to carry amplitude modulated signals must have bandwidths twice as large to pass the same spectral information, because both upper and lower sidebands must be accommodated.

If we recall that the spectra of Johnson and shot noise sources are flat, then the contribution to the total mean square noise from such noise sources will be proportional to the effective noise bandwidth of the system. The bandwidth restrictions imposed by the frequency content of the signals place a lower limit on the permissible narrowing of the noise bandwidth and, therefore, an upper limit on the achievable signal-to-noise ratio.

In many experimental situations, it is possible to reduce the spectral width of the signal by performing one's experiment more slowly. For example, if one is scanning an optical absorption spectrum, one can slow down the scanning rate, thereby taking a longer time to perform the measurement. Since the signal changes more slowly, its bandwidth is lower so that one can use a narrower bandwidth filter to reject noise. The signal amplitude is unchanged, but the noise bandwidth is reduced. Therefore, one ought to be able to achieve improved signal-to-noise ratios by slowing down one's data acquisition rate. Indeed, so long as the noise spectrum is white, this argument holds, and the signal-to-noise ratio can be improved by taking a longer time to perform a measurement and using a correspondingly narrower filter to reduce noise.

When $1/f$ noise is introduced into the picture, this argument falls apart. In Chapter 17, we calculated the mean square noise output of a single-pole filter (with the pole at f_p) driven with $1/f$ noise. We found that for a total time of observation equal to T, the mean square noise was proportional to $\ln(f_p T)$. Let us now think about this result in the suggested context of a slower experiment (longer T) and a narrower filter (smaller f_p). The Fourier components of the signal will occur at frequencies that are proportional to the rate of scanning and, therefore, inversely proportional to the total scan time T. That is, the filter bandwidth can only be lowered by an amount such that $f_p T$ and therefore $\ln(f_p T)$ is constant. This leads to the surprising result that for $1/f$ noise, the *signal-to-noise ratio does not improve with slower scans and narrower bandwidths.* This means that in any experiment, there is a practical limitation to the achievable signal-to-noise ratio. Once the bandwidth has been made small enough so that the noise is $1/f$ in character, there can be no further improvement in signal-to-noise ratio by making adjustments of f_p and T.

The advantage of using chopping and modulation methods in such situations is now clear. By chopping, the spectrum of the output signal no longer is $1/f$ in nature, but has the characteristics of the transducer and amplifier noise around the chopping frequency. One is still limited by the trade-off between the bandwidth and the total time to make an observation, but now one is dealing with a noise spectrum that does not become infinite as the bandwidth is narrowed. Therefore, vast improvements in signal-to-noise ratio can be achieved by reducing the data acquisition rate in connection with chopping methods.

As a final bandwidth issue, we must comment on the bandwidth of a chopper amplifier relative to the bandwidth of a straight dc amplifier. The dc amplifier has a bandwidth limited only by the frequency-response characteristics of the op-amp. The chopper amplifier, however, has a much narrower bandwidth, being in practice limited to frequencies about 1/10 to 1/5 of the chopper frequency. Variations in input signal that are more rapid than this get averaged out in the demodulation process and cannot be accurately amplified. Therefore, the spectral characteristics of the signal being amplified must be known before one can specify the use of a chopper amplifier.

18.3 ANALOG-TO-DIGITAL CONVERSION

18.3.1 The Dual-Slope Integrator

The process of converting an analog signal to digital form is itself a fundamental instrumentation problem, one fully worthy of extended attention. There are many basic methods in current use. One conceptually simple

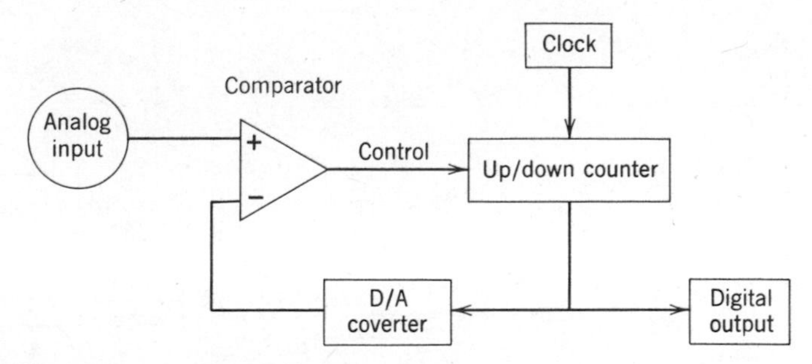

Figure 18.11
Schematic diagram of a possible A/D converter.

scheme is illustrated in Fig. 18.11. The input signal is amplified and compared to the analog output of a digital-to-analog converter (D/A) that is connected to the output of a counter. The counter is of the type that counts up when the control line is HI and counts down when the control line is LO. The pulses that the counter accumulates come from a clock. The output of this converter is stable when a change of one count up or down causes the comparator output to reverse sign.

A somewhat more sophisticated approach is the dual-slope integrator in Fig. 18.12. The analog input is through a contact of a three-position switch. A second contact is grounded, and the remaining contact is connected to a stable reference voltage V_R. The switch position is determined by a digital control circuit. Usually, the switch is made with a network of FETs, and the connection between the logic levels and the switches is accomplished by the types of circuits illustrated in Section 10.4.2. The output of the switch is connected to an analog integrator with time constant τ. The output of the integrator drives a comparator. The control circuit accepts output from the comparator and from the counter, and also controls the counter action. The sequence of these steps involved in a conversion are as follows:

1. *PRESET.* Control circuit sets switch to position A, and presets a number N_0 into the counter. This number N_0 is chosen such that for a fixed clock pulse rate of f_0 pulses per second, the counter will take a fixed time T_0 to count from N_0 down to zero. The control circuit also resets the integrator voltage to zero during this preset time interval.
2. *UP Integrate.* The control circuit releases the zero on the integrator and moves the switch to position B. At the same time, the counter begins to count down toward zero. The integrator voltage grows at the rate of v_S/τ V/s, where v_S is the analog input voltage. After a time T_0, the control circuit detects a zero in the counter, ending this portion of the cycle.

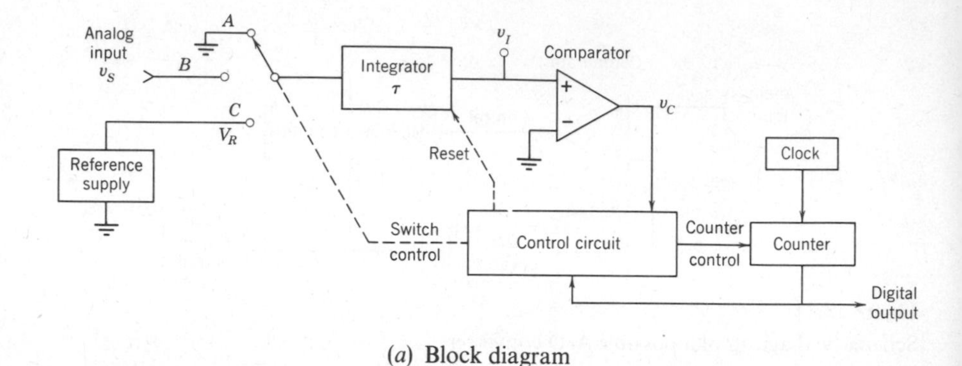

(*a*) Block diagram

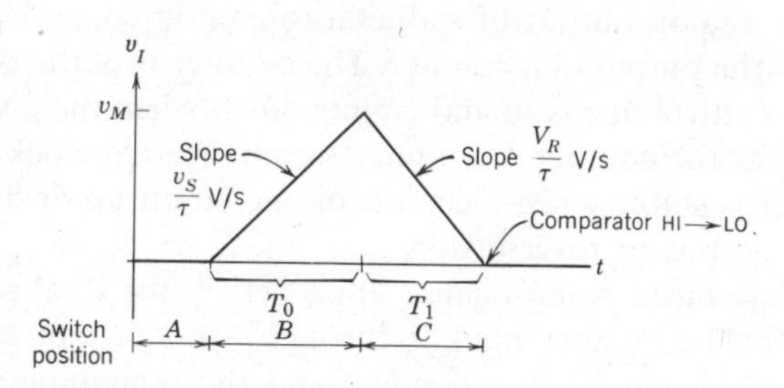

(*b*) Integrator output waveform through one conversion cycle

Figure 18.12
A dual-slope integrator A/D converter.

3. *DOWN Integrate.* Immediately upon detecting a counter zero, the control circuit throws the switch to position *C* and starts the counter counting up from zero. Since the voltage applied to the integrator is now V_R, the integrator voltage drops at the rate of V_R/τ V/s. When the integrator voltage reaches zero, the comparator output v_C switches from HI to LO, at which point the control circuit halts the counter. The count accumulated in the counter at the end of the *C* time interval is accurately proportional to the analog input signal, as the following argument demonstrates.

During the UP integration, the voltage that builds up on the integrator is

$$v_M = \frac{v_S}{\tau} T_0 \tag{18.3}$$

The DOWN integration proceeds at V_R/τ V/s, therefore requiring a time T_1 to reach zero given by

$$T_1 = \frac{v_M}{(V_R/\tau)} \tag{18.4}$$

During the down integration, the counter is accumulating counts at the clock rate of f_0 pulses per second. Therefore, the total number of counts N_1 in the counter at the end of the down integration is

$$N_1 = f_0 T_1 \tag{18.5}$$

Substituting for T_1 and v_M, and using the fact that $N_0 = f_0 T_0$, we obtain

$$N_1 = \frac{v_S}{V_R} N_0 \tag{18.6}$$

That is, the output count depends on the ratio of v_S to V_R and on the preset digital count N_0. It does not depend on the integrator time constant or on the clock frequency. With proper choice of N_0 and V_R, the output count N_1 can be calibrated to indicate an output in any desired unit scale, such as volts, millivolts, or microvolts.

There is an additional feature of the dual-slope integrator that makes it exceptionally useful in environments where interference from the power line is a problem. The time interval for the UP integration T_0 can be chosen to be an exact multiple of the period of the power line. Since the integral of a sinusoid over one period is exactly zero, the dual-slope integrator has the capability of perfect rejection of the line frequency and of harmonics of the line frequency. Thus, in addition to any filtering that may be present, and in addition to the attentuation of high-frequency noise that will automatically accompany the use of an integrator, one can provide particularly strong attenuation at specific frequencies with this conversion method.

The analog portion of an A/D converter can be isolated from subsequent loads by inserting digital isolation devices, such as optoelectronic isolators, in the path of the digital signals. Thus, it is relatively straightforward to guard the analog portion of an A/D converter, include within the guard a high-gain preamplifier and, thus, obtain a self-contained guarded amplifier and A/D converter. Commercial instruments of this type with output scales calibrated in volts are simply called *digital voltmeters* (DVM).

18.3.2 Precision, Speed, and Accuracy

Each type of A/D converter requires a finite time to perform a conversion. There is a trade-off between the speed of a conversion and the precision of the digital output. The longer one takes to perform a conversion, the more precisely one can determine the value of the voltage being converted, and the greater the number of bits of digital output one can obtain. For example, high-speed units capable of performing conversions at rates up to 10 MHz are limited to 8- or 10-bit resolution, which means that the value of the input voltage is determined only to one part in 256 (8 bits) or one part in 1024 (10 bits). In order to obtain as many as 16 bits of resolution (one part in 64,000) much longer conversion times are required. Furthermore, if high precision is required, the input voltage must be examined for a long period of time to obtain an accurate average value. Thus in high precision instruments, the input voltage is strongly filtered to remove analog noise, and the conversion speed is then limited by the time necessary for the step response of the filter to settle down to within the stated precision.

In addition to the trade-off between speed and precision, there is an important question of accuracy. For reasons known perhaps best to psychologists, there is a strong tendency to *believe* the output of a digital instrument simply because it is digital. That is, users of digital instruments often forget that such instruments must be *calibrated* to achieve their stated accuracies. In the dual-slope method, for example, the voltage reference V_R must be precisely known and stable if the output reading is to be an accurate measurement of the input voltage. The clock frequency must be stable in time, and the integrator must not drift. Furthermore, if a preamplifier is used, the gain of the preamplifier must be accurately set, and the amplifier must be free of error-producing offsets and drifts. The user of digital instruments must be continually reminded not to be seduced by the digital output, but rather to maintain calibration standards for digital instruments with the same care one would use for analog instruments.

18.3.3 Sampled and Multiplexed Data Systems

Since an analog-to-digital conversion process occupies a finite time, the output of an A/D converter does not represent a continuous function of time, but rather a series of *samples* of the actual waveform. Figure 18.13 illustrates a sampled waveform. A total of 10 samples are taken, the width of the solid vertical lines representing the time taken for each conversion. Thus, the continuous variation of the analog signal $v(t)$ is represented by a list of digital numbers, each number in the list representing the amplitude of $v(t)$ during one particular conversion interval.

A critical question concerns the rapidity with which these samplings must be repeated. A fundamental theorem, called the *sampling theorem*, states that

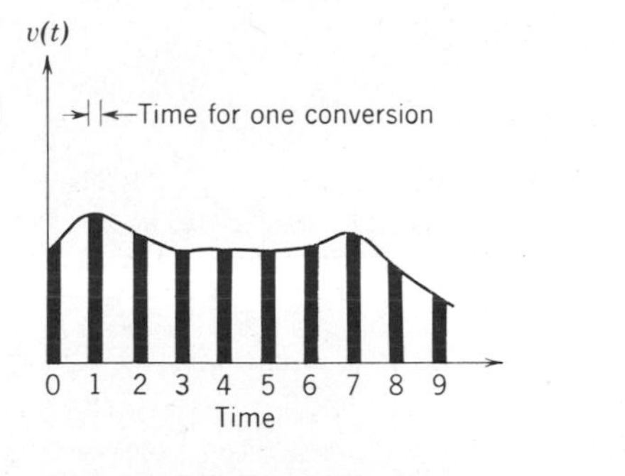

(*a*) Sampled waveform

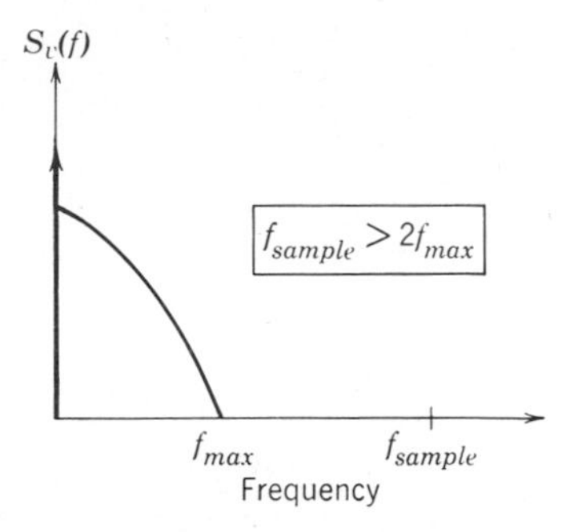

(*b*) Spectrum of waveform

Figure 18.13
The sampling theorem.

if a waveform is sampled at a rate that exceeds twice the frequency of the maximum Fourier component of the waveform, then the complete waveform can ultimately be recovered with proper filtering. That is, if the maximum Fourier component of the waveform being sampled is f_{max}, then a sampling rate of

$$f_{sample} \geq 2f_{max} \tag{18.7}$$

must be used to ensure no loss of information. In practice, sampling rates well in excess of this theoretical minimum are used, partly to simplify the task of recovering the original waveform with filters, and partly to reduce the likelihood of noise-induced errors.

If the time required per conversion is much shorter than the time interval between samples, it becomes possible to use one A/D converter to sample in sequence many different sources of data. Figure 18.14 illustrates the operation of a *multiplexed data system*, in which three analog data channels, v_1, v_2, v_3, are sampled sequentially by a switching network and an A/D converter. The digital output from the A/D is once again a list of numbers, but every third number is associated with a particular data channel. During the *n*th time interval, for example, the first number represents the value of $v_1(n\tau)$, the second represents the value of v_2 at the slightly later time $n\tau + \tau/3$, while the third number represents the value of $v_3(n\tau + 2\tau/3)$. The number of channels that can be multiplexed in such a system can be very large, depending on the speed of the A/D converter and on the rapidity of variation of the analog signals being sampled. The slower the variation of the analog signals, the less often each channel must be sampled, and the greater the number of channels that a single A/D can handle. Data acquisition systems of this general type

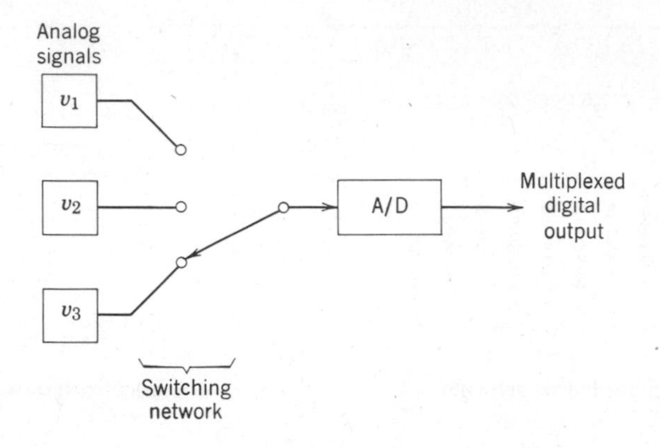

(*a*) Block diagram

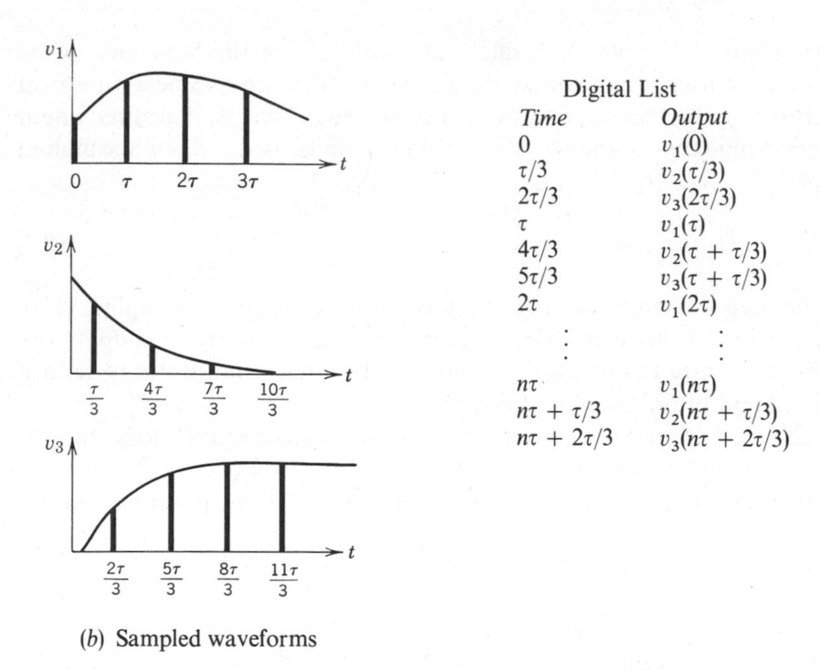

Digital List

Time	*Output*
0	$v_1(0)$
$\tau/3$	$v_2(\tau/3)$
$2\tau/3$	$v_3(2\tau/3)$
τ	$v_1(\tau)$
$4\tau/3$	$v_2(\tau + \tau/3)$
$5\tau/3$	$v_3(\tau + \tau/3)$
2τ	$v_1(2\tau)$
$\vdots$	$\vdots$
$n\tau$	$v_1(n\tau)$
$n\tau + \tau/3$	$v_2(n\tau + \tau/3)$
$n\tau + 2\tau/3$	$v_3(n\tau + 2\tau/3)$

(*b*) Sampled waveforms

Figure 18.14
A multiple-channel data system.

are available commercially in a variety of forms. Some include simple printed output, while others include either a digital tape recorder for storage of the data for subsequent computer processing or direct interfaces to a computer for real-time processing of the data.

18.4 DIGITAL SIGNAL PROCESSING

With the growing accessibility of inexpensive high-speed digital computers, instrumentation practice is shifting increasingly toward the use of digital methods to perform signal-processing tasks. This book has emphasized analog methods extensively, partly because they relate most directly to one's experimental laboratory experience, and partly because a clear understanding of analog signal-processing methods is an important prerequisite to the study of the corresponding digital operations. Our purpose in this final section is to illustrate the simplest kind of digital signal processor, the digital equivalent of a single-pole filter, thereby to point the way toward further study in this important area.

18.4.1 A Simple Digital Filter

Figure 18.15 illustrates a digital signal processor in which a sequence of numbers $\{x_n\}$ (i.e., $x_0, x_1, x_2, \ldots$, etc.) is presented to a digital filter, and a corresponding set of outputs $\{y_n\}$ is obtained. The inputs might be a set of numbers obtained from a sampled analog waveform. Thus, the $\{x_n\}$ series of numbers represents the "input waveform" to the filter and the set of numbers $\{y_n\}$ represents the "output waveform."

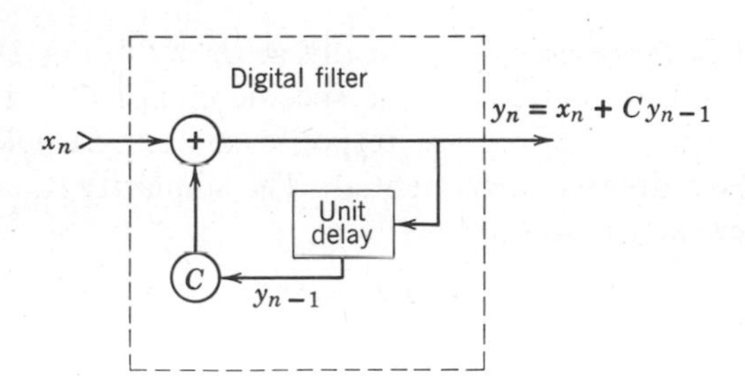

Figure 18.15
A single-pole digital filter.

In the example of Fig. 18.15, the output during the nth time interval y_n is fed back through a unit delay. Thus at the output of the unit delay in the nth time interval is the value of y_{n-1}. This value is multiplied by a constant C, and is then added to the input x_n to form the value of y_n. The algebraic equation that represents this block diagram is

$$y_n = x_n + Cy_{n-1} \tag{18.8}$$

One could easily picture a two-line computer program designed to perform these operations, the verbal equivalent of the computer instructions being:

1. Take the last value of y (i.e., y_{n-1}) and multiply it by C. (18.9)

2. Add to it the next value of x (i.e., x_n) to get the next value of y (i.e., y_n).

These instructions, carried out repetitively, constitute the implementation of this digital filter on a computer.

We shall now examine the response of this filter to both a unit pulse and a step. Consider the unit pulse input first, as summarized here:

$$\begin{aligned} x_n &= 0 \qquad n \neq 0 \\ x_n &= 1 \qquad n = 0 \end{aligned} \tag{18.10}$$

We can solve for the unit pulse response by continuous application of Eq. 18.8. Since only x_0 is nonzero, each y_n after y_1 is simply equal to C times the previous y_n. That is,

$$\begin{aligned} y_0 &= 1 \\ y_1 &= C \\ y_2 &= C^2 \end{aligned} \tag{18.11}$$

with the general term being

$$y_n = C^n \tag{18.12}$$

This output can be represented graphically, as shown in Fig. 18.16*a*. Here the unit pulse response is plotted for the specific choice $C = 1/2$. Notice the outward similarity to the impulse response of a single-pole analog filter (which would be a decaying exponential). The similarity can be made more explicit if we rewrite the solution as

$$y_n = C^n = e^{n \ln C} \tag{18.13}$$

Since C is less than unity, $\ln C$ is a negative number. Thus the set of amplitudes obtained does have an exponential decay, with the quantity $|1/\ln C|$ representing the time constant.

Similar repeated applications of Eq. 18.8 enable us to determine the step response. If we represent the step input as

$$\begin{aligned} x_n &= 0 \qquad n < 0 \\ x_n &= 1 \qquad n \geq 0 \end{aligned} \tag{18.14}$$

then the step response is

$$y_n = \frac{1 - C^n}{1 - C} \tag{18.15}$$

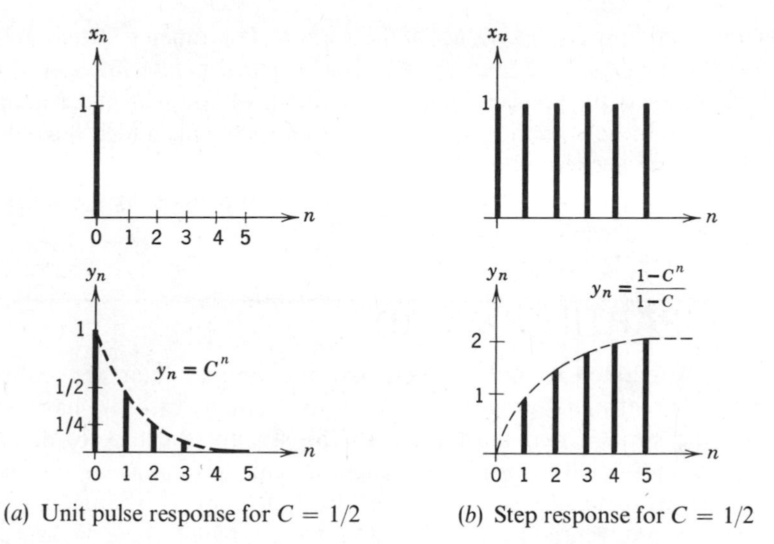

(*a*) Unit pulse response for $C = 1/2$

(*b*) Step response for $C = 1/2$

Figure 18.16

Responses of a single-pole digital filter.

This step response is represented graphically in Fig. 18.16*b* for $C = 1/2$. Once again, the set of amplitudes follow an exponential envelope, with the same time constant determined above. Furthermore, if one takes the difference between y_{n+1} and y_n (the digital equivalent of calculating the derivative of the step response), one obtains the unit pulse response.

$$y_{n+1} - y_n = \frac{(1 - C^{n+1}) - (1 - C^n)}{1 - C} = C^n \tag{18.16}$$

Thus the linear equation (18.8) and its corresponding computer program performs a similar kind of filtering operation on a sampled waveform that a single-time-constant linear circuit would perform on the complete analog signal.

18.4.2 Digital Process Control

Assuming that the example just studied serves as sufficient evidence that it is possible to construct digital algorithms and computer programs that can perform operations on sampled waveforms, one can now conceive of using a digital computer as part of a real-time feedback control system. Sampled data from sensors are fed to a central computer that performs various filtering operations on the data, calculates corrections to any process control parameters, and outputs new values of these control parameters. Because one is working with a digital computer, one can easily perform filtering operations for which there is no precise analog equivalent, or for which the analog equivalent would be extremely difficult to construct. This added flexibility, coupled with the accuracy possible in digital systems, has been responsible for an enormous growth in the use of digital signal processing methods.

18.5 A PARTING WORD

We have attempted to survey electronic circuits and their applications to a variety of real signal-processing situations. In many cases we have only scratched the surface, and must leave to the student the task of delving deeper into individual subjects. Throughout this presentation we have attempted to keep one idea in mind—that the performance of actual devices, the models used to represent them, and the ways in which these devices can be used to perform signal-processing tasks all interact in a significant way. The method ultimately adopted to handle a particular signal-processing task will be a synthesis of "what is desirable" with "what is possible." The final important ingredient is common sense.

REFERENCES

R18.1 Ralph Morrison, *DC Amplifiers in Instrumentation* (New York: John Wiley, 1970).

R18.2 Ralph Morrison, *Grounding and Shielding Techniques in Instrumentation* (New York: John Wiley, 1967).

R18.3 David Bartholomew, *Electrical Measurements and Instrumentation* (Boston, Mass.: Allyn and Bacon, 1963).

R18.4 R. Ralph Benedict, *Electronics for Scientists and Engineers* (Englewood Cliffs, N.J.: Prentice-Hall, 1967).

R18.5 Mischa Schwartz, *Information Transmission, Modulation, and Noise* (New York: McGraw-Hill, 1970).

R18.6 D. H. Sheingold, ed., *Analog-Digital Conversion Handbook* (Norwood, Mass.: Analog Devices, 1972).

R18.7 Jerald G. Graeme and Gene E. Tobey, eds., *Operational Amplifiers: Design and Applications* (New York: McGraw-Hill, 1971).

R18.8 Bernard Gold and Charles M. Rader, *Digital Processing of Signals* (New York: McGraw-Hill, 1969).

EXERCISES

E18.1 Prove the results stated in Fig. 18.2*c*.

E18.2 Plot the waveform $\sin(0.02\pi t) + \cos 2\pi t$ for $0 \leq t \leq 2$ s. Use your graph as a two-second sample of an unknown waveform, and estimate its "dc component" and its "ac component."

E18.3 Repeat E18.2 for the time interval 1 min $\leq t \leq$ 1 min 2 s.

E18.4 Plot the spectrum of the source waveform v_S for the bridge of Fig. 18.2*c* if $v_B = A\cos 200\pi t$, and $\delta = 0.01R_2[\sin(0.02\pi t) + \cos 2\pi t]$.

E18.5 Prove the results stated in Fig. 18.4.

E18.6 Prove the results stated in Fig. 18.5.

E18.7 Estimate the common-mode rejection of the circuit of Fig. 18.6 assuming a 0.001 μF capacitance between *common* and the *guard*, and assuming $Z_C = 1$ kΩ, $R_a = R_b = 1$ kΩ, $R_G = 10\ \Omega$, and $C_G = 0.001\ \mu$F.

E18.8 Prove the results stated in Fig. 18.7.

E18.9 Show in the synchronous demodulator of Fig. 18.9*b* that depending on the phase of the square wave relative to the carrier phase, the output v_O can follow either the positive envelope, the negative envelope, or be anything in between (including *zero* for a 90° phase error). Plot the v_O waveform for several different phase choices to illustrate the effect.

E18.10 A chopped waveform resembles an AM wave, but using a square wave for the carrier. Describe the spectrum in this case.

E18.11 Assume that the ac-amplifier op-amp in Fig. 18.10 has an input offset voltage v_I that varies slowly compared to the chopping frequency. Show by exhibiting the waveforms, that this offset voltage is cancelled out in alternate half-cycles of the chopping waveform.

PROBLEMS

P18.1 One way to sense small motions is to attach one plate of a parallel plate capacitor to the moving object, and leave the other plate fixed. Since the capacitance is proportional to the reciprocal of the plate spacing, position changes appear as capacitance changes. Figure P18.1 shows a circuit that can be used to sense such capacitance changes.

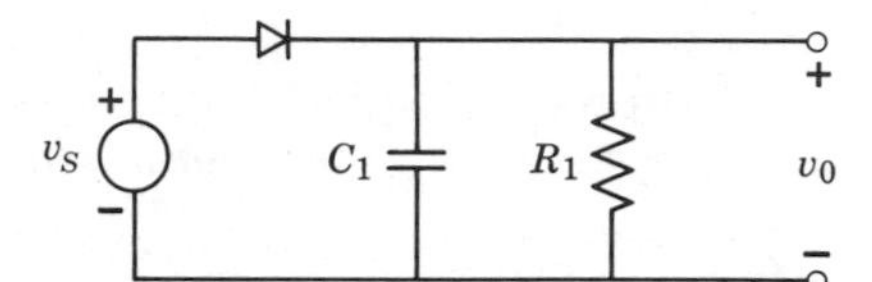

Figure P18.1

Suppose v_S is a 1 MHz, 100 V peak sine wave, and C_1 is a capacitor that varies between 6 pF and 10 pF in response to the motion of an object attached to its upper plate. If the Fourier components of the motion are confined to frequencies below 1 kHz, show that for proper choice of R_1, the voltage v_O, will have a component approximately proportional to the variation in C_1.

P18.2 Given an op-amp with an open-loop gain that might vary between 10^4 and 10^5 because of combinations of temperature changes and supply voltage variations, determine if it is possible, using *one* of these op-amps together with precision resistors, to design an amplifier circuit with a gain of 100.0 guaranteed accurate to 0.1%. Explain your answer.

P18.3 Can the circuit in Fig. P18.3, with proper choice of R_1 and R_2, satisfy the accuracy requirements of P18.2? Explain your reasoning.

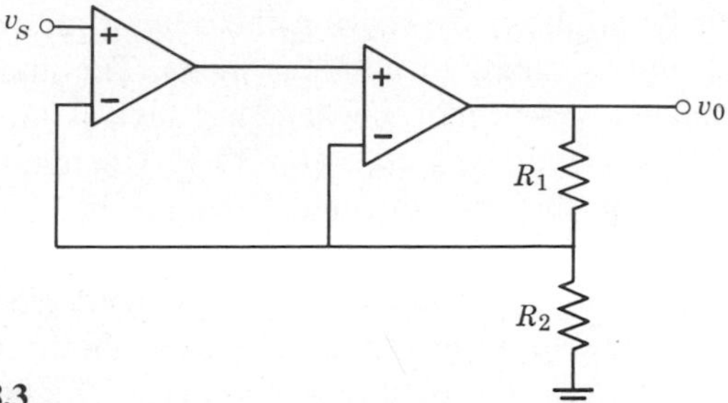

Figure P18.3

P18.4 Suppose in Fig. 18.6 that in addition to C_S and C_G, there is an additional capacitance C_P between common and ground (because of imperfect isolation of the power supply, gaps in the guard, and stray capacitance in any isolation devices). Calculate the common-mode rejection using the values of Exercise E18.7, but assuming $C_P = 10$ pF.

P18.5 Design a three-channel FET multiplexer (see Fig. 18.14), and then design the demultiplexer to re-sort the multiplexed signals into separate sample-and-hold circuits.

P18.6 Devise the digital-filter equivalent of a zero.

P18.7 Devise the digital-filter equivalent of a two-pole system. Can it be made to ring?

P18.8 A psychologist wishes to present two alternating pictures of *equal* intensity to a subject (Fig. P18.8). The brightness of the two lights

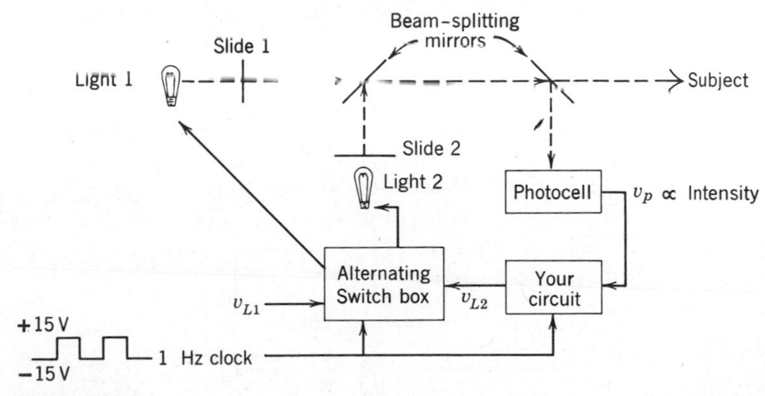

Figure P18.8

can be adjusted by varying the magnitudes of the two voltages v_{L1} and v_{L2} used to light the lights. The alternating switch box applies v_{L1} to Light 1 when the clock is at $+15$ V, and v_{L2} to Light 2 when the clock is at -15 V. The magnitude of v_{L1} is controlled by the experimenter. Your job is to design a circuit that adjusts v_{L2} so that the intensity reaching the subject from slide 2 is identical to the corresponding intensity from slide 1.

A beam-splitting mirror is placed in front of the subject to divert part of the light to a photocell. The photocell produces a voltage v_P proportional to the light intensity striking it. You may assume that the intensity from Light 2 varies linearly with v_{L2}, and that a range of 0 to 10 V provides enough latitude to insure matching the intensity from Light 1.

P18.9 An apparatus provides an analog voltage v_1 (in volts), and a 3-bit binary number N. You wish to generate a voltage v_2 with a magnitude in volts numerically equal to the product of v_1 and N.

(a) Design a system that will perform this multiplication. Your multiplier may be digital, analog, or a mixture, as you prefer. Explain clearly how your system works.

(b) Explain how you would approach this very similar problem: instead of producing the analog voltage v_1, your system responds to a trigger pulse by producing a pulse after a delay of D milliseconds. At the same time, the 3-bit number N appears in an output register. You wish to compute the product DN and place it in a storage register.

P18.10 A typical automobile ignition system uses the interruption of current flowing through an inductor to produce the high voltage necessary to produce the spark that ignites the fuel (Fig. P18.10*a*).

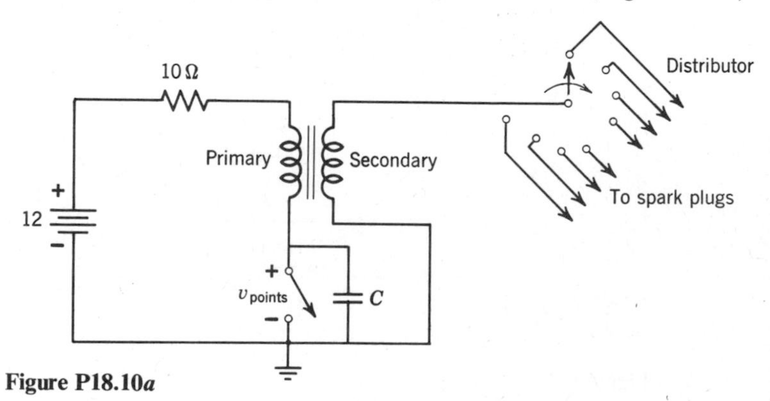

Figure P18.10*a*

The points open and close four times during each engine revolution (for an eight-cylinder engine) producing the voltage waveform across the points shown in Fig. 18.10*b*.

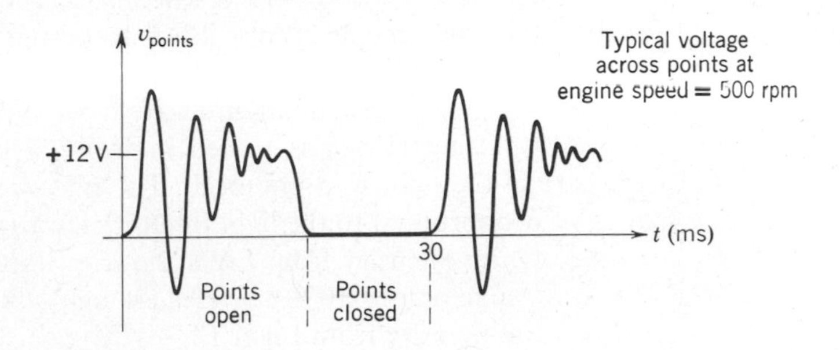

Figure P18.10*b*

You are to design

(a) A *tachometer* that uses the voltage across the points to produce a reading of the engine speed,

and

(b) A *dwell meter* that uses the voltage across the points to produce a reading of the percentage of time that the points are closed The dwell angle is

$$\text{dwell angle} = (\text{fraction of the time closed}) \frac{360^\circ}{\text{no. of cylinders}}$$

INDEX

Boldface page numbers identify major chapter sections dealing with the topic. Page numbers containing "P" refer to pages in the problem section at the end of each chapter. *Italicized* entries indicate cross-references to entries where a more descriptive index for that topic can be found. For example, in the entry – Amplifiers, *common-emitter,* 232, **256,** 306, 434 – page **256** identifies the principal reference to the topic, and the reader is also referred to Common-emitter amplifier, which contains twelve specific indexed items on that topic.